Laboratory techniques in biochemistry and molecular biology

5

LABORATORY TECHNIQUES IN BIOCHEMISTRY AND MOLECULAR BIOLOGY

Volume 5

Edited by

T.S. WORK – *N.I.M.R., Mill Hill, London*
E. WORK – *Imperial College, London*

NORTH-HOLLAND PUBLISHING COMPANY – AMSTERDAM · OXFORD
AMERICAN ELSEVIER PUBLISHING CO., INC. – NEW YORK

Part I

Brian W. Fox

TECHNIQUES OF SAMPLE PREPARATION OF LIQUID SCINTILLATION COUNTING

Part II

P.G. Righetti and J.W. Drysdale

ISOELECTRIC FOCUSING

1976

NORTH-HOLLAND PUBLISHING COMPANY – AMSTERDAM · OXFORD
AMERICAN ELSEVIER PUBLISHING CO., INC. – NEW YORK

ISBN North-Holland – series: 0 7204 4200 1
– volume 5: 0 7204 4219 2
ISBN American Elsevier: 0 444 11216 2

Published by:
NORTH-HOLLAND PUBLISHING COMPANY – AMSTERDAM · OXFORD

Sole distributors for the U.S.A. and Canada:
AMERICAN ELSEVIER PUBLISHING COMPANY, INC.
52 VANDERBILT AVENUE, NEW YORK, N.Y. 10017

Printed in The Netherlands

Editor's preface

Progress in research depends upon development of technique. No matter how important the cerebral element may be in the planning of experiments, a tentative hypothesis cannot be converted into an accepted fact unless there is adequate consciousness of the scope and limitation of existing techniques; moreover, the results may be meaningless or even positively misleading if the technical 'know how' is inadequate.

During the past ten or fifteen years, biochemical methods have become specialized and sophisticated to such a degree that it is now difficult for the beginner, whether undergraduate, graduate or specialist in another field, to grasp all the minor but important details which divide the successful from the unsuccessful experiment. In order to cope with this problem, we have initiated a new series of Laboratory Manuals on technique. Each manual is written by an expert and is designed as a laboratory handbook to be used at the bench.

It is hoped that use of these manuals will substantially reduce or perhaps even remove that period of frustration which so often precedes the successful transplant of a specialized technique into a new environment. In furtherance of this aim, we have asked authors to place special emphasis on application rather than on theory; nevertheless, each manual carries sufficient history and theory to give perspective. The publication of *library volumes* followed by *pocket paperbacks* is an innovation in scientific publishing which should assist in bringing these manuals into the laboratory as well as into the library. In undertaking the editing of such a diverse series, we have become painfully conscious of our own ignorance but have been encouraged by our board of advisers to whom we owe many valuable suggestions and, of course, by our authors who have co-operated so willingly and have so patiently tolerated our editoral intervention.

T.S. & E. Work
Editors

Contents of parts I and II

PART I

TECHNIQUES OF SAMPLE PREPARATION FOR LIQUID SCINTILLATION COUNTING, Brian W. Fox 1
Contents 3
List of abbreviations 10
Introduction 11
Chapter 1. General principles of liquid scintillation spectrometry 13
Chapter 2. Counting systems 41
Chapter 3. Preprocessing techniques: general aims and criteria 86
Chapter 4. Animal tissue processing 107
Chapter 5. Botanical aspects 142
Chapter 6. Cell cultures 149
Chapter 7. Extracts and chromatographic eluates 161
Chapter 8. Macromolecules 173
Chapter 9. Electrophoresis, centrifugation and chromatography on solid supports 184
Chapter 10. Inorganic applications 199
Chapter 11. Quench correction methods, multiple isotope counting and data evaluation 207
Chapter 12. Geophysics and archaeology 244
Chapter 13. Miscellaneous applications and future prospects 254
Appendix I 264
Appendix II 271
Appendix III 280
Appendix IV 284
Appendix V 289
References 291
Subject index 209

PART II

ISOELECTRIC FOCUSING, P.G. Righetti and J.W. Drysdale 335
Contents 337
Chapter 1. Theory and fundamental aspects of IEF 341
Chapter 2. Preparative IEF 377
Chapter 3. Analytical IEF 419
Chapter 4. General experimental aspects 501
Chapter 5. Applications of IEF 527
References 573
Subject index 587

TECHNIQUES OF SAMPLE PREPARATION FOR LIQUID SCINTILLATION COUNTING

Brian W. Fox

Paterson Laboratories,
Christie Hospital & Holt Radium Institute,
Manchester, England

Contents

Preface 7

List of abbreviations 10

Introduction 11

Chapter 1. General principles of liquid scintillation spectrometry 13

1.1. Scintillation mechanism 13
1.2. Quenching 18
1.2.1. Impurity or chemical quenching 20
1.2.2. Colour quenching 23
1.3. Chemiluminescence 24
1.4. The counting vial 27
1.5. Choice of the scintillation spectrometer 31
1.6. Methods of dispensing and cleaning 35

Chapter 2. Counting systems 41

2.1. Homogenous counting systems 41
2.1.1. Toluene-based scintillant mixtures 42
2.1.2. Dioxane-based scintillant mixtures 45
2.1.3. Miscellaneous counting systems 46
2.2. Heterogenous counting systems 48
2.2.1. Disc counting 50
2.2.2. Suspensions and gels 60
2.2.3. Colloid counting 64
2.3. Cerenkov light 76
2.4. Luciferin luciferase assay 82

Chapter 3. Preprocessing techniques: general aims and criteria *86*

3.1. Combustion techniques 87
3.2. Organic solubilizers 95
3.3. Degradative methods 98
3.3.1. Degradation without volatilization of the products 99
3.3.2. Degradation with volatilization of radioisotope 101
3.3.3. Enzymatic degradation 104
3.4. Precipitation methods 105

Chapter 4. Animal tissue processing *197*

4.1. Hard tissue 109
4.1.1. Bone, teeth etc. 109
4.1.2. Hair, lens, skin etc. 114
4.2. Soft tissues 117
4.2.1. Readily homogenised tissues e.g. liver, spleen, etc. 118
4.3. Semi-solid or liquid 121
4.3.1. Faeces 121
4.3.2. Whole blood, plasma etc. 123
4.3.3. Radioimmunoassay 131
4.3.4. Urine and body water 134
4.3.5. Milk and food, etc. 138
4.4. Respired gases 140

Chapter 5. Botanical aspects *142*

5.1. Higher plant tissue processing 143
5.2. Algae, yeasts 146
5.3. Soils and nutritional studies 147

Chapter 6. Cell cultures *149*

6.1. Virus particles and bacteria 150
6.2. Mammalian cells 151
6.3. Subcellular organelles (nucleii, ribosomes, etc.) 152
6.4. Media and broths 154
6.5. Liberated gases 156

Chapter 7. Extracts and chromatographic eluates *161*

7.1. Aqueous extracts 162
7.2. Acid solutions 164
7.3. Alkaline solutions 166
7.4. Salt solutions 168
7.5. Organic solvents 169
7.6. Gradient correction procedures 171

Chapter 8. Macromolecules *173*

8.1. Proteins and amino acids 174
8.2. Nucleic acids, nucleotides, etc. 177
8.3. Lipids and steroids 179
8.4. Sugars and polysaccharides 181

Chapter 9. Electrophoresis, centrifugation and chromatography on solid supports *184*

9.1. Electrophoresis 184
9.1.1. Polyacrylamide gel electrophoresis 184
9.1.2. Electrophoresis on agarose and starch gels 188
9.2. Centrifugation 189
9.2.1. Pellets 189
9.2.2. Caesium chloride gradients 191
9.2.3. Sucrose gradients 192
9.3. Chromatography on solid supports 194
9.3.1. Thin-layer chromatography 194
9.3.2. Paper chromatography 196

Chapter 10. Inorganic applications *199*

10.1. Solvent extraction methods 200
10.2. Precipitation and complex salt formation 203
10.3. Lipophyllic salts 205
10.4. Noble gases 206

Chapter 11. Quench correction methods, multiple isotope counting and data evaluation *207*

11.1. Quench correction methods 207
11.1.1. Internal standard method 209
11.1.2. Sample channels ratio techniques 212
11.1.3. External standardization 220
11.2. Multiple isotope counting 222
11.2.1. Two beta-emitters (e.g. ^{14}C and ^{3}H) 223
11.2.2. Gamma- and beta-emitters together 232
11.2.3. Automatic quench correction (AQC) 233
11.3. Data evaluation 234
11.3.1. Some statistical considerations 234
11.3.2. Computer assisted data handling 242

Chapter 12. Geophysics and archaeology *244*

12.1. Hydrology applications and tritium dating 246
12.2. Fall-out and meterology 249
12.3. Radiocarbon dating 250

Chapter 13. Miscellaneous applications and future propspects . *254*

13.1. Flow cells 255
13.2. Gas chromatography 258
13.3. Analytical applications 260

Appendix I *264*

Isotope tables and decay charts 264

Appendix II *271*

Properties of solutes, solvents and scintillation mixtures 271

Appendix III *280*

Analytical key to sample preparation 280

Appendix IV *284*

Glossary 284

Appendix V, Instruments and chemicals (suppliers) *289*

References *291*

Subject index *309*

Preface

The use of weak beta-emitting radioisotopes in an increasing variety of investigative procedures, particularly in the biochemical field, has stimulated the demand for better methods of assaying those isotopes which are more conveniently incorporated into organic molecules employed as precursors.

The prime object of the present work is to bring together the many different ways that have been designed to prepare the samples in a form suitable for liquid scintillation counting. Although the prime object is to bring the weak beta-emitter into close molecular contact with the primary solvent in the system, the method of doing this is highly dependent on the nature of the sample being measured. The initial sample for assay can exist in any state and associated with many other components which exceed the concentration of the actual beta emitter by many orders of magnitude. Where the level is low, careful thought should be given to the possible means whereby the labelled material itself may be isolated or alternatively the contaminating materials removed from the labelled substance. Occasionally, either one of these processes may be effected by solvent extraction, but before doing so, the sample multiplicity must also be considered. The number of samples may be too great to allow even a simple process of this kind to be conducted within a reasonable time. A decision as to whether or not such a tedious extraction procedure should be contemplated will depend on the level of accuracy needed by the experiment.

Subject index p. 309

Such decisions are often overlooked before assay and many users of liquid scintillation instruments rely solely on the claims of the many commercial scintillation cocktails available, hence incurring unnecessary expense. Often in addition, incorrect application of such cocktails may produce anomalies which are thought to be associated with the experiment itself rather than with the counting conditions employed.

In every counting technique used, some form of 'quenching' (see § 1.2) will be experienced. This reduction of counting efficiency can usually be accurately assessed and allowed for in a number of different ways. However, there are certain situations where these correction procedures cannot be used. It is in these circumstances that anomalous results usually appear and are often unrecognised by the experimenter.

In many, and probably in most biochemical applications, it may not be essential to obtain an accurate assessment of the absolute level of isotope present. Reproducibility of sample counts by careful control of the sample preparation is often the goal to be achieved. The ability to conduct several hundred measurements with a reproducibility of one and often up to 5% between samples is often all that is necessary. The correct initial choice of counting method at the outset will be repaid by the avoidance of many hours of tedious and unnecessary sample preparation, which may otherwise be costly not only in time consumed but also in the preprocessing and scintillation materials used.

Most of the confusion arising from apparently conflicting data in the literature on sample preparation methods, appears to arise from the fact that many authors consistently refrain from stating the reference efficiency of the machine being used. It is thus not possible to determine from the work, whether the increased efficiencies apparently obtained by his method is indeed an improvement or just a demonstration of the acquirement of a scintillation counter with improved engineering. Also since a judicious choice of window settings can appear to confer increased efficiency, open window data should be quoted in making such comparisons for single isotopes. This would avoid giving artifactual increases by quench shifts into lower windows.

It is suggested that for data on the measurement of low specific activity solutions an Instrument-corrected Merit Value (MIV) be used. Where there is a possibility of instability due to phase perturbations such as in heterogenous liquid scintillation counting, an instrument efficiency and stability corrected value (MISQ) be used. These terms are defined in greater detail in § 2.2 and Appx. IV.

Only a passing reference will be made to the scintillation equipment itself, since this has been the subject of several authoritative reviews. The choice of instrument is more often than not dictated by the money allotted to this aspect of the work programme. The instrument expense is usually related more to the cost of data-processing associated with it, rather than to the basic instrument; due consideration should be given to the number of users and samples to be assayed and whether or not some centralization of the facility is envisaged. In practice, an instrument holding 300 samples is a convenient size, since it will allow samples to be counted for 10-min periods over a week-end. Where shorter counting only is required, e.g. 2-min counts, it is clearly an advantage to purchase equipment with a higher potential capacity for samples.

Probably the most important point to bear in mind throughout all the applications of this technique is to be aware of the proportion of disintegrations that are being measured, and to force this to be as high and as constant as possible with the least effort.

List of abbreviations

(excluding those chemicals referred to in Appendix II, table 1)

ANDA	7-amino naphthalene-1,3-disulphonic acid
ATP	adenosine triphosphate
BHT	di-t-butyl-4-hydroxy toluene
cpm(s)	counts per minute (second)
CEA	carcino embryonic antigen
Ci	Curie
DEAE	diethylaminoethyl
DMSO	dimethyl sulphoxide
DNA	deoxyribonucleic acid
dpm(s)	disintegrations per minute (second)
EDTA	ethylenediamine tetra-acetic acid
ESR	external standard ratio
eV	electron volts (KeV = 1000 eV, MeV = 10^6 eV)
FMN	flavin mononucleotide
HeLa	a human cell line
LSC	liquid scintillation counting
NAD	nicotinamide adenine dinucleotide
PCA	perchloric acid
PSD	pulse shape discrimination
RNA	ribonucleic acid
RPH	relative pulse height
SCR	sample channels ratio
TLC	thin layer chromatography
TCA	trichloracetic acid

Introduction

Throughout the book, a simple coding system has been employed to designate the scintillant mixture being used. Such a coding is explained in Appendix II and the detailed compositions are listed in Tables 3 to 5 of this Appendix. The use of this coding system enables the qualitative composition of the scintillant to be recognized. Slight variations in the proportions of the components present are sometimes, but not often, carefully evaluated by the authors who proposed them. Most of the compositions are empirically derived and a rigorous assessment of the relative proportions of primary solvents, scintillants and blenders may reduce the number of really useful compositions even further.

My special thanks are due to my sister, Miss Mary Fox, for the very careful typing of the manuscript and to my wife and parents for checking. Thanks are also due to my scientific friends and colleagues who have suggested additional items, as well as the deletion of some.

Finally, I am particularly grateful to Drs. T. and E. Work, for their considerable help and tolerance.

Subject index p. 309

General principles of liquid scintillation spectrometry

1.1. *Scintillation mechanism*

The basic process of converting the energy derived from a beta particle to photons which can be detected efficiently by photomultiplier systems has been the subject of many detailed and authoritative works. Apart from technical broadsheets prepared by many of the liquid scintillation spectrometer manufacturers, a number of standard works has also been written. Most important of these are by Birks (1964, 1970) and Gundermann (1968). Edited collections of papers delivered at symposia devoted to different aspects of liquid scintillation spectrometry include those edited by Bell and Hayes (1958), Horrocks and Chin Tzu Peng (1971), Bransome (1970), Dyer (1971) and Crook et al. (1972).

Only a broad outline of the scintillation process will be given here, in as sufficient detail as is necessary only to help to understand the ways in which poor sample preparation techniques can introduce error or decrease the efficiency of counting.

Stage 1: The energy derived from a beta particle is first absorbed by the aromatic solvent molecule known as the *primary solvent.* Ninety % of the energy will be dispersed in the primary solvent molecule in exciting the bonding 'sigma' electrons and will be lost from the scintillation process as vibrational energy and heat. About 10 % of the absorbed energy, however, will excite the more fluid and excitable pi electron system which exists around the molecules and raise them to higher excited states. The process is represented by eq. (1.1)

Subject index p. 309

$$\text{beta-emitter } (E) + \text{solvent } (X) \rightarrow \text{excited solvent } (X^{++\cdot\cdot}) \quad (1.1)$$

where E is the energy of the beta particle and X is that of the solvent. The suffix $++..$ represents varying degrees of excited states present in the solvent molecule after the initial excitation energy has been dissipated. The many higher excited energy states of the pi electron system soon dissipate their excess energy ($X^{+\cdot\cdot}$) as *internal conversion energy* and the excited energy descends almost entirely into the first excited singlet state (X^{+}) according to eq. (1.2).

$$(X^{++\cdot\cdot}) \rightarrow (X^{+}) + \text{internal conversion energy} \quad (1.2)$$

The number of such excited molecules, A, is related to the nature of the solvent itself and to the energy of the beta particle by the equation

$$A = sE \quad (1.3)$$

(where s is the solvent conversion factor).

The value of s is such that one excited molecule results from approximately 100 eV of beta particle energy deposited. For the purposes of comparing different solvents from this point of view, it is convenient to regard toluene as having an 's' value of 100. Table 2 Appx. II lists these values for a number of solvents. The data give some insight into the excitability of different solvents and together with the solubility data of a primary solute, can suggest suitable combinations of primary solute and primary solvent. However, there are several other factors to be taken into consideration first.

Stage 2: The energy stored in one primary solvent molecule as excitation energy, does not remain there, but by a combination of thermal movement and other diffusion changes, the excited energy is transferred to an adjacent molecule and forms a short-lived dimer in doing so, a so-called 'excimer' (for a detailed treatment see Horrocks 1971). The energy is very rapidly transferred from molecule to molecule by this *solvent–solvent energy transfer process.* This process is highly efficient and takes place in approximately one picosecond (10^{-12} sec). Provided that the molecules are in close proximity to one another, i.e. no blending or quenching molecules interrupt the process, the transfer is also quantitatively very efficient and can be represented by

$$X_a^+ + X_b \rightleftharpoons (X_a - X_b)^+ \rightleftharpoons X_a + X_b^+ + \text{vibrational relaxation energy} \quad (1.4)$$

Stage 3: The process of solvent–solvent interaction by excimer formation and dissipation continues until contact with a primary solute molecule (Y) occurs, into which energy is preferentially transferred by a solvent–solute energy transfer process. Since the first excited singlet state of the latter is usually slightly lower than that of the former, there is again some energy loss in the transfer process, represented by

$$X_b^+ + Y \rightleftharpoons X_b + Y^+ \quad (1.5)$$

The efficiency of this transfer process is indicated by the solvent–solute energy transfer quantum efficiency (f). This value is the fraction of excited solvent molecules which succeed in transferring their energy to the primary solute and is affected only by the concentration of the primary solute in the solvent. For most of the combinations normally used in scintillation counting the value of 'f' is close to 1.0. The number of excited primary solute molecules (B) arising from 'A' number of excited primary solvent molecules is then given by

$$B = fA \quad (1.6)$$

Stage 4: An excited primary solute molecule, e.g. diphenyl oxazole (PPO), liberates within a few nanoseconds (solute fluorescence lifetime) of its excitation, a fluorescence photon, with a relatively high efficiency (solute fluorescence quantum efficiency = q). The wavelength of the photon emitted will depend on many structural features of the primary solute itself as will also the efficiency with which it is emitted. There is also some evidence to suggest that there is a similar solute–solute energy transfer process involving excimers similar to that between primary solvent molecules, and could be represented as in eq. (1.4).

$$Y_a^+ + Y_b \rightleftharpoons (Y_a - Y_b)^+ \rightleftharpoons Y_a + Y_b^+ + \text{vibrational relaxation energy} \quad (1.7)$$

If this process terminates in light production however, the reaction will be represented as follows

$$Y_b^+ \rightarrow Y_b + h\nu \; (h = \text{Planck's Const.}, \nu = \text{frequency of emitted light}) \quad (1.8)$$

The final number of photons emitted (P) is closely related in amount to the number of excited solvent molecules produced initially, provided that the quenching level is nil

$$P = qB = sfQE \quad (1.9)$$

However, one of the effects of quenching molecules (see § 1.2) is to decrease the number of photons emitted per original excited event and hence the number of photons per event will thereby decrease. Furthermore, it is worth noting at this stage, that the higher the energy of the original beta particle, the greater the number of photons eventually produced, but the effect of quenchers is to reduce the effectiveness of such events in producing photons. Hence the *proportion* of light quanta will be reduced by the same proportional amount, whether the source of the photon emission was a weak or a strong beta-emitter. This fact is used in the 'external standard method' i.e. using a gamma-emitter to produce Compton electrons which in turn are used to determine the degree of quenching of a soft beta-emitter.

Stage 5: The final stage of the liquid scintillation process is a modification of the light spectrum produced to match more closely the spectral sensitivity of the photomultipliers used. The proportion of overlap of the fluorescence spectrum from the primary solute, with the sensitivity spectrum of the photomultipliers is known as the matching factor (m). Owing to an improvement in the design of the photocathodes of the photomultiplier tubes, the degree of overlap has considerably increased, i.e. better matching factors are obtained. In the earlier instruments, the addition of a secondary solute or waveshifter, improved the matching factor by causing the finally emitted light to overlap more completely the less sensitive photomultiplier tubes in use at the time. The secondary solute is a similar aromatic fluorescent molecule to that of the primary solute, but is usually used in about 1/20th the concentration of the latter, and by a non-radiative energy transfer, operating at very low wavelengths (approx. 4 nm), accepts the excited energy state of the primary solute and excites its own electrons to a first excited singlet state. Subsequent light emission is now from the secondary solute and its spectrum will depend on the properties of this molecule. The efficiency of the transfer is usually very high and very little loss of photon emission occurs due to this process alone. In many modern instruments there is little to gain in using such waveshifters, and a simple check with a 0.4% solution of PPO in toluene, compared with and without 0.02% of POPOP (together with a tritium standard) for each of the instruments employed will soon convince one of the necessity or otherwise for its use in any particular machine. This check only applies to homogenous systems; there is some evidence that differences are obtained in heterogenous systems for reasons not yet clearly understood.

Stage 6: The photons from the scintillation solution are emitted in all directions and it is necessary to collect these as efficiently as possible, for them to be seen by the photomultipliers. The position of the scintillation vial, the efficiency of the light collecting system and the design of any light reflecting surfaces used within the vial, all help to increase the fraction of the light collected, the so-called light collecting factor (G), and entering the photomultiplier system. The photons incident on the photocathode (P') are given by

$$P' = GP = sfqGE \tag{1.10}$$

The number of photoelectrons emitted (N) is related to the P' photons by a fraction (mK), the mean photo-electric quantum efficiency

$$N = mKP' \tag{1.11}$$

where 'm' is the spectral matching factor and 'K' is the maximum photoelectric quantum efficiency at the peak of the photocathode spectral response (a value ranging from 15 to 30%).

A useful parameter in comparing the relative efficiency of one scintillant system with another is that of the Relative Pulse Height (RPH). This parameter was first adopted by Hayes et al. (1955) who using a solution of para-terphenyl (TP) in toluene (8 g/l), as a standard (100), were able to compare the efficiencies of a number of different mixtures of primary solute and solvents. The above terphenyl solution produced the same RPH (i.e. 100) as a solution of 3 g/l PPO in toluene, under the original conditions of measurement, and this PPO solution is now used as a standard. For a more detailed survey of modern primary solutes and solvents, see Birks and Poullis, 1972. These latter authors concluded that although the maximum RPH measured is given by a 24 g/l solution of BIBUQ (4,4-bis 2-butyl octyloxy-p-quaterphenyl) in toluene (RPH = 160), under quenching conditions, the counting efficiency was seriously affected. Both PBD (2-phenyl-5-(4′-biphenyly)1,3,4-oxadiazole) and butyl-PBD gave relative pulse heights greater than 150 when dissolved in toluene (12 g/l) in the presence of a number of quenching agents and were therefore more useful for practical use. These authors have also listed a number of factors which determine the magnitude of the relative pulse height in scintillant mixtures. In brief, they are 1) the nature and purity of the primary solvent, 2) the nature and purity of the primary solute, 3) the absorption spectrum and concentration of the primary solute, 4) the nature and concentration of any secondary solute (waveshifter) used, 5) the nature and energy of the ionizing radiation being measured, 6) the concentration of dissolved oxygen which depends on temperature and external atmospheric pressure, 7) the form and concentration in which beta-emitters are introduced (i.e. the effects of additional impurity and colour quenchers), 8) the size and shape of the counting vial, 9) the material of the vial, 10) the nature of the light-collecting system in the instrument, 11) the spectral response of the photomultiplier tubes.

1.2. Quenching

Quenching is a general word indicating that one or several processes are contributing to the suppression of the relative pulse height (RPH). Based on some of the considerations outlined in § 1.1, the factors contributing to quenching which can be modified by alteration in the

TABLE 1.1
Factors contributing to quenching.

	Energy loss could occur due to absorption by	Quench type
Beta-particle source	1) sample itself 2) unlabelled solute associated with sample 3) solid support used with sample 4) colloidal micellar structure	Source
Primary solvent	5) associated energy sink, e.g. dissolved oxygen, etc. 6) associated blending molecules 7) solvent added with sample (e.g. water, alcohol, etc.) 8) physico-chemical modification of the primary solvent	Energy transfer
Primary solute	9) physico-chemical modification of the primary solute 10) too high a concentration of primary solute	Solute
Light emission	11) photon trapping (colour) 12) wavelength shift due to associated waveshifters 13) poor matching factors 14) geometry changes	Photon transfer

method of sample preparation are listed in Table 1.1. Although it would appear to be more logical to discuss each of the types of quenching as indicated, the nature of the quench process and the relative contribution of each of the enumerated factors within a single quenched situation is not adequately known. It is convenient for the purposes of this book to consider the quenching process under two broad groups, impurity or chemical quenching and colour quenching. The former group consists of most of the events outlined whereas the colour

Subject index p. 309

quenching process consists mainly of those energy loss events associated with items 11, 12, and 13 in Table 1.1.

One of the prime objectives of this book is to describe practical methods of living with the phenomenon of quench, so that appropriate steps can be taken to determine the amount of it present, or in most cases, the relative amount present between samples. It is of prime importance in all biochemical work, to know how the quenching varies between samples. Methods of estimating the level of quenching will be dealt with in greater detail in chapter 11.

1.2.1. Impurity or chemical quenching

Those mechanisms of quenching which involve interference with transmission of the beta energy at source or during the excitation states of the scintillation system have been generally classified as impurity or chemical quenching. It will be convenient at this point to examine in greater detail the possible causes of the different forms of this type of quenching.

When the sample source emitting the beta-particle is in complete solution in relation to the phosphor system, the system is said to be *homogenous.* If the sample is separated from the scintillant mixture by a phase boundary, it is referred to as *heterogenous.* It is useful to recognise this difference at this stage as the type of system involved is associated with its own particular problems when a quench correction is required.

The homogenous system can be quenched at source by absorption of energy by the isotopically enriched atom itself. This only applies to very low specific activity material or to certain inorganic applications. In biochemical applications, the actual amount of material containing the beta-emitter in its molecule is usually very small, and does not often constitute a serious quenching hazard in its own right. A more serious contribution however will be the solute and solvent molecules associated with the beta-emitter, and it is the reduction of this parameter to the minimum possible level or its manipulation without excessive and impracticable preprocessing which is the main aim of good sample preparation.

The problem of source quenching is most acute in heterogenous counting of weak beta-emitters (e.g. tritium) on discs (§ 2.2.1). The principle effect of this type of quenching is to convert the counting geometry from the ideal 4 pi to something less than 4 pi. To a certain extent, this problem may be overcome in the case of glass fibre disc counting by disintegrating the discs and suspending the fibres either in thixotropic gels or in the presence of Cab-O-Sil (see § 2.2.2). The latter has the property of absorbing the beta-emitter on to its extremely large surface area, and creating counting conditions nearer to 4 pi.

Interference with the solvent–solvent energy transfer process (eq. 1.4) by solutes and solvents added simultaneously with that of the sample constitutes a far more important cause of quenching. The necessity to add alcohols or other amphiphilic solvents to bring an aqueous phase into intimate contact with a scintillant solution in which it is otherwise insoluble, introduces a marked degree of quenching, and attempts have been made to discover blending agents which will allow for the unification of such a two phase system with minimal quenching. 2-Ethoxyethanol has proved to be one of the most useful blending solvents from this point of view (White 1967).

Dissolved oxygen gas, either from the atmosphere, or from combustion procedures, is an important cause of quenching. An improvement of approximately 33% in counting efficiency can be obtained if the scintillant mixture is flushed with nitrogen (Pringle et al. 1953) or better, argon (Kerr et al. 1957). Ultrasonic degassing has also been suggested (Chleck and Ziegler 1957). By studying the level of oxygen quenching of primary solutes such as PPO and α NPO in xylene in sealed tubes under different pressures of oxygen, it may be deduced that its effect is directly on the solute molecule itself, and not on the solvent.

The relative amount of quenching can be determined by application of the Stern–Volmar equation

$$\frac{Vo}{V} = 1 + \frac{[M]}{[M]_{0.5}} \tag{1.12}$$

where Vo is the unquenched RPH and V the RPH due to the quenching agent. $[M]$ is the molar concentration of quenching agent and $[M]_{0.5}$ is the molar concentration of the quenching agent at which the RPH is one half that of the unquenched RPH. By plotting Vo/V against $[M]$, a straight line with a slope of $1/[M]_{0.5}$ is obtained in most cases examined by Birks and Poullis (1972). Table 1.2 gives a

TABLE 1.2

Half-value quencher molar concentration of various solvents using carbon tetrachloride as quencher (Birks and Poullis 1972).

		Solvent, $[M]_{0.5}$ in 10^{-2} M					
Solute conc. g/l	Solute[1]	Benzene	Toluene	Xylene[2]	p-Xylene	Mesitylene	DN[3]
5	TP	2.74	2.22	2.12	1.91	1.97	—
6	PPO	3.96	2.78	2.64	2.72	2.44	3.25
8	BBOT	5.64	3.82	3.56	3.17	3.09	7.23
8	PBO	6.12	4.50	3.71	3.51	3.42	5.25
10	butyl-PBD	6.31	4.66	4.60	4.22	3.98	6.03
10	PBD	8.17	5.79	5.06	4.83	4.50	6.44
15	BIBUQ	2.98	2.24	2.19	2.46	1.93	2.85

[1] For abbreviations see Table 1 Appx. II.
[2] Mixed isomers (Koch Light Labs. Ltd.)
[3] p-dioxane + 100 g/l naphthalene.

list of half-value quenched molar concentrations derived from the information obtained from a large series of measurements of different solute–solvent combinations. Since the susceptibility of the solvent to quenching is given by the inverse of this value, it can be seen that in general, benzene and dioxane-naphthalene based scintillant mixtures are least susceptible, whereas, p-xylene and mesitylene are most susceptible, with toluene and xylene forming an intermediate range. The best overall combinations i.e. those least susceptible to quenching are 1) 12 g/l PBD in toluene, 2) 12 g/l butyl PBD in toluene, and 3) 7.5 g/l PBO in toluene (Birks and Poullis 1972).

Other causes of quenching include the use of too high a concentration of the solute itself, causing a self absorption phenomenon, which can be readily observed by examining an efficiency-solute concentration curve. A typical curve will first rise and then decrease after reaching a maximum. The point of optimal concentration is usually chosen just as the curve begins to plateau. With blended scintillant mixtures, the optimal level of the solute may be greater than that in the original unblended system, due to the considerably increased solubility of the primary solute itself in the blended scintillant.

The above observations were conducted with carbon tetrachloride as a quenching agent. It is to be expected that different quenching agents will differentially exert their action on scintillator systems. There is also good evidence (Kaczmarczyk 1971) that the Stern–Volmar relationship does not hold at the higher concentrations of quenching agent. Thus there are possibly many instances in biochemical work when quenching is considerable (as in colloid counting techniques) and a different quench relationship from the Stern–Volmar relationship may exist. Under these circumstances, probably the only reliable quench correction technique is the use of a high specific activity standard sample of the beta-emitter being assayed.

1.2.2. Colour quenching

The reduction of photon transmission by absorption due to a chromophore present in solution is often a significant source of quenching in biochemical work. Attempts to bleach the colour often lead to greater impurity quenching, so that nothing is achieved. Cerenkov assaying is of course an exception (see § 2.3).

Coloured impurities can arise from a variety of sources in biochemical work. A common source of trouble is in the incomplete degradation of iron pigments, such as haemin. An approximate correction can be applied by measuring the absorption at 400 nm and by using a standard curve, which can be derived using Sudan Red, the counting efficiency can then be deduced. This however is more useful for the assay of Cerenkov emission quenched with coloured substances.

In general, the effect of coloured impurities in the scintillation mixture is to broaden the pulse height spectrum curve, over that which would have been obtained in a comparably impurity-quenched situation. The exact reason for the broadening is still a matter for discussion. Neary and Budd (1970) suggested that the light from a single photon emission will take several different light paths in a coloured sample, resulting in a broadening as observed. However, although Ten Haaf (1972) has attempted to interpret the data in terms of a mathematical model, good experimental support for the theory is still not available.

Certain highly coloured organic compounds are best combusted in oxygen, ensuring that complete degradation of the material occurs (Yamazaki et al. 1966). For small quantities of blood, the perchloric-hydrogen peroxide technique of Mahin and Lofberg, 1966, is usually very successful, whereas, the Shoniger flask combustion technique is recommended for larger samples (Nathan et al. 1963).

1.3. Chemiluminescence

Many chemical reactions, especially those involving oxidation, emit photons. If the half-life of the reaction is of short duration, i.e. seconds, the process is generally referred to as chemiluminescence, whereas a reaction involving a protracted emission of light energy, i.e. over several hours or even days is generally referred to as phosphorescence. In both cases, the emission of light is most likely due to a continuing chemical reaction rather than the result of a single activating event. Effective coincidence counting of an isotope in a scintillant solution depends on the ability of the instrument to observe photons that have arisen simultaneously from the same multi-photon event. Chemical reactions that are involved in chemiluminescence emit photons as single photon events, but they are nevertheless at such high flux that they are observed as multi-photon events within the coincidence time of the instrument. Thus in the older instruments, where the coincidence delay time was of the order of 1 sec, chemiluminescence was a considerable problem. Modern instruments however possess coincidence times of much shorter duration (10–15 nsec) and thus the chance of observing single photon events are considerably reduced. In fact some modern instruments have a delayed coincidence time, in addition to the normal one in the instrument, to measure the level of chemiluminescence, and by an automated system, to subtract a computed chemiluminescence level from the normal counts. Thus there is an increased sensitivity of assay of an isotope such as tritium, even in the presence of otherwise interfering chemiluminescence.

The problem of phosphorescence is most acute when certain protein solutions in alkali or quaternary ammonium solubilizers are being counted. This phenomenon was early recorded by Herberg (1958) who attempted to relate its extent to a number of physico-chemical parameters within the sample itself, in particular viscosity, against which there appeared to be an inverse relationship. Also it was markedly activated by exposure to light prior to counting. The decay curve usually consists of at least two components, a fast component with a half life of less than 1 min and a slow component, of approx. 15 min half-life. The phenomenon was found to be negligible at concentrations of protein lower than 10 mg/ml and it was observed that acidification of the mixture caused an immediate cessation of the phosphorescence. The problem is particularly acute with scintillant mixtures involving dioxane (Lloyd et al. 1962) especially when serum is being assayed in the mixture (Moriarty 1972). The vial itself makes little contribution to the phosphorescence (Moghissi et al. 1969); dioxane alone will produce light if first excited by strong sunlight, and the kinetics of decay are similar to those produced in the scintillator solutions containing dioxane. However, the reaction is clearly not a simple one, as Kahlben (1967) has pointed out. The phosphorescence due to hyamine in dioxane is considerably enhanced by the addition of naphthalene to the mixture, even though a mixture of hyamine and naphthalene does not show any significant phosphorescence. The reaction is also dependent on temperature, but by lowering the counting temperature from 20°C to 8°C, only a 33% decrease in the counting rate is observed.

Although chemiluminescence has been recognised for many years, it is only recently that it has received serious attention. The physical basis of chemiluminescence has been reviewed by Hercules (1970) who pointed out that not all chemiluminescent reactions involve oxidation. If a reaction is to produce light it must have sufficient energy for excitation and must possess some species capable of forming an excited electronic state following transfer of excitation energy from the emitter.

In order to produce light, the energy required must exceed 47 to 65 kcal/mole. This kind of energy cannot simply be derived by fission of carbon-hydrogen bonds, and there are certain chemical structural requirements. Furthermore there is considerable evidence to suggest that chemiluminescence occurs in energy systems where there is insufficient energy to form excited singlet states, and thus the lower energy requiring, triplet–triplet annihilation reactions have been implicated as a possible source of this phenomenon. Khan and Kasha (1966) have proposed that energy transfers from singlet oxygen dimers to emitters in solution may be a general mechanism for chemiluminescence in liquid scintillation counting. However, there are inconsistencies in the data available for this type of reaction and some form of singlet–triplet transfer mechanism seems to be a more feasible source of the energy (Hercules 1970). No adequate explanation is yet available for light production in quaternary ion solubilization reactions however, but the presence of peroxides in the solvent used would appear to be a major source of this energy light emission.

Addition of a drop of HCl or of 10 % ascorbic acid is usually successful in reducing the level of chemiluminescence where peroxide formation is a basic cause. The light emission exhibited by detergent mixtures, such as with Triton X-100 systems, appears to be of a different origin and is unaffected by acids. A commercial preparation Dimilume™30 produced by Packard, U.S.A., is a scintillant mixture containing an inhibitor claimed to be active in preventing the chemiluminescence exhibited by their own solubilizer and colloid based systems. The product is described as an acidified-emulsion phosphor.

1.4. The counting vial

The early counting vials consisted of glass weighing bottles, 5 cm in diameter and 6 cm high, holding 85 ml. However, it was appreciated by Davidson (1958) that optically clear '5 dram vials', 2.6 cm in diameter and containing up to 15 ml were more suitable, especially when containing about 5 ml of scintillant mixture.

The present day standard vial is shown in Fig. 1.1a. It is constructed of low potassium glass or a suitable plastic, with a screw cap containing either a circle of tin foil or an inner plastic sealing flange, moulded into the cap itself. Glass vials can be washed and re-used, but the plastic caps are best treated as disposable items and discarded after use.

Pyrex glass possesses phosphorescent properties when exposed to fluorescent or strong daylight. It is advisable to employ subdued lighting or sodium-tube lighting in the vicinity of the liquid scintillation spectrometers for this reason and to avoid excessive exposures of the vials to sunlight or other direct light before counting. The phosphorescence of quartz is some 2 to 2.5 times less than that of glass, and quartz is therefore used for very low-level counting such as that required in carbon dating techniques. The low level of potassium 40 present in quartz also reduces the possible background counts from this source. The prohibitive cost of quartz however, will allow its use only in special cases where very low backgrounds are essential. To a large extent, plastic has replaced it. Roughening the glass on the inside of the vial is also recommended as a means of increasing the efficiency of counting in glass vials (Schwertdel 1966).

The use of plastic as a material from which to construct scintillation vials enables the whole vial to be discarded after use, thus eliminating the necessity for the tedious and expensive washing, drying and checking procedures required with glass vials. Furthermore, the translucent nature of such a vial does not interfere with its optical efficiency and indeed, higher efficiencies than in glass vials have been reported (Rapkin and Packard 1960; Rapkin and Gibbs 1963). Polyethylene vials have been especially recommended for use in counting very low

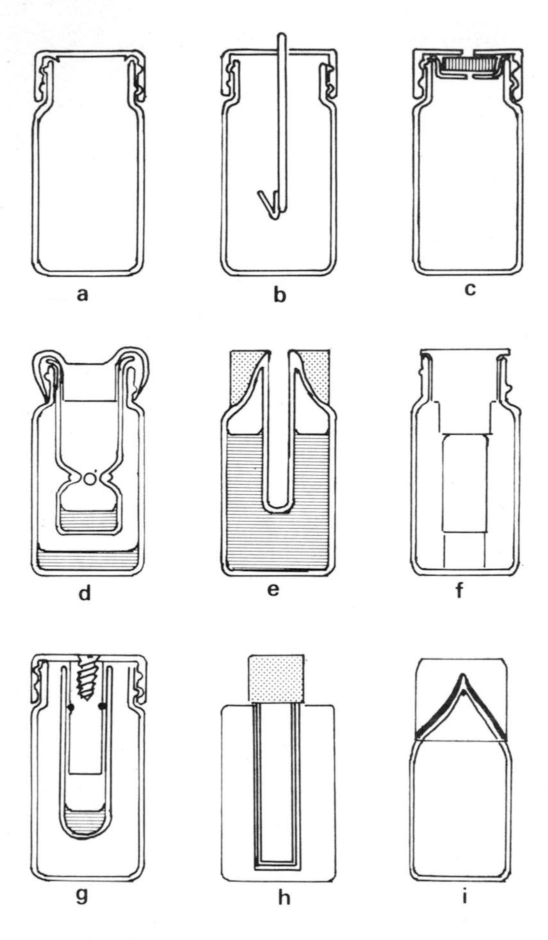

Fig. 1.1. Some examples of types of scintillation vials that have been described. a) Standard vial, b) vial used in small sample combustion, c) vial which allows the introduction of liquid under pressure (Johnson 1972, by permission of Heyden & Son, Ltd., in Liquid Scintillation Counting Vol. 2, eds. M.A. Crook, P. Johnson, B. Scales), d) vial for the estimation of respired $^{14}CO_2$ from enzyme reactions (Slater et al. 1964), e) vial for assaying soft gamma emitters using a lead or tin loaded scintillant in an outer compartment (Ashcroft 1970; by permission of Academic Press), f) vial to estimate thiol levels by measuring the degree of colour quenching (Snyder and Moehl 1971, by permission of Academic Press), g) and h) methods of holding small volumes of scintillant mixture to be used in the standard liquid scintillation spectrometers.

levels of radionuclides under conditions where there is the possibility of contamination of the vial walls with the radioisotope (Davidson and Oliviero 1965; Litt and Carter 1970).

However, there are a number of problems associated with the use of plastic in counting vials and these are dependent on the nature of the plastic used. One of the main problems with vials made of polyethylene is absorption and transmission of toluene, causing the vial to swell, as well as to liberate toluene vapour into the body of the instrument. This process is temperature dependent, being considerably more evident at ambient temperatures. The loss of toluene from such a vial is approx. 150 to 200 mg per day; thus a machine in which 200 such samples are being counted can be filled with some 60 to 80 g of toluene vapour over one weekend. Nylon, on the other hand does not swell with toluene but does distort appreciably when dioxane-based scintillants, containing aqueous samples are used. (For comparisons, see Lieberman and Moghissi 1970; Butterfield and McDonald 1972.)

Polyfluorinated polymers such as PTFE (Teflon) however, do not show appreciable swelling with either type of solvent (Calf 1969). These vials are specially recommended for long-term counting, such as is required in certain hydrological methods, but they are usually too expensive to use for short-term counting as a disposable vial. An interesting effect associated with this plastic is the increased 'figure of merit' (E^2/B) obtained when in contact with scintillant. Even rods of the plastic placed in the scintillant in a glass vial increase the counting efficiency. However, unexplained changes also occur in the pulse height pattern using this material, and thus more careful discriminator settings may have to be made in certain situations.

Artifacts sometimes arise when plastic vials are used in conjunction with external standard ratio techniques for quench correction. There are unconfirmed reports that these artifactual quench correction values are also dependent on the nature of the external standard source (Rauschenbach and Simon 1971). Some artifacts are expected if the primary solute enters the wall of the vial and hence radically alters the geometry of the counting from the external standard source.

During colloid counting in plastic vials, considerable instability of phase structure occurs due to the preferential absorption of one phase into the wall causing an imbalance in the phase composition of the liquid remaining in the vial.

In conclusion, it would appear that if plastic vials are to be used, polyethylene containers are best employed with dioxane-based mixtures, but only PTFE vials can be recommended for any counting which requires a longer storage than 48 hr within the vial.

A number of vials for specialized use have been described and some of these are illustrated in Fig. 1.1. In an attempt to avoid transfer of labelled carbon dioxide in quaternary ammonium hydroxide solution from respiratory enzyme experiments a vial was designed (Fig. 1.1d) to allow the reaction to occur in a flanged tube attached to the cap. The liberated $^{14}CO_2$ passes through holes in the flange and is absorbed in a CO_2 absorbant in the vial (Slater et al. 1964).

Vials have also been designed which house a small sample holder and an electrode in the cap for undertaking automatic combustions of very small samples (Fig. 1.1b). The sample is ignited by applying a high frequency discharge to the projecting terminal on the upper side of the lid (see § 3.1).

Further adaption of scintillation vials include the insertion of a gas chromatography septum into the lid (Fig. 1.1c) in order to allow the introduction of liquid during high temperature digestion experiments without loss of gas pressure and thus of isotope (Johnson 1972). Two further modifications are worthy of note here also. The vial described by Ashcroft (1970) consists of an outer sealed compartment containing an organic tin or lead loaded scintillant (Fig. 1.1e). This enables soft gamma-emitting isotopes (e.g. I^{125}) to be assayed with higher merit values than by conventional liquid scintillation counting alone (chapter 13). Another modification (Fig. 1.1f) described by Snyder and Moehl (1971) is used to determine the level of colour quenching as a means of estimating the levels of thiols present (see § 13.3).

There is clearly scope for considerable ingenuity in the design and modification of counting vials for specific uses and the possibility of reducing the amount of preprocessing.

1.5. Choice of the scintillation spectrometer

It is not the purpose of this book to compare and deal with the different makes of instrument available. There is an excellent review by Rapkin (1972) outlining the development of the instrument to the present day.

In assessing the best instrument for a particular purpose, it is clearly important firstly to consider what proportion of a laboratory budget can be reasonably applied to this particular aspect of the research requirement. In a laboratory situation where the need for the instrument is a real one, the use of the instrument is usually very considerable, and in a research institute concerned with fundamental problems of biomedical research, for example, the need for further equipment is usually soon felt. At this stage, it is worth looking at the real needs of the staff, since the type of radioisotope meaurement undertaken will very largely dictate the type of equipment required. Different instruments offer different services and it is worth examining these real needs in some detail.

In most such laboratories, it is not usually envisaged that fundamental work on the liquid scintillation system will be undertaken, and the primary requirement will be to measure the relative levels of radioactivity in a series of samples, processed by varying techniques. These will usually consist of carbon-14 or tritium labelled samples. Sometimes, other isotopes such as ^{32}P, ^{35}S etc. will also have to be measured. If an occasional other requirement exists however, then more versatile equipment may be necessary. If the levels of radioactivity are going to be low, e.g. less than 2×10^4 dpm per sample, then efficient quench correction becomes important. The level of accuracy tolerated will then determine whether the external standard ratio on the instrument is going to be good enough, or whether it will be necessary to undertake the more tedious but usually more accurate internal standards method.

The price itself will often dictate the level of the computerization acceptable, but serious consideration must be given to the amount of computerised calculation needed. For example, it is really essential to purchase a computer attached to the counting facility, or could a tape readout of the raw data be all that is required to feed into an existing computer system, possibly underused, within the same laboratory?

If many users are envisaged, who require relatively short counting times (say less than 5 min) for a varying number of samples prepared by different methods, it may be necessary to select an instrument having facilities to change the channel settings at the beginning of a row of samples by some preselector device. This facility, which exists on some instruments, is of considerable help in such cases. However, it is necessary to ensure that details of the channel settings (either as a single number code or in detail) are indicated on the print-out itself either with, or ahead of, the samples concerned.

The choice of print-out is often an important consideration if sample information requires to be stored and coded. It is often more convenient if the readout is printed on a single horizontal line for each sample rather than a series of lines, since it is easier to read and also easier to cut up into sections after counting. It also helps if there is some mechanism for the spacing of items at intervals, in order to allow the dissection of the sheet between the items for distribution to different users. In some instruments, the data can be printed on to existing tables, with the information concerning user and experimental details being recorded manually at the beginning of the sheet. However, the paper cost and wastage is usually much greater in these cases.

When the number of users and instruments begin to increase in a laboratory, it is often advisable to consider the economics of maintaining the regular checking, cleaning and general usage of the instruments by a competent technician. He or she may have developed a special interest in the many facets of the technique and with suitable training and experience, would be able to advise users on the state of the machines and to maintain continuity of use and cleanliness of vials placed into the machines. Under pressure situations, 'hogging' of spaces, under pretext of requiring repeated counts, only increases the problems, and a recognition of this and application of suitable controls can prevent the excessive delays in counting that can result.

If a room is set aside for these facilities, it is usually advisable to make accurate records of room temperature, especially at night, to avoid overheating. At the outset, air conditioners should be seriously considered to maintain a constant temperature (say 20°C) and so avoid fluctuations which can lead to sudden unexpected changes in background levels.

There are advantages and disadvantages to operating the equipment at cooled temperatures (e.g. 4–10°C). The main advantages which are claimed are lower background counts and the reduction in chemiluminescence. In modern instrumentation, the former is of doubtful value, since more efficient background reduction can usually be achieved through improved channel settings, more suitable for the working conditions of the experiment itself (see chapter 11). The length of time required for equilibration of the temperature conditions for counting is greater than at ambient and some variation can be expected during the process of sample cooling. The reduction of chemiluminescence is also not complete, and at best only a partial reduction (to varying levels) which persist for longer periods at the lowered temperature, can be achieved. It is always much better to attempt to get rid of the chemiluminescence, rather than to partially sweep it under the carpet by lowering the instrument temperature (see § 1.3). Another disadvantage of the cooled system is that in certain types of counting, e.g. in those systems containing dioxane, freezing of the scintillant mixture can occur, if the water concentrations happens to be low. The colloid system is usually considerably affected by temperature, and can only be used at the lowered temperatures, if the system has already been worked out for that temperature.

Careful consideration should be given to the possible advantages in chosing the right time interval for the counting, the necessity of repeating counts, and the application of background subtraction and low sample count reject systems that are available on most modern instruments. The low sample reject system can result in a considerable saving in time e.g. with a large series of samples from a chromatographic elution. Most of the samples may be approximately background, but a few, possibly in unknown positions, contain varying levels of isotope, some of which may be only 2 to 3 times the background level. If some of the expected peaks also contain 2.5–3 times the background counts, then it is feasible to use the low sample reject system. However, for lower counts than this, the counts need to be measured to similar degrees of accuracy to avoid transient low or high counting rates which can occur under normal low level counting conditions.

A facility for displaying the pulse height spectrum is sometimes incorporated into instruments and this can be a useful means of confirming that the correct channel settings have been made. In any case it is useful to ensure that a suitable output is available on the instrument, to which a multi channel analyser system can be attached, should this be occasionally necessary. In teaching laboratories, it is useful to consider this accessory, since, the basic processes involved in the technique can then be usefully demonstrated.

All instruments should have a facility for regular slight adjustment of the external standards ratio value and suitable regular technical surveillance of this check can often detect a slow deterioration in one of the photomultiplier systems, long before the user is aware of it.

In some instruments, the elapsed time from the beginning of a series of counts can be made. This is of particular value in the investigation of chemiluminescent phenomena, and more particularly in those systems which employ chemiluminescence as a means of analysis, such as luciferin or luciferase assays (§ 2.4), and the detection of low levels of metallic ions (chapter 10).

Most instruments can allow for repeated counts to be made on the same sample as well as repeated cycles of a series of samples. These are useful facilities, especially in conjunction with the elapsed time facility, to detect the presence of some contaminating chemiluminescence or indeed to measure the decay rates of short-lived beta-emitters in inorganic applications.

The future design of the liquid scintiallation spectrometer, in the face of considerable commercial competition, appears to be diverging into two main types. 1) Instruments which can be bought in numbers and provide raw data only, preferably as a tape readout as well as printed sheets, as cheaply as possible. 2) More sophisticated instruments, which employ diverse applications of on line computer calculations to provide finalised data, either as computed lists or as graphs. Data from other sources regarding the samples can also be built into the computations of the final results. It is easy to see that a future development of this feature would be to incorporate the liquid scintillation spectrometer as a member of a more complex, and more comprehensive analytical system, possibly associated with gas chromatographic, mass spectrometric and other sub-micro analytical techniques.

1.6. Methods of dispensing and cleaning

Scintillant chemicals, made up into mixtures ready for counting should always be stored in the dark or in dark bottles. For dioxane-based and blended scintillant mixtures, it is always worthwhile flushing out the storage bottle with nitrogen before stoppering to reduce the formation of peroxides. Anti-oxidants, such as BHT (di-t-butyl-4-hydroxy toluene) can be added without any apparantly deleterious effects on the scintillation efficiency. Dioxane, ethoxyethanol and methoxyethanol are particularly prone to peroxide formation on standing in air for any length of time. Large pieces of granulated zinc placed in the storage bottle will also help to reduce the formation of peroxides.

It is very convenient and desirable for scintillant mixtures to be dispensed automatically and dispensers are available commercially which are designed for use with non-polar solvents. It is advisable to use a dispenser which has only glass–glass or glass–PTFE joints and surfaces. A typical dispenser is shown in Fig. 1.2a. It is not advisable to use such dispensers with dioxane: naphthalene scintillant mixtures however, as the naphthalene readily crystallizes out and blocks the syringe mechanism. An alternative device is the fixed volume tipping dispenser (Fig. 1.2b) which is more useful for these latter mixtures since the inner surface is continually bathed in the vapours of the solvent and tends to prevent the deposition of the naphthalene. It is possible to vary the size of the aliquot measure in the top of the dispenser in some designs.

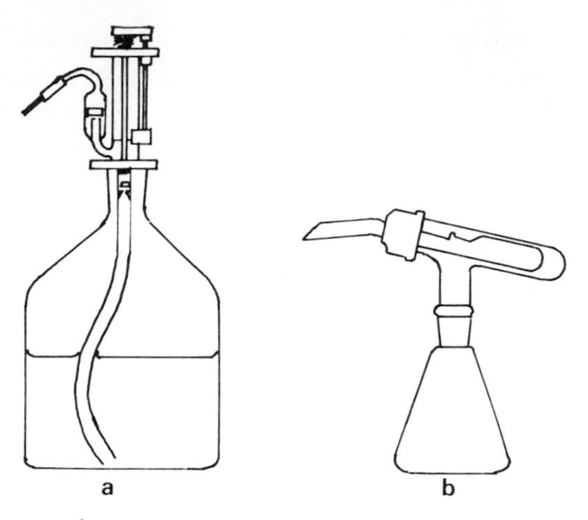

Fig. 1.2. Two basic dispensing techniques. a) All-glass device for the delivery of fixed volumes of scintillant from a stock bottle, ideal for dispensing scintillants which do not contain naphthalene (based on a design manufactured by Jencon, Hemel Hempstead Herts, U.K.), b) a tipping dispenser, suitable for dispensing naphthalene-containing scintillants.

The washing and decontamination of scintillation vials is primarily the problem of removing toluene and water-insoluble scintillant chemicals as completely as possible before the routine washing and rinsing procedures are undertaken. After some scintillation and also some heterogenous counting methods, a film of isotopically labelled deposit may also be present on the walls of the vial. In the latter case, strong detergent and/or sonication procedures may be necessary to remove it.

If disc counting has been employed, there is often no necessity for an elaborate washing system. The disc can simply be removed with a pair of forceps and the scintillant discarded. The vial is then rinsed with methanol and dried. A device described by Harris and Friedman (1969) and illustrated in Fig. 1.3a is of particular value. In the present author's laboratory, the inclusion of an 'O' ring which forms a snug fit with the sides of the vial provides a means of rinsing those vials which may be slightly chipped on the mouth and thus would otherwise

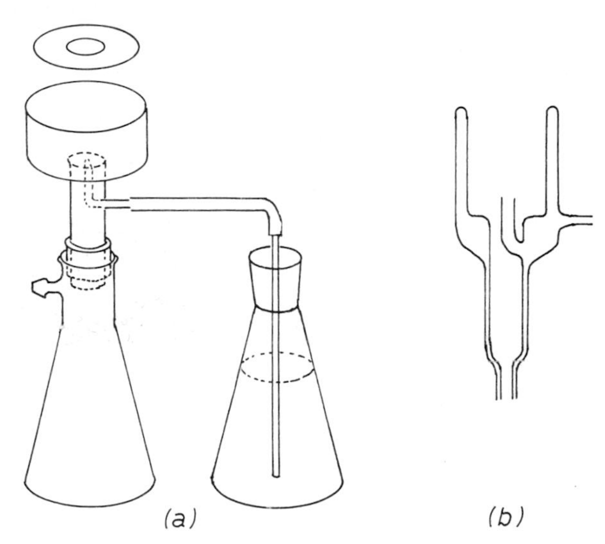

Fig. 1.3. Device for the rapid rinsing of scintillation vials. a) Based on a design described by Harris and Friedman (1969, by permission of Academic Press). The funnel is constructed of brass, with a neoprene ring placed inside the top to ensure close contact between the vial and the base of the funnel. Suction is applied to the Buchner flask from a water pump and the Erlenmeyer flask is filled with a cheap alcohol (methylated spirits). As the vial is pressed on the ring, a reduced pressure in the vial causes the alcohol to wash through the vial and to waste in the Buchner flask. b) A modification made in glass based on the same principle designed by Radin 1973 (Academic Press).

interfere with the suction seal necessary to allow the washing fluid to be drawn into it. A foot-operated two-way tap, which converts an alcohol rinse to a water wash for each vial, accelerates the procedure still more. A cheap grade of alcohol can be used for this purpose. A still further improvement has recently been described (Radin 1973) and is illustrated in Fig. 1.3b. Alternatively a more elaborate device has been described by Drosdowsky and Egoroff (1966) where a continuous extraction of cooled inverted vials with a hot chloroform: methanol mixture is undertaken. The instrument (manufactured under a French patent No. 53760 of Mar 16th 1966) is illustrated in Fig. 1.4.

In the case of homogenous systems employing solubilizers or blenders it is usually necessary to clean by thorough detergent washes

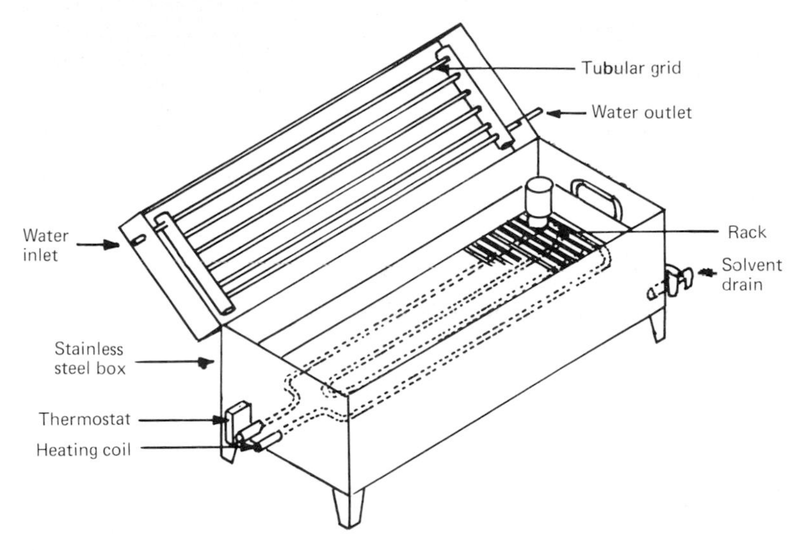

Fig. 1.4. Apparatus for the solvent cleaning of scintillation vials. One litre of a mixture of methanol : chloroform 1 : 1 v/v) forms a 2 cm layer in the base of the apparatus shown. The cooling cover is set in place and the solvent mixture is heated. The thermostat is adjusted so as to allow gentle refluxing to occur on the lid cooling system. The vapour condenses on the cooling coils and returns to the heating area. After 2 hr the heater is turned off, the vials removed with the rack and the whole rinsed in distilled water and dried. A cover is placed over the rack during the rinsing operations. The method is described by Dosdrowsky and Egoroff (1966) and the apparatus is manufactured under a French patent by GRINEX S.A. 43 Rue de Mauberge, Paris (9ème), France. Illust. by kind permission of Academic Press Inc.

and rinses. The vials should be soaked overnight in a strong detergent (after removing the toluene by decantation to waste). Commercial preparations containing detergent are available which can be added directly to the toluene-damp scintillation vials. The bottles are then loaded on to a multi-jet washing system attached to a hot water supply (Fig. 1.5) and thoroughly rinsed, after which they are rinsed twice in de-ionised water and dried in an oven at 120°C, stacked in an inverted position. It has been suggested that heating to high temperatures in 'self cleaning' ovens is also an efficient means of decontamination (Kushinsky and Paul 1969). A temperature of 420°C, maintained

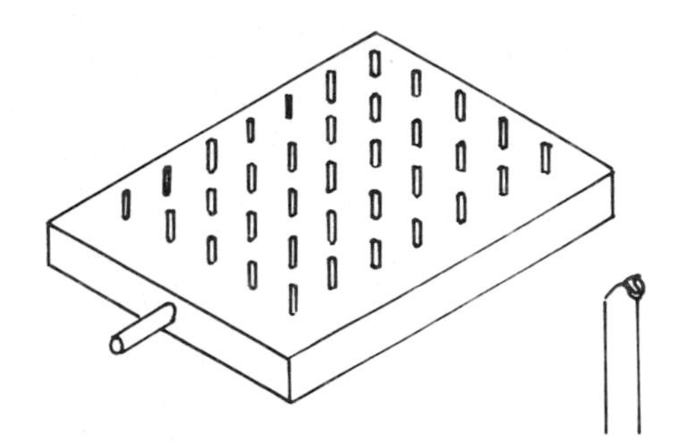

Fig. 1.5. A simple device for the rinsing of vials using the mains hot water supply. Inset, construction of jet to ensure full washing of interior of the vial.

for 3–8 hr, is recommended; this can be used primarily for tritium and ^{14}C samples and would be unsuitable for inorganic salts of isotopes which would not be decomposed at this temperature.

Care should be exercised to avoid excessive chipping of the screw-cap vial, since the shaking necessary in some counting techniques (e.g. colloid counting) will leave traces of scintillant creeping down the outside of the vial and hence contaminating the chain mechanism and the light collector system of the scintillation instrument.

Following colloid counting, the gel is shaken out into a waste jar and the vials are washed directly over a multi-jet washer, since adequate detergent is already available for cleaning in the scintillant itself. Vials should on no account be loaded into an automatic washing machine with toluene or dioxane still present in the vials. Toluene vapours are both inflammable and toxic and sink units where washing is done should be well ventilated and personnel engaged in this duty must avoid both skin contact and inhalation of vapour. If accidental exposure occurs, fresh air should be sought as soon as possible. Medical advice should be sought immediately if nausea and especially vomiting occurs.

Clean vials are best stored inverted in cardboard trays in the dark to avoid excessive exposure to light and hence the light stimulation of phosphorescence in the glass itself. Should vials show evidence of persistant phosphorescence (high background counts after cleaning), heating the vials to 200–300°C will usually destroy the process; for low background work, this procedure is recommended as a routine precaution. Vials should be inspected regularly for small star-shaped cracks in the base – usually an indication of either poor handling techniques or occasionally of a badly moulded batch of vials in which the base is too thin.

Ultrasonic methods have been recommended for the cleaning of vials especially where bulk storage and cleaning is contemplated for multi-user activity. However, great care should be taken that only vials which have contained a similar range of counts be washed together. If some vials are known to have contained excessive levels of radio-isotope (e.g. greater than 10^6 dpm), these should either be washed separately or, if only a few are involved, are best discarded.

The efficiency of both preparing and dispensing scintillant material may often be dramatically checked by occasionally observing the working area in ultraviolet light since diphenyloxazole is highly fluorescent and the spread of the agent can be easily seen. Direct exposure to UV of the scintillant mixtures to be used in counting should of course be avoided. It is strongly advised, and in some working conditions imperative, to carry out both scintillant preparation and dispensing in areas known to be free, or very low in labelled substances. The contamination of a large stock of solvent (e.g. toluene) can cause considerable trouble for some time before the fact is realized. A suspiciously high background in all 'blanks' should be checked for its decay rate and, if possible, its pulse height spectrum, to ensure that it is not due to contamination with an isotope such as tritium. Particular care should be taken to avoid dispensing toluene standards in the region of a laboratory where vial-filling in preparation for assaying is being done. On no account should internal standards in containers other than vials be taken into a room set aside for liquid scintillation spectrometry.

The success of preparation, cleaning and dispensing techniques can only be judged by a very low frequency of vials with an elevated background (say $1\frac{1}{2}$ to 2 times the average) and by a constant low background in the scintillation counter for a number of years.

Counting systems

2.1. Homogenous counting systems

Homogenous counting systems, consisting of one phase only, are the ideal system for assaying beta emitters in a form in which a quench correction procedure can be accurately applied, so that an absolute assessment of an isotope may be made. Many scintillator compositions have been proposed with varying concentrations of primary and secondary solutes as well as complex mixtures of blenders in a variety of primary solvents. The final mixture 'recommended' is often the result of empiricism or serendipity rather than of a rational determination of optimal conditions. Manufacturers have been quick to seize on these variations, giving them a multitude of different names compounded in part from the terms 'phosphor', 'scintillant' and 'fluor'. A disturbing feature has been the introduction of secret recipes of scintillant mixtures, thus adding an avoidable variable into the liquid scintillation counting technique, which is thereby not exploitable by the research worker should an unusual result occur. An example would be that a secret recipe containing 2-phenyl-5(4′biphenyl)-oxadiazole (PBD) would give erroneous results with certain solubilizers (Dunn 1971) without the worker being able to predict such an event. If there is any doubt about the incompatibility of solutes, it is recommended that either scintillant mixtures be made up in the laboratory, or only those commercial mixtures whose composition is accurately known should be employed.

The simplest homogenous system consists of a solution of diphenyl-

Subject index p. 309

oxazole (PPO) in toluene. In order to allow admixture with water or very dilute solutions, an alcohol such as ethanol, or better, 2-ethoxy ethanol, is added to bring the final mixture into a single phase. To increase water-accepting capacity, a solution of naphthalene in dioxane can also be used as a primary solvent system, some of which are listed in Appx. II, Tables 3 and 5.

2.1.1. Toluene-based scintillant mixtures

The most popular toluene-based mixture is a 0.4% solution of diphenyloxazole (PPO) in toluene (Tpp-4, Table 3, Appx. II). In many of the older instruments and in a few modern instruments, photomultipliers are used which are not as sensitive to the wavelengths around the fluorescence maximum of PPO (365 nm) as to the longer wavelengths around 420 nm. It is therefore an advantage to include a secondary solute (see § 1.1) such as 1,4-bis-(5-phenyloxazol-2-yl) benzene (POPOP) or its 4,4′-dimethyl derivative (DM-POPOP) in order to transpose the fluorescence maximum from 365 nm to 415 nm or 430 nm respectively to provide an improved spectral matching with the photomultiplier tubes. However, over the last decade, new photomultiplier tubes with increased sensitivity at the shorter wavelengths have also been introduced into many machines, and it is doubtful if either of these secondary solutes are now required. A simple check at the outset with tritiated toluene standard in a 0.4% PPO solution in toluene and with different levels of POPOP (say 0.005, 0.01 and 0.02%) will soon convince the user of the necessity of this component. If the solute is not required, the resulting saving in time and cost is worthwhile, since these secondary solutes are both expensive and fairly slow to dissolve in the toluene solution.

In its simplest form, this solution (0.4% PPO in toluene with POPOP) is most often used in *heterogenous* counting, such as with discs or in suspension techniques (see § 2.2.1). As a *homogenous* system, it is used to assay pure steroids, esters, ethers, hydrocarbons, lipids and certain gases, especially some of the inert gases. Its greatest use is however, with solutions of proteins as carbamates in quaternary ammonium solubilizers (§ 3.2), which are totally miscible with the solution.

Most samples in biochemical practice however, contain water and dissolved solutes and it is necessary to utilize a ternary combination of the sample, toluene and a suitable 'blender' or 'diluter' such as an aliphatic alcohol, to allow an aqueous solution to form a one-phase system with the toluene and thus provide direct access of the beta-emitter to the primary solute in the electron transfer system. However, blenders also introduce the problem of quenching. There are thus two opposing processes involved in the determination of the eventual counting efficiency. A useful summary of the assessment of the optimal conditions for a few of the commonly employed blending materials was described by White (1967). He concluded that 2-ethoxyethanol was superior to several other alcohols in its blending ability in relation to quenching action. A useful scintillant composition based on this alcohol as a blending agent is referred to as 'Phosphor A' (TNppX4.8, in Table 4, Appx. II). This scintillant mixture will accommodate up to about 1 % of water. For higher proportions of water (i.e. up to 4 %), the naphthalene must be omitted and a larger proportion of ethoxy-ethanol should be added. For water proportions greater than 4 %, dioxane-based scintillants are best employed if homogenous systems are required.

In order to establish the most useful proportion of primary and secondary solutes in the presence of different levels of quenching agent, some form of contour plotting is essential to select the optimal conditions from the variety of different compositions possible. This type of analysis was undertaken by Little and Neary (1971), who examined the effect of nitromethane quenching on toluene, PPO, dimethyl POPOP mixtures. The effect of 0.2 % nitromethane on the equal relative counting efficiency contours is shown in Figs. 2.1 and 2.2. A similar plot for a dioxane naphthalene, PPO, dimethyl-POPOP system indicates that this system is more efficient in that less secondary solute is required to produce the same counting efficiency as the toluene system. In the dioxane-based system also, the counting efficiency is inversely related to the concentration of the primary

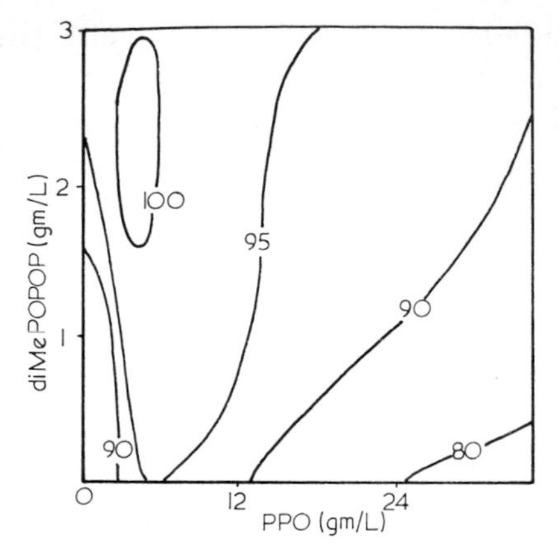

Fig. 2.1. The determination of the optimal composition of PPO and dimethylPOPOP in toluene in the presence of 0.2% nitromethane as quenching agent (Little and Neary 1971, Academic Press, Inc.).

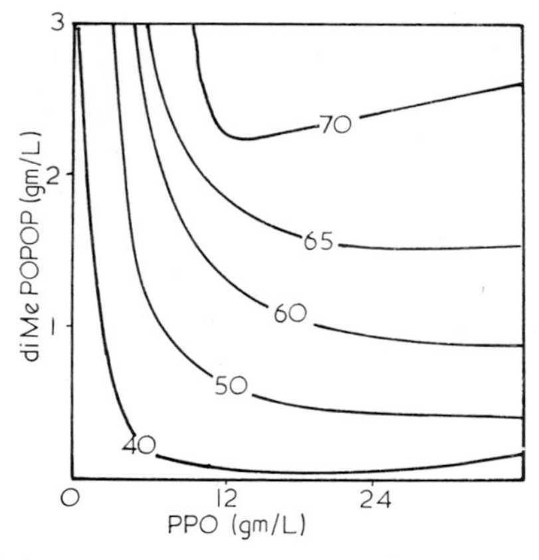

Fig. 2.2. The determination of the optimal concentration of PPO and dimethylPOPOP in a dioxane-naphthalene mixture in the presence of 0.2% nitromethane as quenching agent (Little and Neary 1971, Copyright permission, Academic Press Inc.).

solute, whereas in the toluene system it is fairly constant above 10–12 g/l. Further data based on this analytical approach could give a clearer insight into the nature of quenching and blending agents.

Naphthalene added to a toluene scintillant is a useful secondary solvent in a homogenous system involving hydrocarbon-soluble samples, but it is clearly unsuitable for samples in which there is more than 1% of aqueous sample present. p-Terphenyl is a highly efficient primary solute, but is more sensitive to quenching than PPO, however, it is of considerable value in disc counting, and is significantly cheaper than other primary solutes.

2.1.2. Dioxane-based scintillant mixtures

The use of dioxane as a solvent in place of toluene was suggested by Farmer and Berstein (1952, 1953), who used a solution of p-terphenyl in dioxane (5 g/l) and pointed out that it could accept up to 20% of its volume of water, but counting efficiency was very poor (approx. 3.8% for tritium when the water content was 3%). However, Furst and coworkers (1955) found that considerable enhancement of fluorescence was achieved in such solution by the addition of naphthalene as an energy transporting material and considerably higher efficiencies were thus obtained which enabled larger proportions of aqueous samples (up to 25 to 30%) to be used. Since that time there have been many different compositions described, ostensibly for specialised uses, but which are all grounded on this basic composition.

One of the most successful of these mixtures is that of Bray (1960) who included ethylene glycol and methyl alcohol as blenders in a dioxane solution of naphthalene containing PPO and POPOP. The composition of this mixture (DNppGM) is given in Table 5, Appx. II. A scintillant mixture (DNXpnE) which also accepts a fairly high proportion of water had previously been proposed by Kinard (1957); it contained xylene and ethanol in addition to dioxane and naphthalene. However, Kinard's mixture was mainly used for trapping water derived from combustion experiments (§ 3.2) whereas that of Bray was found to accept higher concentrations of salts and other solutes present in the sample, without precipitation.

It is usually not necessary to use highly purified naphthalene, even commercial grade appears to give entirely satisfactory results (White 1967). Dioxane however, is liable to form peroxides on standing and methods of avoiding contamination involve distillation *in vacuo*

(CARE-EXPLOSIVE!), addition of antioxidants such as BHT (di-t-butyl-4-hydroxytoluene) or maintaining the solvent over granulated zinc. The presence of peroxides in ethoxyethanol and methoxyethanol is probably a greater problem as these alcohols very readily form peroxides. The presence of peroxides in the solvents may be qualitatively and sensitively recognised by the oxidation of iodide to iodine in acid solution. Peroxides should be eliminated as much as possible as they produce considerable and variable quenching.

Method of detecting peroxides in solvents

1) 1 ml of solvent added to 1 ml of potassium iodide solution (10%).
2) Add 0.5 ml of 2 N H_2SO_4 and allow to stand for 2 min. The presence of a brown colour indicates the presence of peroxides. The sensitivity of this reaction is very great and some faint brown colour will always be obtained on prolonged standing, due to the formation of peroxides from atmospheric oxygen in the light.

A drawback in the use of dioxane-based scintillants is the tendency for phosphorescence to occur (§ 1.3). Provided appropriate measures are taken to avoid chemiluminescent phenomena, the dioxane-naphthalene system is ideally suited for ambient temperature counting. Ethyl naphthalene has been shown (Carter and Christophorou 1967) to possess better electron transfer properties and greater solubility than naphthalene itself, features which recommend its use as a more suitable alternative secondary solvent than naphthalene. The ethyl derivate is also less inflammable and volatile and has a higher density, the latter property assisting in its efficiency as an electron capture system. A more detailed study of the electron transfer mechanism involved in naphthalene systems was made by Germai (1970). Scintillation mixtures involving 1-methylnaphthalene have also been described (Goldstein and Lyon 1964).

2.1.3. *Miscellaneous counting systems*

Other solvents have been tried in order to achieve a greater solubility of aqueous solutions, especially those containing inactive and quenching solutes, without excessive quenching of the primary solute itself.

The benzo- and aceto-nitriles have also been suggested (Gomez et al. 1971). Although these are less efficient than toluene as pure solvents for PPO, their real advantage is seen on dilution with quenching blenders or aqueous solutions, when it can be seen that the degree of quenching as dilution proceeds is less than with blended dioxane or toluene scintillants. A 40% solution of naphthalene in benzonitrile would appear to have interesting potentialities as a water-accepting scintillator mixture. An additional interesting feature of this solvent is the fact that it has a fairly high fluorescence yield in the pure state and has been suggested as a useful solvent for use with alpha-emitters, when it can be used without any added primary solute.

An alternative method of detecting particles of higher energy is to use large tanks of scintillating fluid in order to increase the chance of capturing those higher energy particles which would have otherwise escaped from the normal 25 ml scintillation vial. Such a tank could then be used for the detection and measurement of neutrons, deuterons and positrons. The response of mineral oil-based scintillant mixtures was compared by Batchelor et al. (1961) for electrons, protons, alpha particles and recoiling carbon atoms. A number of commercially available scintillants were also compared by Smith et al. (1968) for their response against electrons, protons and deuterons using anthracene as a standard scintillant. Only stilbene (1.21 ± 0.15) showed a response greater than anthracene to 0.525 MeV electrons. The commercially available NE 224 (Nuclear Enterprises, U.K.) is a liquid which is very transparant to its own scintillation emission, and has been used for the detection of neutrons. A Pyrex pipe, ten feet long and $2\frac{1}{2}$ in diameter was filled with the liquid and photomultipliers separated from the liquid by $\frac{1}{8}$ in quartz windows were placed at either end (Berkowitz 1969). This particular detector was designed primarily as a position detector of neutrons.

Large-volume detectors have also been employed for the assay of low levels of gamma-emitters, such as ^{57}Co and ^{60}Co in large volumes of urine (250 to 1000 ml). This is a particularly useful method for the detection of small amounts of labelled Vitamin B_{12} secreted from patients with anaemia (Schilling 1964).

A further elaboration of the scintillation mixture is that described by Ashcroft (1970) who used tetrabutyl-tin-loaded (35% w/v) toluene-based scintillant (Tpp5, Table 3, Appx. II) to increase the cross section capture and hence increase the efficiency of counting of certain soft gamma-emitting isotopes such as Iodine 125. The type of vial used in this tecnique is described in § 1.4 in greater detail.

2.2. *Heterogenous counting systems*

A heterogenous system from the point of view of liquid scintillation counting may be defined as a two phase system in which most of the scintillant is dissolved in one phase and most of the beta-emitting sample is incorporated into the second phase. Such a two phase system can be either solid : liquid or liquid : liquid. In the case of the solid : liquid system, the scintillant can be incorporated into either phase; the sample being incorporated into the opposite phase. There are certain general considerations which can be obtained for the heterogenous system as a whole by studying these two types of solid : liquid systems and it would appear that these considerations may clarify many of the anomalies that are experienced in the more widely used liquid : liquid two phase systems.

Scintillant in the solid phase

The earliest work in liquid scintillation counting was carried out using suspended scintillants. For example, Steinberg (1958) introduced the Pilot B beads, i.e. either beads or filaments of diphenylstilbene packed into a scintillation vial and covered with an aqueous solution containing a beta emitter. Efficiencies for carbon 14 of around 20% were achieved, and of tritium of 0.5%. Three-fold higher efficiencies were obtained when a specially pure preparation of anthracene, known as blue-violet anthracene (Steinberg 1959) was used. The efficiency of the system was still further increased by the addition of a trace of detergent, presumably to increase the effective contact surface of the scintillant with that of the radionuclide-containing aqueous solution. The determination of tritium and carbon-14 in aqueous solution using anthracene powder was described by Schram and Lombaert (1962).

Use of detergent-coated anthracene (Myers and Brush 1962). Blue-violet grade anthracene (10 g) is treated with 10 ml of an aqueous solution of Triton GR-5 (1:333 dilution, Rohm and Haas, Inc.). The anthracene is filtered off and dried *in vacuo* over a drying agent.

The anthracene is loosely packed into a scintillation vial and the aqueous solution is pipetted on to the crystals. Efficiencies of 50 to 100% have been reported for a variety of nuclides and one of the chief advantages of the system is that the counting efficiency is relatively unaffected by solutes which would normally interfere and quench very considerably in homogenous systems. Efficiencies of 29% for carbon-14 and 0.9% for tritium have been obtained, but in view of the fact that the sample can be recovered for further processing and that there is no necessity to allow for chemical quenching, there are probably many instances where this procedure may be very useful, especially in cases where nuclides have given higher efficiencies, e.g. ^{45}Ca (up to 90%).

The use of solid suspended scintillants has been reviewed by Rapkin (1963) and useful details about the parameters which are to be considered in this type of counting were reviewed by Steinberg (1960). The latter author also noted that the maximum efficiency of counting was achieved when the conditions to achieve the highest 'Instrument figure of merit' (i.e. E^2/B) were used. The performance of the suspended scintillant is unaltered by pH, but foaming, if it occurs, does produce some anomalous effects. 75% Ethanol, methanol and acetone solutions can also be used successfully in this system.

Scintillant in the liquid phase

This system is by far the most commonly used of the heterogenous methods, in which the beta emitter is attached to or absorbed in the solid phase and the scintillant is in a solution of the electron-transferring hydrocarbon solvent. The solid phase may be a finely dispersed powder (Suspension and Gel counting, see § 2.2.2) or attached to a sheet of the solid phase as in disc counting (§ 2.2.1). It is also worthwhile considering before-hand whether the sample preparation technique employed results in a heterogenous or a homogenous sample product, as this is not always immediately obvious. A good example of this arises in the assaying of polyacrylamide gels, where an apparent solubilization is in reality a swelling of the gel, and an improvement in the heterogenous association of the scintillant mixture with the beta-emitter attached to the gel support phase (see § 2.2.2).

In order to achieve a high counting efficiency in this system, all materials which would cause quenching in the liquid phase should be avoided, and hence the solid support should contain as little water or toluene soluble quencher as possible in order to achieve a closer, unquenched association of the beta-emitter with the scintillant in a high efficiency, toluene-based scintillant. However, this is not always feasible, and methods of absorbing materials to finely divided surfaces suspended in aqueous rich mixtures of the scintillant are used. However, in general terms, the system should be examined thoroughly to see if the aqueous phase can in some way be removed prior to counting as marked improvement in the counting efficiency can often be achieved.

2.2.1. Disc counting

A logical method of assaying beta-emitters was to regard the photomultiplier tube as a Geiger tube and the photon emission could thus be measured in the same way as in the earlier planchet systems. Early attempts to count the photon emission accurately consisted of laying the scintillant-soaked paper, containing also the radionuclide, directly on to the surface of the photomultiplier tube itself (Roucayrol et al. 1957; Seliger and Agranoff 1959).

Many types of solid supports have been tried, amongst which are filter papers of varying grades, cellulose esters (e.g. Millipore), glass fibre, lens-cleaning tissue, ion-exchange papers and ion-exchange impregnated papers. There are many advantages as well as problems associated with these supports, and an attempt will be made to summarise these since careful choice of support can considerably improve counting technique. A useful review of different types of supports was undertaken by Gill (1967), who concluded that glass fibre and cellulose esters exhibited very similar efficiencies, which were consistently higher than those of any grade of filter paper. In an attempt to reduce the serious self absorption quenching of filter paper, lens-cleaning tissue has also been recommended by Weg (1962).

Cellulose filter paper

A number of commercially available cellulose filter papers were examined by Wang and Jones (1959) who concluded that in general, by counting carbon-14 on a solid support, a reduction of efficiency from 80 to 55% was usually experienced when compared with a homogenous system. This corresponded to an absorption by the support material of some 25 to 30% of the activity in the case of carbon-14. The problem of self-absorption in the case of tritium is of course much greater and these factors will be considered later.

The orientation of the paper with respect to the photomultiplier was discussed by a number of workers, in particular Geiger and Wright (1960) and Bousequet and Christian (1960). They concluded that for carbon-14 this factor was unimportant, but for tritium, the orientation of the paper was important. Better reproducibility, with the sacrifice of some efficiency, was obtained if a paper strip was coiled into a cylinder and placed vertically in the scintillation vial and covered with scintillant (Loftfield 1960). However, it was soon recognised that by cutting the paper into a disc (2.4 cm dia.) and laying it on the base of the vial, equally efficient counting and reproducibility could be obtained, for most isotopes except for tritium (Davidson 1961, 1962). Some idea of the variability that orientation may introduce is suggested by the fact that counting paper discs in a recumbent position will introduce a variability of about 1.5% whereas strips in the vertical position show a variability of about 7.1%.

In order to attain maximum efficiency with this system, the disc should be thoroughly dried, usually by exposure to an infrared lamp; the process of drying is even more important with filter paper than with glass fibre discs and the parameters associated with the drying of this support material are discussed in the following section dealing with glass fibre discs. The discs require to be just covered by a non-polar scintillant mixture such as Tpp4 (Table 3, Appx. II) (Funt and Hetherington 1960); it has even been suggested that soaking the paper with scintillant in a vial is sufficient (Nunez and Jaquemin 1961) and provided the user is satisfied that reproducibility can be obtained under his own special conditions, a considerable saving in the quantity of scintillant used may thus be achieved. The somewhat more energetic beta-emission from sulphur-35 however can be assayed with a greater degree of reproducibility on cellulose filter paper. The efficiency on Whatman No. 1 paper is approx. 45% whereas that on Whatman 3MM, a thicker product often used in electrophoretic techniques, is of the order of 40% (Davidson and Riley 1960a, b). Dual isotope counting on paper discs has been demonstrated for combination of iodine-131 and tritium, where the gamma-emitting iodine is markedly different from that of the tritium (Roche et al. 1962). Dual isotope measurements are considered in greater detail in § 11.2.

Another feature of disc counting, which is more important with glass fibre (Fig. 2.3), is the ability to count several discs in one vial, without serious loss by mutual absorption. Up to 25 mg of potassium gluconate-^{14}C could be counted per cellulose paper disc and up to three discs (corresponding to 75 mg) per vial, before self-absorption effects

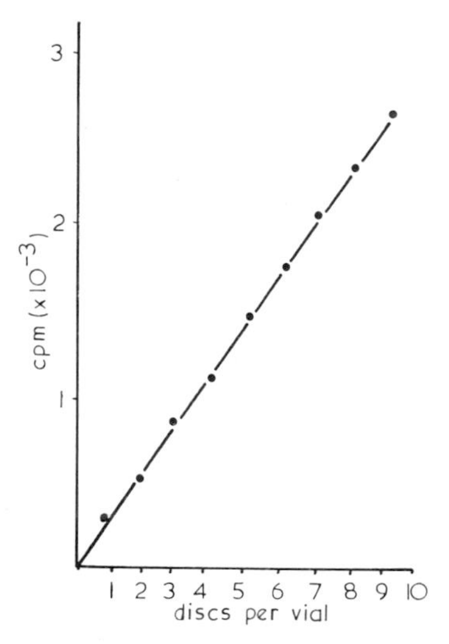

Fig. 2.3. The strictly additive effect of assaying in one vial different numbers of discs on which tritium-labelled thymidine has been equally dispensed and dried.

were evident (Segal and Blair 1962). Some authors have claimed that homogenisation of the paper in the scintillant prior to counting gives better reproducibility (Chakravarti and Thannassi 1971). If the material to be measured is soluble in methanol, it is often more efficient to place the disc in a scintillant which can dissolve the material off the paper. This aspect was considered for tritium by Cayan and Anastassadis (1966). They showed that the effect of extraction can increase the counting efficiency of tritium from 9 to 14%. Some workers advocate the disc technique also on the basis that the non-polar scintillant employed can be re-used, thus effecting a saving. If the system employed is truly heterogenous then there should be no removal of the beta-emitter by the scintillant mixture. However, in view of the fact that only small volumes of scintillant need be used in practice, this saving is fallacious. Although beta-emitter may not be extracted, there is evidence that a quenching agent may be differentially extracted from the paper, thus inducing an unknown and unnecessary variability in the system (Cramer and Arnott 1972) and in the long run may prove to be more expensive. The use of filter paper discs for enzyme kinetic studies was successfully employed by Bollum (1959, 1963) and Mans and Novelli (1961) and was reviewed by Bollum (1966). However, the use of ion-exchange paper has been more widely adopted for this purpose and will be described on p. 56.

Glass fibre discs

Since the introduction of glass fibre in the form of thin sheets, many uses have been found for it in biochemical work. One significant advantage is the fact that no absorption into the fibre itself can occur and the discs can be treated with strong acids in order to dissolve off materials without any deleterious effect on the sheet itself. Glass fibre discs have been shown to be consistently better than cellulose filter paper for the non-selective absorption of material and higher counting efficiencies are obtained. Cellulose paper is only about 60% as effective as glass fibre (Pinter et al. 1963; Gill 1964; Davies and Cocking 1966). Furthermore, up to 25 discs can be counted in the same vial without loss of efficiency due to mutual absorption between the discs (see Fig. 2.3). Consequently, if 1 or 2 discs are insufficient, more can be added until satisfactory counting statistics are achieved; thus if 0.3 ml of aqueous solution is applied to each disc, which is dried and placed in a vial with 16 other similar discs, as much as 5 ml of aqueous solution can be assayed at once. In order to achieve better reproducibility between discs in different vials, it is usually advisable to maintain the same amount of solid material present by adding a standard amount of carrier macromolecule to each disc to ensure that the degree of quenching remains the same throughout. In some cases, glass fibre discs supporting precipitates have been homogenised in the scintillant before counting to increase the reproducibility by allowing for a more constant absorption of the precipitate on to the suspended glass fibres (Johnson and Smith 1963).

Certain precipitates, such as $Ba^{35}SO_4$, collected on glass fibre discs are often homogenised and suspended in a gelling agent to achieve more consistent counting geometry (Gottshalf et al. 1962). Some isotopes, for example ^{89}Sr may be assayed on glass fibre discs with efficiencies of 65% (Creger et al. 1967) and such discs are 90% more efficient than paper discs (Hutchinson 1967).

An alternative procedure, and the method to be recommended if absolute levels of radioactivity need to be determined, is to remove the beta-emitter from the disc by means of a solubilizer (Bransome and Grower 1970, 1971). Tritiated RNA is also more accurately determined in this way (see below), using a solubilizer such as NCS (Birnboim 1970), than if it was assessed directly in a toluene-based scintillant as suggested by Malt and Miller (1967). This method of solubilization is advisable when double isotopes are involved, since the usual quench correction procedures may then be applied to the homogenous counting system so produced. Reduction of any chemiluminescence present by the addition of acetic acid was also suggested by Birnboim as a means of increasing the relative efficiency by decreasing the background.

Method of solubilizing material from glass fibre disc (*Birnbom 1970*)

1) Dry glass fibre disc by suction until the colour goes from gray to chalky white.

2) Place in a scintillation vial and add 0.5 ml diluted 'NCS' (1 part 'NCS' and 3 parts 'Liquifluor' (42 ml/l toluene)).
3) Stand 30 min at room temperature.
4) Add 5 ml toluene: acetic acid scintillant (42 ml 'Liquifluor' and 1 ml glacial acetic acid, per litre toluene).
5) Sample usually clears on standing, count.

The drying of glass fibre discs is easier than that of other solid support materials since the discs may be treated with alcohol, acetone or ether to accelerate the drying process. Several important features of the drying procedure are illustrated in Fig. 2.4. With water still present, i.e. using a very short drying period, the dioxane-based scintillant (DNpp-7) extracts thc bcta-emitter (in this case ^{3}H-alanine) from the disc and counts it efficiently in the scintillant, but with longer drying the reduced amount of water present enables the alanine

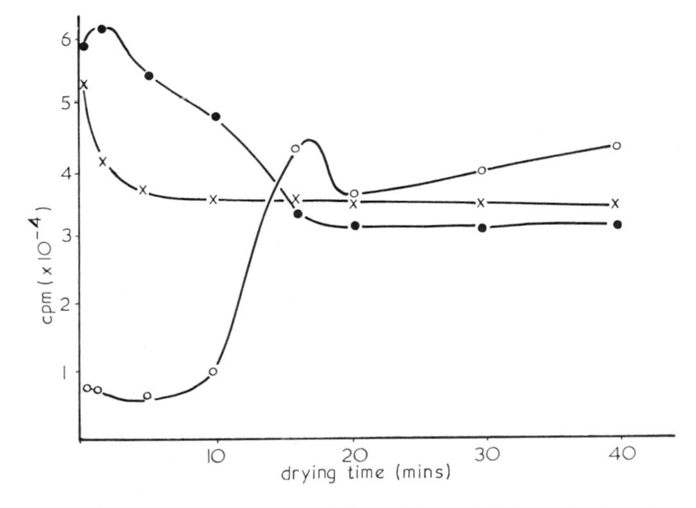

Fig. 2.4. The effect of drying time (by 2 infra-red lamps 18″ from discs) on the relative counting efficiency of glass fibre discs on which an aqueous solution of tritiated alanine has been dried. ●———●, in a dioxane naphthalene-based scintillant (DNpp-7); ×———×, in a Triton X-100:toluene (1:2) scintillant mixture; ○———○, in a blended toluene-based scintillant mixture such as TNppX (see Table 4 Appx. II). (Reproduced by kind permission of Beckmann Instruments Ltd.)

solution to make a homogenous counting system with the dioxane-based scintillant. In the case of the toluene-based scintillant however (Tpp4), after a short drying period the low counts may be due to the water forming a barrier between the beta-emitter and the non-polar scintillant; further drying removes this barrier and the counts increase. It can also be seen that under these conditions of drying (2 infrared lamps approx. 18″ from the discs) the highest efficiency is still not achieved after 40 min of drying. It is therefore useful to know the detailed experimental conditions used by the worker. It is also interesting to note that a Toluene : Triton X-100 scintillant (2 : 1) seems to possess properties of both types of polar- and non-polar systems as far as disc counting is concerned, and gives a count which is less dependent on the degree of drying of the disc. This system could therefore have distinct advantages.

The usual disc diameter is 2.4 cm and it is often less expensive to punch out discs of this size from larger circles or sheets. This is particularly true for glass fibre discs and cellulose acetate membranes, where a cost saving of 5 to 10% (excluding labour costs) may be obtained. A simple brass punch is illustrated in Fig. 2.5. Drying of the discs is best conducted beneath infrared lamps, with minimal contact of the discs with the underlying surface. This can be done by using a cork mat through which pins have been pushed so that the points (approx. $\frac{1}{2}$″ distant from one another) form a 'bed of nails'; the discs lie on top and touch only the points. An alternative is a cork mat into which round-headed mapping pins have been placed; in this case contact is only on the upper spherical surface of each pin. Yet another alternative is to lay the damp discs on to a water-repellent silicone treated glass surface. Marking glass fibre discs is a problem since graphite is easily washed off. A small animal ear-pearcing punch can be used to code the discs (on the edge away from the main filtering area). Glass fibre discs easily disintegrate if pins are used through the disc, and this procedure is not to be recommended.

Ion exchange papers

An important application of disc counting is the use of ion exchange paper, such as DEAE-cellulose, to separate charged from uncharged

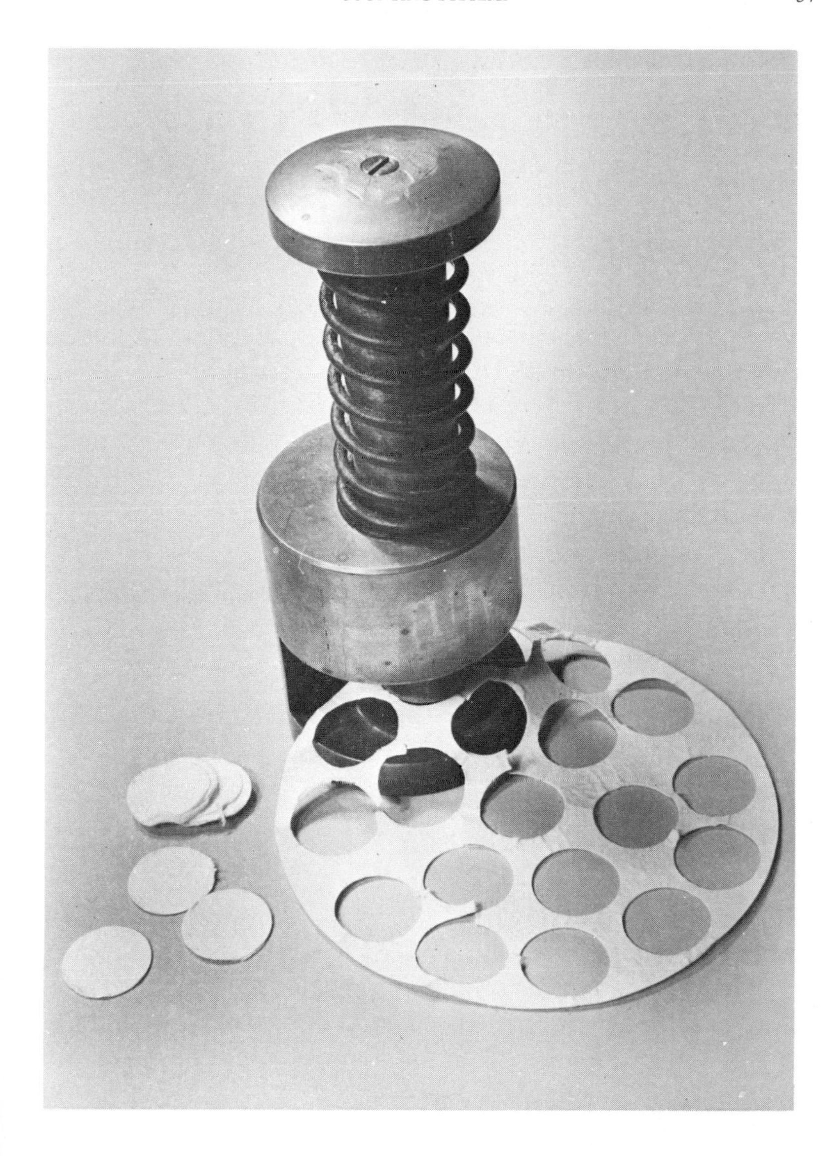

Fig. 2.5. A brass punch suitable for punching out 24 mm circles for liquid scintillation counting.

Subject index p. 309

components of a mixture. This system is particularly useful in enzyme assays which involve the conversion of an uncharged substrate into a charged product. The latter becomes associated with the ionic charges on the disc while the unchanged substrate is washed through. By counting the disc before and after washing, it is thus possible to assay the extent of conversion.

In the estimation of amino acid activating enzyme, a separation of ^{14}C amino acid from ^{14}C hydroxamate was required (Loftfield and Eigner 1963) and Amberlite cation-exchange paper SA-1 (Rohm and Haas, Inc.) was employed. DEAE-cellulose discs have been used for the assay of the products of a thymidine kinase system, and a suitable washing procedure necessary to achieve 98% removal of thymidine while at the same time retaining 100% of the phosphorylated product on the disc was described by Breitman (1963). This procedure is given below.

Method of washing DEAE-cellulose disc free of thymidine in a thymidine kinase assay (Breitman 1963)

1) DE 81 discs (Whatman, W. and R. Balston, Ltd.) added to 0.001 M ammonium formate (20 ml per disc).
2) Stand 10 min.
3) Wash twice with 0.001 M ammonium formate.
4) Wash once with distilled water.
5) Wash once with ethanol.
6) Air dry, add to 5 ml toluene-based scintillant mixture (Tpp4, Table 3, Appx. II).

Alternatively, the discs may be placed in 95% ethanol at room temperature for 5 min with intermittant stirring, blotted on paper towelling and dried (Furlong 1963). A second set of discs which have not been washed with alcohol are dried on a non-wettable surface, to act as control. The enzymic conversion is simply the ratio of the cpm of the discs to that of the unwashed discs.

Similar procedures (using discs) have been described for other enzyme systems, such as galactokinase (Sherman 1965; Gulbinski and Cleland 1968); purine phosphoribosyl transferases (Atkinson and

Murray 1965; Nicholls and Murray 1968); adenosine diphosphate glucose pyrophosphorylase (Ghosh and Preiss 1966); adenosine triphosphate creatine phospho-transferase (Morrison and Cleland 1966) and glycerol kinase and hexokinase (Newsholme et al. 1967; Himms-Hagen, 1968). However, in view of the differential absorption artifacts observed with cellulose filter paper supports, it would be advisable to check the system thoroughly for such artifacts before commencing a series of measurements (Furlong 1970). The different salt concentrations during the enzyme reactions can also seriously alter the level of absorption of the beta-emitter on to the disc (Roberts and Tovey 1970). For example, salt concentrations as low as 12.5 mM can interfere with the assay of alkaline phosphatase. Furthermore, the method of drying the discs is also important in achieving reproducibility, and it is necessary to determine the length of time required for reproducible drying, under the drying conditions used by the worker, as with glass-fibre discs referred to on p. 55. Preliminary checks on these factors would prevent the interpolation of artifacts into the experimental results. Useful reviews of radiotracer techniques as applied to enzyme assays have been given by Reed (1968) and Oldham (1968).

Cellulose ester membranes (e.g. 'Millipore')

Although glass fibre discs have been shown to be highly efficient supports for liquid scintillation counting, cellulose nitrate discs have also been used extensively for this purpose. The weak ion exchange activity has made it particularly suitable for studying hybrid formation in the nucleic acid field (see § 8.2).

However, there are several factors associated with the use of cellulose ester membrane counting which should be recognised and which also apply to other absorptive supports. Materials can absorb *on to* and *into* supporting fibres which can lead to different degrees of quenching by different substances on the same disc (Furlong 1970; Bransome and Grower 1970, 1971). A typical example is the marked discrepancies in assessing labelled adenine and cyclic AMP mixtures, since these exhibit different penetration characteristics into the cellulose ester matrix. Increased efficiencies for counting ^{3}H-DNA over that of the precursors is another effect of such a phenomenon (Furlong 1970; Furlong et al. 1965). Details of parameters controlling adsorption of DNA on to the fibre surface are also discussed in § 8.2.

In some cases, authors have suggested that precipitates on cellulose ester discs should be dissolved before assaying and this is particularly useful with tritium or in mixed isotope studies where homogenous counting is desirable. This can be done by incubating the disc with concentrated ammonium hydroxide as follows.

Method of dissolving RNA and protein precipitates from 'Millipore' discs (Penman et al. 1964)

1) 5% TCA precipitated ^{3}H-RNA filtered (together with 200 mg yeast RNA as carrier) on to membrane disc.
2) Wash twice with 5% TCA.
3) Place membrane in vial and add 1.5 ml 10 N NH_4OH.
4) Stand at room temperature for 60 min.
5) Add 20 ml Bray's scintillant (Table 5, Appx. II).
6) Cool to 0°C for 5 hr. Count.

2.2.2. Suspensions and gels

Hayes et al. (1956) showed that material in suspension could be measured by liquid scintillation counting by finely grinding materials like $Ba^{14}CO_3$, ^{14}C-phenylalanine, ^{14}C-liver tissue, ^{14}C-bacteria, $Ca^{14}CO_3$, ^{35}S-benzidinium $^{35}SO_4$, sodium ^{3}H-acetate etc. and counting directly in a toluene; PPO; POPOP mixture. The sample was shaken and counted several times and the mean of the values taken. Alternatively, the samples were shaken and a count versus time plot was constructed from which extrapolation to zero time would give a reproducible result. Except for tritium, there appeared to be little trouble from self absorption properties for a number of isotopes, provided the grinding of the sample was fine enough. It was noticed in particular that opacity of the sample did not reduce the counting efficiency as much as expected but a trace of colour was considerably more quenching. Early work of this type was reviewed by Helf (1958), Birks (1964) and Schram (1963).

The use of finely divided suspensions has continued to be employed up to the present, and several methods for a more efficient dispersion and grinding of the precipitate have been suggested. In 1967, Alfred suggested that sonication was a useful method of reducing the particles of $Ba^{14}CO_3$ to a form suitable for counting in a toluene-based scintillant. Hoffman (1965) suggested dispersing a similar $Ba^{14}CO_3$ precipitate in the form of a toluene aerosol directly into the toluene based scintillant. Earlier (1963) the same author had proposed a similar procedure for dispersing a tritiated polymer suspension. The very finely divided particles (approx. 100 to 200 μm dia) behave very closely to a homogenous counting system. Mammalian cells on cover slips (Fallot et al. 1965), dried suspensions of tissues and faeces (Handler 1963) and tritiated wool (Downes and Till 1963) have all been assayed directly. ^{14}C in soil has also been measured by direct suspension and the various parameters of weight, colour, sampling time in relation to counting efficiency have been examined (Page et al. 1964).

Thixotropy

Certain colloidal systems show under certain conditions, the phenomenon of thixotropy, where it is possible to liquify a gel by shaking it or by subjecting it to ultra-sound, and on cessation of the mechanical shock, the liquid will reset to a stiff gel. The word is derived from the Greek 'to change by touch'. Funt (1956) showed that addition of aluminium stearate (5%) to a toluene or a dioxane-based scintillant in which a finely divided sample was suspended, caused it to gel and hold the suspension particles apart. However there were technical difficulties in manipulating such gels. This was overcome by employing this property of thixotropy and White and Helf (1956) described the use of thixotropic Baker Castor Oil Co. product known as Thixcin. A mixture of 25 g Thixcin per l of toluene-based scintillant was first homogenised in a Waring blender and the finely suspended sample was then simply added to an aliquot and the whole was shaken. On standing, the mixture assumed a solid gel and the particles were thereby held apart. Approximately 0.5 g of $Ba^{14}CO_3$ may be suspended in 10 ml of this scintillant and assayed between 40 to 50% efficiency. Self-absorption did not appear to arise with this isotope when the original precipitate was less than 60 mesh size. Suspension counting in such gels was up to 10 times more efficient than measuring in a Geiger–Muller end-window instrument by planchette methods.

An interesting effect was observed by these workers when using labelled hexamethylene tetramine. Small quantities showed a high counting efficiency, but addition of further quantities resulted in a degree of quenching which reached a plateau value as further material was added. This suggests that some of the material dissolved in the organic scintillator and quenched the counting system. This has since been observed to be a common feature of heterogenous counting systems in general and suggests that where possible, this technique is best applied to the counting of those substances which are totally insoluble in the scintillant mixture used. Where some solution does occur, the addition of the secondary solute POPOP or its methylated analogue, has been shown to assist in maintaining efficiency of counting (Helf and White, 1957). Materials which normally quench considerably in a scintillator solution e.g. the polynitro compounds, may be counted more efficiently in a heterogenous counting system (Helf et al. 1960), especially if one can be found in which they are totally insoluble. Other isotopes, such as ^{36}Cl, ^{22}Na, ^{90}Sr and ^{133}Ba may also be assessed by this method. Nathan et al. (1958) suggested that $BaCO_3$ gels may provide a more efficient method of determining ^{14}C-CO_2 from combustion techniques, by virtue of the lack of excessive quenching which is a feature of other absorption methods (see § 4.4). Both Cluley (1962) and Harlan (1963) showed clearly that the procedure was indeed superior to Geiger counting.

A further modification of the Thixcin gel technique was described by Shapira and Perkins (1960) who combined this gelling agent with the solubilizing agent, hyamine hydrochloride (see § 3.2). The composition of their scintillant mixture was PPO, 0.4%; POPOP, 0.01%; Thixcin, 3%; Hyamine 10X, 5%; aqueous suspension, 5%. Efficiencies up to 50% were also reported for this system for carbon-14 labelled materials and 10% for tritiated water. These correspond to Merit values (see § 2.3) of 250 for carbon-14 and 50 for tritium. Another material was described by Ott et al. (1959) known commercially as

Cab-O-Sil, M-5 (G.L. Cabot, Inc., Boston, U.S.A.), which consisted of very finely divided (0.015 to 0.02 μ) pure silica in a hydrated form which produces a firm gel with toluene- or dioxane-based scintillants. It is usually employed as a 3 to 5% (v/v) concentration and 10 ml of such a system can support up to 1 g of $Ba^{14}CO_3$. It has been found to be particularly useful in the assay of ^{89}Sr in bone ash preparations. Its use with biological material such as dried algae (Yarbrough et al. 1966) is mainly as a suspending agent. Its useful feature is its ability to absorb beta-emitter on to its very large effective surface area (175 to 200 m^2/g) and reduce the effects of wall absorption of isotope (Blanchard and Takahashi 1961). This will effectively improve the geometry of counting from near 2 pi to something approaching 4 pi. Merit values for carbon-14 and tritium of 420 and 91 respectively were reported by Gordon and Wolfe (1960) when counting 6.5% aqueous solution using 4% Cab-O-Sil. Estimations of carbon-14 in plasma solutions (Spencer and Banerji 1970), traces of $^{137}CsClO_4$, $^{90}SrSO_4$ and $Ag^{131}I$ in rainwater (Germai 1963, 1964) as well as radioisotopes in TLC chromatograms (§ 9.3.1) have all been improved by the concomitant use of this material during suspension counting.

An alternative material is 'Aerosil' (Bush Beach and Segner Bayley Ltd., Marlow House, Lloyds Av., London, EC3) which has been recommended to be used in a similar way to Cab-O-Sil at a concentration of 8% v/v (Eakin and Brown 1966).

High-viscosity methods

One of the problems associated with the production of gels by the use of aluminium stearate was the necessity to heat the mixture in order to produce a gel. This was later overcome by Funt (1961) by the use of aluminium-2-ethyl hexanoate, which did not require heating to produce a gel. Other methods used to increase the viscosity of scintillant mixtures include 25% polystyrene (Hayes 1958) which enabled up to 4 g of bone ash to be suspended in 10 ml of toluene-based scintillant. Methyl methacrylate has also been used to increase viscosity (Shakhidzhanian et al. 1959) and 5 to 8% in a toluene-terphenyl-POPOP scintillant mixture is fluid at 100°C and sets to a gel at room temperature. ^{40}K, ^{90}Sr and ^{137}Cs incorporated into such a system can be assayed at almost 100% efficiency.

A gel structure has also been produced *in situ* by reacting a mixture of commercial toluene diisocyanate (Hylene-TM-65) (du Pont) with a mixture of branched primary amines known as Armeen L-11 (Armour Ind. Chem. Co.) (Bollinger et al. 1967). Alternatively, Benakis (1971) has described resins based on polyolefins derived by polymerization of ethylene, butylene, etc., known as Poly-Gel-B which is soluble in warm toluene and xylene solutions and sets (10%) to a gel on cooling. An efficiency of 44% for carbon-14 using 50 mg dry rat liver powder is claimed for this procedure.

A possible advance in this field would be to employ a finely dispersed solid scintillant in a colloidal form in addition to a scintillator solution, and to take full advantage of the near 4 pi counting conditions that could be achieved.

2.2.3. *Colloid counting*

The interest in colloid counting in biochemistry is derived from the fact that the frequent requirement of the counting of salt or other substrate-rich solutions in large numbers can present a formidable task if the conventional preprocessing techniques of macromolecular precipitation or isolation are undertaken. In this system the importance of the 'Figure of merit' or 'Merit value' should be stressed here. In order to salvage as many counts as possible in low specific activity solutions, it is necessary to find out those conditions where the ratio of the volume of aqueous solution to that of scintillant mixture is optimal to obtain the maximum counts in the scintillation spectrometer. To assess this value, the product of the percentage of aqueous solution and the counting efficiency is assessed. This merit value, corresponding to the 'Figure of merit No. 1' of Paterson and Greene (1965) and to the 'Merit number' of Van der Laarse (1967) will enable such a comparison to be drawn for any particular instrument. To compare such values between instruments, Paterson and Greene (1965) have recommended that this value be divided by the counting efficiency of a standard Toluene, PPO (0.4%), POPOP (0.01%) in the instrument used. The merit values given here are calculated for the instruments on which the work was done, which unfortunately is not known in all cases.

The problem of incorporating larger volumes of aqueous solution into a suitable scintillant mixture was examined by Shapira and Perkins (1960) who noted that by the use of a gelling agent, thixcin (see § 2.2.2) with hyamine hydroxide, they were able to incorporate up to 5 % of aqueous solution and obtained merit values of 50 and 250 for tritium and carbon-14 respectively. In 1962, Meade and Stiglitz reported the use of a non-ionic detergent, Triton X-100 (Rohm and Haas), (iso-octyl phenoxy-polyethoxyethanol, Fig. 2.6), to assist in the

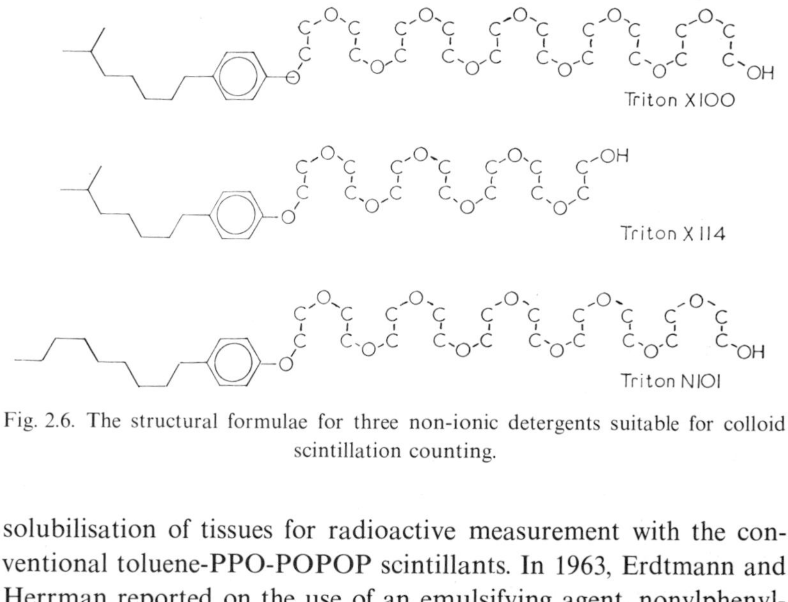

Triton X100

Triton X114

Triton N101

Fig. 2.6. The structural formulae for three non-ionic detergents suitable for colloid scintillation counting.

solubilisation of tissues for radioactive measurement with the conventional toluene-PPO-POPOP scintillants. In 1963, Erdtmann and Herrman reported on the use of an emulsifying agent, nonylphenylpolyglycol ether, which they used in place of hyamine hydroxide. Paterson and Greene (1965) examined the Triton X-100 system further and studied the relationship of the counting efficiency of both tritium and carbon-14 to the proportion of the detergent present; they concluded that a relatively large amount of detergent could be added before there was a significant decrease in the counting efficiency as measured by the merit value. However, these authors only described the use of two Triton X-100 : toluene compositions, tT21 (i.e. toluene : Triton X-100, 2 : 1) and tT76. Turner (1969) re-examined the tT21 system, using D-glucose-^{14}C (U) and tritiated water, the percentage of the mean value obtained by counting in a homogeneous system was plotted against the corresponding sample weight (ranging from 200 mg to 3 g in each case). Although the curves were similar, they did not overlap precisely. Both exhibited a sharp dip around 1.8 g wt of aqueous phase (Fig. 2.7). The counting efficiency for a given channels ratio was also found by Turner (1969) to be higher with Triton X-100 : toluene scintillant (2 : 1) than with Bray's scintillant and differences were also found between the use of oil-soluble (e.g. hexadecane $-{}^{3}$H or toluene ^{3}H) and water-soluble (e.g. water $-{}^{3}$H) standards.

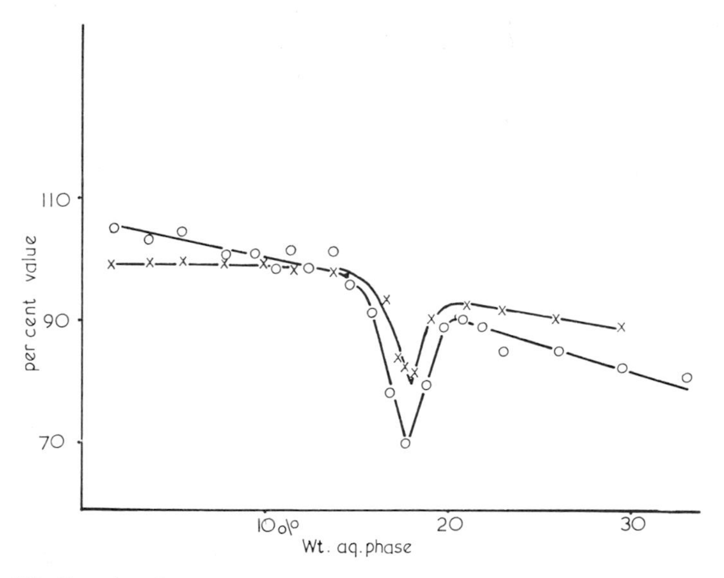

Fig. 2.7. Counting of aqueous carbon-14 (glucose) (×) and tritium (water) (○) samples in a Toluene: Triton X-100 (2 : 1) scintillant mixture, containing 0.4 % PPO and 0.1 % (w/v) POPOP in the toluene phase. The percent value compared with that obtained by measuring in a homogenous system (Turner 1968).

It is thus clear that to understand this system, the whole phase diagram must be studied and not just one, empirically derived, scintillant composition.

A considerably improved method of analysing this system was first described by Van der Laarse (1967) in his study of ground water samples. A phase diagram was constructed based on the appearance of the colloid at 3°C, the operating temperature of the counting instrument used. The complexity of the system was then realised and the reason for the difficulties encountered by the other workers become understandable. By comparing the diagram with the merit values of each point in the phase diagram, one can locate a region where the maximum counting efficiency of that particular sample can be determined. However, one problem in their work that had previously been raised by Benson (1966) was the instability exhibited by the counting system after mixing, as well as the marked temperature dependency of the efficiency of counting. With these conditions in mind the 'phase diagrams' of a number of commonly used solutions in biochemistry were constructed (Fox 1968), together with similar triangular plots of the merit values, from which it was possible to deduce optimal counting conditions where the merit value was highest for a counting stability within a prescribed variance. Although the appearance of the mixture has been used by many workers as a criterion of correctness of composition, this is by no means a valid guide, and the superimposition of merit value data with counting stability over a period of 48 hr is the only valid parameter combination to use in order to obtain the best counting conditions.

Constructing a phase diagram (Example: Triton X-100 : Xylene: Water.)

1) Arrange 36 vials in the form of an equilateral triangle as shown in Fig. 2.8 and marked as in Fig. 2.9.

2) Add 0.5 ml of a concentrated PPO solution (2.67 g PPO, 0.067 g POPOP in 50 ml toluene) to each vial.

3) Add remaining toluene, Triton X-100 and aqueous solution in the volumes shown in Fig. 2.9.

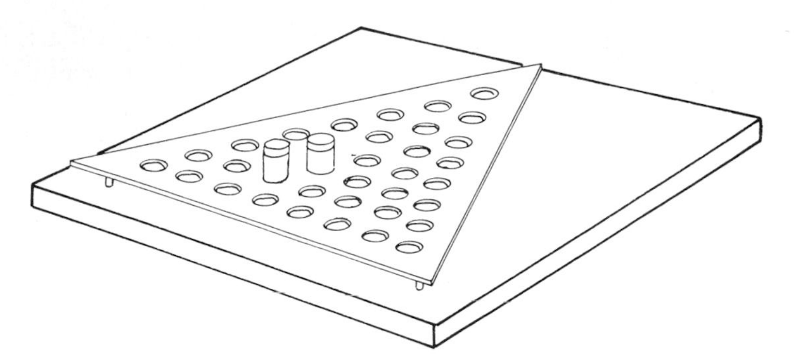

Fig. 2.8. A simple wood and metal support for the construction of a 'phase diagram' to determine the optimal counting conditions using a colloidal counting system.

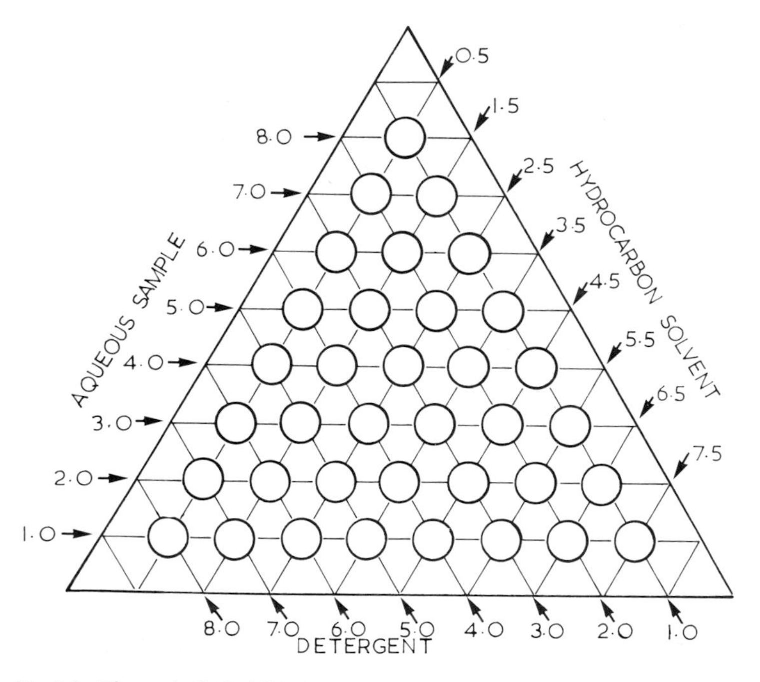

Fig. 2.9. The method of addition of components for the construction of a phase diagram to determine the optimal counting conditions using a colloidal counting system. (Reproduced by kind permission of Pergamon Press.)

4) Shake the vials on a vortex mixer, count to check for background and chemiluminescence.

5) Add 10 μl of either aqueous tritium standard, or ^{3}H-toluene to each vial and shake thoroughly.

6) Count over multiple cycles (preferably over 48 hr if possible) in order to assay each sample 8 to 10 times.

7) Calculate variance on each point and plot the merit values and counting stability data on triangular coordinate graph paper. Construct contours representing increments of merit values of 100 and the variances at less than 0.5%, between 0.5 and 1.0%, 1 to 5% and greater than 5%. It can be seen (Fig. 2.10) that these values also, most surprisingly, follow such contours, and a region of overlap of the

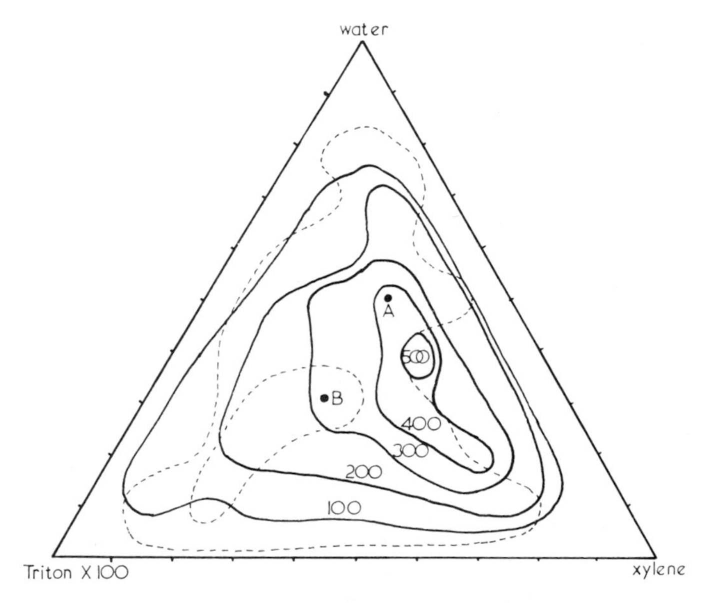

Fig. 2.10. The combined merit value (not instrument corrected) ———, and counting stability - - - - - - - for the system Triton X-100: xylene: water. Point A represents the highest stability nearest the high merit value area and point B the highest stability in the system. Inner stability contour < 0.5% and outer contour < 1.0% coefficient of variation over a 46 hr counting time.

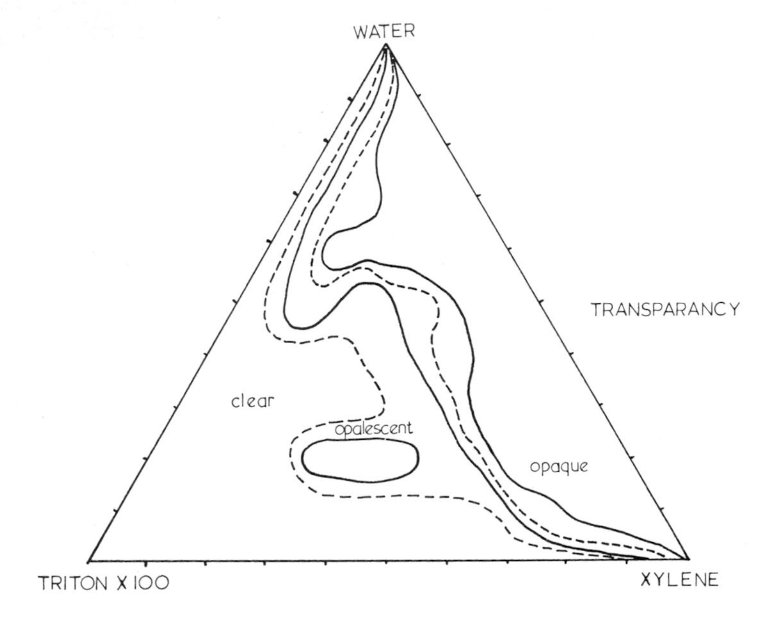
WATER
TRANSPARANCY
clear
opalescent
opaque
TRITON X 100
XYLENE

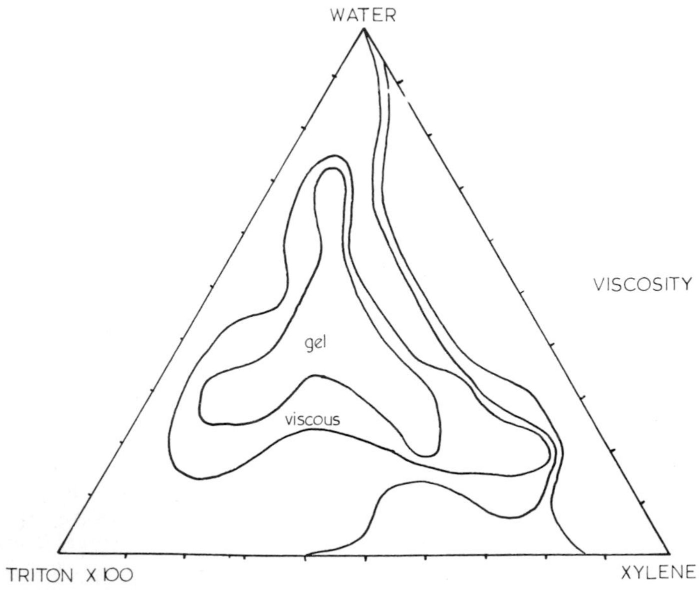
WATER
VISCOSITY
gel
viscous
TRITON X 100
XYLENE

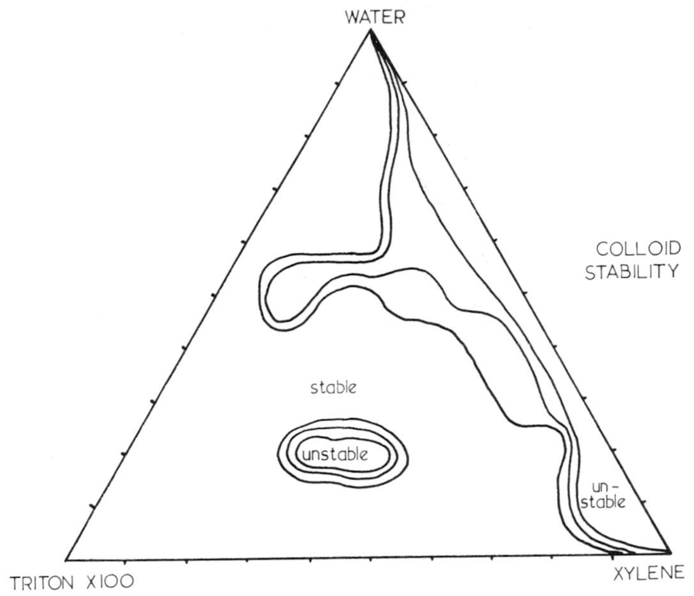

Fig. 2.11. Contours representing the varying degrees (see text) of a transparancy i.e. from clarity to opaque, b) Viscosity, c) Colloid stability (NB, *not* counting stability) for the Triton X-100:xylene:water system. (Reproduced from Liquid Scintillation Counting, Vol. 3. eds. M.A. Crook and P. Johnson, by kind permission of Heyden & Son Ltd.)

highest stability with the highest merit value can then be evaluated (point B). If less counting stability can be tolerated, say for comparative experiments in column fractionation, higher merit values can often be achieved (point A).

In order to describe the visual appearance of the phase diagram for comparative studies, the following arbitrary units are suggested, and these can often be assigned to each sample (preferably by an unbiased assistant after coding and randomization) for the transparancy, viscosity and colloid stability (not counting stability) as follows.

Transparancy: Completely transparent, no trace of opalescence (0); opalescent but transparent (1); opalescent and not transparant, i.e. misty (2); opaque similar to dilute milk (3); opaque, similar to cream or white paint (4).

Viscosity: No significant viscosity beyond that of toluene (0); some viscosity present, intermediate between toluene and pure Triton X-100 (1); similar viscosity to Triton X-100 (2); viscosity similar to thick but flowing cream (3); a gel which retains its position on inversion of the vial (4).

Colloid stability: The mixture retains a similar appearance during the whole of the counting period (0); partially separates in 24 hr but not within 2 to 3 hr (1); partially separates within 2 to 3 hr slowly (2); separates into two layers fairly quickly within 2 to 3 min of mixing (3).

These properties appear to be independent of each other and must be assessed separately if any understanding is to be expected from the appearance. Description based on 'viscous gels' or similar combinations of parameters will show no consistent patterns. However, it can be seen that if the individual assessments are plotted on triangular coordinate graph paper that contours can be drawn, indicating that the parameters are significant representations. Furthermore, it can be seen from the example (Fig. 2.11a, b, c) that the parameters are essentially independent of one another and do not coincide. A further increase in efficiency (merit value) may be obtained by varying the total level of PPO and POPOP in the composition indicated by the choice above. This can easily be achieved by varying the levels of the concentrated PPO scintillant in the mixture and determining the level at which the efficiency curve just begins to plateau (Fig. 2.12). Such optimal values for a number of commonly employed solutions are listed in Tables 7.1, 7.3 and 7.5.

Quench correction

As has already been mentioned, various authors have noted differences in the correction quenching if standards of different oil–water solubility are employed. Channels ratio appears to operate satisfactorily in this system and a direct comparison of the data with that of the non-colloidal scintillant, Bray's, has been made by Turner (1969). However, the external standard ratio cannot be used in this system (Fox 1968). The reason for this is not very clear at the moment, but it would appear that, as Van der Laarse has already pointed out, probably all parts of the phase diagram, even those areas which are optically clear,

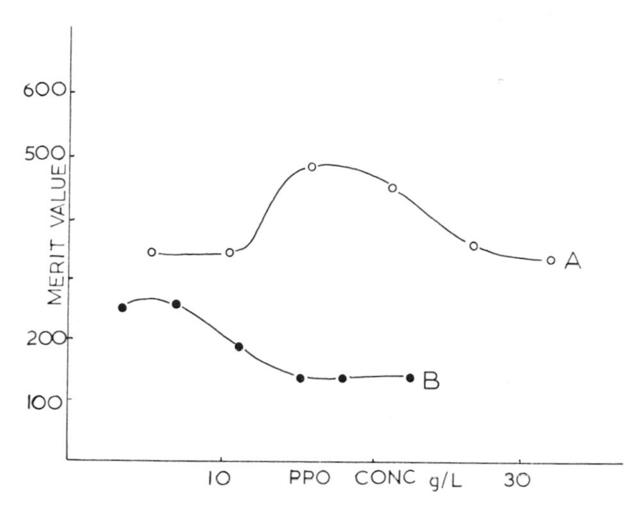

Fig. 2.12. The efficiencies of compositions represented by points A and B in Fig. 2.10 in relation to the concentration of PPO.

are in fact colloidal in nature and hence heterogenous. The micelles in these regions are clearly submicroscopic and probably consist of minute droplets of a highly efficient phosphor system, with a detergent envelope suspended in an environment of lower counting efficiency. The proportion of one compartment will vary relative to the other and to assess the true efficiency of the system as a whole it will be necessary to distribute a standard in exactly the same way as the component to be measured before a true assessment of overall counting efficiency can be obtained. The channels ratio of the actual counts of the sample are probably the most reliable means of measuring the counting efficiency.

The external standard ratio however, measures the average quenching in the whole system, and does not, of course, partition itself as does the material under study. The result is that readings are usually much lower than expected from the counts and bear no obvious relationship to the degree of quenching actually occurring. It is an important point to realise also that many other solubilization techniques may depend on a similar colloid formation for their efficiency and great care should be exercised to see whether the external standard ratio (ESR) can be used in a particular system. This can be done by comparing the ESR values with a series of channels ratio values for a number of differentially quenched samples, before subjecting the unknown samples to this procedure in an instrument in which automatic data processing, based on the ESR, may still further obscure the errors. Many of the errors may not be easily recognisable in their final computed presentation.

A number of other detergents have been investigated by Lieberman and Moghissi (1970) and from these studies, Triton N-101 (Rohm and Haas, Inc., U.S.A.), an alkyl phenyl polyethylene glycol ether (Fig. 2.6) similar to Triton X-100, when combined with p-xylene and PPO with bis-MSB (Table I, Appx. II) as secondary solute was considered to be superior to Triton X-100-toluene system for the measurement of low activity counts in aqueous solution. However, the comparison in this case was made with a toluene system in which sub-optimal conditions were chosen and, due to a phase difference in the Triton N-101 system, this sub-optimal point was closer, but not coincident with, the optimal conditions in the latter system (Fox 1972). This may not be the case with Triton X-114 (Fig. 2.6), which has been reported by Moghissi (1970) to give higher merit values than the Triton X-100 system, for tritiated water. The Triton X-114 system consists of one part Triton X-114 and two parts of a solution of naphthalene, 100 g; PPO, 10 g; POPOP 0.8 g in 1 l xylene. All the counting was done between 0 to 1°C and merit values of approx. 800 are reported. However, it is necessary to take into account the increased efficiency of the instrumentation which they used before a valid comparison can be made with other work.

A similar survey of detergents was carried out by Lupica (1970), who concluded that Igepal CA 720 (General Aniline and Film, U.S.A.) and Sterox DJ (Monsanto, U.S.A.) gave higher efficiencies than Triton X-100. However, again, only a very limited region of the phase diagram was examined and this conclusion may not be valid when the true optimal conditions are determined with these detergents.

Any statement of improved performance must be made at the proven optimal conditions for that system, and without an examination of the whole phase diagram, the results must be regarded as empirical.

It is also necessary to establish the expected counting stability of the so-called optimal point, as it can be seen that the choice of optima is based on the choice of counting stability permitted. The appearance of the colloidal system is no guide to the stability of either the colloid or the counts.

A useful feature of this system, is the unlimited possibility of use in biochemical work where complex solutions are to be measured routinely. It has been extended (Madsen 1969) to the counting of ^{14}C protein by initially dissolving the protein in 1 N NaOH. In the author's laboratory, this method is widely adopted for such measurements, and a technique based on this is described in § 8.1. Merit values for ^{14}C of 2370 (uncorrected for reference efficiency) are usually obtained.

Assessment of the isotope content of urine and plasma has important health physics implications (see § 4.3) and the ability to measure the level of an isotope like tritium present in the urine, without the necessity to distill it, is an obvious advantage. Many other complex solutions such as biological media, chromatographic eluates containing 8 M urea or high salt concentrations, and even milk has been measured in this system. The details of these techniques will be described in the appropriate chapters of this book.

There is clearly a great deal of potential in this particular system, and the inclusion on the market of several proprietar scintillant mixtures containing detergents are listed in Table 6, Appx. II. However, for accurate work, it is advisable to check with the actual samples to be measured, the validity of any claims that are made concerning their versatility of composition and their fitness to be used with automatic external standard ratios, before any routine measurements are conducted.

2.3. *Cerenkov light*

The excitement of the discovery of the misty bluish-white light will always be associated with the Curie's pioneering work of radium salt concentration and isolation in 1910. It was over two decades later that Cerenkov (1934) published his detailed investigations of this light emission which now bears his name. In 1937, Frank and Tamm derived an equation linking the number of photons produced within a specific spectral range with the beta-particle energy and the refractive index of the medium

$$N = 2\pi\alpha\left(\frac{1}{\lambda_2} - \frac{1}{\lambda_1}\right) \times \left(1 - \frac{1}{\beta^2\mu^2}\right)\mathrm{d}x \tag{2.1}$$

where N = number of photons produced within the spectral range λ_1 to λ_2 (nm) in a medium of refractive index μ over a short distance of travel dx of a beta particle of energy E. $\beta = (v/c)$ where v is the velocity of the electron of energy E in the medium. As the beta-particle is emitted from the isotope at the speed of light *in vacuo*, there is a light equivalent of a shock wave produced by a piling up of wavelets which accumulate and interfere with one another. The resulting electro-magnetic emission is recognised as Cerenkov light with a pulse length of less than one nano-second. For a more detailed account of this form of light emission there are many excellent works. General works include those by Jelley (1958), and Marshall (1952), and more detailed accounts relating to the special applications in liquid scintillation counting have been recorded by Belcher (1963), Parker and Elrick (1966, 1970) and by Haberer (1966).

One of the interesting properties of this shock wave is that it is emitted in a cone of light diverging along the axis of travel of the beta-particle as illustrated in Fig. 2.13. The product of the cosine of the angle θ and the refractive index (μ) of the medium is inversely equal to the ratio (β) of the velocity of the beta-particle in the medium (v) the velocity of light *in vacuo* (c): (Cerenkov relation), i.e.

$$\mu\cos\theta = \frac{1}{\beta} = \frac{c}{v} \tag{2.2}$$

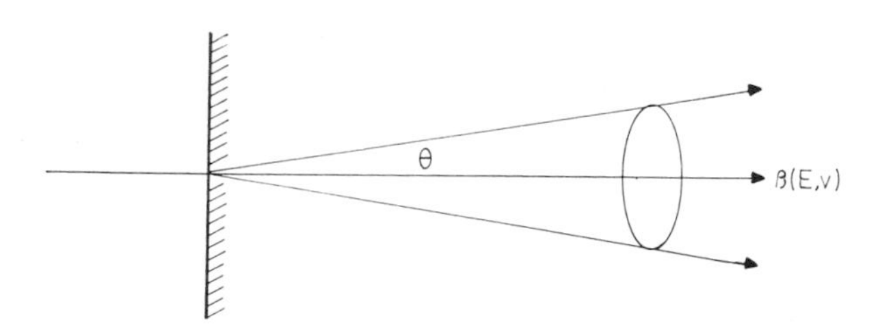

Fig. 2.13. The cone of Cerenkov emission produced by the passage of an electron travelling in a medium at a velocity greater than the speed of light in that medium.

Hence, for any emission of light to occur, a shock wave must be produced and thus θ must be positive and > 0, i.e. $\cos\theta < 1$. It is thus necessary for a beta-particle to exceed a certain energy (and hence velocity) in that medium to achieve shock wave conditions and this has been shown to be 0.263 MeV for water (Kment and Kuhn 1963) but can vary from 0.15 to 0.26 MeV depending on the refractive index of the medium through which it passes. A plot of the Cerenkov threshold (MeV) for electrons and refractive index is shown for a number of common solvents in Fig. 2.14 (Parker and Elrick 1970).

The directional nature of the Cerenkov light and the low photon yield (approx. 40 per disintegration) for ^{32}P and 7 photons per disintegration for ^{36}Cl means that a proportion of the disintegrations will be lost. To some extent, these can be retrieved by using the photomultipliers out of coincidence or singly, but the resulting increase in background due to thermal noise and other events do not usually offer any great advantages. Decreasing the refractive index of the medium will increase the photon yield per disintegration and this was demonstrated by Haberer (1965), who measured the efficiency of ^{40}K (1.32 MeV) and the ^{144}Ce-^{144}Pr conversion (2.98 MeV) in sucrose solutions of varying refractive index from $\mu = 1.33$ to 1.44 and showed even within this relatively narrow range, an increase of efficiency of 7%. However as Parker and Elrick (1966) showed, beta emitters with energies closer to the threshhold value (e.g. ^{36}Cl and ^{204}Tl) tend to benefit most from decreased refractive indexes (i.e. 10% increase per 0.01 decrease in refractive index).

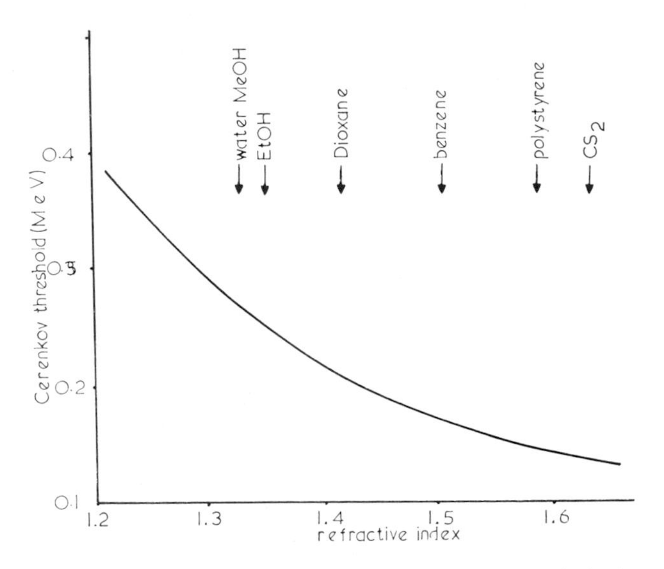

Fig. 2.14. The relationship between the minimum energy threshold of a beta particle allowing for the emission of Cerenkov light and the refractive index of the medium (Parker and Elrick 1970).

The Cerenkov light spectrum ranges from 300 to 700 nm, overlapping the sensitivity range of the photomultipliers. There tends to be a higher photon emission at the shorter wavelengths and in the ultra violet region, this tends to get absorbed by the glass of the vial and by the medium itself. Some plastic vials show improved transparency to these shorter wavelengths and it is probably for this reason that certain vials have been shown to be more efficient for the transmission of Cerenkov light (Parker and Elrick 1966; and Laughli 1969). It has also been suggested however that part of the increased efficiency in polystyrene vials may be due to a weak scintillator action of the plastic itself (Birks 1964; Parker and Elrick 1970).

In order to convert more of the shorter wavelength pulses to the regions of the spectrum more available to the photomultiplier, wavelength shifters have been employed. One of these is the fluorescent indicator, β-methyl umbelliferone (Fig. 2.15c). However, when using

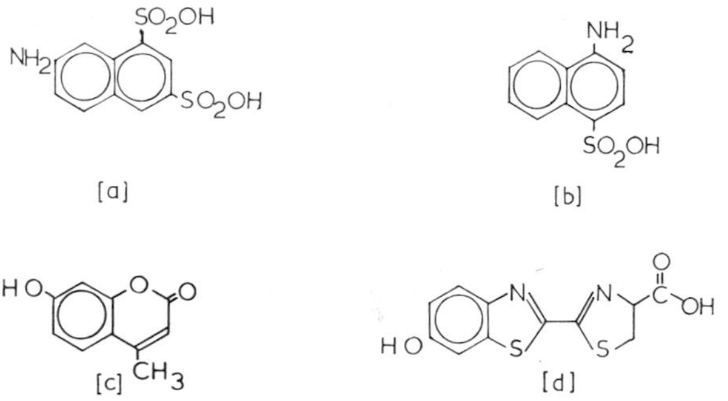

Fig. 2.15. Some examples (a to c) of waveshifters which increase the efficiency of Cerenkov counting, a) 2-amino naphthalene 6,8-disulphonic acid, b) amino naphthalene sulphonic acid, c) β-methyl umbelliferone. Dehydroluciferin, the substrate for the luciferin-luciferase reaction, is shown in (d).

this agent to increase the efficiency of counting it is essential to maintain the correct pH of the system. The naphthylamine sulphonic acid derivatives have proved more useful as they are considerably less pH dependent, and 1-amino naphthalene-4-sulphonic acid was shown by Haberer (1965) (Fig. 2.15b) to increase the counting efficiency of ^{137}Cs (E_{max} = 0.51 MeV) 7-fold. The sodium and potassium salts of 2-amino naphthalene-6,8-disulphonic acid (Fig. 2.15a) have been used by Lauchli (1969) and by Parker and Elrick (1966) as waveshifters. An increase in counting efficiency of ^{32}P from 25 to 50% is reported but the advantages of sample retrieval, characteristic of the Cerenkov counting system is destroyed by the addition of such agents, and the only advantage becomes an increase of the merit values for some of the more energetic beta emitters like ^{32}P.

The technique of counting by means of Cerenkov emission has been described in studies on ribonucleotides (Sebring and Salzman 1964), steroid hormone uptake in tissue slices (Braunsberg and Guyver 1965), strontium 90 in urine samples (Narrog 1965), manganese-56 (de Volpi and Porges 1965), ^{42}K estimations (François 1967; Moir 1971), ^{86}Rb transport in plants (Laughli 1969), ^{32}P labelled phospholipids (Satyaswaroop 1972) or iodine-127 following neutron activation experiments (Hoch et al. 1971).

Counting by means of Cerenkov light is very simple. An aqueous or organic solution is placed in a polypropylene or glass vial, up to 20 ml per vial. The vials are then assayed directly using a channel setting as for tritium. An equivalent volume of water or organic solvent may be used as a blank. If the sample need not be recovered, one of the waveshifters may be added to increase counting efficiency, e.g. for beta energies less than 1 MeV (i.e. ^{36}Cl or ^{137}Cs), 5 mM of 7 amino 1,3-naphthalene disulphonic acid (ANDA) (approx. 15 mg/10 ml solution) may be used. For isotopes with beta energies greater than 1 MeV, one half this concentration should be used. In the case of the neutron activation analysis of ^{127}I, the procedure recommended by Hoch et al. (1971), is as follows.

Following a timed exposure of I^{127} solution to a neutron flux (approx. 2×10^{12} neutrons/cm^2/sec), 25 mg KI in 10 ml water are added together with a drop of phenolphthalein indicator, 8 drops of 12 M NaOH and 10 drops of 5% sodium hypochlorite (laundry bleach). After 60 sec 8 drops of 16 M HNO_2 are added together with 10 to 15 drops of $Na_2S_2O_3$ (0.5 M). In 30 sec, the pale yellow solution is made alkaline with NaOH and the mixture extracted with carbon tetrachloride. The iodine is then extracted from the CCl_4 solution into water and assayed directly by Cerenkov light emission. A time course of the counts are measured for 30–45 min and the level of ^{127}I assayed from standard curves (half life ^{127}I is 25 min).

Several studies have been made on the feasibility of quench correction by the conventional methods available for liquid scintillation counting. An improved method of internal standardization for ^{32}P was described by Vemmer and Gutte (1964) which consisted of comparing an inactive sample of the same composition as the active sample, with and without added standard ^{32}P.

Colour quenching, a serious problem in Cerenkov counting can usually be corrected by the standard channels ratio technique (Stubbs and Jackson 1967; Moir 1971; Wiebe et al. 1971) or by the external standard ratio method (Brownell and Laughli 1969). The two methods were compared by Kamp and Blanchard (1971) who concluded that the channels ratio was the most useful method provided that the counting statistics were adequate. This form of correction however, is only necessary where the bleaching action of such agents as benzoyl peroxide, or sodium bisulphite solutions fail. Due to the relative insensitivity of the system to impurity quenching, colour bleaching is probably the preferred method (Schneider 1971).

However, although very little variation occurs between quenching agents, differences do occur between different solvents. An example of the influence of solvent composition on efficiency of Cerenkov counting of ^{32}P is illustrated in Table 2.1 (Johnson 1969). It was also

TABLE 2.1

The influence of solvent composition of Cerenkov counting efficiency (Eff) of ^{32}P (Johnson 1969).

Solvent	Eff	Solvent	Eff
Water	27	Isopropanol: heptane (1/1, v/v)	35.5
Iso-propanol	34.9	$CHCl_3$: isopropanol (2/1, v/v)	35.7
Heptane	37.2	$CHCl_3$: methanol (2/1, v/v)	34.1
Isopropanol: water (1/1, v/v)	31.1	Formic acid (98 %, w/v)	25.9
Ethyl acetate: water (4/1, v/v)	32.1	Formic acid: butyl acetate (5/1, v/v)	27.1

observed that the external standard ratio values departed from the true value when highly quenched, low volume samples were used and Kamp and Blanchard (1971) concluded that scintillator volumes should be made greater than 10 ml if possible.

A detailed examination of the Cerenkov counting in aqueous and organic media was also conducted by Haviland and Bieber (1970). He related the efficiency not only to the volume of solvent and its composition but also to the vial material. They derived an equation which described the relative efficiency of counting as follows

Subject index p. 309

$$\frac{\eta_1 a_s + \eta_2 a_v}{\eta_1 a_w + \eta_2 a_v} \tag{2.3}$$

where η_1 and η_2 are geometric factors dependent on the volume of the sample and a_s, a_v and a_w are the Cerenkov efficiencies for the solvent alone, vial alone and water alone.

The isotopes amenable to assay by the Cerenkov light method are listed in Table 14.1, efficiencies are quoted for aqueous solutions. However, it can be seen from Fig. 2.14 that if a solvent could be used with a refractive index greater than 1.559 (using eq. 2.1) it should be possible to count carbon-14 (a_v 0.159 MeV) by this technique, and vastly expand its potentialities. Ross (1970) examined this point and has shown that a surprising accentuation of counting efficiency (from a theoretical 1.3% to an observed 12.3%) is obtained in the highly refractive α-bromonaphthalene (refractive index, 1.6582). This was explained by anomalous refractive dispersion effects. Although the choice of monobromonaphthalene would severely limit its potentialities as a practical method of counting carbon-14, the results with this solvent suggest that Cerenkov light may play a more important part than is realised in certain conventional liquid scintillation systems.

Another potentially important application of the detection of Cerenkov light is in the flow monitoring of column eluates and of nuclear power station effluents. Rippon (1963), showed it to be a sensitive method for the latter application but with poor isotope discrimination capabilities when compared with ion-exchange systems. However there appears to be no reason why the method should not be applicable to single isotope analysis in chromatographic systems. An interesting application of the continuous radiometric analysis of an isotope isolated by automatic analytical procedures was described by Colomer et al. (1972).

2.4. *Luciferin luciferase assay*

In theory, any chemical reaction involving the production of light, will be a candidate for quantitation in the liquid scintillation spectrometer. The method is particularly suitable for the detection and quantitation of very low levels of light.

Two important biological systems whose light emission has been exploited in this manner, are those of the *Photobacterium fischeri* luciferase, which uses reduced nicotinamide adenine dinucleotide (NADH) and flavin mononucleotide (FMN) as an energy source and the Firefly (*Photinus pyralis*) luciferase which employs adenosine triphosphate. By selection of the appropriate conditions of reaction and light measurement, levels of these co-factors down to 10 femtomoles (10^{-14} moles) may be quantitatively assayed.

The Firefly luciferase system may be represented as follows

$$\text{Luciferase} + \text{dehydroluciferin} + O_2 \xrightarrow[\text{pH 7.4}]{\text{ATP}} h\nu\ (\lambda\ \text{max. 562 nm}) \tag{2.4}$$

The photobacterium system

$$\text{NADH} + H^+ + \text{FMN} \xrightarrow{\text{dehydrogenase}} \text{NAD} + \text{FMNH}_2 \tag{2.5}$$

$$\text{FMNH}_2 + O_2 + \text{Luciferase} + \text{long chain aldehyde (e.g. tetradecanal) pH 7.0} \rightarrow h\nu\ (\lambda\ \text{max. 485 nm}) \tag{2.6}$$

As with Cerenkov counting, some sensitivity advantage may be gained by using photomultipliers out of coincidence, provided the thermal and other noise characteristics of the instrumentation is minimal.

The purification of the Firefly enzyme was described by Green and McElroy (1956) and the substrate for this enzyme is dehydroluciferin (Fig. 2.15d) which can be purified according to the method of Seto et al. (1963).

Method of assay of ATP using Firefly luciferase (*Schram 1970b*)

1) Dissolve content of one vial of firefly extract (Sigma, FLE-50) in 5 ml distilled water (this gives a final concentration of buffer, 0.05 M K_2HSO_4:0.02 M $MgSO_4$; pH 7.4) Store, −20°C.
2) Dilute an aliquot of this enzyme solution 10-fold with the same K_2HSO_4:$MgSO_4$ buffer and leave overnight at 4°C before use (to deplete endogenous ATP).

3) Bring enzyme solution to 20°C prior to use (enzyme prep.).
4) 10 μl sample solution added to 100 μl enzyme.
5) Wait 10 min exactly, then count for 1 min or more.
6) Standard curve obtained using ATP solns from 0.1 to 10 picomole (10^{-12} M) per vial.

Count rates of approximately 2 to 5 × 10^4 cpm are obtained for levels of ATP ranging from 1 to 4 picomoles.

Method for the assay of ATP using Firefly luciferase (*Stanley 1971*)

1) Firefly luciferase preparation: two dessicated firefly lanterns homogenised with 1 ml cold arsenate buffer (100 mM sodium arsenate, 40 mM $MgSO_4$ adjusted to pH 7.4 with H_2SO_4 and Millipore filtered).
2) Centrifuge at 20,000 *g* for 30 min at 2°C.
3) Clear, straw coloured supernatant stored at 0°C (for further details of the aging of the enzyme preparation see the original text).
4) Scintillation vials, acid washed and kept in oven at 100°C until required.
5) Add 1 ml arsenate buffer to vial.
6) Equilibrate vial at 20°C for 10 min in sample changer of liquid scintillation counter.
7) Add 50 μl of ATP sample.
8) Add 20 μl of firefly enzyme solution, cap and count for 0.1 min, 8 sec after adding enzyme.

There are clearly many possible application of this counting system for the analysis of low levels of ATP, such as the assay of ATP-sulphurylase (Balharry and Nicholas, 1971). The assay of NADH and FMN was described by Stanley (1970). A semiautomatic microtransferator (Fig. 2.16) and a special cell for conducting these types of assay have been described by Schram and Roosens (1972).

Applications, such as amino acid activation studies in *Bacillus stearothermophilus*, periodicity of photophosphorylation in *Acetabularia mediterranea* and the development of amphibian eggs, have been explored using this technique.

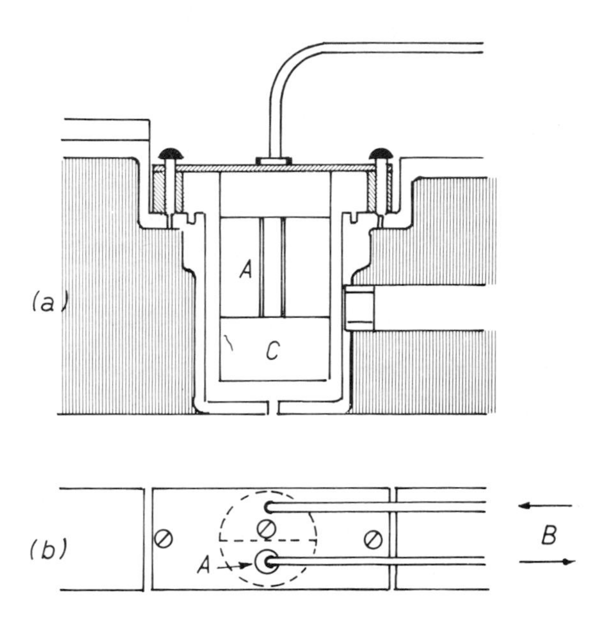

Fig. 2.16. A semiautomatic microtransferator for the assay of NADH and FMN by the luciferin-luciferase system. The cell is built into the sample drawer of the Intertechnique SL20 counter. The body of the cell is made of quartz tubing (A) and occupies an eccentric position, closer to the photomultiplier used for the measurements. Motor driven syringes are used to pump the solution in and out (B) of the cell housing (C), an excess of air being used to expel the last drops remaining during the emptying process. Usually no rinsing is required but a single rinse with distilled water is adequate if some tailing is experienced. Side view (a), top view (b) (Schram and Roosens 1972). (Reproduced by kind permission from Heyden and Sons Ltd., Liquid Scintillation Counting, Vol. 2, eds., M. Crook, P. Johnson and B. Scales.)

An interesting study of White et al. (1971) has indicated that it is possible to employ a number of analogues of the substrate luciferin in this reaction, and the quality of the photon emission is clearly dependent on the structure of the substrate. This would suggest that further exploitation of this aspect may lead to light emitters even more compatible with the photomultiplier systems in use and hence will still further increase the sensitivity of the system.

Preprocessing techniques: general aims and criteria

The prime object of preprocessing a sample is to convert it by degradation into a form which allows the beta-emitting component to achieve the closest electronic association with the primary solvent and solute. The length of time required to achieve this result will depend on the nature of the sample, the level of accuracy required commensurate with the number of samples to be measured, and the level of counts expected in the sample itself.

The ideal end-product of preprocessing would be a beta-emitting component completely soluble in a high efficiency scintillant mixture such as a toluene solution of PPO and POPOP (e.g. Tpp4, Table 3, Appx. II); the necessity for POPOP will depend on the type of photomultipliers used (§ 2.1.1) or whether the final sample-scintillant mixture is heavily quenched. The majority of biochemical samples however, fall far short of this ideal, due to the presence of biopolymers, salts and water.

One should first consider the possibility of complete extraction of the labelled component from the tissue by simple solution techniques. In the case of lipids (see § 8.3) this may be the quickest method of assay. The *total* radioactivity present in a *tissue* is probably best assessed by a combustion technique, for which there have been many modifications. Such methods are particularly suitable where an absolute assessment of radioactivity is required, but are limited by the time required to undertake the combustions. If large numbers of tissue samples are to be measured, of which at least 20 mg of each is available for analysis, it may be more convenient to employ either a chemical degradation procedure, such as a wet oxidation technique, or a solubilizer of the quaternary ammonium type.

A number of alternative procedures are also available for special circumstances, which allow a large number of samples to be prepared quickly with good internal reproducibility. The choice of quench correction method used will also largely depend on the technique employed, bearing in mind that only homogenous systems can be simply and accurately corrected for quenching and that only an approximate quench correction can be made in heterogenous counting, hence limiting its use to situations where absolute assessments of radioisotope content are not necessary.

Particulate material derived from tissues either directly, e.g. ribosomes, nucleii etc., or following precipitation, can be measured by counting the sample directly on a disc in a high efficiency scintillant mixture (§ 2.2.1). However, for tritium, there may be too much coprecipitated material and excessive quenching will occur due to the compact nature of the residue in the filter disc. The complete solution of such material or its complete oxidation to tritiated water or to ^{14}C-carbon dioxide is often advisable for accurate comparative measurements.

3.1. Combustion techniques

The main object of the combustion procedure is to convert a tissue or an organic compound into carbon dioxide and water as completely as possible, an appropriate solvent can then be used to absorb them quantitatively in a form suitable for admixture to a toluene or dioxane based scintillant mixture. At the same time, those materials which would normally quench the system have themselves been degraded.

The classical method of converting an organic compound or piece of tissue into its elementary components for quantitative analytical estimation of the constituent elements is by means of a combustion train based on the original Pregl-Roth procedure (Pregl-Roth 1958). In the scintillation counting application, the gases are absorbed in solvents which are miscible with toluene-rich mixtures. The relative temperature difference between a solid carbon dioxide trap (i.e. solid CO_2 dissolved in *n*-propanol or a chloroform : carbon tetrachloride mixture) and a liquid nitrogen trap is most frequently employed initially to separate water from carbon dioxide, the latter freezing only in the liquid nitrogen trap. An alternative method of trapping $^{14}CO_2$ or $^{35}SO_2$ is to use a strong alkali or a quaternary ammonium base. A methanolic solution of the base traps $^{14}CO_2$ as the carbonate which is readily soluble in toluene-rich scintillant mixtures, and efficiencies for ^{14}C between 30 and 60 % are obtained depending on the proportion of base in the scintillant mixture. Bases such as Hyamine 10X (Passman et al. 1956) phenylethylamine, Primene 81R and ethanolamine have all been recommended at various times. Hyamine 10X was originally manufactured by Rohm and Haas as a germicidal agent and is described as *p*-(diisobutylcresoxyethoxyethyl) dimethyl benzyl ammonium hydroxide. It is normally supplied as a 1 M solution in methanol. However, although a number of such agents have been described for the trapping of CO_2, a solution of KOH is still considered by some authors (Dunscombe and Rising 1969) to be superior. An ingenious system for the collection of gases directly into scintillation vials (Fig. 3.1) was described by Smith and Phillips (1969). Both $^{14}CO_2$ and $^{3}H_2O$ can be collected by this method by cooling the whole apparatus in liquid nitrogen. The CO_2 is trapped in a vial containing a solution of ethanolamine (60 ml) in methyl alcohol (360 ml), 0.5 ml of such a solution will absorb as much as 10 ml of carbon dioxide at atmospheric pressure. When the vial is brought to room temperature, a scintillant such as Tpp-4 (Table 3, Appx. II) is added (5 ml) together with 1 ml approx. of methyl alcohol (necessary to blend the mixture into a single phase). The simultaneous assay of sulphur-35 and tritium by a combustion procedure was described by Roncucci et al. (1968) who used phenyl-ethylamine at $-80°C$ to trap the sulphur dioxide formed.

Several scintillant mixtures have been recommended for the absorption of tritiated water. Kinard's solution (see Table 5, Appx. II) was designed for this purpose. Bagget et al. (1965) suggested the use of a

Fig. 3.1. A simple trapping system for $^{14}CO_2$, using standard laboratory glassware. The vial contains 0.5 ml of a solution of 60 ml ethanolamine in 360 ml methanol and the whole is cooled in liquid nitrogen (adapted from Smith and Phillips 1969).

dioxane : toluene : naphthalene mixture for the absorption of tritiated water, recoveries of 94 to 97% being reported. Conway et al. (1966) have recommended the following two scintillant mixtures. Ethoxyethanol (200 ml), toluene (800 ml), PPO (6 g) and POPOP (0.2 g) which allowed for recoveries of 96 to 98% of tritiated water and ethoxyethanol (400 ml), ethanolamine (120 ml), toluene (480 ml), PPO (6 g) and POPOP (0.2 g) for $^{14}CO_2$. A mixture for $^{14}CO_2$ absorption employing phenylethylamine was claimed by Peterson (1969) to give 97% recovery, it consisted of toluene (430 ml), methanol (300 ml), phenylethylamine (270 ml), PPO (5 g) and dimethyl POPOP (0.5 g).

One of the chief problems with such combustion systems is the low counts often obtained by excessive quenching of the counting samples due to the dissolved oxygen present. This quenching action of oxygen was recognised by Pringle et al. (1953) who suggested that the problem may be overcome by flushing the system with nitrogen in order to displace the oxygen. An improvement of about 30% was achieved by this procedure. These authors also suggested that photo-oxidation and dimerization processes were probably responsible for this quenching action. The solubility of oxygen in toluene is 240 mg/l at 180°C and under normal atmospheric conditions, toluene will contain 30 to 40 mg/l. Argon has a greater density than air (1.38 with respect to nitrogen) and its efficiency in displacing oxygen is greater than that of nitrogen. An increase in pulse height of some 21 % is achieved (Ott et al. 1955) by flushing for 10 min with argon using a sintered glass gas disperser. An 18 kC, 200 watt ultrasonic generator will also degass scintillator solvents sufficiently in 6 min at 60°C (Chleck and Ziegler 1957).

A useful summary of the problems encountered in the measurement of low levels of ^{14}C and ^{3}H by the combustion method was recorded by Ober et al. (1969).

A problem associated with the combustion procedure is the relative slowness of the assays, since the oxidation train has to be flushed out and checked before another sample is introduced. For a detailed account of these early procedures, readers are referred to the review by Jeffay (1962). Although more sophisticated designs of oxidation trains have been described (Knoche and Bell 1965; Smith 1969), these are not often employed.

An alternative procedure is to burn the sample in oxygen within a closed vessel and then to condense the oxidation products in an appropriate cooling train or within the combustion flask itself. This principle has been widely accepted as probably the most quantitative and accurate means of assaying both ^{14}C and tritium in biological materials. Jacobsen et al. (1960) combusted 5 to 25 mg of the sample with a mixture of copper and copper oxide in a small sealed tube. The standard Parr bomb technique has also been used (Sheppard and Rodgher 1962), when up to 3 g of tissue may be combusted, and is capable of detecting as little as 0.0004 μCi tritium or 0.0001 μCi ^{14}C.

The addition of the absorbants directly to the combustion flask after the oxidation has taken place is the simplest but potentially hazardous procedure. Many modifications of the combustion flask procedure have been developed in an attempt to speed up and simplify the process of oxidation especially for this application to liquid

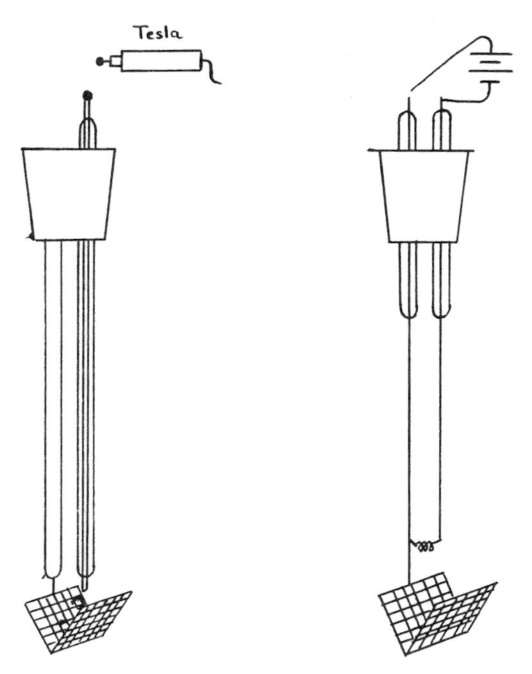

Fig. 3.2. Two designs for a platinum basket for use in a large Buchner flask combustion system (Kelly et al. 1961). (Reproduced by kind permission of Academic Press.)

scintillation counting. Kelly et al. (1961) used a large (2 l) Buchner flask into which a platinum basket constructed as in Fig. 3.2 is placed. The flask is flushed with oxygen and the electrode containing the sample is then inserted. The tissue is ignited by passing a current to heat the filament or by means of brush discharge from a Tesla coil. The flask is cooled and 10 ml of 1 M hyamine hydroxide in methanol is added. After absorption has taken place, 3 ml is removed and assayed in 10 ml of toluene based scintillant (e.g. Tpp4, Appx. II). If *tritiated water* is to be measured, the flask is placed on a mixture of solid carbon dioxide in chloroform: carbon tetrachloride for 0.5 hr and then treated with 20 ml of a blended toluene scintillant such as TppE (20%) (Appx. II) to absorb the water condensed as ice on the base of the flask, which is then transferred to ice water to allow for slow thawing. An aliquot of the scintillant mixture is then taken for counting.

There are several problems associated with the procedure as described by Kelly et al. but nevertheless the procedure has been widely exploited and several useful modifications have been devised. A number of serious explosions occurred with these original designs, due to a sudden expansion of gases during the oxidation, especially violent if traces of washing solvent had remained behind. In one design, a second side arm with an expansion vessel was introduced (Martin and Harrison 1962). An aluminized polythene sheet (0.07 mm thick) was stretched over an open neck to allow for such sudden expansions and prevent the breakage of the vessel itself. Oliviero et al. (1962) used an infrared system for the ignition of the sample. Dobbs (1963) described a series of modifications employing standard jointed glassware and designed to reduce the possibility of an accident to negligible levels (Fig. 3.3a). The modification (Dobbs 1966) (Fig. 3.3b) employing the combustion flask in the horizontal position is probably the best laboratory constructed device described. A number of different materials have been employed in the construction of baskets for these combustion methods, such as kanthal resistance wire (Franc et al. 1965), nichrome (Conway et al. 1966), platinum (Yamasaki et al. 1966) and platinum (10%):iridium (Davidson and Oliviero 1968).

In the method described by Dobbs, a solid sample is weighed on to a filter paper (2 cm sq) which has been left in the balance case for several hours to equilibrate its moisture content. The sample is wrapped up into a 1 cm square package within the filter paper and placed on the combustion platform. (Liquid samples are weighed in a small gelatine capsule which is placed on a small strip of filter paper to assist ignition of the capsule.) The combustion head is secured with elastic bands and the flask flushed thoroughly with oxygen. A slight negative pressure is maintained following flushing (about 5 cm of mercury) to facilitate the introduction of scintillant mixture following combustion. Following ignition by means of a Tesla discharge, when the flask is cool it is removed from the gas line. Liquid scintillator (10 ml) is introduced by syringe into the flask through the serum cap

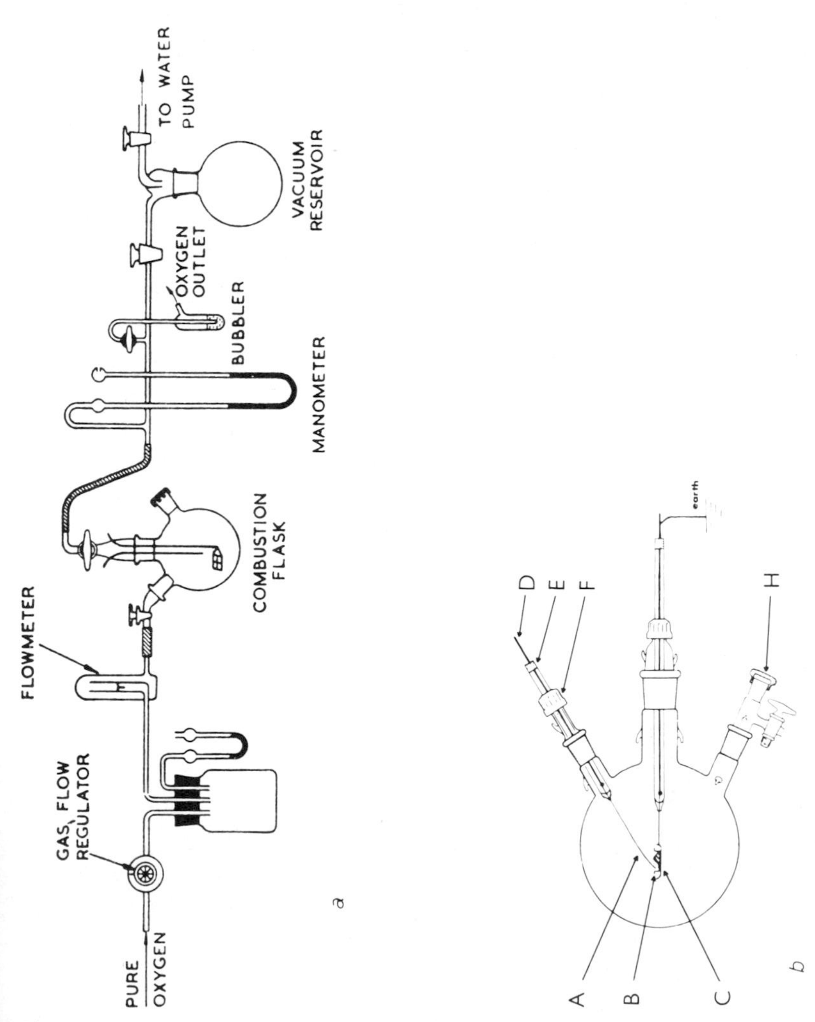

Fig. 3.3. A) combustion system devised by Dobbs (1963a) possessing many of the safety features to allow for sudden expansion during combustion. A clip-cap on the combustion flask is tight fitting but would be rejected in an explosive expansion. In B (later design, Dobbs 1971) the flask is used in a horizontal position and incorporates an expansion system involving elastic bands. A) Platinum ignition electrode, B) sample in cachet, C) platinum gauze platform, D) copper, E) epoxy resin, F) screw cap seal, G) earth, H) serum cap. (Reproduced by kind permission of Academic Press.)

and swirled around in the flask. For tritium, Kinard's scintillant was used (see Table 5, Appx. II) and for carbon-14 or sulphur-35, a phenylethylamine, 270 ml : methanol, 300 ml : toluene, 430 ml : PPO 59, dimethyl POPOP, 0.59 scintillant mixture was used. The flask is left for 15 min with occasional shaking, and an aliquot (9 ml) is removed and assayed in a scintillation vial after bubbling nitrogen through the scintillator at 0.05 l per min for 3 min. For samples combusted in gelatine vials, it is recommended that the absorbing scintillant is allowed to stand for one hour with the products of combustion before assaying.

A novel approach to the combustion flask technique was made by Gupta (1968) who used plastic bags as combustion vessels. A modification of this method has been described by Lewis (1972) (Fig. 3.4). The

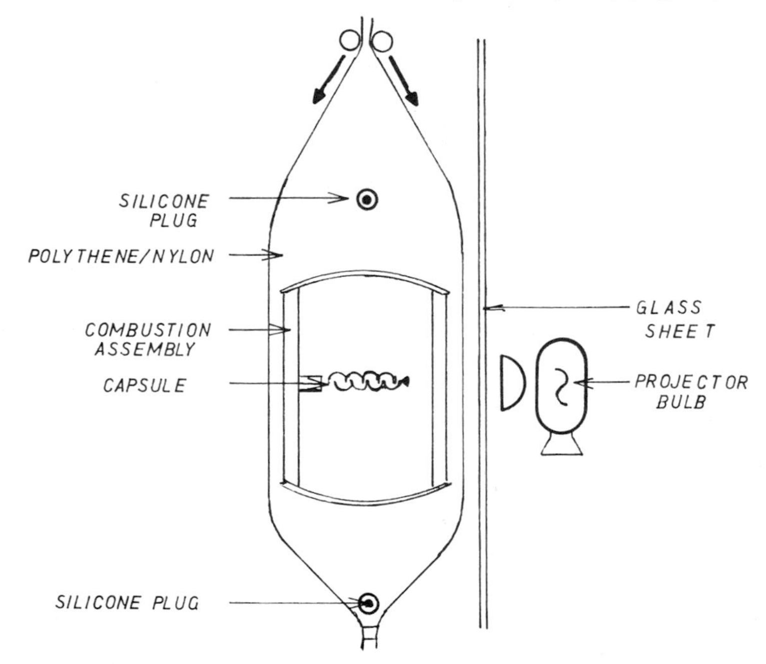

Fig. 3.4. A plastic bag combustion system (Nylon-Polythene laminate 'Synthene 5023', produced by Smith and Nephew Plastics, Ltd., Welwyn Garden City, England) described by Lewis, 1972. (Reproduced by kind permission of Pergamon Press.)

method as originally described by Gupta consisted of appropriately bent platinum sample holder attached to a watch glass by means of a silicone rubber plug. The whole is placed in a polyester plastic bag and sealed after flushing with oxygen. After combustion, initiated by means of a focussed infrared beam, an absorbing scintillant mixture is injected through a small silicone bead previously fixed into the side of the bag. Lewis's modification attempted to overcome the occasional melting of the upper regions of the bag by deflecting the ball of heat as it is formed by means of upper and lower shields in a combustion assembly. About 150 mg of tissue is weighed on to the assembly. During the combustion, a stream of cold water is directed around the sides of the bag to keep it cool. After combustion, absorbing scintillant is injected through the upper silicone bead, and after a suitable time to allow for absorption, an aliquot is removed from the lower bead. The use of a disposable bag clearly eliminates the necessity for the laborious decontamination of the vessel following combustion, and the modifications described also reduce the chance of a serious explosion occurring.

Most of the methods described above may be undertaken with materials which are readily available in the average laboratory. Commercial firms have been interested in this system as an adjunct to scintillation counting and a number of oxidisers consistent with different degrees of automation are now available commercially. Some of these are based on the sophisticated combustion system described by Kaartinen (1969) where the combustion of the samples as well as the subsequent absorption and dispensing of scintillant components are undertaken automatically. Over 99 % recoveries are reported for these procedures which have been reviewed by Davidson et al. (1970).

3.2. Organic solubilizers

The problem of solubilizing proteins and other macromolecules into a form miscible with toluene or dioxane-based scintillant mixtures has been the subject of much work in the radioisotope field since the introduction of liquid scintillation counting.

Formamide. One of the earliest methods of dissolving tissue in a solvent suitable for liquid scintillation spectrometry was that of Lahr et al. (1955) who used formamide as a protein solvent and ethanol to blend the formamide complex into the toluene solution for counting. Pearce et al. (1956) also reported the use of this reagent as a tissue solvent.

Hyamine 10-X (Rohm and Haas, Inc.). Vaughan et al. (1957) recognised the value of this base as a solubilizer of amino acids and proteins and described a method of counting by combining the hyamine solution with methanol-blended toluene scintillant. This system was compared directly with the formamide–ethyl alcohol system by Kinnory et al. (1958) who were able to show that the latter system possessed certain distinct advantages. The formamide–ethanol mixture dissolved proteins at a faster rate than hyamine, and the counts were reproducibly stable. The counts in the hyamine solutions increased over several days until they reached a constant value already achieved by formamide. The hyamine solution, although capable of accommodating more sample, caused considerable quenching most often from the presence of coloured products. Internal standards should be used routinely with all samples, since they invariably suffer widely varying degrees of quenching, even amongst samples of the same type. A further problem associated with the use of hyamine hydroxide as a solubilizer of proteins, is the frequent production of chemiluminescence which leads to spurious counts and high backgrounds. This phenomenon was studied by Herberg (1958), Dulcino et al. (1963) and Kalbhen (1967). The chemiluminescent reaction appears to be associated with the presence of peroxides in the scintillation solvents used, and alkaline solutions of dioxane are particularly troublesome (§ 1.3). In the case of hyamine digests, it occurs at a higher intensity at higher temperatures and decreasing the temperature from 20 to 8°C was shown by Kalbhen (1967) to reduce the photochemical reacton by only 30%. The removal of peroxides would thus be the most satisfactory way of eliminating the effect of this reaction on the counts. The peroxide effect occurs at alkaline pH, and hyamine solutions can be neutralised with two to three drops of 12 N HCl, or more efficiently with 10% ascorbic acid (McClendon et al. 1971), prior to adding to the scintillant mixture, without serious loss of counting efficiency. Although ultrasonic disintegration has been advocated for hymaine digests by Wang (cited in a review by Rapkin 1961), Hansen and Bush (1967) did not find any significant advantage of ultra-sound over digestion between 35 and 50°C for a few hours. Digestion over 70°C however produces excessive quenching.

'*NCS*' (Nuclear Chicago Corporation). Several other complex bases have been developed for this type of solubilization, amongst which is the material, NCS of 'molecular weight within the range 250 to 600 and containing a straight-chain alkyl group with an average length of 12 carbon atoms'. The base is normally employed as a 0.6 M solution in toluene. A number of different tissues can be dissolved in this system including muscle, liver, blood and plasma and the digest has been compared directly with those using hyamine by Hansen and Bush (1967). It was shown by these authors that NCS gave higher figures of merit than hyamine hydroxide in their studies and that with both bases, both internal standards and channels ratio could be used to determine the degree of quenching.

Two 'solubilizers' have been produced by Beckmann Instruments, namely the Bio-Solve™ series, BBS-2 and BBS-3. However, these agents apparently contain surface-acting agents which lead to the formation of colloids and should therefore be regarded as heterogenous systems (see § 2.2.3).

Soluene 100 is a Packard Instruments formulation of a quaternary ammonium hydroxide, reported to be a dimethyl dodecylundecyl ammonium hydroxide of molecular weight of 386 (Laurencot and Hempstead 1971). It is normally used as a 0.5% solution in toluene and blended into a BBOT, toluene, naphthalene scintillant mixture with 2-methoxyethanol. This mixture has been particularly recommended for the radio-assay of whole blood (§ 4.3.2). However, it should be noted that this solubilizer cannot be used with butyl- or isopropyl-PBD as fluor since considerable quenching, considered to be due to a direct interaction of the solubilizer with these fluors, has been reported to occur (Dunn 1971). Although this material is considered to be a superior solubilizing agent, it is liable with certain proteins to induce a very considerable chemiluminescence when mixed with a toluene-PPO scintillant mixture. This chemiluminescence can be reduced by addition of HCl or 10% ascorbic acid solution. Soluene 100 has however been shown to be of significant value in the solubilization of polyacrylamide gels (see § 9.1.1). The presence of a small proportion of water appears to be essential for the successful application of most of these solubilizers (Moorhead and McFarland 1966).

Method for the determination of protein dissolved in Hyamine or NCS solubilizers (Schmuckler and Yiengst 1968)
It is possible to determine the protein solubilized in hyamine hydroxide or in NCS (and probably in other quaternary bases as well) by a modification of the Biuret reaction.

1) Protein sample (approx. 10 mg) diluted to 1 ml with NCS or Hyamine.
2) Add 4 ml freshly prepared Biuret reagent (225 mg $CuSO_4 \cdot 5H_2O$ dissolved in 97 ml abs. MeOH, add 1 ml ethylene glycol followed by 2 ml of tetramethyl ammonium hydroxide in methanol, 24%).
3) Heat mixture 15 min at 50°C in a water bath.
4) Cool and determine optical density at 550 nm. The usual controls are included and bovine serum albumin or rat liver protein are used as standards.

3.3. Degradative methods

The process of chemically degrading biological tissue to smaller molecular weight species to achieve solubility of isotopically labelled fragments is a parallel process to that of combustion. Three basic procedures are available here; in one, the gaseous products are flushed out with an inert gas, or are allowed to diffuse out of the oxidation system. This method will of course, avoid quenching by water and the oxidising chemical (degradation with volatilization of the isotope, § 3.3.2). In another method, the products are retained within the mixture and are assayed in the presence of the oxidation system (degradation without volatilization of the products, § 3.3.1). Thirdly, enzymatic degradation (§ 3.3.3) of an insoluble macromolecule gives products which are soluble in a suitably blended scintillant mixture.

3.3.1. Degradation without volatilization of the products

Where the level of radioactivity in the tissue is reasonably high so that a moderate level of quenching will still leave sufficient counts to be statistically meaningful, e.g. 1000 cpm, it is possible to oxidise an organic compound or a piece of tissue with an oxidising agent such as nitric or perchloric acid and to retain the oxidation products within the oxidising mixture. In this case, the oxidation can be carried out in the scintillation vial and no additional collecting apparatus will be required. For multi-sample situations, where a rapid assessment of relative uptake is required, this can often be the preferred method of assay. It is not clear why the method is so successful as it would be expected that serious losses of isotope may occur by volatilization of the oxidation products, unless the solubility of $^{14}CO_2$ in strong acid solutions is unusually and unexpectedly high. Nevertheless, O'Brien (1964) described a useful method of oxidation of small samples of tissues with concentrated nitric acid, by heating the mixture in a sealed vial for a short period (30 min at 70°C) and cooling. Apparently, low quantities of nitric acid (i.e. less than 0.5%) did not seriously modify the counting system. One of the main advantages of the use of these wet oxidation methods is that little if any chemiluminescence is induced. The procedure is found to be particularly useful in the assay of insect cuticle. Counting efficiencies of 2.8% for tritium and 30% for ^{14}C, have been claimed. An improved scintillant mixture was described by Pfeffer et al. (1971) for use with nitric acid oxidation which consisted of a dioxane, naphthalene; butyl PBD (see DNb10, Table 3, Appx. II).

Assay of labelled tissue by a wet oxidation method (*Pfeffer et al. 1971*)

1) For every 1 g minced tissue add 5 ml colourless concentrated nitric acid in a capped scintillation vial (no inside foil should be present).
2) Heat the mixture 2–5 min at 70°C in a water bath.
3) Dilute 1 to 10 ml with 0.75 M Tris.
4) Add 1 ml of this mixture to 10 ml of the scintillant mixture DNb10 (Table 3, Appx. II).

A number of different tissues can be assayed by this procedure, e.g. rat brain, skeletal muscle, packed blood cells, whole blood, stomach, small intestine, caecum, large intestine, eye, testes, spleen and skin. With fat, brain and testes, it was recommended that 0.2 ml of Biosolv 3 be added to 10 ml of the Tris neutralised digest before adding 1 ml of the mixture to the scintillant (10 ml), 'in order to prevent lipophyllic materials forming micelles which would otherwise render some of the materials uncountable'. The more probable interpretation is that colloidal micelles were formed after addition of the Biosolv enabling the system to be assayed more efficiently (§ 2.2.3). Faint yellow and pink colourations were produced in digests of liver and blood respectively. The recoveries of glycine-1-^{14}C and glycine-2-^{3}H subjected to this procedure were reported to be $99.3 \pm 2.9\%$ and $97.8 \pm 3.7\%$ respectively.

A similar oxidation system for the assay of ^{35}S consisted of oxidising the sample with nitric acid and perchloric acid containing also magnesium nitrate in glycerol (Jeffay et al. 1960). ^{35}S was quantitatively converted to magnesium sulphate which was dissolved in a mixture of dimethylformamide and toluene. This solution could then be assayed in a scintillant with 49% efficiency. Again, an advantage of this procedure is that many samples can be done at the same time since the reaction may be carried out in a number of open scintillation vials.

The wet oxidation method has also been used for the assay of iron isotopes in certain blood turnover studies; these methods will be described in greater detail in § 4.3.2.

Method of oxidising tissue by perchloric acid: hydrogen peroxide (Mahin and Lofberg 1966)

1) Tissue (0.2 ml or 100 mg) treated with 0.2 ml perchloric acid (60 %) in a scintillation vial.
2) Add 0.4 ml hydrogen peroxide (30 %, '100 vol').
3) Incubate 70 to 80 °C for 30–60 min until the tissue dissolves.
4) Cool, and add 5–6 ml ethoxyethanol and 10 ml Tp6 (no POPOP).

Even bone can be readily assayed by this procedure, about 200 mg being used. A 2.4 cm paper disc may also be oxidised. It is recommended that POPOP should not be used as a yellow colour develops which causes a great deal of quenching. The following efficiencies are achieved; tritium, 10–11 %; ^{14}C, 75 %; ^{35}S, 75 %; ^{45}Ca, 80–85 %.

Of the wet oxidation procedures described, the latter method is probably the most widely used for animal tissues and can be adapted for the assessment of a number of biological materials, including protein and nucleic acid preparations. Provided the temperature of incubation does not exceed 80 °C (explosive chlorine peroxide is formed at too high a temperature!) there is surprisingly little loss of labelled ^{14}C and ^{3}H by volatilization. The method however is not to be recommended for labelled carbohydrates and thus is not as useful as volatilization techniques in work with labelled plant material (Fuchs and De Vries 1972).

3.3.2. Degradation with volatilization of radioisotope

The oxidation of organic compounds in aqueous solution may be undertaken by the application of oxidising agents, the products of such oxidation, e.g. $^{14}CO_2$, $^{35}SO_2$ etc. being liberated either by simple diffusion, or removed by the passage of nitrogen or other inert gases such as argon. The trapping of radioactive gases has already been described (§ 3.1).

One of the earliest procedures for the chemical oxidation of organic materials into CO_2 was described by Van Slyke and Foch (1940) for the elemental analysis of organic compounds. A mixture of fuming sulphuric acid, periodate and chromic acid was used for the oxidation;

the mixture was boiled for 1–2 min in a stream of nitrogen and the liberated CO_2 was trapped in an alkaline solution. For liquid scintillation counting the carbonate so formed could be precipitated as barium carbonate and counted by suspension counting § 2.2.2) or the alkaline carbonate solution could be counted in appropriately blended toluene or dioxane based scintillant mixtures. This apparatus was further modified by Fuchs and De Vries (1972), (Fig. 3.5). Another wet-oxidation procedure was described by Katz et al. (1954) in which a mixture of sulphuric acid and solid potassium persulphate was used with silver nitrate as a catalyst to commence the oxidation of the

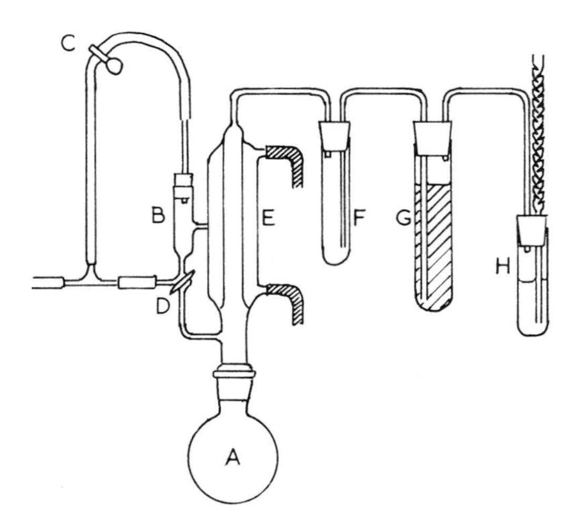

Fig. 3.5. A method of trapping $^{14}CO_2$ from a wet oxidation as described by Fuchs and de Vries (1972). Up to 100 mg of freeze-dried material is dried to constant weight in a 50 ml round bottomed flask (A). The van Slyke combustion system (mixture is used as described in the text and the SO_2 formed is removed from the gas mixture by passing through 4% $KMnO_4$ in 2 N H_2SO_4 in tube G. A mixture of ethanolamine: 2-ethoxyethanol (1:2 v/v, 25 ml) is placed in trap H to which a Vigreux column 60 cm long is attached to increase the efficiency of the extraction system. The whole apparatus is continuously flushed with nitrogen. After washing out the Vigreux column with methanol, the final trapping solution was made up to 50 ml and aliquots of 1 ml to 5 ml were used for counting. The condenser (E) and safety trap (F) are used to prevent carry-over of spray. (Reproduced by kind permission of Pergamon Press.)

organic compound in aqueous solution. The oxidation reaction was carried out in a conical flask at 70°C for 30 min. The liberated $^{14}CO_2$ was abosrbed into sodium hydroxide in a beaker placed in the centre of the base of the flask. After absorption had taken place, the solution was counted in a blended toluene-based scintillant mixture.

A modification of this procedure was described by Baker et al. (1954). A non-spillable ink bottle was used, and the oxidation procedure was carried out in one compartment and the absorption in another, the whole vessel being adiabatically enclosed. Using a larger 'Mason' jar, Walker and Lougheed (1962) used a similar oxidation procedure to deal with larger samples of organic material.

The use of a mixture of nitric acid and perchloric acid was described by Belcher (1960) for tissues and solid organic compounds and is useful for both ^{14}C and ^{3}H, 2–5 g of material can be oxidised at once by this procedure. The method is conducted in a conventional Kjeldahl flask, first oxidising with nitric acid and then with perchloric acid. In the case of tritium, the products of the oxidation are distilled off and the distillate is counted. Carbon dioxide is trapped in KOH solution and the carbonate solution is then counted in an ethanol-blended Tpp4 scintillant.

The use of hyamine hydroxide in a Warburg flask for the trapping of $^{14}CO_2$ following oxidation has also been described by Snyder and Godfrey (1961).

Many of the procedures described for the estimation of ^{14}C by the oxidation procedure cannot be used for the estimation of tritium because the oxidations are usually carried out in aqueous solution and the dilution of the tritium by the water would be too great to allow for the accurate estimate of the tritium. This is not the situation where dry materials or whole tissue is used with a concentrated acid, containing only a minimal quantity of water. An ideal method would be to convert the tritium present in the original compound to another organic compound containing tritium which was soluble in toluene-based scintillants. Such a method was described by Tamers et al. (1962) who incorporated the tritium into acetylene by reacting with calcium carbide and then to benzene by polymerization (§ 12.1.1).

This reaction is clearly of considerable value where low activity tritium is required to be measured.

3.3.3. Enzymic degradation

This method is based on the fact that most macromolecules, like deoxyribonucleic acid, ribonucleic acid, proteins and polysaccharides are insoluble in blended scintillants but their monomeric or oligomeric hydrolysates are usually soluble in these scintillant mixtures. Although many heterogenous systems have been described for assaying radioactivity in these materials it is often desirable to bring the molecules into homogenous solution so that accurate quench assessment can be undertaken, especially when there is more than one isotope involved. Although solubilization techniques are useful (see § 3.2) in this regard, it is often more convenient to degrade small quantities of macromolecular solutions to monomeric or short-chain products by means of enzymes. The protein of the enzyme can usually be made so low as not to interfere with the counting system. Although the technique does not appear to have been described very often in the literature it has proved to be very useful in the author's hands. An example of such a method is described below for DNA, but there appears to be no reason why other macromolecules may not be assessed in this manner, especially if the sample that requires to be measured is in the form of a relatively high specific activity solution of the macromolecule in buffer.

Method of assay of isotopes in nucleic acid

1) The nucleic acid solution (up to 5 mg/ml, 0.2 ml) is dispensed into clean conical centrifuge tubes of 10 ml capacity.
2) Aliquots (0.05 ml) of pancreatic DNase solution (1.0 mg/ml of 3 m $MgCl_2$ 0.05 M Tris, pH 7.3) is added and the mixtures incubated for 2–3 hr or overnight at 37°C.
3) The tubes are treated with a suitable blended scintillant mixture (10 ml) such as DNXppE (Table 5, Appx. II) and the mixture is poured into a vial for counting.

In the case of RNA, RNase may be used with shorter incubation times; and for proteins Pronase B has been found to be the most useful enzyme. Other enzymic methods are described in the appropriate sections of chapter 8.

3.4. Precipitation methods

The formation of a precipitate between the beta-emitter and an inorganic or organic ion, such that a specific component of the mixture is quantitatively separated from the bulk, is a useful means of concentrating and selecting that species. There can be many other labelled species present in the original mixture, and it is often, for example in certain enzyme reactions, convenient to isolate one labelled species from another in order to determine the extent and rate of the reaction. Although other methods involving selective absorption on to ion-exchange celluloses are used (§ 2.2.1) there are some reactions in which a precipitate can be formed in a quantitative manner, often from large bulk of solution. The precipitate, now in small bulk, can be assayed on a filter disc or by solubilization within the scintillation vial.

There are only a few instances where an inorganic cation is used to precipitate a labelled organic anion as a method of concentration and radiometric analysis, although it is difficult to understand why this technique has not been employed more frequently. The simplest example would be the precipitation of a labelled oxalate with calcium ions, but there appears to be no record of such a use as applied to liquid scintillation counting, even though the assessment of ^{45}Ca activity by this method is very well established and frequently employed.

The precipitation of a macromolecule by means of dilute perchloric or trichloracetic acids is however extensively employed (chapter 8). In order to concentrate a labelled organic species, an organic ion could be used, with which it selectively forms an insoluble precipitate. The only important criteria for success seem to be that the precipitate has low solubility in the reactant solution and that it should be colourless. It is worth noting that the lower the solubility of the precipitate, the greater is the volume of solution from which the trace component can be isolated. If the precipitates are not soluble in toluene based scintillant mixtures, for routine comparative measurements, there seems to be no reason why they should not be filtered on to discs and dried and counted.

One of the most useful applications of a selective precipitation is in the isolation and assay of blood glucose $-^{14}C$ (§ 4.3.2). The glucose is precipitated as the strongly coloured phenyl-glucosazone which is then converted into the colourless glucozotriazole for counting (Steele et al. 1957; Jones and Henschke 1963).

A still further use of an organic precipitant is seen in the assay of labelled lactate in body fluids by conversion to aldehyde and precipitation as the very insoluble dimedone complex (§ 4.3.2).

Selective precipitation has also been widely used in the field of nucleic acids (see § 8.2). These are known to form complex insoluble salts with certain high-molecular weight quaternary ammonium salts. Typical examples of this reaction are described by Trewavas (1967) and Sibatini (1970) using cetyl trimethyl ammonium salts in 2-methoxy ethanol and methyl alcohol in the presence of EDTA.

There appears to be considerable scope for the wider application of specific precipitation in the field of liquid scintillation counting and with the advantages to be gained in reduced sample preparation time.

Animal tissue processing

The assay of radioactivity in animal tissues presents a wide variety of problems, derived in most cases from the insolubility of body components and from coloured products which accompany the degradation reactions. The best approach is to degrade the material as far as possible to small molecular weight species which would be readily soluble in homogenous blended systems. If the tissue has a medium specific activity e.g. an expected count for tritium of approx. 1000 cpm per 20 to 30 mg wet weight, then the application of a blended homogenous system is the method of choice. The problem arises when the amount of tissue needed to produce an adequate number of counts exceeds the capacity of the blended scintillation systems and an alternative method of radioactivity concentration is needed.

A frequent requirement in drug distribution studies is to estimate the total concentration of labelled material present in the whole animal at a given time after treatment. Although it is possible to estimate the body content by difference between the counts administered and those which are excreted in the urine, faeces and respired gases, it is more accurate to obtain directly an assessment of the total activity still present in the animal.

Radioassay of whole mouse (tritium, ^{14}C, ^{35}S and ^{32}P)

This method is useful where the isotope would be volatilised in procedures involving ashing of the carcass.

1) Place dead mouse in a 25 ml silicone rubber stoppered measuring cylinder and just cover with 12 N NaOH.

Subject index p. 309

2) Agitate gently at intervals over 2 to 3 days to dissociate the fragments. The carcass usually dissolves in this period at room temperature. Record the final volume.
3) Mix 0.6 ml of the solution with 1.5 ml distilled water and neutralise carefully (ice cooling) with 12 N HCl (approx. 0.5 ml needed). Make up to 3.0 ml with distilled water. The final salt concentration is approx. 1.2 M and this volume can then be mixed with 7.0 ml of a Triton X-100 : toluene scintillant (1 : 1), containing 8 g PPO per l, shaken thoroughly and counted using a water-soluble standard to correct for quenching.

If the radioactive component needed to be assayed is stable to combustion, such as metallic and alkaline earth ions, then it is usually more accurate to remove as much of the non-radioactive components of the tissue as possible before handling and to concentrate the label to a residue. Although it is more accurate to count these residues in a homogenous liquid scintillation counting system, it is often less time consuming to count the residue on a glass fibre disc, especially with the more energetic beta-emitters. The following method of assaying the total body calcium-45 is based on this approach.

Radioassay of total body calcium-45 (modified from Cramer and Ross 1970)

1) Ash the whole animal carcass by heating it in a muffle furnace at 600°C overnight.
2) Dissolve the ash in the minimum quantity of 3 N HCl (approx. 2 ml per 20 g). Measure the total volume.
3) Place 100 μl of the acid solution on to a glass fibre disc, dry and assay in 4 ml of Tpp4 (Table 3, Appx. II).

The counting efficiency of calcium-45 under these conditions is between 88 and 90%. If there are insufficient counts present in this single disc, it is possible to add more discs to the same vial as described in § 2.2.1. If there are still insufficient counts present in up to 2 ml of this solution (approx. 20 discs), then a procedure involving the isolation and purification of the calcium salts is necessary (§ 4.1.1).

Any other isotope that does not volatilize on ashing can be measured by a similar procedure.

4.1. Hard tissue

This type of tissue includes bone and cartilage, teeth, horn, hair collagen and skin.

4.1.1. Bone, teeth etc.

Radioassay of bone and teeth is usually concerned with the measurement of calcium-45, strontium-89 and 90, phosphorus-32 and carbon-14.

Methods of assaying calcium-45 in bone were first described by Lutwak (1959) who suggested two alternative procedures. The first was to establish a scintillant mixture made up of a ternary system of toluene, absolute alcohol and water which would incorporate calcium chloride solutions without precipitating the salt. The mixture suggested however was accompanied by a considerable level of quenching and thus a second method was described. Certain calcium salts (acetate and lactate) as well as calcium hydroxide were converted into the 2-ethyl hexanoate salt, which was soluble in non-polar scintillant mixtures and could thus be assayed as a high efficiency homogenous counting system. A significant advantage of this method was the fact that the hexanoate salt was sufficiently soluble in toluene to be extractable by this solvent from aqueous solutions, and thus could be assayed with high efficiency in an appropriate non-polar scintillant mixture. However this method was limited by the fact that the salt could not be precipitated from the more usual chloride, citrate or phosphate solutions and it was necessary to undertake some tedious preliminary conversion. An additional problem was the fact that the level of calcium present had to be kept low to avoid interfering precipitation.

An alternative procedure is to dissolve the ashed bone in nitric acid and inject the solution through a spiral capillary of plastic scintillator. It is claimed (Pickering et al. 1960) that the decreased self absorption artifacts from the absence of precipitated calcium salts as well as the increased recoverability of the sample were significant advantages. These could be overriding factors in some assay requirements which could make this technique useful.

Alternatively, the calcium chloride solutions obtained by solution in hydrochloric acid may be dried by evaporation at 130°C and the residue assayed as a suspension in a dibutyl phosphate: toluene scintillant mixture (TpdD, Table 4, Appx. II). Approximately 500 mg $CaCl_2$ may be supported in 15 ml of this scintillant mixture and assayed at between 75 and 80% efficiency (Hardcastle et al. 1967).

In order to ensure that only calcium or strontium salts are measured, it is necessary to employ an intermediate oxalate purification step (Carr and Parsons 1962; Sarnat and Jeffay 1962). The first-mentioned authors converted the oxalate into carbonate by heating the dried precipitate to 500 ± 25°C, dissolving the resulting carbonate in 1 N HCl and assaying the solution in an ethyl alcohol-blended scintillant mixture in the presence of Cab-o-Sil to absorb any precipitate which may be produced. By converting the oxalate into nitrate, this salt may be radioassayed in a scintillant system already containing ethylene glycol and nitric acid (TpEGN, Table 4, Appx. II). In the latter technique, the whole of the preprocessing can be conducted in a sintered glass crucible and hence reduce any losses due to unnecessary transfers. In both methods it is possible to assay the specific activity by a gravimetric estimation at an intermediate step. For less than 50 mg calcium salt, the oxalate may be converted into the perchlorate by treatment with 60% perchloric acid, and the resulting solution assayed in a 20% tributylphosphate solution in a toluene-based scintillant (Tpp3, Table 3, Appx. II) as a near homogenous scintillation system (Humphres 1965).

These different procedures were reviewed by Lerch and Cosandry (1966) who maintained that with most of the apparently homogenous systems, some precipitation of the calcium salts occurred and introduced varying degrees of heterogeneity into the counting system. They recommended that both calcium-45 and phosphorus-32 in bone should be assayed as the pure dried salts on filter paper, hence ensuring that reasonably constant (albeit heterogenous) counting conditions were achieved. In fact, Cramer and Ross (1970) suggested that the even simpler process of applying 100 μl of the HCl solution of the calcium salt to glass fibre discs and drying and assaying in Tpp4 (see Table 3, Appx. II) was adequate for most purposes. Glass fibre discs are 90% more efficient than filter paper for ^{45}Ca (Hutchinson 1967) and the problem of shedding of the glass fibres referred to by these authors is only a disadvantage if the discs are needed to be stored for reference purposes. It is of interest to note that although the technique for the assay of calcium-45 in urine described by Turpin and Methune (1967) appears to be potentially useful for the assay of this isotope in bone samples, its use in this respect does not appear to have been described. The method consists of precipitating the calcium ion as the acetone dimethyl acetal complex and counting the precipitate as a suspension (see § 4.3.5).

Assay of specific activity of calcium-45 in bone

The method is compounded from data described by a number of authors, but is primarily based on the original method described by Carr and Parsons (1962)

1) Ash the bone sample at 600 to 800°C for 8 hr in a muffle furnace.
2) Dissolve ash in the minimum amount of 1 N HCl[a].
3) Treat an aliquot (containing up to 100 mg) with excess 4% oxalic acid in a hard glass tube and adjust the pH to 3.5 to 4.5 with ammonium hydroxide (suitable indicator bromocresol green).
4) Heat solution to 70°C for 20 min, cool, centrifuge.
5) Wash precipitate with water (to which a drop of ammonium hydroxide has been added), centrifuge and decant off the supernatant.
6) Dry the precipitate at 140°C[b].
7) Heat the tube to 500 ± 25°C for 4 hr or overnight.
8) Cool, weigh (as carbonate).
9) Add 1 N HCl to dissolve the precipitate.

10) Heat the solution to dryness at 140°C by gradually raising the temperature to avoid spitting.
11) Dissolve the residue in 2 ml ethyl alcohol and add 10 ml of Tpp4 (Table 3, Appx. II[c]) and transfer to vial for counting.

Note (*a*). A rapid comparative assay can be carried out at this stage by applying 100 μl of the sample to a glass fibre disc, drying and assaying in Tpp4 (Table 3, Appx. II).

Note (*b*). Where an accurate assay of mixed isotopes is to be made, or the specific activity of the calcium present needs to be measured, the following additional steps require to be made. The calcium oxalate precipitate is first converted into perchlorate by treatment with 60% perchloric acid. An aliquot containing 200 to 500 μg calcium is then transferred to a white porcelain dish, Glycine buffer (pH 12.66, 5 ml) is then added together with approx. 10 mg of a dry indicator mixture containing 0.2 g calcein, (fluorescein-3,3′-biomethyl imino diacetic acid, Hopkin and Williams), 0.12 g thymolphthalein and 20 g KCl, mixed and ground to a fine powder. A standard solution of EDTA (0.93 g disodium EDTA diluted to 1 l) is run into a magnetically stirred mixture until the endpoint (green to pink) is achieved. Blanks and standards are used in the usual way.

Note (*c*). An alternative procedure at this stage is to use 15 ml of TpdD (Table 4, Appx. II) and to assay as a suspension.

Strontium-89 may also be assayed by the procedure described above using the glass fibre disc modification (Note a). This method can also be applied to the assay of egg shells and faeces (Creger et al. 1967). For the disc counting, quench correction is best undertaken by the internal standard method, using a standard ^{89}Sr salt prepared in the same way (§ 11.1.3). Phosphorus-32 may also be assayed by this technique, using solid supports (Lerch and Cosandry 1966).

The assay of strontium-85 in the presence of calcium-45 can be undertaken at the ethanolic solution step in the above scheme (step 11). The channels settings for this particular isotope combination are shown in Fig. 4.1. The separation of $^{90}Sr(^{90}Y)$ and ^{45}Ca is similarly shown in Fig. 4.2.

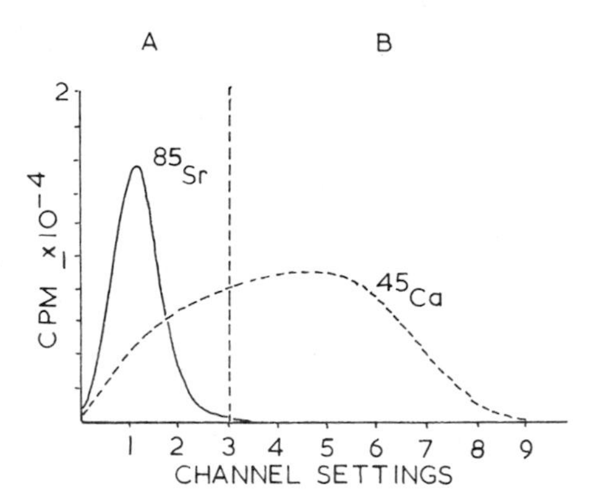

Fig. 4.1. The spectra of ^{45}Ca and ^{85}Sr. The data obtained using a 2% window scan of a system containing 2.0 ml ethanol/10 ml BBOT scintillant (4 g/l in toluene) plus 20 mg calcium^{++} as $CaCl_2 \cdot 2H_2O$ containing approximately 0.05 μCi ^{45}Ca and 0.1 μCi ^{85}Sr. Gain 9.0 A + B represent the approximate positions of the upper and lower channel settings (Carr and Nolan 1967). (Reproduced by kind permission of Beckmann Instruments Ltd.)

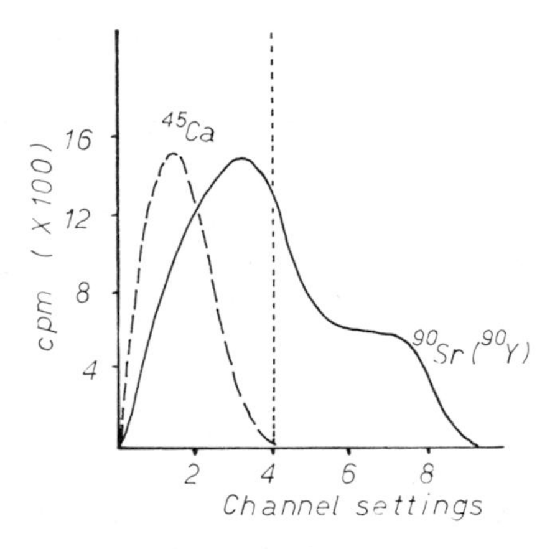

Fig. 4.2. The spectra of ^{45}Ca and ^{90}Sr (^{90}Y). Lower gain settings than those required by ^{85}Sr in Fig. 4.1 are needed the approximate position of the discriminator required to separate the 2 isotopes. (Nolan 1972) (In Liquid Scintillation Counting, Vol. 2, eds., M. Crook, P. Johnson and B. Scales, by kind permission of Heyden and Son Ltd.)

If the level of activity in the bone is relatively high, then the more rapidly applied wet combustion procedure of Mahin and Lofberg (1966) (§ 3.3.1) can be successfully applied to samples up to 20 mg weight, and efficiencies of the order of 10–11 % for tritium, 75 % for carbon-14, sulphur-35 and 80 to 85 % for calcium-45 may be achieved. This simpler method is clearly only suitable for samples of bone where the usual ashing procedures would lead to isotope loss, and where the level of isotope is high enough to permit measurements on such small samples. For isotope distribution measurements in small rodents, such as the distribution of labelled drugs, it is a very suitable method.

Solubilizers, such as the Beckman product (BioSOLV 3) have been suggested (McClendon et al. 1971) in combination with an ashing procedure for the assay of ^{45}Ca in skunk toe nails. After ashing the toe nails in a muffle furnace for 12 hr at 1200°F, 15 to 30 mg of the ash is placed in scintillation vial with 0.5 ml concentrated HCl. After heating the mixture on a sand bath at 100°C, the vial is shaken gently and the excess acid allowed to evaporate to dryness. The residue is dissolved in 2 ml of dilute HCl and after the residue is dissolved, 1 ml of (BiSOLV) is added and allowed to stand for approx. 5 min. A scintillant mixture containing 8 g/l butyl PBD, 0.5 g/l PBBO, BioSOLV (10 % v/v) in toluene is added. For routine assays of this type, it is most likely that an increased merit value could be obtained by careful choice of conditions in a colloid counting system as described in § 2.2.3.

4.1.2. Hair, lens, skin etc.

The radioisotope content of collagenous materials can usually be assayed by using a suitable combustion procedure (see § 3.1) or by employing a wet oxidation procedure (§ 3.3.1). However, several satisfactory techniques have been described which exploit the various chemical and physical properties of the material to be measured.

Hair, wool and similar material is usually sufficiently divided to be assayed by a heterogenous counting system, such as by suspension counting directly in a toluene based scintillation mixture such as

Tpp4 (Table 3, Appx. II) (Downes and Till 1963). Up to 500 mg of tritiated wool can usually be assyed by this means in a standard vial and when the results are compared with those from the same samples obtained by a combustion technique, the counting rates are apparently proportional to fibre mass up to about 200 mg. Above this level, absorption artifacts are evident. Indeed, the absorption effects have been studied in some detail by Downes and his co-workers, and methods of accurate assay of such isotopes as ^{3}H, ^{14}C, ^{35}S and ^{203}Hg have been described.

An ingenious application of these studies of the relationship between fibre diameter and the counting efficiency of ^{14}C incorporated into the fibre has led to a good quantitative assay of the mean fibre diameter by initial absorption of ^{14}C-formic acid into the wool and relating the counting efficiency, following stabilization of the counts, to absorption of the beta particle from the isotope. A known radioactive scintillant solution (0.5 ml) consisting of ^{14}C formic acid (100–250 Ci/g) 50 μl together with 50 μl of Triton X-100 in a scintillant mixture of butyl PBD (5 g/l) and POPOP (0.5 g/l) in toluene is added to a known mass of wool (100 mg) and the vial is sealed. The counts are assayed over an extended period until they decrease to a stable value. The counting rate eventually attained is related to the mean wool diameter according to Fig. 4.3. The counting efficiency of ^{203}Hg (E_{max} 0.21 MeV) is also shown, where the higher energy radiation is less affected by the fibre diameter than in the case of ^{14}C.

Skin samples are most accurately assayed by the combustion technique (§ 3.2). Wet oxidation techniques using nitric acid (O'Brien 1964) or perchloric acid-hydrogen peroxide (Mahin and Lofberg 1966) have been specifically used with this tissue. The quaternary ammonium salt solubilizers (§ 3.2) can also be readily used. A novel method of examining the regeneration of the epidermis has been described by Hennings and Elgjo (1970). Adhesive (Scotch') tape is used to remove the surface layers (squames) of the skin at different times after labelling with tritiated thymidine. The tape is assayed in a toluene-based scintillant (e.g. Tpp4), and a study of the kinetics of epidermal development can thus be undertaken.

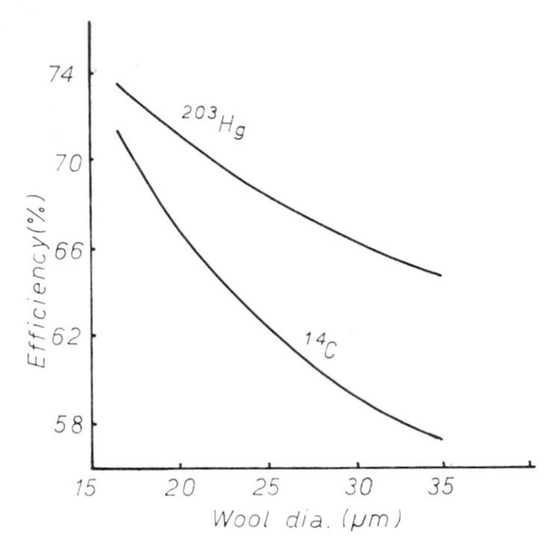

Fig. 4.3. The relationship of the counting efficiency of ^{14}C (formic acid) and ^{203}Hg in wool fibres of different diameters. It can be seen that the influence of fibre thickness is inversely related to the energy of the isotopic radiation (Downes 1971). (Reproduced by kind permission of Academic Press Inc., from Organic Scintillators and Liquid Scintillation Counting edited by Horrocks, D.L. and C.T. Peng.)

The formation and degradation of collagen has been studied using scintillation counting techniques. A rapid assay system to study the extent of incorporation of ^{14}C-hydroxylysine into collagen was described by Blumenkrantz and Prockop (1969). The level of incorporation is determined by first allowing synthesis of labelled collagen in 10 day old chick tibia and then degrading the product using the enzyme collagenase. After extracting the solution with toluene, it is oxidised with sodium metaperiodate and the ^{14}C formaldehyde so formed is trapped as a dimedone complex, which is soluble in toluene-based scintillant. An estimate of the collagenolytic activity of a homogenate of bone cells was also described by Woods and Nichols (1963). The technique here consists of first decalcifying bone chips after removing protein and lipids. The bone cell homogenate is then incubated for varying lengths of time and ultrafiltered to remove solubilized collagen material. Collagen itself readily dissolves in quaternary hydroxides such as NCS (Hansen and Bush 1967). The optimal sample composition is 420 mg collagen with 4.0 ml NCS and 0.7 ml water. An efficiency of 13.4% for tritium was recorded. Hyamine hydroxide appears to be less efficient in its solubilization properties for collagen, efficiencies of only 5.1% being reported.

Eye lens suffers the same problems as other collagenous materials in being difficult to dissolve. A method of radioassay of tritium-labelled protein from rabbit lens was described by Gerweck et al. (1972).

Twenty four hr after labelling the lens protein by incubation of the excised lens in medium TC 199 containing labelled amino acids, the lenses are removed, blotted, weighed and homogenised in 2 ml of 10% trichloroacetic acid. The homogenates are centrifuged and the pellets washed with absolute ethanol and once with ether before being dried at 100°C and weighed. Hyamine hydroxide (0.0167 ml/mg protein) or NCS (Amersham Searle product) (0.0125 ml/mg protein) together with 0.1 ml of water is added to the precipitate. Digestion is carried out from 24 to 72 hr at 50°C. An aliquot (0.2 ml) of this digest is then added to 10 ml of TDNpdM or Tpp4 (Tables 5 and 3, Appx. II).

4.2. Soft tissues

The problems encountered in the radioassay of soft tissues are concerned not so much with the insoluble nature of the material to be studied but with the quenching associated with coloured materials carried over by incomplete degradation. The most efficient method is undoubtedly the complete oxidation of the tissue by combustion in oxygen which has been fully described (§ 3.3). However, where a large number of samples need to be assayed and compared it is often possible to develop a specific degradative technique which is more rapid, gives reproducible efficiencies and which is more directly compatible with the experimental conditions of the biochemical work being undertaken.

The tissues are best considered in groups based largely on the methods that have been found to be the most suitable. The final method chosen for the assay of any particular soft tissue will depend on, a) the level of the radioactivity present in the tissue, b) the nature of the isotope being measured, c) the presence or absence of coloured impurities, such as blood and bile pigments, and d) whether the whole or part of the organ need be assayed.

In general if the tissue is of high specific activity, e.g. greater than 5000 cpm/100 mg wet weight, then a wet oxidation system based on 20 mg wet weight will give a counting level which is satisfactory for the majority of requirements. However, should the counts be well below this figure, then the combustion technique is best employed. However, should a large number of samples of low specific activity be required to be assayed, then a solubilization technique is probably the best approach. It should be noted however, that the newer oxidation equipment that is commercially available may analyse up to 60 samples an hour and represent a real challenge to the analysis of multiple samples by any other procedure. However the automated equipment available tends to be expensive and unless a lot of routine combustion measurements are anticipated, they may not be an economical proposition for the average biochemical laboratory. Hunt and Gilbert (1972) conducted a comparative study of digestion and combustion techniques for a number of soft tissues; poor recoveries were recorded from the digestion of lung and spleen but unusually high values were obtained for muscle. As expected, combustion procedures produced consistently high recoveries.

4.2.1. Readily homogenised tissues e.g. liver, spleen, etc.

The tissue in this case may be homogenised in a suitable saline solution and an aliquot taken for further processing. Tissues of this type can usually be solubilised either by the wet oxidation procedure or by the use of a suitable solubilizer. The decision will usually be one of economics, the solubilizer technique being more expensive than that of the peroxide-perchloric acid method. The solubilization method however is capable of accommodating a greater quantity of tissue and is thus more useful when the specific activity of the tissue is expected to be low.

Solubilization of liver preparations with hyamine hydroxide or other similar quaternary salts often results in a phosphorescence which persists for a considerable period. This luminescence appears to be similar to the peroxide-induced luminescence and is diminished by the addition of a suitable radical scavenger such as 10% ascorbic acid or by acidification with acetic acid or HCl.

Although the most usual requirement in assaying a tissue of this type is the measurement of the total radioactive content, it is sometimes convenient to modify the preprocessing to give a product in which the isotopically labelled component is more concentrated than in the original sample. If ashing of the sample is not feasible due to the volatility of the isotope, extraction of the tissue homogenate with a suitable solvent, e.g. ether or a higher secondary alcohol, or precipitation of the macromolecular components of the tissue with a suitable precipitant may be considered. Such procedures are particularly suitable when an assessment of the location of the isotope in the tissue is needed, for example, in pharmacological investigations using labelled drugs.

In isolating the tissue from the animal, the animal is first exsanguinated and the liver or spleen removed, rinsed in clean physiological saline, blotted on filter paper and placed immediately on tin foil lying on top of crushed CO_2 or alternatively dropped into liquid air. This allows for a rapid freezing of the tissue so that degradative enzyme activity can be checked and hence reduce the possibility of a change in the chemical state (and possible volatility) of the labelled material present. The tissue can then be stored at solid CO_2 temperatures or in liquid air. Even at $-20°C$, there is some reduction in the moleculaar weight of DNA due to nuclease activity. When the tissue is solid, it can easily be broken and weighed. For a truly representative sample however, it is necessary to homogenise the whole tissue and remove an aliquot of the homogenate for processing and counting.

A method of cleaning the small intestine, is to remove the tissue from the animal and place it directly into a beaker of crushed ice in water. The intestine stiffens slightly under these conditions, and by forcing iced saline from a 20 glass syringe (without a needle attached) through it, the faecal contents can be ejected. The intestine can then be blotted on a piece of filter paper and frozen as for the tissues above.

Should the epithelial lining be required, it may be removed by cutting with sharp scissors along its length, placing on a filter paper with the epithelial cells upwards and the upper layer scraped gently off in short lengths with a blunt scalpel. In order to avoid excess cells attaching themselves to the filter paper and becoming very difficult to remove, a long microscope cover slip can be placed beneath the intestine at the point at which scraping is being done.

Where labelled components are to be measured, it is an advantage to remove as much water from the tissue as possible to reduce the level of unlabelled water diluting any tritium oxide produced in combustion or wet oxidation procedures. Care must be taken however to ensure that the tritium to be measured is not already present as tritiated water in the aqueous compartment of the cells. In this unlikely event, the water may be distilled off and assayed in a blended homogenous counting system such as DNXpnE (Table 5, Appx. II). If unchanged drug is to be assayed, then some property of the drug, such as extractability in lipophyllic solvents or precipitability with specific precipitants, should be exploited first. It may also be necessary to separate the components by column, paper or thin layer chromatography before an assessment of the isotopic content can be made due to the considerable excess of unlabelled material present. These techniques are described in chapters 7 and 9.

Assay of ^{35}sulphur in soft tissue.

A wet oxidation procedure can be used. Pirie's reagent* (1 ml) with 250 mg wet weight of tissue is heated slowly (1–2 hr) on a sand bath at 260–280°C. The residue is cooled to 150°C and dissolved in 1 ml glycerol at approx. 100°C. The cooled solution is dissolved in 6 ml ethanolic solution of NN-dimethylformamide (1 : 3 v/v) and mixed with 10 ml of a toluene-based scintillant e.g. Tpp3 (Table 3, Appx. II) (Jeffay et al. 1960). This procedure can equally well be employed with

* Pirie's Reagent; prepared by adding 3 vol of concentrated nitric acid to one vol of 60% perchloric acid; take one fourth of this mixture and saturate with magnesium nitrate (approx. 50–60 g/100 ml). This saturated solution is then added to the remainder of the acid mixture.

other isotopes such as ^{32}P and ^{45}Ca, but not where the risk of volatile products is likely, as with ^{14}C.

4.3. Semi-solid or liquid

This section deals with those body fluids and excretory products which do not require any significant mechanical effort to disintegrate other than mixing or suspension. Often it is necessary to measure the rate of excretion of isotope from the whole body after administration of a labelled drug; an efficient means of separating faeces from urine, and a trap for expired carbon dioxide are required.

Many designs of metabolic cages have been described and some of these are available commercially. An apparatus suitable for one or very few mice or rats is shown (Fig. 4.4).

4.3.1. Faeces

The difficulties associated with the radioassay of faeces are similar in many respects to those encountered in the radioassay of bone and whole blood. Colour quenching is usually an important consideration.

The variation in water content of the faeces between species and between different collections following drug treatment often makes collection difficult. Variable results will be obtained if presented as a specific activity related to the wet weight of the sample. It is thus useful to dry the faeces where possible, by heating or by standing in a dessicator and heating the sample as a dry homogenate. A preliminary test to ensure that the isotope is not present in a volatile component may be conducted by placing a small sample of the faeces in a closed Petri dish on a warm plate with a piece of solid CO_2 on the lid. Some water will rapidly condense on the undersurface of the lid and an aliquot (e.g. 0.2 ml) can be assayed in a blended scintillant (e.g. 10 ml TppE, Table 4, Appx. II). If the isotope is in the dry faeces, the whole collection should be processed together and thoroughly homogenised before sampling. It is often tempting to sample animal faeces by selecting pellets for assay, but there will be a considerable variation between them, depending on the time and order of evacuation.

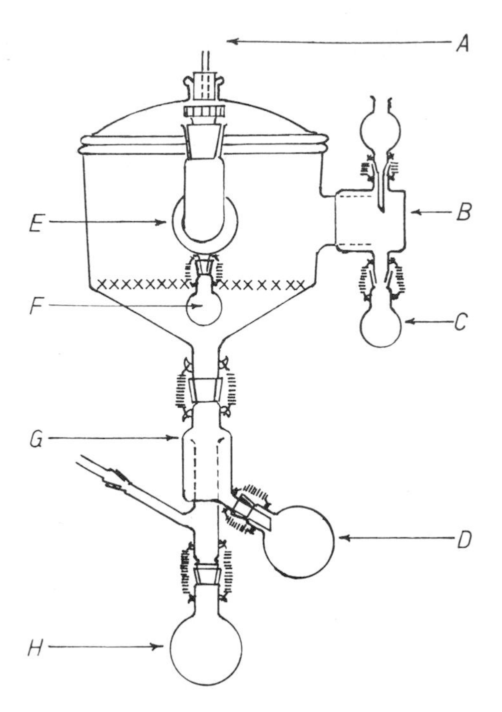

Fig. 4.4. A metabolic cage suitable for a small rodent (Jencon Scientific Ltd, Herts). A) air intake, B) water and feeding area, C) unconsumed water trap, D) urine collecting vessel, E) solid food receptacle, F) unconsumed food trap, G) faeces urine separating zone, H) faeces trap.

If it is necessary to sample a large quantity of faeces it is often more convenient to homogenise in a Waring blender and to assay an aliquot of the homogenate by solubilizing or by a suspension technique, depending on the penetrability of the beta emission.

Both ^{45}Ca and ^{32}P can be assayed by a similar procedure to that used in the assay of these isotopes in bone (§ 4.1.1). The faeces are first ashed and then the residue is dissolved in hydrochloric acid (Nolan 1972).

In order to derive a suitable 'impurity quench curve' for the assay of faeces, a synthetic faecal ash has been described (Carr and Parsons

1962) and consists of the following mixture: NaCl (15.2 g), KCl (44.8 g), saturated $MgCl_2$ solution (204 ml), Fe_2O_3 (1.22 g), H_3PO_4 (88%, 92.0 g), H_2SO_4 (98%, 20.5 g), and $CaCO_3$ (80 g). Sudan IV can be used to produce a suitable 'colour quench' curve. Using these correction procedures, faeces with as little as 40 pCi ^{45}Ca/mg can be assayed.

The dry oxidation of faeces by the bomb oxidation technique is not recommended, as very low recoveries (70%) have been reported (Burns and Glass 1963).

4.3.2. *Whole blood, plasma etc.*

The direct assay of radionuclides in whole blood presents several problems in sample preparation. Proteins are readily precipitated by alcohol-blended scintillant mixtures and introduce heterogeneity and poor reproducibility in the counting system. Furthermore, short chain peptides, salts and other low molecular weight species present in blood also introduce impurity quenching and varying levels of haem and other porphyrins which absorb light strongly in the visible regions and varying degrees of colour quenching.

Early work on the solubilization and decolourization of whole blood samples has been reviewed by Houtman (1965). Solubilization was effected by using hyamine hydroxide or alcoholic potassium hydroxide. The haem pigments were persistent in these preparations and hydrogen peroxide (10 drops of 30%) (Herberg 1960) or sodium borohydride (Fales 1963) was used to decolourise.

The following method given below was described by Hansen and Bush (1967).

1) Add 0.4 ml whole blood to a scintillation vial.
2) Add 1.4 ml freshly made saturated solution of benzoyl peroxide (6–7 g of benzoyl peroxide in toluene (30 ml) at 60°C, cool rapidly to 18°C, stand 1 hr, filter).
3) After decolourization has occurred, add 4.5 ml NCS and heat for 30 min at 50°C or until no trace of solids remain.
4) Add 10 ml Tpp12 (see Table 3, Appx. II).

Other commercial solubilizers have been recommended for whole blood assays. Laurencot and Hempstead (1971) used the Beckmann Instrument product, BioSOLV-3. Care should however, be taken in the interpretation of results obtained from detergent-based 'solubilizers' when the external standard ratio method has been employed for quench correction (see § 2.2.3).

Assaying of whole blood by the wet oxidation technique of Mahin and Lofberg (1966) is particularly successful, since decolourization occurs through the nature of the technique itself. The procedure normally used is exactly as described in § 3.3.1 using 0.2 ml of whole blood.

The method described by Lindsay and Kurnick (1969) using Triton X-100 and the Nuclear Chicago (NCS) to 'facilitate the solution of tritium-containing tissue' should be used very carefully. It is not surprising that chemiluminescence was found to be excessive and the recommendation that the 'external standard may be used' may lead to entirely spurious results. In this case, only internal standards of similar solubility character to the radionuclide containing material under investigation can be relied upon.

In many cases, it is often unnecessary to combust or solubilize the whole sample of blood and the possibility of a limited preprocessing to concentrate and if possible, to convert the beta-emitter into a form which is soluble in a homogenous counting system should be considered. A typical example of the specific type of preprocessing which leads to a cleaner product for homogenous counting, is that of the estimation of blood ^{14}C-glucose. In one method, the glucose is first converted to potassium gluconate by the method described by Blair and Segal (1960). An aliquot of the supernatant after removing protein is then dried on discs and assayed (Segal and Blair 1962). Alternatively ^{14}C-glucose can be assayed after deproteinization of laked blood with Somogi's reagent (to 6 ml laked blood add 2 ml 0.3 N $Ba(OH)_2$ and 2.0 ml of 5 % $ZnSO_4 \cdot 7H_2O$); glucose is converted initially into glucose penta-acetate by treating the glucose solution with sodium acetate and acetic anhydride (Jones 1965). The penta-acetate of glucose is a particularly useful derivative as it is soluble in a toluene-based scintillant mixture such as Tpp4. ^{14}C-labelled glucose-1-phosphate on the other hand can be treated with potassium iodate which converts the carbon-14 of this compound into ^{14}C formic acid which can be stabilized as bromophenacyl-^{14}C-formate which is also soluble in toluene and hence in a form suitable for radioassay (Gabriel 1965).

Haemin crystals, containing carbon-14 can be assayed by adding 2 drops of methanolic hyamine hydroxide (1 M) and after a short period, 1 ml of 2-butoxy-ethanol and 2 drops of 30% hydrogen peroxide. On warming the vials to 60 to 70°C, the intense colour of the haemin disappears. The resulting solution can then be blended into a toluene-based scintillant (2 ml), using more butoxyethanol if required. The toluene-based scintillator solution recommended in this case is a twenty-fold dilution of the commercial product 'Liquifluor' (Table 6, Appx. II) in toluene (Murty and Nair 1969).

Labelled gases such as $^{14}CO_2$ and the inert gases, such as ^{85}Kr, can also be assayed by the liquid scintillation counting procedure. Krypton-85 in blood can be assayed by adding 0.5 ml whole blood to a scintillation vial containing a scintillation mixture consisting of toluene: Triton X-100 (6:4 v/v) containing PPO (5 g/l) and dimethyl POPOP (0.5 g/l). The cell debris settles to the bottom of the vial after vigorous shaking and the dissolved gases escape into the upper toluene-rich phase (Smith et al. 1970). Removal of the cells appears to be unnecessary in most counters, presumably due to the improved geometry in modern instruments. The technique as described is highly heterogenous and should be used with care; only an internal standard of ^{85}Kr would be reliable with this method.

An ingeniously simple technique for the measurement of dissolved $^{14}CO_2$ in whole blood consists of liberating the labelled gases inside a scintillation vial with a mixture of sulphuric and lactic acids added to the whole blood sample. The liberated $^{14}CO_2$ is then trapped on to a glass wool pad soaked in ethanolamine and attached to the inside of the cap by a support similar to that illustrated in Fig. 1.1c. After an appropriate length of time to allow complete absorption of the liberated gas, the cap is removed and replaced on a clean vial containing a toluene-based scintillant and shaken to allow dissolution of the ethanolamine carbonate (Hagenfeldt 1967).

The relative rate of absorption of iron from the gastrointestinal tract into the blood plasma needs to be known in certain iron deficiency diseases. The two isotopes, ^{55}Fe and ^{59}Fe can be used to study this parameter in humans (Peacock et al. 1946). A technique for the assay of these two isotopes in the same sample of blood was described by Hallberg and Brise (1960). Iron-59 has two energetic gamma emissions (1.1 and 1.3 MeV) and two beta emissions (E_{max} 0.27 MeV and 0.46 MeV). Iron-55 on the other hand decays by an electron-capture mechanism, emitting X-rays of 0.0059 MeV with a fluorescence yield of 25% and hence the latter isotope is detected with rather more difficulty than the former.

The earliest methods of sample preparation for the assay of these two isotopes consisted of boiling whole blood with a mixture of fuming nitric acid, and perchloric acid and later with sulphuric acid. The cooled solutions of iron salts were then treated with ammonium hydroxide to precipitate the $Fe(OH)_3$. The latter was then dissolved in oxalic acid : ammonium oxalate mixture (pH 4.5) and electrolysed to deposit the iron. This deposit could then be dissolved off in perchloric acid as ferrous perchlorate and assayed in a blended toluene solvent in which it is soluble. These methods were described in detail by Dern and Hart (1961a, b), Jenner and Obrink (1962), Katz et al. (1964). The overall method was modified by Perry and Warner (1963) who reduced the ferric hydroxide in the early stages of the method to ferrous salts by means of an ascorbic acid–hydrochloric acid mixture.

The method has been modified on several occasions since, in an attempt to cut down the amount of preprocessing required to produce the counting sample. The most extreme of those is to add an aliquot of whole blood directly to discs of filter paper and to assay on the dry discs. This technique is obviously of low accuracy and reproducibility but may suffice in some cases where only approximate comparative assessments are required (Lockner 1965). Another method which has gained considerable popularity is to convert the initial ferric hydroxide to a white ferriphosphate by treatment with phosphoric acid, ammonium chloride and ethanol, this can then be assayed as a suspension using Cab-O-Sil as a suspending agent (see § 2.2.2) (Eakins and Brown 1966). Efficiencies of 19.4 and 38.4% are reported for ^{55}Fe and ^{59}Fe respectively. A deficiency in this method is the inability to estimate the degree of quenching accurately, due to the heterogeneity of the technique. An alternative mixture consists of Triton X-100: toluene (1 : 2) and is assayed as a colloid system. Counting efficiencies for ^{55}Fe and ^{59}Fe counted simultaneously were 3.2% and 40.2% respectively. It is possible that more efficient counting of these isotopes could be achieved by a more rigorous examination of the phase diagram characteristics as described in § 2.2.3. An alternative heterogenous system was also described by Graber et al. (1967) in which ferric benzene phosphonate is prepared as a white powder, and assayed as a gel suspension in a toluene scintillant following ultrasound homogenisation of the precipitate. These authors suggest a procedure for the conversion of 2 ml of whole blood within the same vial to a counting system after a series of wet oxidations, evaporations, furnace heating and solution procedures which take approximately two days to complete.

Of all these procedures, it would still appear that the conversion of iron to the toluene-soluble ferrous perchlorate results in the best final product for assay as a homogenous system, capable of being corrected for quenching by the usual procedures possible for homogenous systems (see chapter 11). If only the total iron of a blood sample needs to be assayed however, there appears to be no basic reason why the method of solubilization described by Hanson and Bush (1967) described above for ^{14}C content, should not be adapted for this purpose, since it is possible to correct for quenching much more easily within this system.

The details of administration of the ^{55}Fe and ^{59}Fe salts to the patient have been described by several authors. Campbell and Powell (1970) recommend that 5 μCi of ^{59}ferric citrate be ingested with a breakfast meal (containing also 5 mg food iron) and that 25 μCi ^{55}ferric citrate be incubated with plasma and injected 1 hr after the breakfast. Blood is then removed 12 to 14 days later for assay. The percentage absorption of ^{59}Fe from the intestine was therefore given by eq. (4.1)

$$\% \text{ Absorption of } ^{59}\text{Fe} = \frac{^{59}\text{Fe act in blood}}{^{59}\text{Fe given orally}} \times \frac{^{55}\text{Fe given i.v.}}{^{55}\text{Fe act in blood}} \tag{4.1}$$

The reverse administration of the isotopes had earlier been described by Pitcher et al. (1965) followed by the assay of the whole blood in 10 days.

The selection of optimum counting conditions is usually achieved by counting ^{59}Fe essentially free from overlap from ^{55}Fe by setting the upper discriminator of the lower (^{59}Fe) channel to a point where ^{55}Fe overlap is least. In the Cab-O-Sil method described by Eakins and Brown (1966), with a setting of 400 to 1000 with a gain of 12.5 %, ^{59}Fe was assayed free from ^{55}Fe; and by using a gain of 100 % with a channel width of 50 to 400, ^{55}Fe counts with a 19% contamination by ^{59}Fe were obtained.

It is interesting that these latter authors, working in the United Kingdom observed background levels of 0.75 nCi/l of whole blood of ^{55}Fe, similar to the levels (0.96 nCi/l) detected by Palmer and Beasley (1965) in the United States. The latter authors attributed the levels observed to the results of nuclear weapons testing. The minimum detectable level for ^{55}Fe in blood by the liquid scintillation counting system appears to be around 0.2 pCi/ml (Eakins and Brown 1966).

Another technique for measuring the specific activity of ^{55}Fe in human blood was described by Cosolito et al. (1968). This method relies on the conversion of the iron to the toluene-soluble di(2-ethylhexyl) phosphate salt. They recorded normal human blood levels of 10.55 ± 0.58 pCi ^{55}Fe per ml, which contains 5.30 mg of stable iron (i.e. sp act 1.99 ± 0.11 pCi/mg).

Plasma

The levels of ^{55}Fe and ^{59}Fe in plasma can be measured by the techniques outlined for whole blood above.

A useful summary of the methods available for the study of beta-emitters in plasma and serum is given by Parmentier and Ten Haaf

(1969), in a general review. Since only small quantities of pigment are present in plasma and serum, procedures for the direct assay of the sample can usually be considered.

Miller et al. (1969) have described such a method for plasma as follows:

1) Plasma (1 ml) pipetted into scintillation vial.
2) Add 4 ml freshly-prepared methanolic hyamine hydroxide 10X (Hyamine 10X (66 g) gradually added to 30 ml spectral quality methanol in a 250 ml Erlenmeyer flask with constant magnetic stirring).
3) Add 15 ml scintillator fluid (Tpp5, see Table 3, Appx. II).
4) Shake bottle and count.

When assaying low levels of beta-emitters in plasma and serum, it is useful to be able to accommodate as much of the sample as possible in the scintillant system in order to achieve a high merit value in counting. The combination of a commercial scintillant mixture (Nuclear Enterprises, NE 213) (10 ml), hyamine hydroxide (2 ml) and n-octanol (1 ml), can accommodate up to 1 ml of sample. The highest merit value recorded was 41 for tritium (5.76% efficiency for a 7.1% plasma sample in scintillant). Spencer and Banerji (1970) found that plasma readily dissolves in dimethyl sulphoxide (DMSO), and described a method of counting where 0.5 ml ^{14}C-labelled plasma solution (26% plasma in DMSO) was assayed in 10 ml of scintillation fluid (dioxane-naphthalene system containing Cab-O-Sil) i.e. merit value of 750 for ^{14}C if 1 ml is used.

The measurement of tritium in plasma by direct assay in a toluene: Triton X-100: hyamine hydroxide system was described by Whyman (1970); 1 ml of plasma was incorporated into 10 ml of the scintillant (toluene: Triton X-100, 55:25 v/v) and 0.1 ml hyamine, and a merit value of 140 for tritium was achieved. Chapman and Marcroft (1971) also described a Triton X-100: toluene (1:1 v/v) system, this time without added solubilizer, and up to 18% plasma (v/v) in the scintillant mixture could be employed. Merit values of 1145 were obtained for plasma ^{14}C but there were no values for tritium, and no indication of counting stability. A detailed study of this system at ambient temperatures indicated that higher merit values and increased stability could be obtained by a more careful choice of Triton X-100: toluene ratios (Fox, unpublished data). From Fig. 4.5 it can be seen that a composition indicated by position (2) would have a merit value for tritium of 177 (instrument reference efficiency 45 %) and a stability of 1 % over a period of 48 hr. At the position (1) a merit value of 176 is obtained, with a higher counting stability (coefficient of variation of 0.3 %). The composition of the latter mixture is a 2:7 basic mixture of Triton X-100: toluene, which contains 6 g/l of PPO. A merit value of 1500 for carbon-14 is also obtained at this point. No information is available however, about the possible variation

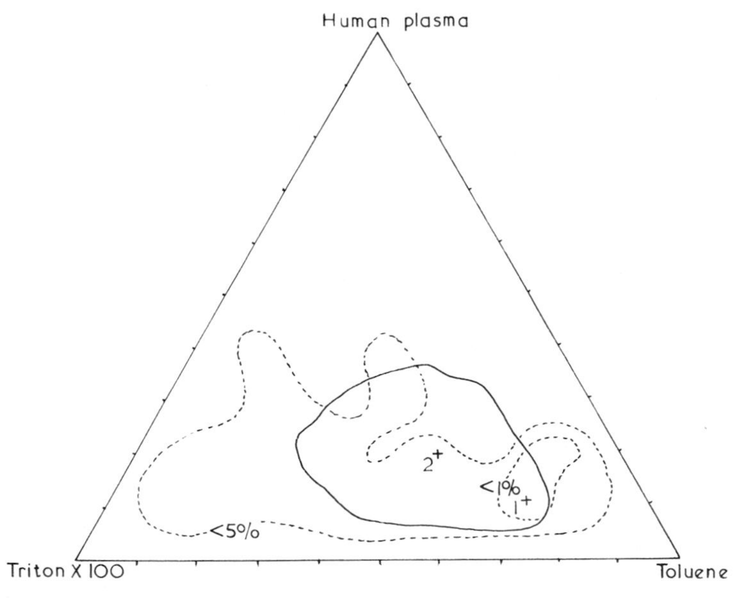

Fig. 4.5. A merit value-counting stability diagram for a sample of human plasma: Triton X-100:toluene system containing tritiated water. The concentration of PPO used to determine the diagram was 6 g/l at all points. The final concentration of PPO for the selected point would be determined by varying the concentration of PPO for this point until a maximum merit value is obtained.

between samples of plasma, and great care should be exercised in extrapolating from one experimenter's findings to another in this particular system.

The assay of plasma tritiated water was described by Moss (1964) who separated the tritiated water from plasma by freeze-drying. The lack of contaminating proteins and pigments in the water so obtained enabled him to assay the samples at an efficiency greater than 5 times that of other current systems involving protein removal.

A double isotope study of the distribution of adrenaline and noradrenaline in plasma samples was described by Siggers et al. (1970) who initially separated the two compounds by electrophoresis. The estimation of ^{45}Ca in plasma (58% efficiency) was described by Turpin and Bethune (1967); 1 ml of plasma was mixed with 10 ml of a scintillation mixture consisting of toluene:ethanol (1:1) containing 10 g PPO/l and the whole was shaken with the suspending agent, Cab-O-Sil. Oxby and Kirby (1968) used a 1:1 mixture of toluene: Texafor FN11 (Glovers, Chemicals Ltd., Leeds 12, U.K.), and reported that up to 8 ml of plasma can be added to 12 ml of the counting mixture (i.e. 40% v/v). The preferred composition is 4 ml plasma and 4 ml water with 12 ml of the scintillant mixture and ^{45}Ca is assayed at 45% efficiency at 5°C. This corresponds to a merit value of 900. Nadarajah et al. (1969), using a mixture of Triton X-100:toluene (1:2 v/v) at 4°C with plasma proportions from 7 to 25% recorded merit values of 770 with 2 ml samples added to 15 ml scintillant.

Methods of quench correction are described in a later chapter (chapter 11). However, an approximate assessment of quenching in clear plasma samples under a given set of counting conditions was described by Toporek (1960) who used a standard quench curve based on the protein concentration which was simultaneously assayed. Within the sort of variation expected between plasma samples, a good correlation can be obtained.

4.3.3. Radioimmunoassay

Radioimmunoassay techniques have enabled the detection and analysis of picogram levels of drugs and naturally occurring components of whole blood, and considerable advances are taking place in the application of this technique.

The technique requires that the material being measured is an antigen, or a hapten which will combine with a protein constituent to form an antigen. The assay normally consists of incubating the antigen being measured (usually an aliquot of plasma), with a known amount of labelled antigen and with a fixed amount of serum antibody. The labelled antigen can be employed as a labelled hapten (drug, hormone etc.) or the protein component of the antigen can be labelled with a protein label such as ^{125}I. In either case, the labelled and unlabelled antigens compete for the limited antibody sites in the antiserum added (ideally, sufficient to bind approx. 50% of the labelled antigen). At equilibrium therefore, the labelled and unlabelled antigens will be partly bound and partly unbound to antibody. The unbound components are usually absorbed on to an absorbent such as charcoal or resin and the level of bound radioactivity (present in the supernatant fluid), is measured by the scintillation counting technique. In the case of ^{125}I labelled antigen, the level can be measured by gamma counting methods as well. In practice, a standard curve obtained by varying the levels of antigen is prepared and the sample results extrapolated directly from it.

Many commercial houses have been quick to seize on the potential market involved in producing antisera, labelled antigens etc. for a variety of hormones and drugs, and have produced kits which incorporate the necessary ingredients. Much original work has also been undertaken and new assays are being prepared.

One of the earliest uses of this technique was in the assay of the level of digitalis glycosides in the serum of patients undergoing therapy for congestive cardiac failure and supraventricular tachyarrhythmias, where a correct level of the administered glycoside is essential for success and could prove fatal if it is only slightly exceeded. The production of a kit to assay the level in serum was described by Smith et al. (1969). The method was later modified for post-mortem specimens to understand the causes of failure, where red cell lysis is also more common (Phillips and Sambrook 1972). In the latter case, the colour is bleached with a commercial bleach preparation containing sodium hypochlorite and a colloid system based on Triton X-100: toluene system was used for the assay of the supernatant fluid. Both digoxin and digitoxin antisera are available commercially as well as the labelled antigens. For ^{125}I counting, the efficiency of counting directly in scintillation mixtures is approx. 90%, whereas, using crystal-well gamma counters, the efficiency is approx. 70%. The vial described by Ashcroft (1970) (see Fig. 1.1e) which employs a 30% solution of tetrabutyl tin (more recently, 10% tetrabutyl lead has been used) is approximately as efficient as the well counting method and like the latter, allows for the recovery of the liquid should this be required. A vial cap, containing a central hole suitable for the insertion of a micro centrifuge tube is produced commercially which simplifies the recovery procedure.

In view of the particular value of the ^{125}I technique, it is not surprising to find that the assay of thyroid hormones has been described by this procedure, and is also available in kit form. In the latter case, the absorbing surface is usually an ion exchange resin, either as a dry resin or as an ion exchange strip (Murphy and Pattee 1964).

The radioimmunoassay technique can assay levels of vitamin B_{12} (Wide and Killander 1971). The unbound vitamin B_{12} is absorbed on to intrinsic factor which is itself attached to a polysaccharide such as cellulose or Sephadex. In this case Vitamin B_{12}-^{57}Co is used as the 'antigen'.

Carcinoembryonic antigen (CEA), a tumour-specific antigen present in gastrointestinal tract tumours, has been used in an ingenious double-isotope radioimmunoassay developed by Egan et al. (1972). Anti-goat immunoglobulin labelled with ^{131}I is used to monitor the preparation of horse anti-goat sera which is used, in turn, to precipitate the ^{125}I-labelled – CEA – bound antigen–antibody complex, derived from a sample of the patient's serum. The levels of this antigen in patients which different cancers were reported by Reynoso et al. (1972).

The measurement of plasma corticosteroids by a competitive radioassay procedure using ^{3}H-cortisol has also been described by Murphy (1967), Nugent and Mayes (1967); and human placental lactogen (HPL) was measured by (Saxena et al. 1968).

The whole field of radioimmunoassay is rapidly expanding and a large number of papers are appearing describing its use for a number of naturally occurring substances as well as the distribution of drugs. A useful review of the methodology as applied to peptide hormone heterogeneity is by Yalow (1973), and reviews edited by Kirkham and Hunter (1971) should be consulted. An authoritative monograph by Chard will appear in one of the future volumes of this series.

4.3.4. Urine and body water

Urine is one of the most convenient body fluids for an assessment of the body contamination by radioisotopes and a sensitive means of estimating beta-emitters, which at the same time requires minimal processing, would be of considerable value to health physicists, pharmacologists and medical research generally. It is not surprising therefore to find that much effort has been concerned with scintillation systems designed to estimate very low levels of tritiated water in a distillate of urine.

Only a small quantity of urine needs to be distilled as a rule, and a simple method is to place a piece of solid carbon dioxide on the lid and a few ml of urine on the lower plate of a warmed Petri dish. This will allow sufficient water to be condensed for an assay to be made. Alternatively a simple device was described by Simpson and Greening (1960) which consisted of a small distillation flask with a bent side arm (Fig. 4.6). A small volume of urine was placed in the flask which is lowered into an oil bath at 110°C. The distillate is collected in the bend by draping over it a wet cloth. The glass wool plug prevents frothing and subsequent contamination of the distillate. The change in isotope concentration on distillation (isotope effect) was found to be negligible. A scintillant composition able to accept relatively large volumes of distilled water was a simple dioxane-naphthalene scintillant containing PPO and POPOP (Table 3, Appx. II) (Butler 1961). The mixture (13 ml) could accept 3 ml of distilled water. Greater sensitivity could also be achieved by an improved low background counter designed

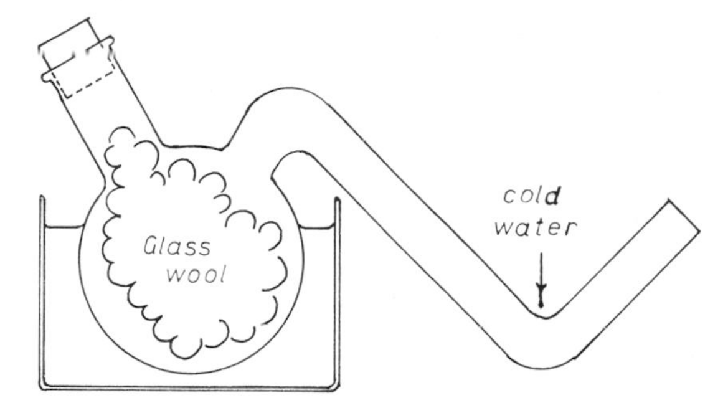

Fig. 4.6. A glass distillation flask for the rapid recovery of tritiated water from urine. The distillate is recovered by cooling the bent portion of the tube with a wet cloth strip (Simpson and Greening 1960). (Reproduced by kind permission of Macmillan Ltd.)

for the assay of tritiated water (Boyce and Cameron 1962).

However, in some instances, distillation may not be employed, such as in the assessment of labelled materials which do not distill off. If these isotopes have greater beta energies than 1 MeV and the Cerenkov effect can be measured, then this would be a most efficient way of measuring the isotope since the presence of salts would not interfere significantly with the sensitivity of the measurement. On the other hand the presence of coloured impurities would introduce serious quenching problems which could be overcome by the use of bleaching agents; but removal by absorption on charcoal (Francois 1967) is not to be recommended unless it is absolutely certain that the isotope itself is not absorbed.

One of the most useful methods for the examination of radioisotopes in urine, assuming, as is usually the case that there is plenty of material available for the examination, is to use a colloid system. The composition recommended by Moghissi et al. (1969) is Triton N101 : p-Xylene (1 : 2.5) containing 7 g PPO and 1.5 g bis MSB per l; this scintillant (17 ml) is mixed with 8 ml urine. Chapman and Marcroft (1971) also suggested the use of a Triton X-100 : toluene scintillant (1 : 1) for the assay of carbon-14 in urine samples. A merit value of 663 was obtained using 2 ml of urine and 10 ml of scintillant mixture. Nadarajah et al. (1969) used Triton X-100 : toluene (1 : 2) for the assay

Subject index p. 309

of calcium-45 in urine using 2 ml of urine and 15 ml of scintillant mixture; obtaining a merit value of 770. Unless these values are corrected for the reference efficiency of the instrument used they are not comparable.

In all these cases however, the optimal counting conditions were not assessed systematically and hence the counting stability data are not known. It is essential, especially in the case of unprocessed biological material like urine, to determine the counting stability to avoid artifacts arising from the varying times of assay after preparation of the colloid system.

In an attempt to find the correct conditions for the counting of urine taking these factors into consideration, the present author (1972) examined the phase diagram of the ternary system, Triton X-100: toluene: normal human urine. The optimal region for this sample is a 1:1 mixture of Triton X-100: toluene as recommended by Chapman and Marcroft but the ratio of scintillant to sample is 6:4 v/v. An instrument corrected Merit Value (MIV) of 965 for tritium is recorded for this sample. However, the properties of urine can change considerably during disease states and under different physiological stress conditions and it is necessary to determine such stability data for the type of urine under test at the time. If the counting stability can be shown to be good by repeated counting over a 48 hr period, the method is the most sensitive direct assay of urine.

The excretion of water by urination, perspiration and exhalation against its retention within tissues constitutes an important balanced process in normal healthy individuals. Certain disease states, especially those involving muscle, show marked perturbations of this water balance and in the distribution of body water. The assessment of the nature of these perturbations can be of an important investigative and diagnostic value, especially in certain types of muscular paralysis following injury (McTaggert and Cardus 1971).

Following intravenous administration of tritiated water in physiological saline (sp. act. 1 mCi/10 ml), venous blood samples are withdrawn at approx. 2, 24 and 48 hr. All the urine excreted during the period is collected, but a fresh sample is laid aside at the times of blood withdrawal. Distillation of a small quantity of both the plasma and urine samples will provide sufficient water for the assay of the label present. An aliquot of the water (0.2 ml) is usually added to 15 ml of TDNpdM (Table 5, Appx. II). The total body water (TBW) is thus obtained by application of eq. (4.2)

$$TBW_t = \frac{I_c \exp - \left[\frac{1}{n}\sum \frac{1}{t_{i+1} - t_i}\left(n\frac{D_{t_{i+1}}}{D_{t_i}}\right)\right] t - t_c}{D_t} \tag{4.2}$$

where I_c = total microcuries of tritiated water to be injected (expressed as total disintegrations per unit of time) *less* that remaining in the syringe *less* also that lost in the urine during the initial equilibration time ($t_c - t_0$). The tritiated water specific activity is given by D for a constant volume of sample and is expressed as dpm. In the first interval of time ($n = 1$), i.e. from $t_i = 2$ hr and $t_{i+1} = 24$ hr, the factor in square brackets is the mean fractional loss of tritium per hour while $t - t_c$ gives the number of hours from the initial equilibration (t_c) to the time (t) at which the tritiated water specific activity (D_t) is determined.

By measuring the total body water of a number of male patients with different degrees of trauma-induced paralysis, McTaggert and Cárdus (1971) concluded that, on the average, the total body water was inversely related to the extent of paralysis.

The rate of tritiated water loss from the plasma can also be determined for each interval (rate constant, λ). The rate of urinary excretion (rate constant v) for each interval is thus determined from eq. (4.3)

$$v = \frac{1}{t_{i+1} - t_i} \ln\left(\frac{\text{recovered}}{\text{original}} - 1\right) \tag{4.3}$$

The ratio recovered/original is the TBW recovered during the interval considered, divided by that calculated to be present at the beginning of the interval. The remaining rate of loss by excretion, exhalation, perspiration etc. (rate constant β) is obtained by difference. A knowledge of these rate constants thus allows for a study of the perturbations of intracellular fluid and plasma movements during chronic disease states.

4.3.5. Milk and food, etc.

There are very few published examples of the liquid scintillation counting of milk samples, although the assay of isotopes in this material must be an important contribution to the knowledge of the fate and potential danger of environmental pollution resulting from fall-out.

Both combustion and wet oxidation techniques have been employed for the assay of the isotopic content of milk, and these would appear to be the obvious method of dealing with this material. However, due to the fact that the sample is itself an emulsion, it would seem feasible to assay certain types of more energetic isotopes directly. For such isotopes as ^{45}Ca and ^{90}Sr (^{90}Y), a suspension technique as described by Gordon and Wolfe (1960) would seem to be appropriate, using either whole milk or a suspension of a lyophilisate. Cab-O-Sil (4 %) is usually sufficient to maintain the suspension whilst counting in a xylene-dioxane-naphthalene scintillant such as DNXpnE (Table 5, Appx. II). There is some difficulty however, as is usual with such heterogenous systems, in assessing the amount of quenching.

The assay of ^{45}Ca can be undertaken in much the same way as is used to assay that isotope in bone (Sarnat and Jeffay 1962) (§ 4.1.1). However, the tedious amount of preprocessing required by this method, may not be desirable in many instances, where the level of isotope present is reasonably high.

An alternative procedure is to combine milk with Triton X-100: toluene directly, and as shown in Fig. 4.7 there are certain compositions where the combination of these two colloid systems produce a single colloid structure, which in some regions is clear. A suitable composition for the assay of cows milk is Triton X-100: toluene (3 : 5 v/v). The milk sample (2 ml) is shaken with 8 ml of the scintillant mixture to provide the necessary counting system. However, the resulting mixture should be checked for both counting stability and quenching using water-soluble standards. In the case of the cows milk sample used in Fig. 4.7 (Fox, unpublished) an instrument corrected merit

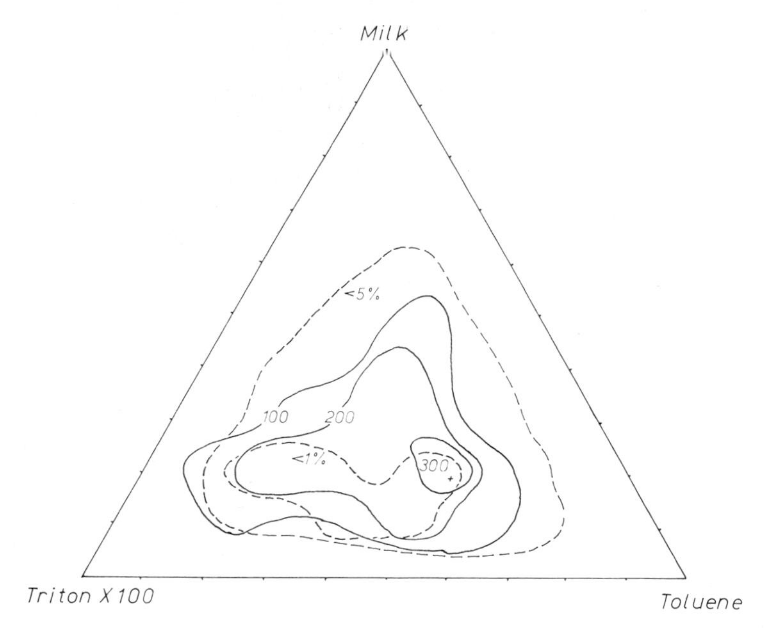

Fig. 4.7. The colloid system, cow's milk: Triton X-100:toluene. A MIV of 664 was obtained for tritium and a max MISQ value of 307. Further improvement in efficiency can be obtained by modifying the PPO concentration at this composition.

value of 644 was obtained and a counting stability of 1.1 %. A rigorous assessment of different milk samples has not however been made and there is likely to be considerable variation in their properties, however, for routine work on a number of similar samples it would be worthwhile to determine the optimal conditions (§ 2.2.3).

The assessment of the level of isotopes in food are similar to the methods described for tissues in the earlier part of this chapter. However, there are reports of specific assays of iron isotopes (Eakins and Brown, 1966) and phosphorus-32 (Ellis et al. 1966) which have been assayed in foodstuffs and which should be consulted. The latter isotope was precipitated as the phosphomolybdate and assayed as a suspension. The concentration of isotope produced by this procedure allows about 10^{-5} μCi to be detected.

4.4. Respired gases

The trapping of moist CO_2 from animals that have been treated with ^{14}C-labelled drug is best effected by the use of single animals. The method of Godfrey and Snyder (1962) is simple and straightforward and the layout is shown in Fig. 4.8.

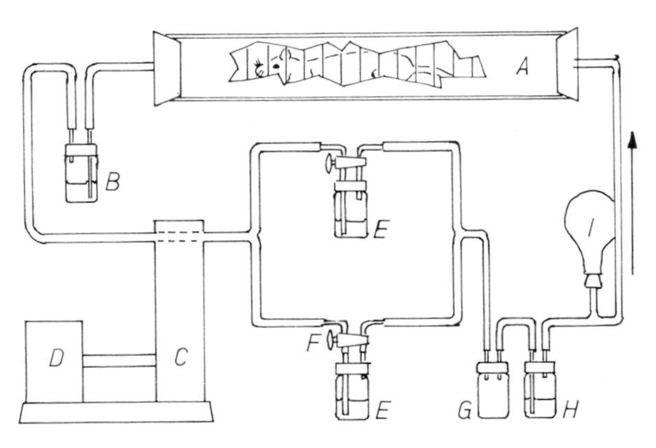

Fig. 4.8. A method for the assay of respired $^{14}CO_2$ from a rodent. For details, see text (Godfrey and Snyder 1962). (Reproduced by kind permission of Academic Press.)

The labelled CO_2 is liberated from the animal in the tubular cage A and is pumped around the system by means of a circulating pump (C, D) operating at about 185 ml/min. The respired gases are first dried in a concentrated sulphuric acid scrubber (5 ml, B). The stream is then split, and led into each of the vials E which contain 4 ml of 1 M hyamine hydroxide solution in methanol. Following a further scrubbing in 5 ml of concentrated sulphuric acid (H) to remove any hyamine vapours, the air again travels through the animal cage. The balloon (I) is filled with oxygen, which enters the system as the oxygen is used up in respiration. The hyamine solution is then assayed in Tpp4.

Where the levels of labelling of the respired CO_2 is very low, a technique has been described by Kearns (1969) for the use of a commercial benzene synthesiser, to convert the CO_2 to benzene. However, this procedure, although very sensitive (0.1 pCi/gC) is time consuming and expensive, about a week being required per sample. An alternative procedure for low specific activity material was described by Moghissi (1970). In this case, the CO_2 is trapped in lithium hydroxide and the resulting lithium carbonate is precipitated by ethanol. Due to the large differential solubility of the hydroxide and carbonate, the latter is separated and represents a considerable concentration of the carbon present. The precipitate is dried thoroughly by desiccation overnight and is assayed (up to 12.5 g) in 15 ml Tpp4 (Table 3, Appx. II). The efficiency is around 55% and the sensitivity of the system is about 1.8 pCi/gC. A similar procedure employing suspension counting can be employed to avoid the use of hyamine and quaternary bases in an enclosed system as described by Godfrey and Snyder (above). A 2.5 N sodium hydroxide solution may be substituted for the quaternary base and following absorption of the gas, a 0.5 ml aliquot of the NaOH is placed in a vial containing loosly packed Cab-O-Sil and then 15 ml TppE (Table 4, Appx. II). The sodium carbonate precipitates out on the Cab-O-Sil surface and by using sodium carbonate standards, quite reliable data can be obtained (Harlan 1963).

Amongst the quaternary bases, 2-phenylethylamine is not recommended for this type of trapping as the faster flow rate needed results in some losses of CO_2. A solution of ethanolamine in methanol (20:80 v/v) is a useful trapping agent and 11 ml of this mixture may be added to 10 ml of Tpp4 for assaying dissolved ethanolamine carbonate.

CHAPTER 5

Botanical aspects

There are considerably fewer reports of the use of liquid scintillation counting in the assay of beta-emitters in plant material, when compared to the literature on animals. Many of the reports that have been written employ either direct plating of dried residues of leaves or other parts of the plant on to planchettes and assay by the Geiger–Muller end window technique or by proportional counting. It is difficult to see why this has been so, since many of the techniques that have been worked out for the animal systems are directly applicable to the botanical field. There are certain exceptions which are due to the higher levels of polysaccharides in plants compared with the mammalian field.

Many of the isotope applications that have been described have employed radioautographic methods for the tracing of the progress of labelled herbicides and insecticides in the tissues of leaf and stems of plants. Other studies include the transport of phosphorus-32 as phosphates from soil to plant and the reverse. Some of these studies have been carried out on small fragments of leaf tissue, or leaves and stems. In either case, it is most convenient to measure this isotope by Cerenkov counting, to avoid the problems of correcting for impurity quenching. A bleaching of the chlorophyll and carotenoid pigments is usually all that is necessary.

In some cases, such as the unicellular algae and yeasts, or the filamentous organisms, suspensions can often be assayed; but, unless the species being measured is tritiated water or other volatile material, it is usually advisable to remove the water by freeze-drying or oven-drying to improve counting efficiency.

5.1. *Higher plant tissue processing*

One of the most convenient ways of converting plant tissue that has been labelled with ^{14}C-labelled herbicides or pesticides is to combust the sample, either by burning in oxygen (§ 3.1) or by the wet combustion method of Van Slyke (see § 3.3.1, Fig. 3.15). This method has been compared (Fuchs and de Vries, 1972) with the hyamine solubilization technique of Veen (1972), the wet oxidation method of Mahin and Lofberg (1966), and the combustion technique of Gupta (1966). It was concluded that the Mahin and Lofberg procedure was unreliable as a means of assaying total ^{14}C activity in plant tissue, and the hyamine solubilization was accurate provided the method as outlined by Veen was strictly adhered to. In this latter method the problem was sometimes very considerable, especially in the case of large samples. The wet oxidation technique of Van Slyke however, gave reproducible and accurate results in their hands (for samples up to 100 mg dry wt.). If speed was important, the combustion method based on the Shoniger flask method (see § 3.1) proved to be better than the others.

Although Fuchs and De Vries (1972) have conducted this critical comparison, their main criticism of the highly convenient Mahin and Lofberg procedure as applied to plant tissue is the loss of ^{14}C during the digestion when ^{14}C-glucose distribution was being studied (as much as 23 % in the data quoted). The method adopted was to incubate the peroxide–perchloric mixture for 40 min between 70–80°C. However, the data for phenylalanine ^{14}C uptake was as good as with the other methods.

In the case of the hyamine solubilization method of Veen (1972) the procedure is as follows: 40 mg lyophilised plant material is transferred to scintillation vials together with 1 ml of 1 M hyamine hydroxide (10X, Rohm and Haas, Inc.) in methanol. Allow solubilization to take place in the dark at 60°C for 20 hr. Add hydrogen peroxide (small drop) to decolourise. After cooling the sample, add 0.5 ml glacial acetic acid to the suppress the chemiluminescence. Add, at 2°C 12 ml of a scintillation mixture consisting of 48 g naphthalene, 9.6 g PPO, and 0.48 g POPOP per mixture of dioxane (800 ml) and methoxyethanol (160 ml). Reproducibility is good up to 10 mg dry weight, and above this value tends to lose this stability presumably due to incomplete solubilization and decolourization.

A useful procedure for the storage of labelled plant material was described in an earlier work of Fuchs and De Vries (1969); the samples (in this case shoots of tomato plants) were frozen in liquid nitrogen, ground in a mortar with solid CO_2 and freeze dried. The dry powder could then be stored under nitrogen at 2°C until required.

Several methods of application of radioisotopes have been described. The most straightforward is to place a leaf blade and its freshly cut petiole directly in the labelled solution for a fixed length of time. Alternatively, the material may be injected in some species, e.g. 5 μCi of ^{14}C-betaine hydrochloride in 100 μl solution injected into a wheat stem about 9 cm above the top node. The wheat plants in this study (Bowman and Rohringer 1970) were then subjected to 16 hr light per day, and the fresh shoots were removed 48 hr to 10 days after. In this case, the formation of formate was being studied, and consequently the shoots were boiled with water and the cooled extract assayed in a toluene : Triton X-100 (2 : 1) scintillant mixture.

The presence of sulphur-35 in plant tissue has been studied by Grundon and Asher (1972) using a combustion procedure undertaken in a simple structure shown in Fig. 5.1. Following combustion, the sulphur dioxide is allowed to absorb in 5 ml of deionised water for 16 min, followed by washing the flask and basket with about 80 ml water. A mixture of toluene : Triton X-100 : water in the ratio 1 : 1 : 2 was used containing 5 g PPO/l as scintillant. The sulphur was assayed at 73% and very high recoveries were recorded.

The assaying of calcium-45 in plant tissue is conveniently done by the method of Retief et al. (1972). The plant material is first wet ashed with $HNO_3 : HClO_4$ (10 ml : 2 ml (60%)) per g dried matter. The mixture is heated until white fumes of $HClO_4$ become visible (5 min).

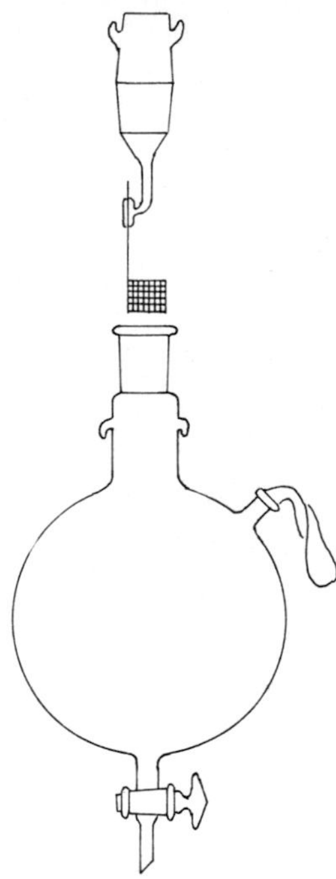

Fig. 5.1. The simple combustion system suitable for the assay of ^{35}S as SO_2 (Grundon and Asher 1972). The 3 l 'PYREX' flask is flushed with oxygen. The sample with an ignited wick is lowered into the flask on a 48 mesh platinum basket, 2 cm high, via a 45/40 cone and socket joint. The flask is also fitted with a flaccid balloon which helps to prevent unnecessary pressure within the flask from a sudden expansion of the gases. For $^{35}SO_2$, water (approx. 5 ml) is present during the combustion, in order to trap the gases. (Reproduced by kind permission of the American Chemical Society.)

Dilute to 25 ml and filter. Add 1 ml carrier $CaCl_2$ (approx. 30 mg Ca), to 3 ml of the wet ashed material, together with 20 ml of 99% ethanol and 1 ml of 5 N H_2SO_4. Calcium sulphate is rapidly formed, and the mixture is shaken centrifuged, and the supernatent decanted. The residue is treated with 2 ml 99% ethyl alcohol and 8 ml of a scintillant containing 6.25 g PPO and 0.125 g dimethyl POPOP per l of toluene (Tpd).

Several of the isotopes used in plant ion transport studies are Cerenkov emitters, such as ^{86}Rb, ^{36}Cl and ^{32}P. In these cases, the plant material to be measured need not be totally in solution. The chopped roots of barley can be suspended in water for counting. Lauchli (1969) used 2.5 mM 2-amino-6,8-naphthalene disulphonic acid (ANDA) and obtained an efficiency of 60% for ^{86}Rb using a 1.0 g sample of barley root. A counting efficiency of 13% was obtained for ^{36}Cl under these conditions. Leaf discs (10 mm dia cork borer) placed directly into the vial containing water or hexane can also be successfully used for the assay of these isotopes; such discs can also be incubated with phosphate solutions for 3 days prior to measuring, to assay uptake of the ion by the leaf tissue (Averbuch and Avnimelach 1970). If greater quantities of plant material require to be assayed by this method, the plant (up to 2 g) can be wet-ashed in 2 ml of a $HNO_3 : HClO_4$ mixture as used for ^{45}Ca assay. To avoid glass surface absorption, 4 ml of a solution of KH_2PO_4 (not less than 30 ppm) is added and the whole diluted to 25 ml for counting.

5.2. Algae, yeasts

The method of Yarborough et al. (1966) for the assay of ^{45}Ca and ^{89}Sr in algal cells exemplifies the fact that for a number of energetic beta-emitters, it is necessary only to dehydrate the cells by sequential desiccation in a series of alcohols followed by concentration and measurement as a suspension in Tpd5 using Cab-O-Sil as a suspending agent. The same procedure could be adopted for yeasts and filamentous algae. However, for carbon-14 and tritium, a wet oxidation procedure or a combustion procedure would appear to be desirable. Internal quenching by the algal cells of the more energetic isotopes occurred above 75 mg/ml.

For routine measurements of the tritium and carbon-14 content of these cells, the method of Veen using hyamine hydroxide (§ 5.1) would appear to offer the most convenient system, since the major problem would be the presence of chlorophyll pigments, and in this procedure, these are effectively decolourised. The measurement of photosynthetic activity by liberation of labelled $^{14}CO_2$ can be conducted in the way applied to mammallian cells (§ 6.5) and essentially this procedure was employed by Schindler (1966) for the study of carbon-14 uptake by algae during photosynthesis.

5.3. *Soils and nutritional studies*

The principal beta-emitters in normal soil are ^{14}C, ^{40}K, ^{87}Rb, ^{232}Th and other thorium isotopes, and ^{238}U and other uranium isotopes. Amongst these isotopes, there are α, β and γ emitters, and the assay of the isotope content of soils is therefore complicated. In fact the measurement of beta activity is not usually undertaken because of the complexity of this emission. The γ spectrum provides a much better analytical technique for the isotopes present.

The radioactive composition of surface soil was given by Talibudeen and Yamada (1966) whose data is reproduced in Table 5.1. In order to assay for ^{40}K in the soil, it is possible to interpose a 0.15 mm thick brass (126 mg cm^{-2}) to allow the more energetic beta-emission from this isotope and stop the majority of others.

However, the assay of soil for the presence of ^{14}C-labelled herbicides and pesticides, is probably best accomplished by the Parr bomb technique of Hilton et al. (1972). The procedure consists of mixing air-dried soil (0.2 g) with 0.1 g cellulose powder in the ignition cup of the Parr bomb assembly. 2.5 N KOH (20 ml) is placed in the beaker to absorb the CO_2 produced. After the combustion, 1 ml aliquot of the KOH solution is added to a vial, followed by 10 ml of 2 M NH_4Cl in 2 M $BaCl_2$. Following precipitation of the $BaCO_3$, the vial is centrifuged (using the normal centrifuge rubber cups in reverse), the precipitate is washed twice with acetone and dried at room temperature. 10 ml of Tpp4 scintillant (Table 3, Appx. II) together with 2.5% Cab-O-Sil are then added.

For the main bulk of the soil and nutritional studies, extracts are used; these can be treated in the same way as those derived from mammalian tissues (§ 3.7).

TABLE 5.1

In the case of thorium and uranium, the isotopes of the radioactive decay series producing a complex series of beta energies are grouped as Th fam and U fam respectively. The half value thickness data is also given for beta-emission (mg cm^{-2}) and for gamma-emission (g cm^{-2}). (From data described by Talibudeen and Yamada 1966.)

		Emission energy (MeV)			Composition surface soil per 100g		Approx. ½ thickness	
Isotope	Half life years	α	β	γ	Content	Total radioact.	mg cm^{-2} β	g cm^{-2} γ
^{14}C	5.8×10^3	—	0.16	—	2.0	14	6	
^{40}K	1.3×10^9	—	1.32	1.46	1.4	1135	108	30
^{87}Rb	5.0×10^{10}	—	0.28	—	1×10^{-2}	229	13	
^{232}Th	1.4×10^{10}	3.95 to 4.01	—	0.06	6×10^{-4}	80	—	9
Th fam		3.95 to 8.76	0.01 to 2.26	0.04 to 2.62	—	1080	0.2 to 216	7 to 43
^{238}U	4.5×10^9	4.18	—	0.05	1×10^{-4}	40	—	8
U fam	—	4.18 to 7.68	0.03 to 3.15	0.05 to 2.45	—	636	0.3 to 354	8 to 40

CHAPTER 6

Cell cultures

Isotopically-labelled precursor uptake experiments in cell cultures often result in the necessity of measuring the radioisotope content of the whole cell as well as the degree of uptake of the labelled material into the subcellular and macromolecular components of the cell. The radioassay of these latter components are considered in more detail in chapter 8. The purpose of this chapter is to describe ways of measuring the total radioisotope activity of virus particles and bacteria (§ 6.1) and mammalian cells (§ 6.2) in culture and the amount of isotope that may be excreted into the medium surrounding it or liberated as gas.

In many cases the relative assessment of the uptake of labelled species between different aliquots of cells is all that is required. In other experiments, it may be necessary to determine the absolute levels of radioactive material present, in order to provide a balance sheet for subsequent fractionation.

In the first type of experiment, i.e. where comparative measurements only are normally required, a constant level of quenching is desirable and this can often be achieved more successfully and with less dependence on individual sample preparation accuracy by operating at medium rather than at high counting efficiencies (§ 1.2). This can be achieved by using a heterogenous counting system or by the use of solubilizers. In the latter application however, an efficient solubilization method or a combustion technique may be necessary, especially if the number of counts available is low.

Subject index p. 309

6.1. Virus particles and bacteria

The methods adopted for the assay of beta-emitters in viruses and bacteria will be dependent on the nature of the problem being studied and the isotope being used. The diameter of a bacterium (approx. 0.5 μ) is close to that of the half value thickness for a tritium beta particle in matter of unit density (0.4 μ), thus a number of beta particles originating from tritium atoms in the centre of a bacterium would be expected to be absorbed to a limited extent. However, since there is likely to be a more uniform distribution of the label in the cell, the effect of self absorption can be considered to be negligible. Membrane filters with pore sizes capable of retaining virus particles (less than 0.027 μ) or bacteria (less than 0.22 μ) are suitable supports (see Table 6.2). In the case of bacteria it would need approx. 2×10^9 organisms to cover a disc (2.4 cm dia) one organism deep. These assay systems are most suitable for some relative uptake studies, since between-sample variation is more likely to be due to support differences than to differences between cells.

An alternative method is to centrifuge the organism suspension to a pellet and to assay in the same way as a protein precipitate (§ 9.2.1). Formamide has been suggested as suitable solvent Neujahr and Ewaldsson (1964) and amounts up to 10 mg dry weight are considered to be reasonable. However, most organic solubilizers are capable of dissolving pellets of these organisms very effectively.

Bacteria and presumably virions labelled with ^{32}P may be assayed efficiently with or without added waveshifters by direct measurement of the Cerenkov emission (§ 2.3). Alternatively, the use of the bioluminescent properties of the Luciferin reaction has enabled the ATP content of bacterial cells to be ascertained (10^{-4} to 10^{-3} picograms per cell), and this value has been used in the assessment of the number of bacteria in soil, fresh water and in sea water (Holm Hansen and Booth 1966). The level of bacterial contamination in infected urine has also been estimated by this procedure (Jones cited by Stanley 1971).

6.2. *Mammalian cells*

An estimation of the tritium content of mammalian cells by direct assay of the dessicated cells in suspension or attached to a solid support is subject to gross inaccuracies and to considerable variability. The method has been suggested as an additional procedure to give absolute counts in autoradiographic methods, however, in view of the fact that the dimensions of the average mammalian cell (8 to 15 μ) is several fold greater than the half value thickness of tritium beta particles, only a small proportion of the total content of tritium (possibly 10 to 15%) will be emitted from the cell.

The efficiencies of 8.7 to 14.8% for tritium obtained by Fallot et al. (1965) for the direct counting of whole cells are higher than would be expected and counting variation was, not surprisingly, reported. In most studies with whole cells, if accurate data are to be achieved, it is important to assay the cells following solubilization of an oxidative degradation product and to count in a homogenous scintillant system.

For cells that normally grow by attachment to the glass surface, such as HeLa, chinese hamster or V79 cells some help can be achieved in the handling processes by growing the cells on cover slips (in sterile scintillation vials for example). Following removal of the media and labelled precursors present, the coverslip may be treated with a solubilizer and incubated at an appropriate temperature for the solubilizer concerned (see § 3.2). A toluene-based scintillant added directly to the coverslip–solubilizer mixture is then used to assay the activity. The coverslip does not need to be removed since no significant quenching occurs from this source.

Assay of the level of nucleic acid synthesis in HeLa cells (*Ball et al. 1972*)

1) A cell suspension (1 ml, 0.25 to 4 × 10^5 cells) is pipetted into a sterile scintillation vial. Vial flushed with 5% CO_2 in air. Capped.
2) The suspension is incubated for 2–4 hr to allow attachment of the cells to the glass walls of the vial.
3) Medium (1 ml) is added, containing 1 Ci/ml of either 6[^{3}H]thymidine (~ 10 Ci/mM) or 5[^{3}H]uridine (~ 15 Ci/mM) for the assay

Subject index p. 309

of DNA and RNA respectively, and the mixture is incubated at 37°C for the appropriate time.

4) Ice-cold saline (10 ml) is added to stop the metabolism and the saline and medium are carefully decanted off.
5) Ice-cold perchloric acid (1.5%, v/v) is added to fill the vial. Care must be taken not to dislodge cells.
6) Allow to stand 1 to 2 min and repeat the washing procedure twice.
7) Fill the vials with ethanol and stand at room temperature for 10 min to remove lipids and excess water from within the cells.
8) Decand off the alcohol and add 1 ml perchloric acid (5%, v/v).
9) Heat in an oven 80°C for 40 min.
10) Cool, add 10 ml of a Toluene: Triton X-100 scintillant. The composition recommended by the authors is Toluene (2 l); Triton X-100 (1000 g); PPO (16.5 g); diMePOPOP (300 mg).

Mammalian cells that do not attach to glass may not, of course, be treated in this manner. However, they may be filtered on to glass fibre discs following incubations and saline washing in suspension and centrifuging. The cells may then be washed with ice cold perchloric acid as described in the above procedure for cells on glass coverslips. Wenzel and Stohr (1970) compared the efficiency of counting such cells on cellulose filter paper by both scanning procedures and liquid scintillation counting of the dried discs. They concluded that the scanning procedure was 62 to 70% and 60 to 68% as efficient as LSC for tritium and carbon-14 respectively. No counting efficiency for tritium was however recorded and it is suspected that it was lower then would have been obtained if the cells were collected on glass fibre discs and solubilized in a suitable quaternary base solubilizer and assayed in a toluene-based scintillant.

6.3. *Subcellular organelles (nucleii, ribosomes, etc.)*

The direct assay of tritium in organelles other than nucleii is analogous to the problem of assaying similar isotopes in bacteria. Provided that a suitable membrane filter can be found, the organelle may be assayed on a solid support and counted by either a heterogenous disc counting technique (§ 2.2.1) or by dissolving the material off the disc and assaying by a homogenous counting procedure. For absolute radioactivity data it is essential to use the latter procedure.

In Table 6.1 are listed the approximate sizes of the different subcellular organelles. Table 6.2 lists some of the filtration characteristics of a number of commercially available filter which could be used for the specific filtration of these organelles. Although this application of filters does not appear to have been exploited, there appears to be no reason why some specificity of organelle measurement should not be obtained. A combination of filters in a step-wise filtration should be able to separate several labelled species of organelles of different sizes within cellular homogenates.

The more usual procedure is to centrifuge the organelle to a pellet and treat as in § 9.2.1.

TABLE 6.1

Approximate sizes of subcellular organelles and cells (diameter in microns).

Subcellular organelles		Mammalian cells	
Ribosomes	0.015	Red blood cell (dia)	7.2
Glycogen granules	0.03	Red blood cell (thick)	2.1
Ferritin granules	0.01	Megaloblast	14 to 19
Mitochondria	0.4 to 1.5	Late erythroblasts	10 to 14
Nucleii	3.0	Normoblast	7 to 10
Lysosomes	0.2 to 0.3	Myeloblast	11 to 18
Megaloblast nucleus	12 to 13.5	Myelocyte	11 to 16
		Neutrophil	10 to 12
Other organisms		Eosinophils	10 to 12
		Basophils	8 to 10
Polio virus	0.027	Lymphoblast	15 to 20
Bacteria	0.2 to 0.5	Large lymphocyte	12 to 15
Yeasts	3.0 to 5.0	Small lymphocyte	6 to 7
Pollen	12 to 50	Monocyte	16 to 22
Fungal spores	8 to 10	Megakaryocyte	40
Algae (unicellular)	20 to 30	Blood platelets	2 to 3

Subject index p. 309

TABLE 6.2

Mean pore size and the variation expected in some commercial membrane filters.

Make*	Type	Mean pore size (μ)	Var $\pm\ \mu$	Make	Type	Mean pore size (μ)	Var $\pm\ \mu$
Millipore	SC	8.0	1.4	Polyvic	BC	6.0	2.0
	SM	5.0	1.2				
	SS	3.0	0.9	Mitef	LC	10.0	2.0
	RA	1.2	0.3		LS	5.0	1.5
	AA	0.8	0.05				
	DA	0.65	0.03	Sartorius	SM	12.0	
	HA	0.45	0.02			8.0	
	PH	0.30	0.02			3.0	
	GS	0.22	0.02			1.2	
	VC	0.10	0.008			0.8	
	VM	0.05	0.003			0.6	
	VF	0.01	0.002			0.45	
						0.3	
Microweb	WS	3.0	0.9			0.2	
	WH	0.45	0.02			0.15	0.03
						0.1	
Duralon	NC	14.0	3.0			0.05	
	NS	7.0	2.0			0.027	0.007
	NR	1.0	0.3			0.015	0.005
						0.008	
Celotate	EA	1.0	0.1			0.0075	0.0025
	EH	0.5	0.05			0.005	
	EG	0.2	0.05				

* Millipore, Microweb, Duralon, Celotate, Polyvic and Mitef are produced by Millipore (U.K.) Ltd., Heron House, 109, Wembley Hill Rd., Wembley, Middlesex; Sartorius from Sartorius-Membranfilter GmbH, 3400 Gottingen, W. Germany, P.O. Box 142.

6.4. Media and broths

The problems associated with the assay of these solutions are closely allied to those which occur with plasma and urine (see § 4.3.2 and § 4.3.5 respectively). The need for radioassay usually arises when cells are known to be excreting radioactive products of metabolism and an assessment of the level of such activity requires to be made. Owing to the presence of proteins such as 'peptone' or serum, as well as a wide variety of both inorganic and organic nutrients, severe chemical and colour quenching can occur if an attempt is made to radioassay directly in a homogenous counting system.

In order to assay potentially low levels of radioactivity, especially if produced by secretion from mammalian cells or bacterial cultures, it is necessary to use the largest amount of the sample possible whilst at the same time reducing the total preprocessing to a minimum. The colloid counting system can usually be used with advantage. The varying compositions of the broths and media require different colloid counting conditions. Some typical examples of the Triton X-100 : toluene composition as applied to a number of commonly employed media are given in Table 6.3. These compositions should only be used as a

TABLE 6.3

The composition of Triton X-100:toluene counting mixtures recommended for optimal counting of a number of commonly used culture media. The Merit Values given are for tritium and are not corrected for counting stability.

Culture medium	Scintillant composition		Counting mixture		Merit value
	Triton X-100: Toluene (v/v)	PPO (g/l)	Scint. (ml)	Medium (ml)	(^3H)
Fischer's + 20% horse serum	7:9	6	8	2	456
Tryptone: Yeast: glucose (TYG)	1:1	10	8	2	536
Nutrient Broth	1:1	4.5	6	4	448
M9	1:2	4.5	6	4	1038
Eagle's MEM	1:1	3.3	8	2	313

guide since the actual proportions used should be determined at the outset for the particular medium needed to be assayed (§ 2.2.3).

Bray's scintillant (Table 5, Appx. II) can accept amounts of such media up to approx. 5 to 10% of the counting volume. If the final scintillant appearance is clear and one phase, i.e. the sample has dissolved completely in the system, this is probably the best method of counting, since the normal procedures of external standard ratio can be applied without fear of artifacts arising from the heterogeneity of the colloid system. The scintillant however can only reliably be used where the specific activity of the sample is sufficiently high, since the efficiency of counting tritium can be low as 5% for these volumes.

As with urine and plasma samples, the possibility of extracting the labelled components should be seriously considered as this procedure could lead to counting conditions which will be less subject to the severe quenching brought about by the large quantities of protein and salts present in the original solution. Such concentration can often be effected on ion exchange columns or by means of molecular sieves. Freeze drying does not usually help, since the level of salts and proteins present interfere not only with the freeze drying itself, but also with the scintillation counting.

6.5. Liberated gases

In some cell culture and enzyme experiments, the need arises to trap and assay liberated gases, in particular labelled CO_2, SO_2 or H_2S, as a measure of the extent of an enzymic or metabolic reaction.

The well known assay technique employing the Warburg respirometer can be used for the absorption of CO_2 into a suitable absorbent such as a quaternary ammonium base or a solution of alkali. Methods of trapping these gases have already been described with reference to the combustion and wet oxidation techniques (chapter 3). It is however, useful here to describe some of the other apparatus designs that have been recorded to absorb gases liberated from cell and tissue culture systems.

The use of the Warburg flask in liquid scintillation counting. An example of this procedure in enzyme assays was described by Kobayashi (1963). Two ml of the enzyme histidine decarboxylase in buffer and 30 μg of pyridoxal phosphate are placed in the main body of the flask. In one side arm, place 1 mg of ^{13}C-1-histidine and in the other 0.3 ml of 1 M citric acid. Hyamine hydroxide solution (1 M in methanol) is then placed either in the centre well of the flask or on a piece of Whatman filter No. 1 paper (7.5 × 25 mm). Before placing the paper in the centre well, the edges are treated lightly with paraffin wax to prevent the hyamine wetting the sides of the flask or well. After a suitable reaction time (2 hr at 37°C) the enzyme reaction is stopped by the addition of the citric acid. The hyamine or paper is removed from the well which is washed out with hyamine if necessary. Hyamine and washings (up to 1.5 g hyamine) are added to 15 ml of DAppD (Table 5, Appx. II).

It is also possible to use a scintillation vial in which to liberate the CO_2 and to trap the liberated CO_2 on a pad treated with hyamine or ethanolamine fixed to the underside of the lid. After allowing for absorption, the cap is transferred to a clean vial containing some Tpp4 scintillant and the vial inverted to allow the absorbed base carbonates to admix with the scintillant. An example of this type of vial is illustrated in Fig. 1.1.

In place of the quaternary base on the filter paper, a solution of alkali may be used, and a simple trapping system was described by Runyon et al. (1967) to undertake this on the top of a screw cap culture vessel (Fig. 6.1).

If the carbon dioxide requires to be liberated from the carbonate in solution, a fully enclosed device which will allow both the liberation of the gas and its absorption to take place would be useful. Moss (1961) described a simple means of doing this (Fig. 6.2). Alternatively, a system constructed from standard ground glass joint equipment was described by Roberts et al. (1965) (Fig. 6.3). Diffusion of the carbon dioxide is complete in 30 min at room temperature, but routinely, 90 min is allowed for the completion of the process. In this case, 1 ml of 1 M hyamine hydroxide (in methanol) is used as the adsorbant in the scintillation vial. Tpp4 is mixed with the resulting hyamine solu-

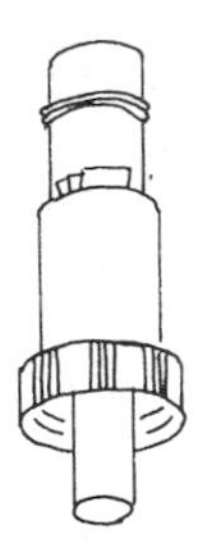

Fig. 6.1. A simple device built up from a standard screw cap dropper assembly and a serum bottle stopper, to trap $^{14}CO_2$ from enzyme systems. A small piece of filter paper in the shape of a fan is placed in a glass tube inserted into the rubber part of the dropper as shown. A small volume (0.1 ml) of NaOH is injected on to the filter paper from a syringe through the serum stopper, in order to absorb the CO_2 liberated. The system can be autoclaved (Runyon et al. 1967). (Reproduced by kind permission of Elsevier.)

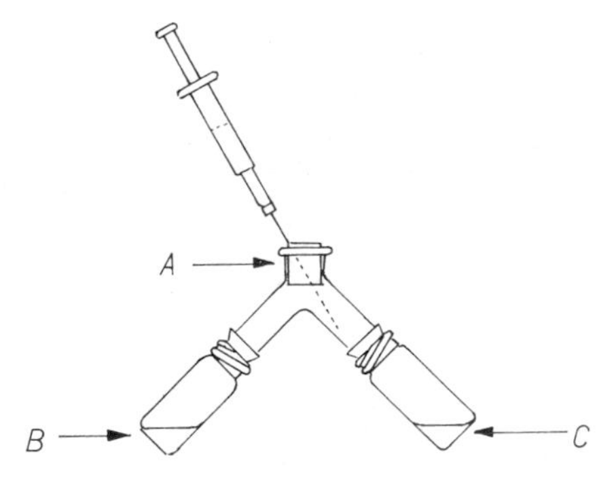

Fig. 6.2. A method for liberating $^{14}CO_2$ from plasma and trapping directly into a scintillation vial. One ml of 85% (10 milli equiv.) lactic acid is injected from the syringe through the penetrable stopper (silicone) (A) into the vial C containing 25 ml (0.5 milli equiv. CO_2) of plasma. The second scintillation vial (B) contains 0.5 ml of a quaternary ammonium solubilizer (in this report, Primene 81R, 8N). The whole is then placed in an agitater for 24 hr at 2°C. Scintillation fluid (15 ml, Tpp4) is added to vial B and the mixture is assayed (Moss 1961). (Reproduced by kind permission of Pergamon Press.)

tion. For a larger number of such routine assays, a rotating diffusion system as described by Aures and Clark (1964) will be found to be useful (Fig. 6.4). This system was used for the investigation of inhibitors of the enzyme histidine decarboxylase in mouse mast cell tumour lines.

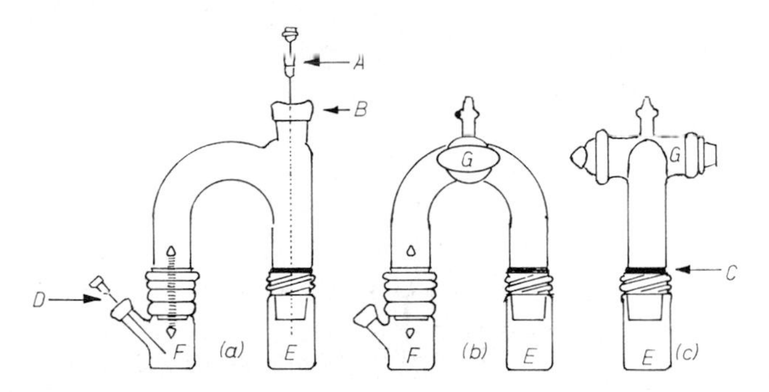

Fig. 6.3. The transfer of liberated $^{14}CO_2$ may be carried out in ground-glass joint systems. a) is suitable for experiments in which bicarbonate is not involved. The desired gas atmosphere is established by allowing water-saturated gas to flow through the 22 gauge needle (D) inserted into the rubber serum bottle stopper attached to the flask F (25 mm dia, 25 mm high up to the ground glass joint). The gas passes out through the spinal needle (A) with the stylet removed. The vial E contains 1 ml of 1 M Hyamine 10X in methanol. Incubation is conducted at 37°C for 90 min. b) and c) are side and end views of an alternative design for bicarbonate buffers. The tap (G) is used to allow gassing of the flask F without entering the vial (E). Following absorption of the $^{14}CO_2$, 15 ml of a toluene solution of PPO (0.3 %) and POPOP (0.01 %) is added to assay (Roberts et al., 1963). (Reproduced by kind permission of the Plenum Publishing Co.)

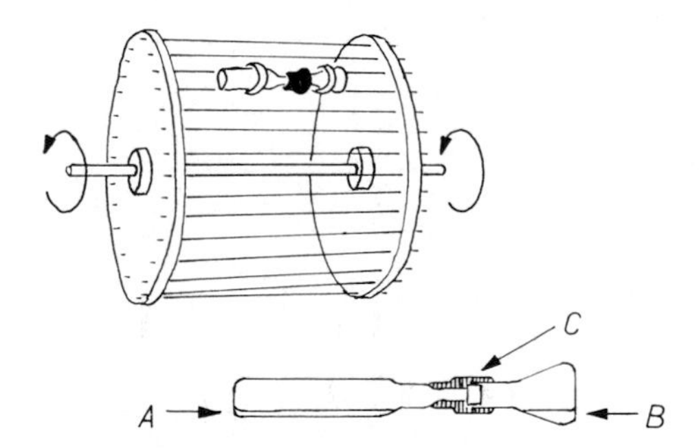

Fig. 6.4. A device for the assay of a multiple series of respiration experiments liberating $^{14}CO_2$. The ampoule-conical flask combination is held to a rotating cage as shown, and the gas liberated in ampoule (A) is mixed with a quaternary base in the small conical flask B. The rubber cap (C) is to provide a safe connection between the two vessels (Aures and Clark 1964). (Reproduced by kind permission of Academic Press Inc.)

Barium hydroxide has been suggested as an efficient absorbant of carbon dioxide by a number of workers (e.g. Cluley 1962). The resulting carbonate precipitate may either be assayed as a suspension, in which case it is necessary to sonicate it (25 Kc/sec/2 min) to reduce the size of the aggregates, or better it may be dissolved in a solution of EDTA (0.4 ml, 10 %) in Tris buffer (1 M, pH 9.0). A scintillant consisting of Triton X-100 : toluene (1 : 2, v/v) is used to assay this solution, and depending on the quantity of barium carbonate solution, efficiencies of 50 to 70 % can be expected. Sonication of the precipitate is reported to increase the efficiency of counting of carbon-14 from 31.1 to 58.9 %.

Extracts and chromatographic eluates

The radioassay of an isolated extract or component of tissue following injection of a labelled drug or biochemical precursor is a common requirement in biochemical pharmacology and in medicine. The physico-chemical nature of the agent, especially after its initial metabolism in the body, will largely determine the manner in which the radioassay will be carried out. It is possible, for example, that the products of interest may be easily extracted with solvents which are readily miscible with high efficiency scintillant mixtures such as Tpp4 (Table 3, Appx. II). However, more often than not, such tissue extracts are aqueous or acidic solutions. In each case there is usually an efficient means of making the most from the few counts in the extract. The main problem is thus to ensure that the beta-emitter of interest has entered solution completely. Since in these types of experiments, one is usually concerned with a particular product of drug metabolism and not so much with the radio-activity of the whole tissue, it is necessary to make sure that the technique employed has extracted *all* the material.

Having established the detailed extraction procedures for the tissue to be studied, it is assumed that the final extracts will be one or more of several types and these are best considered in terms of their ionic strength, pH and salt content. The same problem will usually arise with chromotographic eluates and they can be considered together with extracts.

The considerable variety of chromatographic techniques requires both the accurate assessment of the radioisotope concentration and

Subject index p. 309

a means of sensitively locating peaks of activity. There are several different flow devices available for the direct monitoring of the isotopes as they pass from the column. These consist of two basic designs, a) a sampling device, which either continuously or intermittently diverts a constant proportion of the eluate stream, which after mixing with an aliquot of scintillant mixture is counted in the usual way, or b) a suspension of a solid, insoluble scintillant is interposed in the eluate stream in the form of a flow-through cell. Both these designs will be considered later in more detail (§ 13.1).

As with most aqueous solutions where the volume available does not present an embarrassment, it is worthwhile, at the outset, to decide what order of magnitude of counts are to be expected from the system. If the eluate is likely to be of relatively high specific activity, and relatively low in salt components, then any of the water-accepting scintillant mixtures can be used, such as dioxane-naphthalene and blended toluene systems. The more usual situation however is that the amount of radioactivity present is very low, and the full resources of a system in which high merit values are an integral feature of the method are usually essential to obtain statistically meaningful counts. If the eluted material is a labelled macromolecule that is precipitable with trichloracetic or perchloric acid then the soluble material may be separated from the macromolecule by filtering on to glass fibre discs (§ 2.2.1). The macromolecule is thereby concentrated and may be measured at approx. 15% efficiency (tritium) in Tpp4 scintillant (see § 2.2.1); the soluble washings may also be assayed in a dioxane-naphthalene system. If the specific activity is approx. 1000 cpm per 0.5 ml or less, a colloid system could be used (§ 2.2.3). Some specific applications of these systems will now be considered.

7.1. Aqueous extracts

An aqueous extract of tissue could contain unlabelled proteins and other high molecular weight materials which would precipitate in either a dioxane or blended toluene scintillant mixture. It is therefore necessary first to determine if these components of the extract can be removed without interference with the levels of labelled components present. There are several protein precipitants available, but the most convenient is probably ethanol which will allow the final aqueous solution, freed from high molecular weight proteins, to be blended more readily into scintillant mixtures.

If the macromolecular component itself is labelled high specific activity solutions may be quickly measured by disc techniques, by simply adding an aliquot to a disc and drying. It is advisable that glass fibre discs be used so that if an absolute measure is required, the protein can be easily removed from the disc by the application of a solubilizer or a proteolytic enzyme. However, for low specific activity solutions, a colloidal technique is more useful. Some useful Triton X-100: toluene compositions are illustrated in Table 7.1.

TABLE 7.1

Some useful mixtures for the colloidal scintillation counting of a number of solutions used in biochemical procedures. The merit values quoted (MIV) are instrument corrected but are not corrected for stability of counting.

Aqueous solution	Scintillator mixture		Counting mixture		Merit value (^{3}H)
	Triton X-100: toluene (v/v)	PPO (g/l)	Scint. (ml)	Aq. soln. (ml)	
Water	1:1	8	6	4	1231
8 M urea	1:1	8	6	4	1142
5% sucrose	2:3	6	5	5	989
2 M NaCl	7:3	8	7	3	989
0.03 M ammonium-formate	2:3	5	5	5	778
1.0 M ammonium-formate	3:4	8	7	3	736

If the macromolecule is a protein, it is often convenient to use a proteolytic enzyme to degrade the molecule, and so render it more soluble in blended-toluene or dioxane-based scintillant mixtures and thus enable the material to be measured by means of a homogenous scintillation counting method. Similarly, DNase and RNase enzymes may be used for DNA and RNA respectively (chapter 8).

In the eluation of oligonucleotides, a high concentration of urea (usually 7 to 8 M) is often used to reduce the hydrogen bonding and limit intermolecular interaction. However, the eluate in these cases cannot normally be assayed directly by either dioxane-naphthalene or blended toluene systems, since immediate precipitation of urea and of naphthalene occurs causing excessive quenching and counting heterogeneity. Although disc counting methods can be used, short-chain oligonucleotides behave in a similar manner to urea and it is difficult to remove the latter and leave the former. The colloid system offers an excellent means of overcoming this problem, since a 7 M urea : toluene : Triton X-100 composition can be employed which can accept 4 ml of such a solution in a total volume of 10 ml, as a clear stable gel with a high merit value.

The more usual situation in biochemical work however, is that the extracting or eluting solvents contain acids, alkali or salts and these additional components of the solution to be assayed will influence the counting technique used, especially in low specific activity solutions.

7.2. Acid solutions

For blended scintillant mixtures based on toluene or dioxane-naphthalene, acid solutions are usually readily assayed in much the same manner as dilute aqueous solutions. However, in general, they produce more quenching problems which vary in extent from acid to acid. Perchloric acid causes less quenching than trichloracetic acid, and where either can be used, for example in the analysis of labelled nucleic acid and proteins in tissue, the former should be considered first. Its lower absorbance in ultraviolet is also a very considerable advantage. The relative quenching activities of a number of commonly used acid solutions are indicated in Table 7.2.

As a rule in homogenous counting systems, acid solutions do not stimulate chemiluminescence, but in fact reduce it if it already occurs.

TABLE 7.2

The half molar quench values of different acids. 0.5 ml of a graded series of the acid solutions were assayed in DNXppE (TABLE 13, Appx. II).

Acid solution	$M_{0.5}$*
Trichloracetic	1.48
Perchloric	1.30
Hydrochloric	1.94
Sulphuric	3.63
Nitric	1.90

* Molarity of acid present in scintillant mixture which quenches scintillant by one half.

The action of the acid is to delay peroxide-induced chemiluminescence, and hence an elevated background count with an acid solution could be due to a suppressed chemiluminescence reaction and will persist for a long time.

Although the problem of chemiluminescence is a minor one with dilute acid solutions and dioxane-based scintillant mixtures, this is not the case with strong acid solutions used in conjunction with colloid systems where it can present a severe problem. No method has been recorded for the elimination or reduction of this form of chemiluminescence, and it is advisable to abandon the colloid technique if this should be excessive. Strong acid elution from Dowex columns can often result in spurious peaks of chemiluminescence, which unless one is aware of it could be thought to be due to tritium. The effect may be reduced by raising the lower discriminator threshold, sacrificing some tritium counting efficiency in the process. For carbon-14 measurements however, this may be a worthwhile procedure as a routine. Continuous flow-through monitoring where chemiluminescence is expected, is an essential adjunt to ensure that the peaks obtained originate from a radioactive isotope and not from an artifactual light-emitting reaction.

Due to their considerable sensitivity for low specific activities however, colloid systems are still very useful for dilute acid solutions. Stronger acids could first be diluted or treated with sufficient alkali to allow the aqueous solution to be regarded as a salt. Fewer photochemical anomalies are usually encountered under these conditions. Reducing agents, such as ascorbic acid, do not appear to have any effect on this type of chemiluminescence. A commercial material 'Dimilume' (Packard Instr. Ltd.) has been recommended for luminescent artifacts that occur with 'Instagel', a commercial colloid system. The effect of this product is said to shorten the time necessary to decrease the counts to a tolerable background level, but this level is usually still higher than expected. A list of useful scintillator compositions for the more dilute acid solutions is given in Table 7.3.

TABLE 7.3

Colloid scintillation counting mixtures for a number of acid solutions used in biochemistry. The merit values quoted (MIV) are instrument corrected but are not corrected for counting stability.

	Scintillator mixture		Counting mixture		
Acid solution	Triton X-100: toluene (v/v)	PPO (g/l)	Scint. (ml)	Aq. soln. (ml)	Merit value (^3H)
5% TCA	13:7	10	8.5	1.5	662
5% PCA	3:1	3	6	4	1148
0.1 N formic	6:11	10	8.5	1.5	706
1.0 N HCl	2:5	8	7	3	1030
3.0 N HCl	5:11	5	8	2	748

Acid solutions do not interfere with the Cerenkov emission and may readily be used in the assay of ^{32}P and similar Cerenkov emitters. Care should be exercised however to limit the actual strength of the acids used in the liquid scintillator to prevent release of acid vapours which may damage electronic components of the instrument.

7.3. *Alkaline solutions*

There are many accounts of the direct assay of small volumes of sodium and potassium hydroxide solutions (less than 0.2 ml) in a water-accepting scintillant such as Bray's Mixture (Table 5, Appx. II) or one based on the dioxane-naphthalene system. However, in the latter case, there are also many reports of excessive chemiluminescence produced when used in conjunction with alkaline solutions. The phenomenon appears to be associated with all three components (§ 1.3) and is usually considerably reduced by initial neutralisation with acids.

Since the use of alkaline solutions is often a convenient and rapid method of digesting proteins and small amounts of tissue, there is frequently a requirement to radioassay such solutions, especially in situations where organic solubilizers are not readily to hand. If the level of potassium hydroxide used is high, then the presence of ^{40}K in the naturally occurring potassium ion (approx. 0.012%) may give rise to slightly elevated background levels but usually this may not matter too much. Sodium hydroxide may be used in most cases. In general, it usually is preferable to neutralise the alkaline solution with acid before attempting to assay it, and also to avoid as far as possible the photochemical artifacts associated with alkaline solutions. In any case, concentrations greater than 2 N should always be diluted to avoid damage to the scintillator itself. Acidification is, of course, not possible where alkaline solutions have been employed to trap $^{14}CO_2$. In this case, an aliquot of the alkaline solution may be added to loosely-packed Cab-O-Sil, followed by 15 ml of TppE as described in § 4.4.

Lithium hydroxide solutions appear to produce excessive chemiluminescence even after neutralization when used in colloid systems. Ammonium hydroxide solutions also produce chemiluminescent artifacts in dioxane-naphthalene systems and thus blended toluene-based scintillant solutions should wherever possible be used for ammonia. Care should be taken to avoid the accidental possibility of measuring ammonia and HCl solutions in the same counter as loose-fitting bottle caps could result in considerable damage to the surface of photomultipliers and light collecting surfaces through deposition of ammonium chloride.

Solutions in hyamine hydroxide and other quaternary ammonium bases should be assayed in a simple toluene-based scintillant such as Tpp4. Any blending of this mixture is best carried out with methanol. Dioxane-based scintillants are to be avoided for organic bases, due to the chemiluminescence induced.

7.4. Salt solutions

These solutions are probably the most frequently encountered systems in biochemical work and comprise a large proportion of different chromatographic eluants. Sodium chloride solution, in particular, is one of the most frequently employed and an examination of the acceptance volume of sodium chloride solutions in Bray's scintillant reveals that potential errors can easily occur if this is not recognised Fig. 7.1 shows the effect of adding different concentrations of sodium chloride solution to Bray's scintillant; it can be seen that the level of salt that it is possible to add is in fact very restricted owing to precipitation.

To overcome this difficulty, especially in column chromatographic applications where the specific activity of the eluate (i.e. cpm/ml) is often low, and the salts present are in varying concentration, the

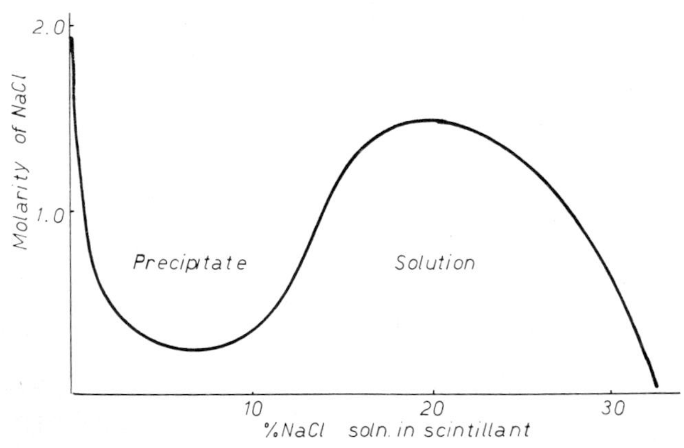

Fig. 7.1. The diagram shows the effect of adding sodium chloride solution of different molarities to Bray's mixture (DNppGM, Table 5, Appx. II). The line represents the boundary between precipitation (upper portion) and complete solution (lower portion).

colloid system is the most appropriate. However, before embarking on this method of radioassay, it should be emphasised that for high specific activity samples (e.g. greater than 1000 cpm/0.2 ml), a homogenous system or even disc counting may be more convenient. The advantage in the use of the colloid system is that a small manipulation, such as a neutralisation, can be incorporated into the routine mixing of sample aliquots with scintillants and providing the correct sample volume is maintained in each case, highly reproducible results may be achieved. Some examples of optimal conditions using the Triton X-100: toluene colloid system are given in Table 7.1. It can be seen that different salts require different Triton X-100: toluene combinations for optimal counting. However, sub-optimal, but nevertheless still efficient counting (i.e. in terms of merit value) may be achieved with fewer basic Triton: toluene mixtures.

The buffering of salt solutions, as in the hydroxyapatite separation of nucleic acids, does not seriously modify the system required for assays in the presence of sodium chloride, the concentration of salt playing the dominant role in determining the optimal counting conditions. Ammonium formate solutions do require different optimal conditions, depending on the concentration of the salt for maximum merit values, but for counting a gradient concentration of the salt, a point can be chosen which is constant throughout the gradient (see § 7.6).

7.5. *Organic solvents*

Organic solvents are often used in the isolation of lipids and steroids from aqueous extracts. These solvents vary widely in their quenching capacities, some of which are shown in Table 7.4.

In general, the halogenated paraffins are powerful quenching agents and if possible, should be avoided in counting systems. Ethers and the alkoxy alcohols have a tendency to produce peroxides which in turn induce chemiluminescence, but storage over granulated zinc is often sufficient to reduce and even eliminate this. If an aromatic solvent, such as benzene, toluene or xylene can be used for the extrac-

TABLE 7.4

Half molar quench values for some organic solvents in DNXppE (TABLE 13, Appx. II).

Solvent	$M_{0.5}$*
Carbon tetrachloride	~0.03
Chloroform	~0.4
Acetone	0.7
Methanol	9.5
Ethanol	10.0

* The molarity of solvent present in the scintillant mixture which quenches the scintillant by one half.

tion or elution, then these are very efficiently incorporated into toluene-based homogenous counting systems. Acetone is a more effective quenching agent than either ethyl or methyl alcohols, and should also be avoided if possible. Most organic solvents however can be admixed with a toluene based scintillant such as Tpp4 (Table 3, Appx. II), and assayed efficiently as a homogenous system, to which all the methods of quench correction may be employed. When the isotope is not present in the water phase, it is an advantage if possible to dry the organic extract thoroughly over anhydrous magnesium or sodium sulphate or magnesium perchlorate, since this procedure could considerably enhance the counting efficiency.

Counter-current systems use two immiscible layers in the analysis of substances having different partition coefficients. After the appropriate number of transfers, it is often more convenient to render the two immiscible phases into a single phase for the purpose of routine assay of the components present, rather than attempt to separate and analyse the individual phases. This is sometimes achieved by addition of ethanol or similar amphiphilic solvent and then the solution is usually compatible with a dioxane-naphthalene or a toluene-based scintillator. With phosphorus-32 labelled materials, there is no problem with the radioassay by means of Cerenkov emission. The effect of different organic solvents on the efficiency of measurement of this emission is shown in Table 2.1, page 81.

7.6. Gradient correction procedures

In order to avoid the necessity of a laborious calculation of the degree of quenching of each sample taken from a gradient salt or pH elution from a chromatographic column, several methods can be suggested. With eluates containing a large number of counts, homogenous counting is usually adeuqate. A small aliquot of the eluate is added to a suitably blended scintillant mixture, either manually or automatically and an external standard ratio or sample channels ratio can be used for automatic correction, often on the instrument itself.

However, the problem arises when low activity samples are to be measured and a colloid system is being used successfully. The degree of quenching must usually be determined either by sample channels ratio or more usually, by the use of internal standardization. The former suffers the disadvantage that a high enough level of sample counts are needed to provide satisfactory counting statistics and the latter is a tedious and time consuming process.

An alternative procedure is to attempt to make the samples of the whole run quench to the same degree, so that for the usual purpose of determining the relative magnitude of peaks, this is adequate. When macromolecules are being separated, for example from centrifugation procedures employing sucrose or caesium chloride, the macromolecule is precipitated on a disc which is washed and thus the relative quenching of each fraction is thereby made the same, independent of the gradient in which they originally existed. This equating of the quenching is made still further comparable between fractions by employing a carrier macromolecule such as calf thymus DNA or bovine serum albumin, in order to mask, by a relatively constant quench, the varying quench that would be obtained from the varying levels of macromolecules present from the fractionation procedure. For small molecules, such as nucleotides, provided the eluting substances will adhere to the disc e.g. DEAE cellulose discs, then after washing the discs will be, from a counting point of view, identical.

However, the colloid system does afford a sensitive means of measuring a gradient elution of isotope by exploiting the special properties of the phase diagram, in that regions can be found where the counting efficiency (in terms of merit value) is similar at both ends of the gradient sequence. This enables a series, albeit at slightly lower merit values than optimal to be measured without correction for efficiencies. Where very accurate assessment is necessary, it is usually essential in any case to combine the fractions below a peak and determine the specific activity of the combined eluates.

If the same gradient elution system is to be used frequently and a homogenous counting system be employed, it is probably more convenient to plot a standard curve of gradient salt concentration (or fraction number) against counting efficiency in a given set of elution conditions. The correction factor can then be read off as the appropriate fraction number. In practice, this is usually found to be necessary where a homogenous system is used with small aliquots of eluate as samples, but is usually found to be unnecessary in colloid systems, as the counting efficiency does not vary a great deal throughout the sequence. The compositions recommended for two gradient combinations often used in fractionation procedure are illustrated in Table 7.5.

TABLE 7.5

Colloid scintillation counting mixtures which are designed to allow a gradient of concentrations to be measured in the same mix with a merit value which varies only by approx. 2% along the gradient. The merit values quoted are not corrected for stability and are not optimal but are instrument corrected (MIV).

Gradient solution			Scintillator mixture		Counting mixture		
Component	from	to	Triton X-100: toluene (v/v)	PPO (g/l)	Scint. (ml)	Soln. (ml)	Merit value (^3H)
Ammonium formate	0.03 M	1.0 M	1:1	4	7	3	850
HCl	1.0 N	3.0 N	1:2	4	7.5	2.5	710

Macromolecules

The difficulties usually associated with the radioassay of macromolecules containing beta-emitters are due to, a) the insolubility of the material, even in dilute alcoholic solution, and b) the presence of salts, sugars etc. which are often associated with the macromolecules when they are required to be assayed. These problems are best solved by consideration of the environmental conditions in which they are found, since the technique eventually employed, may depend more on the state of its solution environment than of the state of the macromolecule itself.

If the labelled precursor enters the macromolecule relatively selectively, e.g. in cell culture experiments or in ascites cells, it is often possible under certain conditions to measure the total cellular radioactivity as an index of macromolecular labelling. This can only be done when the cell pool radioactivity is very low or negligible compared with that in the macromolecule. These conditions may often be achieved by a) judicious choice of precursor to be used, or b) 'chasing' out the labelled pool components with a higher concentration of the unlabelled precursor to dilute the pool label to an infinitesimal level, or c) by waiting until the pool concentration of the labelled precursor has achieved a very low level compared to that of the macromolecule itself. Preliminary experiments to achieve this differential between the levels of labelling of the precursor as compared to the macromolecule will often save many hours of unnecessary preprocessing. More often however, levels of macromolecular uptakes and pool labelling require

Subject index p. 309

to be assayed simultaneously and preprocessing and fractionation has to be undertaken.

In assaying the conversion of monomeric precursors to macromolecules or the reverse (degradative changes), liquid scintillation spectrometry can provide a means of rapidly assaying the relative proportion of the components present, usually by absorbing the macromolecule on to a solid support and washing off the monomeric units with suitable solvents. The potential application of this principle is considerable, but it appears to have been exploited only in the nucleic acid field to any considerable extent and somewhat less so in the field of protein synthesis. Thus some preliminary careful thought given to the details of the preprocesssing could enlarge the scope of the radiochemical analysis within the experiment.

8.1. Proteins and amino acids

One of the most frequent requirements in biochemical work is to assay protein, either in solution of buffers or salts often of high ionic strength, or as a solid having been freeze-dried or as a pellet obtained by centrifugation. The most efficient method is of course to degrade the protein to carbon dioxide and water by combustion procedures (see § 3.3). However, either for numerical reasons or (more often) due to the presence of inorganic salts, the preprocessing required for these combustion methods is often impractical and tedious. If the protein is already in a solution of low ionic strength and its assay is a regular requirement for a large number of samples, it is worthwhile determining optimal counting conditions for a colloid scintillation procedure (§ 2.3). However, if only a few determinations are to be made then several alternative methods are available.

If of fairly high specific activity (say greater than 1000 cpm/0.2 ml), an aliquot may be taken (approx. 0.2 ml) precipitated and added directly to a glass fibre disc. If salts are present, these can usually be washed away by means of ice-cold 5% perchloric or trichloracetic acid. Less quenching is observed when perchloric acid is used and residual traces will not seriously quench the final count. The disc can be dried and counted in a standard Tpp scintillant composition (Table 3, Appx. II). Solubilization of the protein precipitate in the minimum quantity of a quaternary ammonium solubilizer will enhance the counting efficiency for tritium containing proteins and, to a lesser extent, for carbon-14 (Pillarisetty et al. 1971).

However, if the solution is of low specific activity, and the requirement is for the assessment of its specific activity, then an aliquot would be analysed for its protein content and a second aliquot could be precipitated with a protein precipitant and the pellet so obtained solubilised in a quaternary ammonium hydroxide solubilizer or, depending on the amount present, assayed by the perchloric peroxide wet oxidation method of Mahin and Lofberg (1966) (see § 3.3).

In the routine assaying of carbon-14 in protein solutions it is sometimes convenient to plot a standard quench correction curve based on the known amount of protein present. This method, suggested by Toporek (1960), does require that estimation of the protein is being done and that there are no coloured impurities present. A plot of log efficiency against protein concentration is approximately a straight line over a short range (e.g. up to 80 mg).

It is possible to undertake limited enzyme kinetic studies directly on the disc and estimate protein synthesis. The method is described by Mans and Novelli (1961), and consists of incubating an enzyme extract with the appropriate amino acids as substrate directly on the filter paper. The reaction is stopped by addition of 5% TCA which also precipitates the protein on to the disc and washes away the free amino acids. After drying it can be assayed directly in Tpp4 (Table 3, Appx. II).

The use of 2-phenylethylamine has been recommended by Francis and Hawkins (1967) for the assay of both tritium and ^{14}C labelled solutions of proteins. A solution of PPO and POPOP in the aromatic base alone will count, but some improvement is obtained by addition of some toluene to the system e.g. PPO (0.4%), POPOP (0.005%) 2-phenylethylamine (4.5 ml) and toluene 0.5 ml. This system will tolerate up to 60% of its volume of water, containing up to 20 mg protein/ml. The absolute efficiencies recorded were 2 to 5% (merit values 120 to 300) for tritium and 25–45% (merit values 1500–2700) for carbon-14.

It is often desirable to assay the levels of radioactive label present in amino acid fractions produced from an automatic amino acid analyser following interaction of the amino acid with the ninhydrin colour reagent. The radioassay for such a product was described by Olsen et al. (1968).

Where the level of the radioisotope in the solution is very low, it is clearly of interest to attempt to use a method which would realise the highest merit value and thus make maximum use of the counts available. The colloid system was suggested by Madsen (1969) for the assay of ^{14}C-labelled protein dissolved in 1 N NaOH. In their method, 1 ml of this solution (or hydrolysate) was mixed with 15 ml of a toluene : Triton X-100 system (2 : 1) containing 4 g PPO/l and then neutralised with 0.1 ml concentrated HCl. Although this technique compares favourably with thixotropic methods, it is desirable to avoid large pH changes in the presence of the primary scintillants and here there is no reason why the solution in NaOH should not first be neutralised and the resulting salt solution of the protein be added to the scintillant mixture. In this case, used by the present author, the merit values (not corrected for instrument efficiency under these conditions) lie within the range 540 to 570 for tritium and 2370 for ^{14}C which are comparable or slightly better than those of Francis and Hawkins (1967) who use a method based on homogeneous systems which are easier to correct for quenching. However, if the optimal point for assaying salt solutions in Triton X-100 : toluene systems be carefully chosen, much better merit values could be obtained.

A direct comparison of the efficiency of counting paper strips by the scanning method using proportional counting and liquid scintillation counting suggestst that the former method is some 60 to 70% as efficient for both ^{14}C and tritium measurement as the liquid scintillation technique (Wenzel and Stohr 1970).

The extreme importance of avoiding recycling of amino acid tracers in measurement of turnover rates of proteins is emphasised by the work of Schinke and Doyle (1970).

8.2. Nucleic acids, nucleotides, etc.

The considerable effort that has been put into the study of these substances in the last two decades has relied very heavily on the use of tritium and carbon-14 labelled precursors as well as ^{32}P for the labelling of the basic phosphodiester chain of the molecule. At the stage at which the radioassay is required, the macromolecule may be in solution as a trichloracetic or perchloric acid extract or together with salts or other solutes as well as other macromolecules such as proteins and polysaccharides. In addition, buffers, polyacrylamide or agarose gels may be present, or the nucleic acid may be a precipitate, or mixed with high concentrations of caesium or rubidium salts or sucrose in fractions from sedimentation centrifugation studies. The special radioassay procedures will be described in conjunction with the technique itself.

If a high molecular weight nucleic acid is added directly to Bray's or similar water-accepting homogenous scintillant, it will be precipitated as fine strands, in some cases, invisible to the naked eye and could be regarded superficially as in solution. For ^{14}C or ^{32}P assay, this factor may not be very important, except that it is not possible to employ the external standards ratio technique for quench correction due to the heterogeneity of the system.

The most efficient method of absolute counting of the nucleic acid is, as with other organic materials in a suitable state, by combustion procedures (§ 3.1). However, for large numbers of samples and where there are other contaminating solutes present, this is cearly an unnecessarily tedious procedure. A more convenient method is often to degrade the macromolecule to short-chain oligonucleotides and nucleotides by an enzyme, e.g. pancreatic deoxyribonuclease for DNA and ribonuclease for RNA.

The main problem however often originates not from the nucleic acid but from the associated solute molecules. These can either be removed by washing the precipitated nucleic acids on a membrane or glass fibre disc or they may be included by employing a colloid scintillation procedure (§ 2.2.3). This latter method can only be usefully employed where relatively large volumes are available such as those derived from column chromatography. In the case of tritium-labelled nucleic acids, the material is best brought into homogenous scintillation conditions, and digestion methods may be used for quantities up to 10 mg. A procedure recommended by Hattori et al. (1965) is as follows: 0.1 to 0.3 mg of the labelled sample is placed in a scintillation vial and 0.5 ml of water or 2 M NaCl is added together with 0.04 ml of concentrated HCl (final concentration of the HCl 0.9 N). Following incubation at 50°C for 20–24 hr in silicone-stoppered vials, 1 ml of hyamine is added together with 10 ml of a dioxane-based scintillation mixture (100 g naphthalene, 10 g PPO, 250 mg POPOP in 1 l dioxane).

An interesting phenomenon described by Paus (1972) was that certain macromolecules, such as RNA, absorb water on to the surface of the molecule and thereby reduce the efficiency by which the beta radiation penetrated into the scintillant environment. This absorption of water is enhanced by the hydrophobic nature of the scintillant system itself.

It has been realised since 1966 that water is essential for the proper action of quaternary ammonium solubilizers. Moorhead and McFarland (1966) showed that the addition of some water to the NCS : PPO : POPOP system in toluene, stabilises and enhances the counting efficiency by increasing the total number of emissions recorded, the greatest increase being at the lower energy part of the spectrum.

Precipitates should be removed from glass fibre discs before assaying. Using a quaternary ammonium solubilizer, 98 to 99% of the RNA and DNA on the disc can be brought into solution and may be assayed by a homogenous technique, when all the conventional quench correction may be used.

An alternative method for the assay of nucleic acids is to precipitate selectively with cetyl trimethyl ammonium salt, which is soluble in 2-methyoxyethanol and methanol. Complete recovery has been claimed by first precipitating the macromolecule on a glass fibre disc, and subsequently dissolving off the precipitate in either of these solvents prior to assay in a blended toluene scintillant.

Cerenkov counting can be employed for the assay of ^{32}P-labelled nucleic acids (§ 2.3). Not only is the wastage negligible by this procedure, but the 'merit value' is very high due to the fact that the whole sample can be employed for measurement; later it can be recovered for further processing if necessary.

8.3. *Lipids and steroids*

The radioassay of lipids and steroids is usually simplified by the fact that they are mostly readily soluble in hydrocarbon solvents and thus provide ideal homogenous counting conditions. The most useful scintillator solutions would thus be the non-polar solution of PPO in toluene (for details of lipid separation see Kates, this series vol. 3).

Lipids are normally isolated and fractionated by such solvents as chloroform, carbon tetrachloride or acetone. These are also well known powerful quenching agents and thus should be removed as far as possible by distillation or preferential absorption before attempting to count the lipid sample. The presence of as little as 3% chloroform in the final scintillator volume will reduce tritium counting efficiency by 50% (Table 7.4).

The measurement of lipid radioactivity derived from thin-layer chromatography presents no problem. Direct assay of samples containing also up to 300 mg silica in a blended toluene-based scintillant mixture such as TNppX or TppE (Table 4, Appx. II) is usually adequate, and in most cases it is possible to apply an external standard ratio method as a means of quench correction.

Phospholipids labelled with ^{32}P are best measured by Cerenkov counting methods (§ 2.3). This is a particularly convenient method for those phospholipids that have been extracted by solvent mixtures containing chloroform, which would normally quench in a conventional liquid scintillation counting system.

For thin layer chromatographic separations of phospholipids on silica gel, a 1 : 1 mixture of chloroform : methanol containing about 10% water has been recommended (Webb and Mettrick 1972) as suitable solvent. The final assay is achieved by suspending the gel, together with methanol and Cab-O-Sil (4 % w/v) in a toluene-based scintillant, such as Tpp4 (Table 3, Appx. II).

Assay of the specific activity of a phospholipid spot on a thin layer plate (Schneider 1971).

1) The spot on the thin layer chromatogram is located by autoradiography or by spraying with molybdate spray, and is scraped into a test tube.
2) The scrapings are digested with 0.4 ml of 70 % perchloric acid and 0.02 ml of 4 % ammonium molybdate at 180°C for 30 min.
3) Mixture cooled and centrifuged, and 2.5 ml of 0.22 % ammonium molybdate solution is added followed by 0.1 ml of the Fiske–Subbarow reagent (Grind together in a mortar 30 g anhyd. sodium bisulphite ($NaHSO_3$), 1.0 g anhyd. sodium sulphite (Na_2SO_3) and 0.5 g 1-amino-2-naphthol-4-sulphonic acid. Dissolve 1.58 g of this mixture in 10 ml water just before use).
4) Heat in a boiling water bath for 7 min.
5) Cool and centrifuge off the silica gel.
6) Take 2.5 ml of the supernatant and add to cuvette for an estimate of absorbance at 830 nm. (A blank is prepared by using a similar weight of silica gel preferably from a blank region of the same plate.)
7) For radioassay, the aliquot in the cuvette is added to a polythene scintillation vial containing 2 ml of a 1 : 1 mixture of 1 M ammonium persulphate and 10 ml NaOH. About 7.5 ml of water may be used for the transfer of material from the cuvette, digestion tubes and other equipment used to make a final counting volume of 12 ml. Bleaching is complete in 2 hr.

Although, not strictly speaking a liquid scintillation counting procedure, an interesting application of the light source offered by liquid scintillation fluor in the presence of a radioisotope has been used in the estimation of the colour resulting from the action of sulphuric acid on lipid chromatographed on a thin layer. The vial used (Snyder and Moehl 1969) is illustrated in Fig. 1.1f. It is probable that this technique may have wider applications in other areas, where the automation and data presentation methods on modern instruments can be used with advantage in multiple measurements of certain unlabelled materials capable of producing a colour reaction.

Although many of the methods employed for the assessment of phospholipids can also be applied to the assessment of steroids due to their solubility in non-polar solvents, there are some additional problems. When low levels of steroids are assayed there is a danger of absorption of the steroid on to the surface of the vial resulting in a change in the geometry of counting. In order to avoid this, it is recommended (Kandell and Gornall 1964) that at least 20 to 50 μg of inert steroid is also included in each assay as a carrier molecule to reduce this phenomenon.

The assay of steroids in a variety of media has however radically changed over the last few years due to the advent of radioimmunoassay procedures, and steroids have been very widely studied by this procedure. For most low level assays therefore this procedure is probably the most specific and accurate (§ 4.3.3).

When introducing a tritium or ^{14}C atom into a steroid, in order to trace its metabolism or distribution, the assumption is usually made that no change occurs in its chemical properties. However, these effects do exist, and one net result of this is to cause the differently labelled forms of the same substance to behave differently during fractionation procedures. This effect is particularly important in the steroids and this subject has been well reviewed by Klein (1970). It is mainly of concern in those situations where isotope dilution assays are employed for the assessment of low levels of steroids.

Steroid conjugates however are usually water-soluble and not toluene-soluble and in these cases a water-accepting scintillant system such as Bray's composition or a dioxane naphthalene system is recommended (Table 5, Appx. II).

8.4. Sugars and polysaccharides

The problems associated with the assay of sugars and polysaccharides are in many ways similar to those which are experienced with nucleic acids and their precursors. For the pure materials, the combustion procedure is the best method of assay, provided that the system used allows for adequate oxygen for the size of sample used. Due to the high carbon content of these substances, a greater volume of oxygen will be needed. This is a particular problem with plant tissue or any other high cellulosic material.

The wet oxidation procedure of Mahin and Lofberg (1966) often results in losses when sugars are being assayed. Kim and Rohringer (1969) suggested that these losses were due to the long heating required in the originally described procedure, and used a short heating time of 2 to 3 min. Fuchs and De Vries (1972) on the other hand also examined plant material and suggested that 40 min heating resulted in the maximal recoveries, which in their hands was approx. 90%. The Van Slyke procedures however (§ 3.3.2) gave higher recoveries, approaching 100%.

Aldose sugars are often precipitated as osazones during their characterisation and isolation, but these are not suitable derivatives for liquid scintillation counting due to their bright orange and yellow colours and to their insolubility. Steele et al. (1957) suggested the conversion of the derivative into the colourless glucotriazole derivative, but this material is also fairly insoluble. In order to increase the solubility in the types of scintillant mixtures used, Jones and Henschke (1963) suggested that the boric acid complex of the triazole be used, since solubility in blended scintillants, especially those of the Bray's type, was much greater. The composition suggested was as follows: naphthalene, 60 g; PPO, 4 g; POPOP, 0.2 g; methanol, 100 ml; ethylene glycol, 20 ml; p-dioxane, 880 ml; and boric acid 25 g; up to 100 mg of the triazole may be dissolved in 5 ml of this scintillant. Solubilizers can also be successfully employed with these carbohydrate derivatives (§ 3.2).

Using a number of milder oxidation procedures however, it is possible to remove one labelled portion of the sugar. For example, perchloric acid will oxidise the terminal carbon to formaldehyde, which may then be trapped as the dimedone complex and measured as a precipitate on glass fibre discs. Alternatively, the terminal group may be further oxidised to the formic acid and trapped as the bromo phenacyl formate (Gabriel 1965). There are probably many ways in which selective enzyme systems could be used to remove specific parts of sugar molecules before radioassay, but they do not seem to have been exploited.

Electrophoresis, centrifugation and chromatography on solid supports

9.1. Electrophoresis

9.1.1. Polyacrylamide gel electrophoresis

Polyacrylamide gel electrophoresis for the separation of proteins and nucleic acids is very frequently employed in biochemical procedures both as an analytical and a preparative tool (see this series Vol. 1 by Gordon and Vol. 4 by Gould and Matthews). Although several methods of staining and subsequent dosimetry have been successful, the measurement of beta-emitters especially in gels which contain more than 5% of polyacrylamide, has caused some difficulty owing to the considerable insolubility of the polymerised material.

The gel is usually formed *in situ* by polymerisation of acrylamide monomer and methylene bis-acrylamide in the ratio of about 20:1. The resulting polymer concentration may vary from 1 to 25% in an aqueous medium. The physical consistency of such gels ranges from fragile jellies to hard rigid solids. The more dilute gels, i.e. less than 5%, will dissolve in hydrogen peroxide solution on prolonged standing. Solubilizers of the quaternary ammonium type only succeed in causing swelling of the gells, rather than actively dissolving the polymer; this process is used with advantage (after extrusion), however, in solubilizing the material present within the gel and assisting its release from the swollen slices.

The radioassay of gel slices by heterogenous counting techniques is usually accurate enough if only a qualitative position of the label within the gel is required. The resolution of the peaks containing the labelled material will depend on the thickness of the slices taken from the gel in relation to the total length of the gel, as well as on the level of counts per unit thickness within the peak area. The compromise decision between large numbers of counts per slice (shorter counting times but less definition) or a larger number of slices (longer counting times per slice, but better definition) will have to be made. Simple gel fractionators are illustrated in Fig. 9.1, and may be adequate for sectioning into approx. 1 mm slices. However, similar sectioning procedures as those employed in histology can also be used, and a microtome cutting procedure following embedding and dehydration was described by Gray and Steffansen (1968). In this case, slices from 50 to 250 μm were prepared.

The slices as prepared by one of the procedures outlined above, can be placed individually on to glass fibre discs, dried overnight at 50°C and assayed in Tpp4 (Iandolo 1970). Alternatively the gel may be treated by one of several ways, advisedly after macerating the gels as finely as possible by forcing through a stainless steel gauze (200 mesh). The resulting macerate, dispersed in water (3 ml), can then be heated (41–42°C) with a colloidal scintillant such as 'Instagel' (Packard) (Leon and Bohrer 1971) (7 ml). Following cooling and keeping in the dark for a few hours, the mixture is assayed as a fine, almost colloidal suspension. A commercial gel fractionator is available (Maizel Gel Fractionater, AGD-30A, Savanth Instruments Inc., Hicksville, N.Y.) based on a design by Maizel (1966). This system has been combined with a flow-through scintillator system based on admixture with a dioxane-based scintillator, for a finer and more sensitive radioactivity tracing procedure in polyacrylamide gels (Bakay 1971).

A few commercially available solubilizers were examined by Paus (1971) for their ability to 'solubilize' polyacrylamide gels. Soluene 100 (Packard Instruments Ltd.) and 'NCS' (Nuclear Chicago Ltd.) had the best 'solubilization' properties for gels up to 2.5% at 60 to 65°C. Gels up to 5% appeared to swell only, but the gel contents dissolved on addition of the scintillant. An important factor in this process appears to be the need for a small proportion of water to be present.

Subject index p. 309

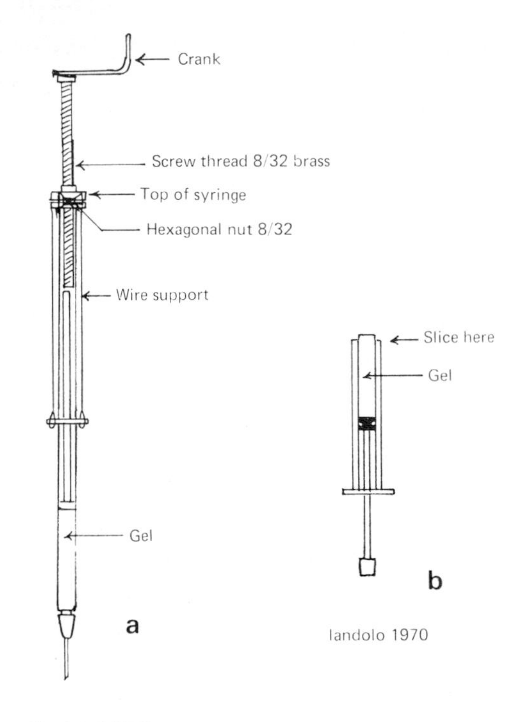

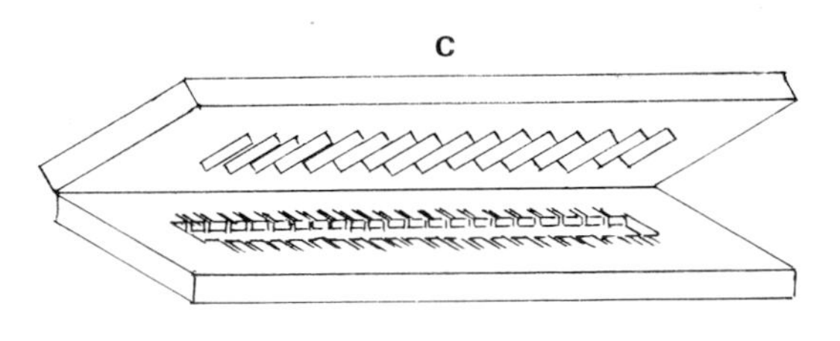

Fig. 9.1. Polyacrylamide gel fractionators readily made from materials available in the laboratory. A) Device described by Ward et al. 1970, made from two syringe barrels (Academic Press Inc.), B) a simple slicer made from a single syringe barrel (Iandolo 1970) (Academic Press Inc.), C) a slicing device described by Cain and Pitney (1968) which simultaneously slices a gel into equal slices (Academic Press Inc).

A recommended combination for 'NCS' for example appears to be, water: NCS (1:9, v/v). The use of hydrogen peroxide (Tischler and Epstein 1968) or sodium hydroxide solution (Alpers and Clickman 1970) has also been recommended in addition to the solubilizer; 1 mm slice of gel is incubated with 0.1 ml of hydrogen peroxide at 50°C for several hours and 1 ml of 'NCS' is then added followed by 10 ml of Tpp3 (Table 3, Appx. II). Alternatively, the slices can be placed directly into a scintillation vial and allowed to dry overnight at room temperature; the resulting dry gels can then be dissolved in a mixture (0.25 ml) of concentrated ammonium hydroxide: 30% hydrogen peroxide (1:99, v/v, stored at 0°C) by incubation at 50°C for 4 to 8 hr. After cooling to room temperature they may be treated with a colloid based-scintillant mixture such as Biosolv, BBS-3 (Beckmann): toluene-based scintillant (1:5, v/v, 10 ml) and assayed (Goodman and Matzura, 1971).

A somewhat tedious heterogenous technique described by Boyd and Mitchell (1966) consists of 'fixing' the macromolecule within the gel with phosphotungstic acid, and replacing the water phase with the scintillant. This is usually achieved by washing the slices with a succession of solvents ranging from a mixture of glacial acetic acid: ethoxyethanol (1:1 v/v) to toluene. It is possible to see however, that an auto-histological system could probably be adapted to this procedure.

One of the most useful advances in this field is the design of a gel which does not modify the electrophoretic efficiency of polyacrylamide but nevertheless is capable of easier dissolution. Two such developments have been described. Ethylene diacrylate (2 g) together with acrylamide (30 g) in 100 ml buffer produces a gel which dissolves in a mixture of hyamine and 1 M aqueous piperidine (1:9, v/v) within 1–4 hr; this mixture can then be diluted with a dioxane-based scintillant mixture such as Kinards solution (DNXpnE) (Table 5, Appx. II) (Choules and Zimm 1965). An even better modification would appear to be the use of N,N′-diallyltartardiamide (DATD) as a mole to mole replacement for methylene-bis-acrylamide as cross linking agent. This results in a polymer with characteristics almost identical with the acrylamide product, but differing in that it is readily soluble in 2% periodic acid (Anker 1970).

The preparation of DATD

1) Add 2.5 moles of allylamine (142.5 g) to 1 mole of diethyl tartrate (206 g) in 10 vol ether.
2) Reflux mixture overnight.
3) Cool. Filter and wash crystals with 10% ethanol.
4) Dry *in vacuo.*

It is often desirable to measure the specific activity of protein bands in polyacrylamide gel separations, following staining. A method for undertaking this was described by Alpers and Glickman (1970). The gels are first fixed in 10% TCA and stained with Coomassie Blue. An unstained duplicate is segmented and allowed to stand for 8–12 hr in 50 to 100 ml of 5% TCA. The slices are removed and rinsed in water. They are then dissolved in 0.3 to 0.4 ml of 1 N NaOH. The protein present is estimated on a 0.1 ml aliquot by the method of Lowry et al. (1951). A 0.2 ml aliquot is added to 1 ml of NCS (Nuclear Chicago) solubilizer and heated at 65°C for 1 hr in a scintillation vial. Ten ml of a toluene-based scintillant such as Tpp4 (Table 3, Appx. II) is added and the mixture assayed. Absolute counts may be assessed by using internal standards or channels ratio techniques.

A similar procedure was described by Cain and Pitney (1968) and consists of using a colorimetric assessment of the dye by first dissolving the slice in 1 N ammonia.

9.1.2. Electrophoresis on agarose and starch gels

Agarose and starch gels in the electrophoretic separation of proteins has been applied to the study of isoenzymes and serum components. Most work has been concerned with stained gels, but the system is clearly upon to the study of proteins which have been labelled with beta-emitters.

A convenient system of assaying both types of gel is to add a slice (up to 5 mm thick) to a total of 4 ml of 5% perchloric acid and heat the mixture at 60–70°C for 20 to 30 min to allow solution, which is usually complete in this time. The resulting mixture is cooled and the solution is added to 6 ml of a toluene: Triton X-100 system (3 : 1) containing 3 g PPO/l. The whole is shaken and after standing in the dark for approximately 1 hr, is assayed in the usual way, using internal polar standards for quench correction.

The assay of radioactivity in agar blocks was described by Lagerstett and Langston (1966), in connection with the study of plant growth regulators. There seems to be no reason however, why the scintillant which was systematically derived by these authors, should not be used for agar generally. Their method was to place a 3 mm tube (of volume approx. 0.03 ml) in a polyethylene vial with 0.5 ml formamide as solvent. Scintillator consisted of 20 ml 1 : 4, ethanol: Tpp5 (Table 3, Appx. II).

9.2. *Centrifugation*

9.2.1. *Pellets*

Many biochemical experiments end up by the necessity to assay a pellet of denaturated protein, calcium phosphate or carbonate, macromolecules, glycogen etc. for a beta-emitter. Probably one of the commonest situations is a protein pellet, 100 to 200 mg wet weight, following a 5% TCA or 5% PCA digestion.

The following method is a convenient procedure for assay of such material, where the whole pellet is available for the radioassay.

1) Pellet treated with 1.5 ml 2 N NaOH in a 10 ml conical centrifuge tube.
2) Heat to approx. 80°C in a water bath for 15 to 30 min or until completely dissolved.
3) Cool and neutralise with 2 N HCl (1.5 ml).
4) Add the mixture to 7 ml of toluene: Triton X-100 (3 : 7) scintillant (Table 7.1) and shake thoroughly.
5) Count after standing in the dark for 1 hr and use a polar internal standard such as tritiated water.

If the pellet consists of a macromolecule (e.g. RNA, DNA or protein) which is soluble in saline-citrate solution and it is necessary to estimate both the total amount present and the radioactivity, a minimum volume of saline-citrate (0.15 M NaCl, 0.015 M Na citrate) can be used to dissolve the material and then the following method adopted depending on the type of macromolecule present.

1) Add 50 μl to a scintillation vial.
2) Add directly to each aliquot, 20 μl of either pancreatic RNase (500 μg/ml SSC) for RNA pellets (X hr); or DNase (500 μg/ml $MgCl_2$ buffer pH 6.4) for DNA pellets (2 hr) or Pronase (1 mg/ml phosphate buffer, pH 6.4) for protein pellets (2 hr).
3) Incubate at 37°C for the time indicated.
4) Cool, add 10 ml of phosphor DNXppE or a blended toluene-based scintillant such as TppE or TpC (Tables 4 and 5, Appx. II). Internal standard sample channels ratio or external standard ratio may be used for quench correction.

The level of protein present due to the enzyme, is usually too low to produce any significant heterogeneity in the system, and since all of the material being measured is usually in low molecular-weight species, it is usually counted as a homogenous system.

For pellets that are difficult to dissolve, such as cartilaginous material, bone residues etc., the procedure of Mahin and Lofberg (1966) can usually be used with success. For application to such pellets the following procedure is advised.

1) Pellet is drained thoroughly or preferably dried, if possible in the scintillation vial itself.
2) Add 0.2 ml of 60% perchloric acid to saturate the pellet.
3) Add 0.4 ml of 30% hydrogen peroxide (100 vol) to the wet pellet.
4) Close the vial and heat not higher than 80°C until dissolved.
5) Cool and add 15 ml of blended toluene scintillant (TpC 3.7, Table 4, Appx. II); use 2-ethoxyethanol as a blending solvent to bring the scintillant mixture into one phase, if this does not occur on the initial addition of the scintillant mixture. All the standard methods of quench correction can again be employed to determine the dpm.

9.2.2. *Caesium chloride gradients*

Several methods of assaying the label present in caesium chloride fractions have been described. The method that has been most commonly employed when DNA separations are being undertaken is to precipitate all or part of the fraction with trichloroacetic acid or perchloric acid under ice cold conditions, usually in the presence of a carrier of bovine serum albumin or of DNA itself (e.g. 50 to 100 μg) and filter either on glass fibre discs or on Millipore filters such as HAWP (0.45 μ pore size). The advantage of glass fibre discs have already been referred to (see § 2.2.1). The fact that they can usually be washed with solvents such as alcohol and ether, will assist in the speed of drying.

However, since an average gradient centrifugation often ends up with 40–50 fractions, a six-place centrifuge head may often necessitate the radioassay of some 250 to 300 fractions. In order to simplify the procedure, the following method is suggested based on the Triton X-100 : toluene system:

1) Fractions (10 to 12 drop) are collected in the usual way into small disposable plastic capped vials such as are available for blood sampling (total volume collected approx. 0.2 ml).
2) Water, 1.8 ml, is added to each vial, except those that will eventually be used to establish the shape of the refractive index (and hence density) of the gradient (e.g. every 10th tube in a 50 tube sequence).
3) After a UV absorbance reading at 260 nm or 280 nm, the contents of the vial are added to a Triton X-100 : toluene scintillant mixture (8 ml, 1 : 1, v/v containing 8 g PPO and 0.3 g POPOP per l) shaken thoroughly and allowed to stand for 1 hr before assaying.

For alkaline solutions, such as from alkaline caesium chloride gradients, 1.8 ml of a dilute HCl solution, sufficient to neutralise the amount of NaOH present in the gradient fraction is added to the vial, before admixture with the Triton X-100 : toluene scintillant as above.

In these colloid systems, the difference between the counting efficiency of the first and last samples from the gradient is usually negligible, and therefore it is unnecessary to make any gradient corrections. If accurate radioactivity measurements are required however, it is advisable to use internal standards of having a water-toluene partition comparable to that of the sample itself. In most cases water-soluble standards will usually suffice.

If the level of beta-emitter is high enough it may be convenient to take 10 to 20 μl of each sample and to assay using disc counting methods (see § 2.2.1). Using fraction splitting devices (§ 13.1), it is possible to collect an appropriate number of drops separately, either directly into the scintillation vials themselves or on to a strip of filter paper (Whatman 3 MM) as described for sucrose gradients (§ 9.2.3).

9.2.3. Sucrose gradients

The assessment of beta-emitters in sucrose gradient fractions is essentially the same as the problem in caesium chloride in that it is desirable either to dilute the sucrose levels to tolerable non-quenching levels or alternatively to remove the sucrose before counting. The final choice of the method will depend largely on whether all or part of each fraction can be used for assay. For example, in the radio-assay of DNA strand rejoining, first outlined by McGrath and Williams (1966), only a few labelled mammalian cells are used per tube (approx. 10^5) and insufficient DNA is obtained from the lysate to allow for the normal absorption measurements of the DNA to be made. The total fraction is then used for the radioassay of the isotope present. A simple technique for the fractionation of such a sucrose gradient was described by Carrier and Setlow (1971); a long strip of Whatman 3MM or 17 paper is used to collect the drops on premarked sections along its length. The strip is then submerged in ice-cold trichloracetic acid and after drying and cutting into squares is assayed in a non-polar scintillant such as Tpp4 (Table 3, Appx. II). Alternatively, the fractions are collected in separate test tubes in a fraction collector, and calf thymus DNA or bovine serum albumin (0.1 ml of a 1 mg/ml solution) is added to each tube as a carrier followed by ice-cold 5% trichloracetic acid; the tubes are left overnight (or at least 1 hr) at 2°C. The fine precipitates are then filtered on to glass fibre discs and assayed as described in § 2.2.1.

Occasionally it may be necessary to assess the level of labelled DNA in relation to the amount of labelled precursor also present, e.g. in the radioassay of pool to macromolecule ratios. The pool precursor molecules would sediment in the upper layers of the sucrose gradient and would normally be washed out by the above technique. The Triton X-100: toluene system is very suitable for the assay of the whole sucrose fraction and is carried out as follows:

1) 10 to 20 drop fractions of a 5 to 25% gradient are diluted to 5.0 ml by adding an appropriate volume (approx. 4.2 ml) of either distilled water for neutral gradients, or dilute HCl, calculated to neutralise the NaOH present for alkaline sucrose gradients. The total final volume must be 5.0 ml.
2) Add 5 ml of a Triton X-100 : toluene colloid system (2 : 3) and shake vigorously. Assay after allowing one hour to equilibrate (see § 2.2.3).

RNA and its precursors may be similarly assayed.

For RNA separations in sucrose gradients, a method of extraction of the molecule from the gradient was described by Paus (1970). This consists of extracting a solution of RNA (approx. 200 g/ml, approx. pH 6) with Soluene : pentanol (1 : 4, v/v) mixture from a solution. This procedure extracted about 95% of the RNA present. The phases separated spontaneously within 5 min to 6 hr depending on the concentration of buffers used. A 0.8 ml aliquot of the extracting layer was then added to 15 ml of a scintillant mixture such as DXpdEA or TpdA (Tables 4 and 5, Appx. II). Good counting efficiencies were claimed.

The commercially available BioSolv has also been employed for the measurement of RNA in fractions from a sucrose gradient centrifugation (McClendon et al. 1971). Sucrose gradients containing 5 to 40% sucrose in a buffer containing 0.1 ml NaCl, 0.1 ml sodium acetate and 1 mM EDTA adjusted to pH 5.0 were studied in this case. To work with such high concentrations of sucrose it is usually essential to add water to reduce its concentration. The procedure involves adding 1 ml of this solution to 10 ml of a non-blended toluene scintillant (Tb6) containing 10% v/v of Biosolv-3 (Table 6, Appx. II). As in the Triton X-100: toluene systems however, great care should be exercised when attempting to interpret the level of quenching in the system using the external standard ratio method.

9.3. Chromatography on solid supports

9.3.1. Thin-layer chromatography

The use of beta-emitters in thin layer chromatographic analysis has been widely used. The radioassay usually consists of drying the plate in the usual way and adding the scrapings from the plate directly into a non-polar scintillant such as Tpp4 (Table 3, Appx. II). The problems of assay by this technique are identical with those outlined in suspension counting methods (§ 2.2.2). Where possible, a scintillant should be chosen which actively extracts the labelled material from the scrapings, since a true solution is thereby achieved and all the usual quench correction techniques may then be applied. The efficiency of extraction can be readily determined by counting with the scrapings present and following centrifugation and washing of the scrapings, recounting.

Apart from the simple method of scraping the solid support with a spatula or more commonly with a strip of chamferred plastic, a number of other ingenious methods have been described. A useful pouter designed for this purpose is illustrated in Fig. 9.2b. The fraction is collected in the base of the pouter and assayed in a non-polar scintillant such as Tpp4. Alternatively, thin layer sheets with plastic base may be cut into segments with a scissors or a scalpel and the segments placed in the vial for counting in the usual way. An automated fraction scraper has been described by Snyder (1970) (Fig. 9.2a) and commercial instruments are now available.

Lipids do not present as great a problem as those materials which are not readily soluble in the toluene-based scintillant mixtures. Krichevsky and Malhotra (1970) have described a method for the analysis of neutral lipids on silica gel plates using a hexane: ether:

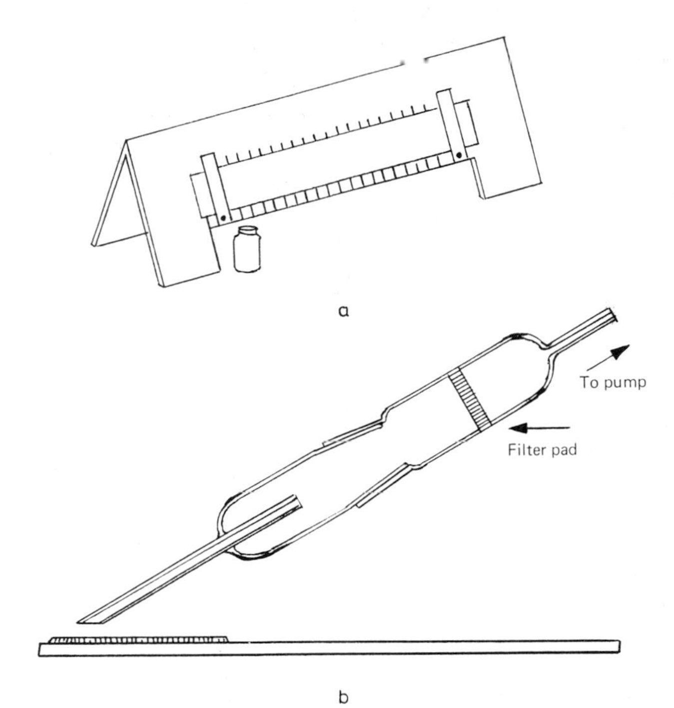

Fig. 9.2. Two devices that can be used for the removal of thin-layer surfaces for liquid scintillation counting. A) A simple construction made in plastic (Snyder 1970). The thin layer plate is held on the sloping surface by clips and a vial is placed below the section to be scraped with a spatula. B) A device for sucking thin-layer samples from a small area on to a filter pad placed on the glass sintered disc.

glacial acetic acid (90 : 10 : 1) mixture and using a colloidal scintillant in which to suspend the silica scrapings. A similar method for the separation of phospholipids was also suggested using $CHCl_3$: methyl alcohol : 7 M ammonia (230 : 90 : 15) and counting the scrapings as before. The scintillant not only extracts the lipid material from the solid support but also assists in suspending the silica particles themselves.

Material which is labelled with ^{32}P may be assayed directly on thin layer material by suspending in organic or aqueous media and counting in a scintillation counter. Counting efficiencies in toluene are approx. 80 %, and in water approx. 50 %. The parameters which determine the best counting conditions for this isotope were studied by Havilland and Bieber (1970).

An alternative method for assaying material absorbed to silica gel was described by Shaw et al. (1971) in which the gel itself was dissolved in hydrofluoric acid; a Triton X-100: toluene system was used for assaying both carbon-14 and tritium. The procedure is as follows:

1) The silica gel scrapings are added to polypropylene vials.
2) Add 0.3 ml of water and, whilst rotating the vial, 0.4 ml of hydrofluoric acid. The gel and its absorbed material dissolve.
3) Add 10 ml of a toluene: Triton X-100 (2:1) scintillant mixture (containing also 5.5 g PPO and 0.125 g diMePOPOP per l.

The maximum efficiency recorded for this method for tritium was 32 % and 87 % for carbon-14. Many stains that may have been used in the original plate for the location of the spots are also decolourised and degraded by this procedure, and thus the method has a potentially wide application. Useful reviews of techniques involved in the radioassay of thin layer chromatograms have been given by Snyder (1968, 1970).

9.3.2. Paper chromatography

The problems associated with the counting of beta-emitters on filter paper have already been discussed in some detail in connection with disc counting (see § 2.2.1). Additional difficulties associated with the counting of paper chromatogram strips are usually due to the presence of stains used to locate the spots in the first instance and of buffers and other solvents used in the process which may interfere with the counting.

It is often required to measure the exact proportion of beta-emitter present in different spots following a scanning procedure; and integration methods, based on the data achieved by this procedure, are usually tedious and unreliable. When assaying the proportion of label present in spots it is usually advisable to base the dissection of the strip on crude auxillary data derived from scans or autoradiography in order to achieve as clean a separation as possible between the spots and not to divide the activity present. When two spots are close together however, it is often an advantage to cut so that a small section of the paper separates the two, this acts as a blank and also confirms that there is an area of low activity separating the spots. However, caution should be exercised in interpreting data derived from this method of dissection of the chromatograms and it is better to dissect a strip into equal portions as often as possible, but at the same time attempting to ensure that a complete spot occurs within such a division.

In view of the variability experienced in counting different shapes and sizes of filter paper by a heterogenous method, solvent extraction of the material being assayed should as often as possible be attempted first, so that a homogenous system can be used for counting. However, failing this, the strips themselves should be washed free of associated salts where this is possible without removing the beta-emitter.

The problem of orientation of the filter paper experienced by earlier workers has been described in § 2.2 in relation to disc counting. These problems are of course, more acute when strips from filter paper chromatograms are used. The problem of colour quenching when dyes are present is particularly acute, and in such a case, a combustion procedure is probably the best means of assaying. The modification suggested by Baxter and Senoner (1964) of spraying the filter paper first with trinitrobenzene-1-sulphonic acid (2 mg TBNS Na in 100 ml dry acetone) not only helped to locate the spots but also considerably improved recoveries during the combustion of the paper chromatographic spots of amino acids. Combustion procedures also avoid problems associated with differential absorption of materials into the fibres of the cellulose itself and hence creating variation in the tritium count amongst different chromatogram spots.

For accurate work therefore, with paper chromatography, a combustion procedure should be employed, especially if the material being measured does not elute with a convenient solvent. Where, the sample is easily eluted, the strip may be clamped between two microscope slides as indicated in Fig. 9.3; the eluting solvent is allowed to soak into the wick and thence into the strip, delivering the eluate in a scintillation vial. In order to avoid excessive eluting solvent, this procedure is preferred to that of dipping the paper into the solvent.

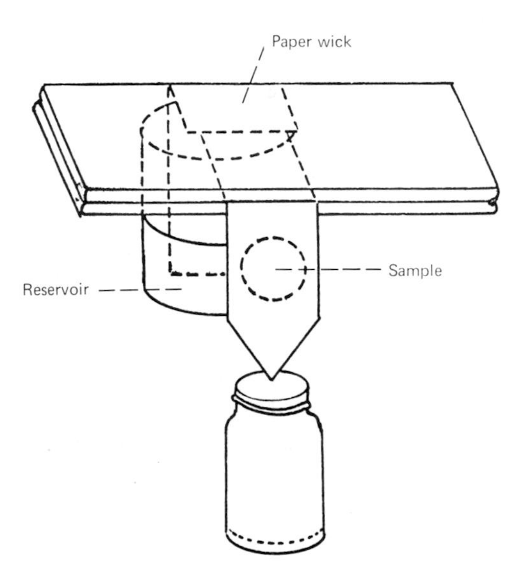

Fig. 9.3. The method of elution of spots on filter paper using the capillary action of a solvent from a reservoir; the eluate is collected directly from a strip clamped between two microscope slides.

Inorganic applications

Over 50 isotopes may be assayed with varying degrees of efficiency in the liquid scintillation spectrometer. The emission of light however, may be derived from a number of different causes and the varying proportion of these utilizable sources of energy within each isotope will determine the efficiency with which the isotope may be measured. Such energy sources include gamma, positive electron, alpha, electron capture events, Auger and Cerenkov light. The quenching actions of different solvents also vary depending on the nature of the energy source, and hence each isotope will behave in a characteristic manner from this point of view. However, for many of the isotopes, the main contribution will be derived from the beta emission contribution much as in the case of tritium and carbon-14. Variation in efficiency will occur depending on the mean energy of the beta emission.

Alpha-emitting isotopes have been studied in depth by Horrocks (1964). These include ^{217}At, ^{220}Rh, ^{232}Th, ^{233}U, ^{236}U, ^{236}Pu, ^{239}Pu and ^{242}Cm. In addition, Basson (1956) has studied ^{211}At.

Other isotopes that are predominantly beta-emitters and can be assayed by liquid scintillation include ^{35}S, ^{55}Fe, ^{63}Ni, ^{22}Na, ^{24}Na, ^{32}P, ^{60}Co, ^{90}Y, ^{106}Ru, ^{109}Cd-^{109}Ag, ^{95}Zr-^{95}Nb, ^{113}Sn-^{113m}In, ^{137}Cs-^{137m}Ba, ^{151}Sm, ^{241}Pu and ^{252}Cf and have been examined by Horrocks and Studier (1961), Horrocks (1963), Steyn (1966), and Ludwick (1960). Isotopes capable of being studied by this technique are listed in Table 1, Appx. I.

The usual object of any inorganic application involving metals is to

Subject index p. 309

convert the inorganic ion into a form which is soluble in non-polar scintillant mixtures, in order to create as far as possible the conditions required by a homogenous counting technique. This is particularly essential in many inorganic applications, since quench correction techniques are more sensitive to changes in the structure of the counting system.

The most usual procedure is to combine the metallic ion with an organic molecule to form a toluene-soluble ion-association complex. A colourless product is preferable but not always possible. A useful review of the complexing and solubilization methods for a large number of inorganic ions is by Dyer (1972).

An alternative procedure is to employ an organic salt of the ion, in which the lipophyllic character is dominant and hence assists the solubility into a toluene based system. Octanoate and *n*-caproate salts are particularly suitable for this purpose.

Provided the energy source is sufficiently penetrating, a suspension of a finely divided salt, held in gel suspension, can often be used and particular success has been achieved using this technique for those isotopes that can be concentrated by precipitation from very dilute solutions, as in the determination of ^{60}Co in environmental sources as the highly insoluble thio-cyanate complex (see chapter 12).

In this chapter, only those isotopes that have been studied in relation to biochemical and biological problems will be considered, and which have not been included in the remaining parts of the book.

10.1. Solvent extraction methods

Plutonium-241, which is produced from plutonium-239 by neutron capture in nuclear reactors, decays to americium-241 with a half life of 13.2 year and is accompanied by a predominantly β^- emission. Since these isotopes are being used in ever increasing quantities in nuclear reactors, possible contamination of biological systems become more and more likely. The detection of ^{241}Pu was early described by Horrocks and Studier (1958) who extracted the isotope in dibutyl phosphate and assayed the solution in a xylene-based liquid scintillation counting system. Using a similar procedure, Ludwick (1961) determined the isotope in urine to a minimum detectable level of 2.2 pCi using a 24 hr counting time.

As the level of β^- activity diminishes, the level of α activity will increase, since the half life of americium-241 is 458 years. By separating these two activities on the spectrometer and assessing the ratio at any given time, the time of origin of the isotope can be determined.

Similar solvent extraction methods are also used for an ingenious assessment of the level of thorium in an Yttrium solution. The technique consists of extraction of thorium, from the mixed solution of ions, by means of a solution of di-2-ethylhexylphosphoric acid (20% v/v) in a solution of p-terphenyl (3.0 g) and dimethyl POPOP (0.1 g) in toluene. By adding known amounts of thorium a straight line can plot the amount of activity extracted, as expected from the partition rule between immiscible solvents. When the curve is extrapolated to the ordinate, the amount of thorium in the original solution can thus be determined (Fig. 10.1). It is claimed (Kim and McInnis 1972) that amounts as low as 10 ppm can be assessed by this technique.

It is often not necessary to remove the aqueous layer from the scintillation vial when a scintillant extractant is used. For example, an aqueous sample containing trivalent transplutonium actinides can be extracted with a 0.01 M solution of an organic-soluble extractant dissolved in a Tpd-type scintillant mixture (Table 3, Appx. II). The organic extractant used is chosen to be most suitable for the appropriate ion. In the series described by McDowell (1972) 1-nonyldecylamine sulphate was most generally suitable, but di-(2-ethylhexyl) phosphoric acid is also useful in certain cases. The aqueous layer, after vigorous shaking, forms small droplets on the side of the vial but does not usually interfere with the counting efficiency.

This latter solvent can also be used for the extraction of ^{147}promethium from urine samples (Ludwick 1964). The method is as follows

1) Urine (250 ml) is treated with concentrated nitric acid (20 ml) in an Erlenmeyer flask and boiled to near dryness in a fume cupboard.

Subject index p. 309

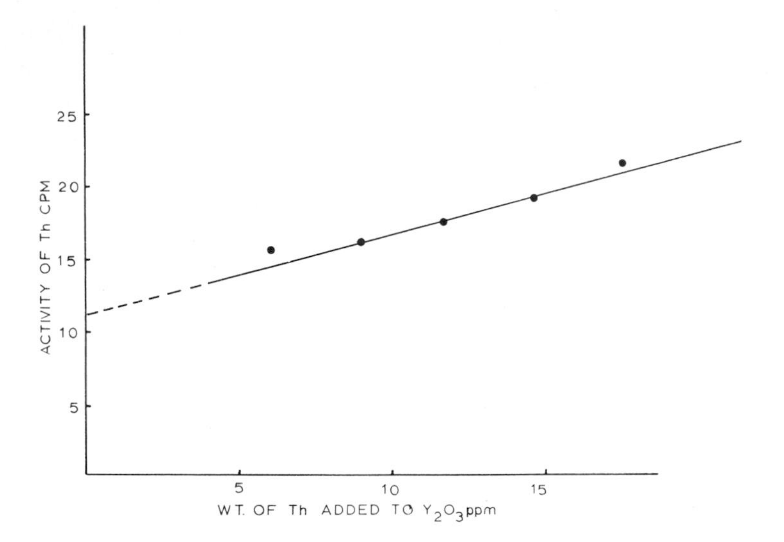

Fig. 10.1. Method of estimating the level of thorium present in a sample of yttrium oxide by solvent extraction following known amounts of added thorium and extrapolation to zero (Kim and McInnes 1972) from 'Organic Scintillators and Liquid Scintillation Counting' (Academic Press Inc.).

2) Several more additions of nitric acid are made and boiled off until the mixture is colourless.
3) The wet ash residue on completion of near drying procedure is dissolved in approx. 15 ml of dilute nitric acid and transferred to a 25 ml volumetric flask.
4) pH is brought to 3.4 with dilute NaOH, and the volume further adjusted with water.
5) The solution is extracted twice with 250 μl portions of 7% HDEHP, di(2-ethylhexyl) phosphoric acid, in Tpd4 scintillant (Table 3, Appx. II). The upper organic layer is removed with a micropipette. A final extraction with 100 μl of the HDEHP/Tpd4 mixture is made and the organic layers combined and transferred to a vial and assayed. Approximately 94% of the isotope can be removed by this procedure.

10.2. Precipitation and complex salt formation

The process of complex salt formation is closely allied to the solvent extraction procedures referred to in § 10.1 since the formation of a complex salt of the metal is probably required before it becomes soluble in the complexing agent itself. However, whereas in the latter case, the complexing agent is the solvent by which the complex is isolated, in the present system, the complex is a precipitate which is either insoluble and is assayed in a heterogenous system or is soluble in a different solvent and thus assayable as a homogenous system.

Both ^{241}Pu and ^{239}Pu have been assayed simultaneously in tissue, by co-precipitating the two elements as the ferriphosphate complex as follows (Eakins and Lally 1972).

1) The plutonium isotopes are isolated by acid elution from an anion exchange column, the eluate is evaporated and treated with concentrated acid and again evaporated.
2) The residue is dissolved in water in 10 ml centrifuge tube.
3) 2 mg of an iron salt carrier (ferric ammonium sulphate) is added.
4) Ammonia is then added to precipitate the ferric hydroxide.
5) The precipitate is washed with water and the residue is dissolved in 0.25 ml orthophosphoric acid.
6) Ethanol (containing ammonium chloride, 0.01 M in absolute ethanol) is then added to precipitate the ferriphosphate complex.
7) The precipitate is centrifuged and the residue is washed with ethanol.
8) The complex salt is then slurried with a dioxane-based scintillator together with Cab-O-Sil and assayed as a suspension.

The gain spectra for these two isotopes is shown in Fig. 10.2 and it can be seen that the separation that can be achieved is very good.

^{236}Pu and ^{238}Pu can also be assayed by this procedure and nearly 100% efficiency can be obtained with each of the isotopes. Some isotopes, e.g. ^{147}Promethium, behave very similarly to calcium; in one laboratory (Moghissi 1970), 10 mg praesodymium is used as carrier to co-precipitate this isotope as the oxalate. This salt may then

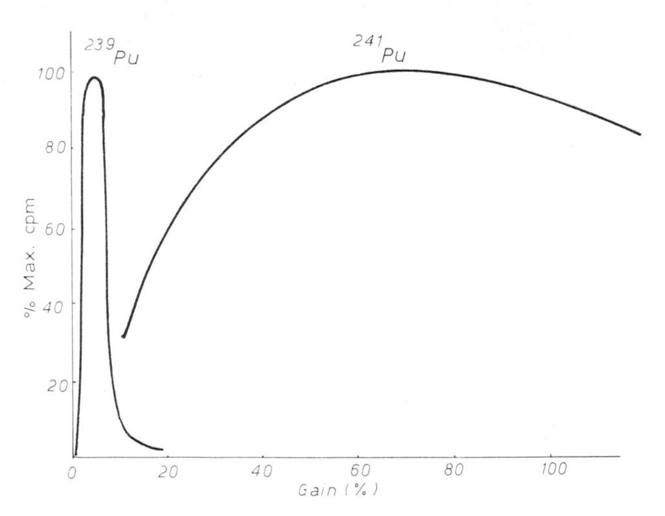

Fig. 10.2. Gain spectra of ^{239}Pu and ^{241}Pu (Eakins and Lally 1972). (From Liquid Scintillation Counting Vol. 2 Crook, M., Johnson, P. and Scales, B. eds., Heyden and Sons, Ltd.)

be dissolved in EDTA solution and assayed in a colloid system. Efficiencies as high as 85% are reported with Y values of 1.8 pCi/sample.

Chlorine-36, present in gaseous chlorine, has also been assayed by the conversion of chlorine to $SiCl_4$ which can then be added up to 55% by volume to a toluene-based scintillant mixture. Y values of from 0.25 to 0.4 pCi/g chlorine were recorded by Ronzani and Tamers (1966). However, the conversion of chlorine-36 to sodium chloride and precipitation with ethanol and drying allowed about 11 g of sodium chloride to be added to a vial and assayed in Tpp4 with an efficiency of 95%. Under optimal conditions, Moghissi (1970) recorded a Y value of 0.43 pCi/g Cl.

The use of stainless steel cladding to encase the enriched uranium oxide in advanced gas cooled reactors results in the exposure of nickel in the stainless steel to neutrons and the conversion to nickel-63. This isotope has a soft beta emission (E_{max} approx. 67 KeV) and a half life of 120 year. Liquid wastes from such reactors could result in the production of a potentially biologically important waste product which is difficult to detect. The isotope may however be complexed with pyridine and ammonium thiocyanate as tetra pyridine nickel dithiocyanate $[Ni(C_2H_5N)_4]$ $(CNS)_2$ which is soluble in a dioxane-based scintillant mixture (Harvey and Sutton 1970).

The use of chelates labelled with carbon-14 to estimate inactive metal ions of the transitional type has been described (Chiriboga 1962), and provided that a suitable selective precipitating or complexing agent can be found for an appropriate metal ion there appears a considerable scope for analysis of both radioactive and non-radioactive metal ions by this technique.

10.3. Lipophyllic salts

Many inorganic metallic ions form salts with long chain fatty acids, and these salts have the capacity of dissolving in polar scintillant mixtures and hence allow an otherwise insoluble material to be assayed in a homogenous system.

There are clearly two ways in which the method can be used. Either for the estimation of labelled metal ions by synthesising the fatty acid derivative or alternatively by labelling the fatty acid and assaying the metal salt as an indirect assay of the concentration of the metallic ion. The octoate and *n*-caproate salts have been frequently employed for this purpose. Amongst the ions that have been assayed are caesium-177/barium-137m (Horrocks 1964), rubidium-87 (Flynn and Glendennin 1959) samarium-147 (Wright et al. 1961), plutonium-239 (Flynn et al. 1964), and nickel-63, using the *n*-caproate salt (Gleit and Jumot 1961).

The estimation of fatty acids themselves can be made by using nickel-63 (Ho 1970). The free fatty acids in serum have been estimated by this procedure, and only 10 μl of serum need be used for the assay. The resulting nickel fatty acid salt is soluble in Bray's scintillant mixture which is used for the radioassay.

10.4. Noble gases

The measurement of low levels of ^{85}Kr in biological fluids and in gaseous phases is facilitated by the relative solubility of the gas in toluene-based scintillant mixtures. The ratio of the concentration of the noble gas in a toluene solution to that in the space above it also decreased as the temperature increases. Horrocks and Studier (1964) showed that at $-15°C$, the ratio for Xe = 5, Radon = 32 and Kr = 0.9. At 27°C, the ratio for Xe = 3 at atmospheric pressure.

The earlier method was to pre-evacuate the vial and inject the gas and scintillant. In the method adopted by Curtis et al. (1966) pressures are usually limited to 25 mm Hg or less, using a cathetometer to measure the exact volume added. The scintillator used in this case is Tpp7 (Table 3, Appx. II) and the counting efficiency is 92.5%.

An alternative, more sensitive, procedure is to pump the gas from a krypton container into the vial at approx. 600 mm Hg. Since the level of krypton in dry air is of the order of 1.14 ppm, a cubic metre of air will be required to produce 1.14 ml. The scintillant used may be either scintillator plastic shavings (20–40 mesh) or toluene-based scintillant (the plastic shavings are an inexpensive by-product of Pilot Chemicals Inc., Watertown, Mass., U.S.A.). In the first case, Sax et al. (1968) reported an efficiency of $94.4 \pm 0.7\%$ and a detection level of 1 pCi/m^3 of air sample. In the latter method, described by Moghissi (1970), an evacuated glass vial fitted with a luer stopcock is first filled with krypton to a pressure of 500 to 600 mm Hg. After filling, the stopcock is closed and by means of a 50 ml syringe, a toluene-based scintillant is allowed to enter the vial, dissolving the gas. The vial may almost be filled if the scintillant was previously efficiently deaerated. The krypton remains in solution, even though the vial is left open for a few minutes. The Y value by this procedure is reported to be 0.14 pCi/ml Kr.

Quench correction methods, multiple isotope counting and data evaluation

11.1. Quench correction methods

In all liquid scintillation counting, there is some degree of reduction of efficiency of counting due to an interference with energy transfer processes leading to the production of a photon or to absorption of the photon by coloured or other light-absorbing substances present. The different mechanisms by which the process of quenching may occur have already been described (§ 1.2). The skill of sample preparation lies in the ability to reduce this phenomenon to a minimum without increasing the effort to do so.

The alternative is to find methods of calculating the degree of quenching present under the conditions of counting found to be most suitable. For a single isotope, this procedure is normally straightforward. However it must be remembered that the only efficient internal standard is a standard sample of the beta-emitter being measured. In most cases, especially in homogenous systems, an alternative standard may be used such as labelled hexadecane or toluene. In heterogenous systems alternative standards should match closely the solubility and polar characteristics of the sample being measured. For multiple isotopes, the problem becomes more complex, but not impossible. The advent of computerised techniques for tedious repetitive calculation of raw data, has led to the incorporation of this facility 'off' or 'on' line to many instruments, hence simplifying such calculations.

Subject index p. 309

The influence of quenching is felt to the greatest extent with the weaker beta-emitters, such as tritium and to a lesser extent with carbon-14 and sulphur-35. It is therefore useful to work out methods for minimal quenching for tritium in order to achieve the best results for other isotopes. In colloid systems this may not necessarily correlate however. In biochemical experiments involving uptake of labelled macromolecular precursors, it is usually better to use the more expensive carbon-14 labelled material in heavily quenched samples, although a costing based on the efficiency of the sample preparation method, usually repays consideration at an early stage of the work.

The most direct method for the determination of the degree of quenching in a sample, is by the use of an internal standard. This technique is probably the least vulnerable to hidden errors which may not be recognised in the final computation of absolute radioactivity. However the method is very tedious where large numbers of samples are involved. The sample channels ratio (SCR) method, initially described by Baillie (1960) and subsequently exploited extensively, has gained considerable populariy. The external standard, a gamma source which induces Compton electrons within the vial, can be used either as an external equivalent of the internal standard or as a convenient source with which to measure a channels ratio, when it is then known as an external standard ratio (ESR) technique.

The different methods have individual advantages and disadvantages and will be separately discussed. The ultimate choice of the quench correction method used will depend largely on a) the number of dpm's expected b) the nature of the beta emitting sample material c) the nature of the scintillation mixture used and d) the number of samples to be assayed. As a general guide line, if the counts present in a homogenous counting system are high, e.g. > 5000 cpm, the sample channels ratio (SCR) will give accuracy without the requirement of any additional handling procedures except the computation of the existing data. With less than 5000 cpm, the external standard ratio is most valuable provided only that the system is strictly homogenous. For a heterogenous system the internal standard or in some cases the channels ratio methods are the only reliable procedures.

Different quench correction procedures are reviewed and compared by Peng (1970) and Rogers and Moran (1966). The errors involved in data processing by these procedures will be briefly outlined in § 11.3.1.

11.1.1. Internal standard method

Homogenous counting techniques

The most frequently used standards employed in toluene-based scintillant mixtures are tritiated and 14-carbon labelled hexadecane. Although for precise absolute assessments of quenching, hexadecane is probably the best material to use, it suffers the disadvantage of having a melting point around ambient temperature and is difficult either to weigh or measure by volume. The use of tritiated toluene is now becoming more popular, since it is considerably cheaper than hexadecane, and for routine use in multiple samples, it is easier to dispense or 'spike'. A 25 μCi/ml standard solution of tritiated toluene can be made by diluting the high specific activity tritiated toluene obtainable commercially, with purified toluene. Thus 10 μl of this solution, using a fine bore, bulb pipette will then add about 500,000 dpm for tritium standard. The temperature dependence of toluene volume is shown in Table 11.1 but this factor is not very important at room temperature, especially if only comparative data are required. Between 18 and 26°C, a variation of $\pm$ 0.43% in toluene and of $\pm$ 0.08% of water may be expected. An alternative method of dispensing has been described by Thomas et al. (1965) based on a mechanical push-button device for the 'Hamilton' microsyringe; the reproducibility is said to be better than 1%. The calibration of a National Bureau of Standards Tritiated Toluene was described by Garfinkel et al. (1965).

A useful, comparable carbon-14 standard was suggested by Marlow and Medlock (1960) which consisted of a solution of benzoic acid-7-^{14}C in toluene. However, ^{14}C-toluene is now readily available, and a similar standard to that of ^{3}H-toluene described above may be made and dispensed in a similar manner.

Although in toluene-rich scintillant mixtures, a toluene or similar

TABLE 11.1

To correct the internal standard of the toluene or water for temperature, multiply the dpm per 10 μl at 22°C by the appropriate factor to obtain the true dpm at the temperature of counting.

Temp.	Toluene		Water	
°C	density (g/ml)	factor	density (g/ml)	factor
10	0.87489	1.0128	0.99973	1.0019
12	0.87305	1.0107	0.99952	1.0017
14	0.87120	1.0086	0.99927	1.0015
16	0.86935	1.0064	0.99897	1.0011
18	0.86750	1.0043	0.99862	1.0008
20	0.86564	1.0021	0.99823	1.0004
22	0.86379	1.0000	0.99780	1.0000
24	0.86193	0.9978	0.99732	0.9995
26	0.86007	0.9957	0.99681	0.9990
28	0.85821	0.9935	0.99626	0.9985
30	0.85635	0.9914	0.99567	0.9979
32	0.85448	0.9892	0.99505	0.9972
34	0.85261	0.9870	0.99440	0.9966

standard can be used, a better standard in heavily water-quenched systems is either tritiated water itself or a water-soluble tritiated material. Tritiated water is used successfully, 10 μl of a solution of 25 μCi/ml being of the right order of magnitude of specific activity to provide a quick assessment of quenching in the sample. An aqueous carbon-14 standard however, presents somewhat of a problem, but in the authors hands, a solution of oxalic acid in water, containing a droplet of chloroform has lasted for considerable time. The oxalic acid concentration should be high enough to prevent bacterial growth, e.g. near saturated conditions, 9–10 g/100 ml. Sodium ^{14}C bicarbonate solutions can also be used, but care must be taken to ensure that the pH of the sample is alkaline before addition. The use of ^{14}C as a standard for the assessment of the quenching of ^{35}S in solution has been recommended by Buckley (1971).

The calculation of dpm from the sample counts per minute using the internal standardization method is given by eq. (11.1)

$$\text{dpm sample} = \frac{N(n_1 - bkd)}{n_2 - n_1} \qquad (11.1)$$

where n_1 = cpm of sample alone; n_2 = cpm of sample + standard added; N = dpm of standard added.

It is to be noted that it is not necessary to subtract background counts (*bkd*) from each of the cpm values in the denominator, since this will automatically be removed in taking the difference.

The method described above is suitable for most homogenous counting systems involving those isotopes that are likely to be encountered.

Heterogenous counting systems

A two-phase counting system, where the sample is present in a different phase from that in which the scintillator itself is dissolved, presents a greater problem in the exact measurement of the quenching present. In the case of filter disc methods (see § 2.2.1), if the material containing the beta-emitter is retained on the disc during the counting, then for carbon-14 any similar material of known specific activity will probably be an adequate standard for the measurement of quenching. Whatever is used as a standard however must closely resemble the solubility properties of the beta emitting sample itself. For a material like DNA, which may be measured as a thin film on the surface of these discs, it is usually advisable to combust or solubilize a small number of the samples after placing on the disc so that an assessment can be made of the relative degree of quenching under the conditions of the experiment itself, and provided the variation between the selected samples is small enough, the assessment of the counting efficiency can usually be extrapolated for the remaining samples without any further checking. It must be carefully assessed whether or not the beta-emitter is or is not dissolved from the disc by the scintillant solution itself. If the disc is removed and the counts appear in the supernatant scintillation fluid, then it can be assumed that solution has occurred and that the internal standards can be used as in the homogenous system described above.

In the case of suspension counting, the same principle applies. If a good comparative value of dpm is needed from the result of multiple suspension assays, a few should be assayed by an alternative homogenous counting procedure in order that an accurate assessment can be made for quenching. In the case of colloid counting systems, provided the standard has similar water-toluene partitioning characteristics to the sample material itself, the internal standards technique is probably the most accurate way of assessing the quenching in this system.

A glass rod coated with ^{14}C-labelled barium carbonate as a reusable quench correcting device has been suggested (Dobbs 1963), but its use has found limited application and its suggested use also for tritium correction has been criticised (Bloom 1963).

11.1.2. Sample channels ratio technique

This method of quench correction relies on the fact that the pulse height spectrum is depressed to lower values during quenching. Thus, since the instrument has a fixed window system, less pulses will appear in the higher window settings and more will be observed in the lower. The ratio of pulses in the upper windows to those in the lower will thus be related to the level of quenching activity present.

In its simplest form, a series of quenched samples are measured simultaneously in two different channels which may or may not be overlapping. The ratios of the counts in these two channels are plotted against the known counting efficiency determined by internal standardization. This curve is then used to determine the efficiency by extrapolating from the channels ratio of the sample counts in the two channels selected. The method has the advantage that in a multichannel instrument, all the information required to compute quenching is obtained in the single counting period. The chief disadvantage of the procedure is the fact that many samples have insufficient counts present to allow for a good statistical level of counting and thus to provide an accurate enough ratio figure in a reasonable time. However, in those conditions where sufficient counts are present in the sample, the method is a particularly good one, especially where computational facilities are available on the instrument itself. Another useful advantage is the fact that the sample itself could be recovered if necessary to carry out repeat counts or pulse height analysis without contamination by excessive counts due to added internal standard. Furthermore, the method is not subject to pipetting errors, as is the case with the internal standards measurement.

Much of the subsequent work in this field has been concerned with a suitable choice of window settings for the evaluation of a series of quenched samples whilst at the same time retaining as many of the counts available as possible to allow for good statistical counting conditions. Clearly, overlapping windows have the advantage that more counts will be observed in each window than would have been the case had they been adjoining or separated. However, the sensitivity to quenching may not be as great.

When constructing a quench curve, it is often worthwhile using a set of standards whose count level and degree of quenching is of the same order as that used in samples of unknown quench, since the parameters involved in determining the calibration curve will then be similar to those involved in the assay itself. For routine use, a set of quenched standards for periodic checking of the channels ratios is often useful.

For unblended toluene-based scintillants, small increments of carbon tetrachloride, chloroform or acetone will produce large quenching effects (Table 7.4). The primary alcohols, such as methanol, ethanol and particularly 2-ethoxyethyl alcohol, will produce less quenching per aliquot and in certain instances may be more useful. For dioxane-napthalene scintillant mixtures or blended toluene mixtures, water or alcohol are suitable quenchers. For very efficient quenching of aqueous solutions, 5–10% trichloracetic acid can be used. Hyamine hydroxide can be used as a quenching agent in either blended or non-blended scintillants.

A number of dyes can be used to simulate biological colours, such as Sudan and Methyl Red as a colour simulant for haemin. If such colour-quenched samples are to be measured in any number, it is often possible to relate the optical density at a given wavelength to counting efficiency. An example of this procedure is in the assay of hyamine hydroxide solutions of pigmented hair (Bersagnes 1963) where a simple assessment of the absorbance at 430 nm can be related to the counting efficiency of the sample. This value of 430 nm is a compromise between the maximum absorption peak of the colour and the sensitivity peak of the average photomultiplier system. The optical density is linearly related to quenching up to 50% quench.

Construction of a sample channels ratio calibration curve (SCR) (adapted from Bush 1963)

1) The scintillant mixture to be used (10 ml) is placed into each of eight scintillation vials.
2) For a) non-blended toluene scintillant, add the following volumes (ml) of acetone: 0, 0.1, 0.2, 0.5, 0.75, 1.0, 1.5, 2.0.
   or b) blended toluene or dioxane-naphthalene scintillant, add either volumes of water, or (better) unlabelled sample in 7 or 8 steps, up to (but not including) the point at which phase separation occurs.
3) With a full window setting, check that the vials are not contaminated.
4) Set the instrument to amplification or gain settings appropriate for the isotope to be measured, using an unquenched standard in the appropriate scintillant (see instrument manufacturers information).
5) Put in an unquenched 'background' sample (i.e. not containing any radio-activity). Set the upper discriminator of both channels to the top value on the instrument (infinity on some instruments). Set the lower discriminator of both channels (channels A and B) to a point (D_1) where the background (dotted curve) does not interfere. (see Fig. 11.1).
6) Remove the reference background sample and run through all the samples of the quenched series to assess background levels (5 min counts are sufficient at this stage).
   Background counts should be insignificant compared with the 2

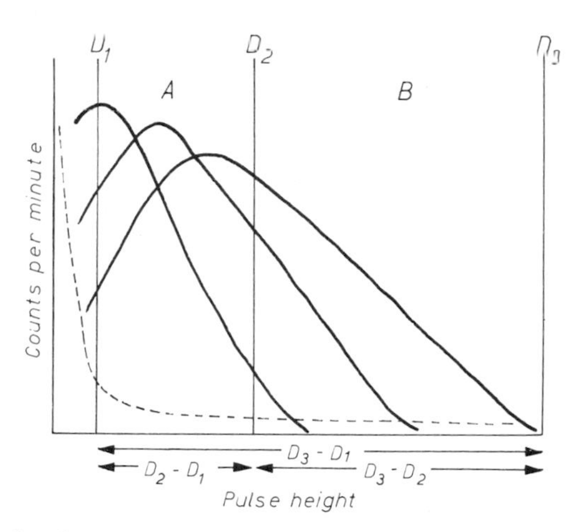

Fig. 11.1. A series of quenched pulse height spectra and the position of three discriminators to create two non-overlapping windows.

to 5×10^5 dpm expected from the standard and can usually be ignored for this purpose.

7) An aliquot of the standard is carefully pipetted into each vial, e.g. 10 μl of ^{3}H-toluene (25 μCi/ml), or ^{14}C-toluene (10 μCi/ml). Shake.
8) Put in the unquenched sample containing the standard and proceed as follows depending on the isotope.

TRITIUM:

a) lower the upper discriminator of channel B to a point D_3 so that any further decrease would result in a significant loss of cpm.
b) raise the lower discriminator of channel B to a point where the cpm are approx. 3/5 of the original cpm (D_2).
c) lower the upper discriminator in channel A to this same value.

The instrument is now set so that the ratio of channel B/channel A counts is approx. 1:5, so that as the scintillant is quenched, the value will fall through 1.0 to fractional values. This value of unity will thus fall within the region expected of the quenched samples and offers the least statistical uncertainty for a given number of total counts. A calibration curve as shown in Fig. 11.2 may be drawn from which efficiencies for tritium may be derived.

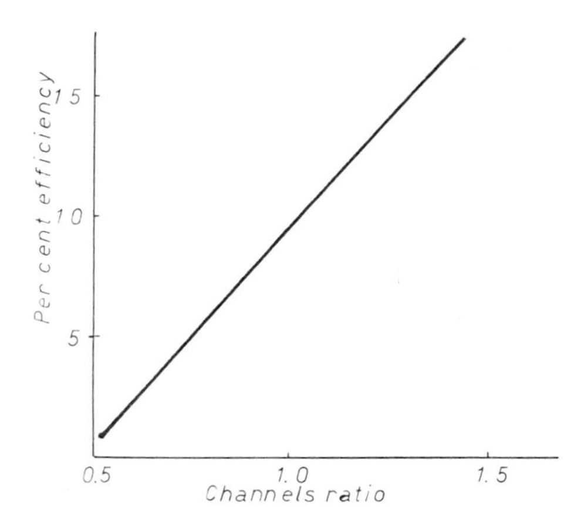

Fig. 11.2. A channels ratio quench correction curve for tritium (see text).

CARBON-14; SULPHUR-35

a) lower the upper discriminator of channel B to a point D_3 where any further decrease will result in a significant decrease in the count rate.

b) lower the upper discriminator of channel A to a point D_2 where the level of counts is approx. $\frac{1}{4}$ of the original counts.

The instrument is now set so that the ratio of channel A/channel B counts is approx. 0.25, and as the scintillant is quenched and the efficiency falls, this value will increase in a non-linear manner (Fig. 11.3) and from which the sample efficiency may be extrapolated.

Bush (1963) demonstrated that the same quench calibration curves could be used for a wide variety of quenching agents, including coloured materials. For the usual laboratory requirements, this has been shown to be substantially correct, however, as indicated in chapter 1 there are differences between quenching agents which have been the subject of several fundamental studies on the general mechanism of the quench process.

Carbon-14 quenching is much less affected by temperature differences that tritium. The pulse height spectrum of tritium decreases as

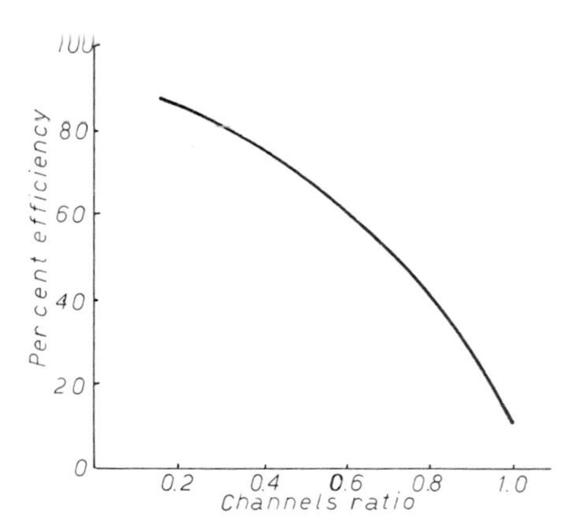

Fig. 11.3. A channels ratio quench correction curve for ^{14}C (see text).

the temperature increases linearly between +23°C and −25°C. As the quencher concentration increases, the relative increase in counting efficiency obtainable on cooling is greater (Kaczmarczyk and Ruge 1969).

The background counts can be considered to be due to two main groups of components, some of which quench in the same way as above, and the rest, due to instrument and thermal noise do not. Thus at extreme quenching conditions, the background level of counts will approach that of the instrument and the thermal noise component.

A further refinement of the sample channels ratio technique was described by Caddock et al. (1967), where the ratio of square roots of the counts rate in each channel is taken rather than the ratio of the count rate itself. This modification allows for an evaluation of the precision of the value of dpm obtained. From Fig. 11.1, the net counts (less background) can be fitted to eq. (11.2) (Channel A = $D_2 - D_1$, Channel B = $D_3 - D_2$).

$$\sqrt{A} = \alpha + \beta\sqrt{B} + \gamma B \tag{11.2}$$

where α, β, and γ are calibrated constants. A typical calibration curve derived from a set of quenched standards is shown in Fig. 11.4. The

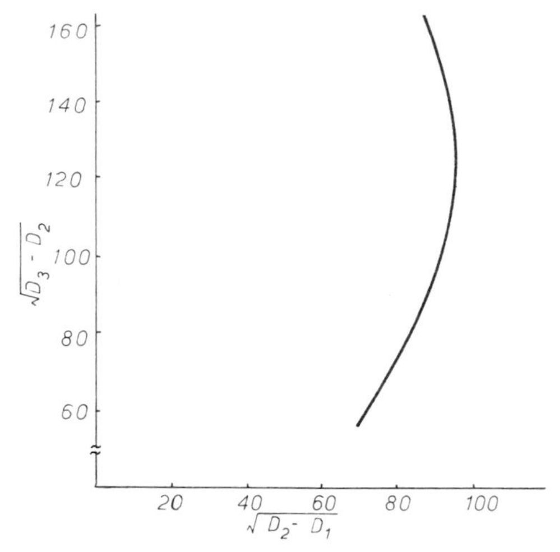

Fig. 11.4. The method of channels ratio quench correction devised by Caddock et al. (1967). (Copyright permission, Pergamon Press.)

counts observed in channel A will then be in error from the calibrated curve by $\sqrt{A} = (\alpha + \beta\sqrt{B} + \gamma B)$.

The Channels ratio method may also be used successfully in estimating the level of colour quenching when Cerenkov emitters are being assayed. The settings for an isotope such as ^{42}K are similar to those of carbon-14 procedure described above. In a 1000 scale unit instrument, the lower discriminators of each channel are set to 100 and the upper discriminators of Channel A are set to 300 and that of B to 1000. An amplification factor of 25% can be used and the sample is diluted until the colour intensity causes the ratio of A/B to fall within the range of 0.35 to 0.55, being the most linear region of the efficiency channels ratio curve (Moir 1971). Alternatively, a split window similar to the tritium isotope setting can be adopted with the upper discriminator setting to 1000 and that of A being set so that the ratio of B/A was approx. 1.0 at an amplification of 40% (Kamp and Blanchard 1971).

A comparison of the efficiencies of Cerenkov counting in various solvents is given in Table 2.2.

Balance point quench correction technique (*Ross 1964*)
As the pulse height spectrum shifts to lower energy levels during quenching, the number of pulses increase in the lower windows and decrease in the upper ones. There is thus a region where the curves intersect or nearly so. If the upper and lower discriminators are thus positioned on either side of these cross over points (Fig. 11.5), then

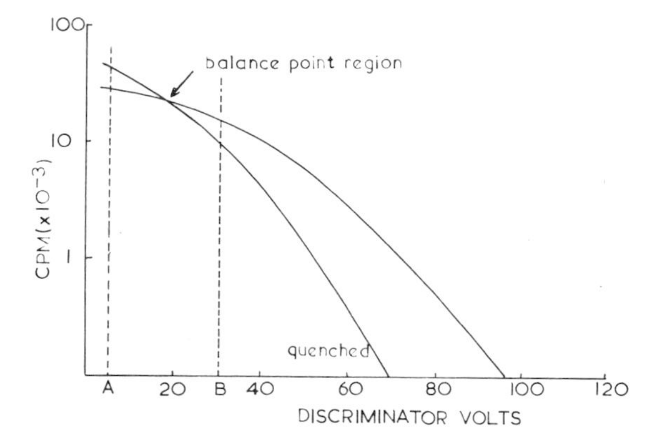

Fig. 11.5. The balance point quench correction method of Ross (1964). The discriminators are placed on either side of the balance point region so that the counts lost above it by quenching are regained below it, ensuring that constant counts remain within the window.

counts lost in the upper half of the window caused by quenching, will be compensated by a gain in the lower half of the window, and hence the total number within the window stays constant. The effectiveness of this method clearly depends on the accuracy with which the user can locate the region of the cross-over points, and can then prove that the window chosen will retain the same level of counts over a range of samples quenched within defined limits.

This method relies on sufficient counts being present to allow a relatively narrow window to be used. Furthermore, the selection of the correct window width is essential for the success of the technique. The window width chosen is the result of a compromise between being wide enough to have sufficient counts to be statistically valid and narrow enough to equate gains and losses on either side of the cross-over region. As the window width around the balance point is increased, the ideal centre voltage setting slowly shifts to higher values; by a systematic increase in window width, a suitable compromise between adequate counts and lack of variability is achieved. Efficiencies in the region of 55–57% for carbon-14 can be obtained in such a window.

11.1.3. External standardization

By 1964, facilities became available on commercial instruments to undertake quench correction by means of external standardization. This technique (outlined in 1962 by Higashimura et al.) consists in bringing a small pellet of a gamma-emitter to the base or the sides of the scintillation vial and assaying the level of Compton electrons produced within the scintillant in the vial. There is also some contribution from the vial wall itself, and this has been calculated to be about 2% of the total when a vial of approx. 120 ml is used.

The Compton spectrum is complex and results from a variety of recoil and conversion electrons generated within the scintillant. The number of these electrons produced will depend on the stopping power of the scintillant components and the vial wall thickness. The absolute level of counts derived from this procedure will therefore be very sensitive to differences in vial geometry and thickness and to the volume of the scintillant used. These factors were studied in detail by De Wachter and Fiers (1966). By measuring the absolute level of counts in a channel setting, using a method similar to the balanced quench correction method (§ 11.1.2) a linear relationship between the log efficiency and the log quencher concentration can be obtained which follows the equation

$$\log E = f \log N + E \qquad (11.3)$$

where E is efficiency, N quench concentration, f and E are constants which depend on the instrument settings and the nature of the isotope used.

However, by measuring of the counts from an external standard source in two channels and assaying the level of quench as a channels ratio similar to the sample channels ratio described for ^{14}C (§ 11.1.2), a value less susceptible to geometry and scintillation volume changes is obtained (External standard ratio method ESR). This ratio is now the most commonly employed form of quench correction, simply because of its ease of operation on automated equipment and because of its independence of the sample count size, ensuring therefore that the statistics of counting is maintained constant between samples (Fig. 11.6).

When assessing published data, it is useful to know how the external standard is sited with respect to the sample during the assay. In some older instruments, a ^{133}Ba source was employed beneath the vial. In

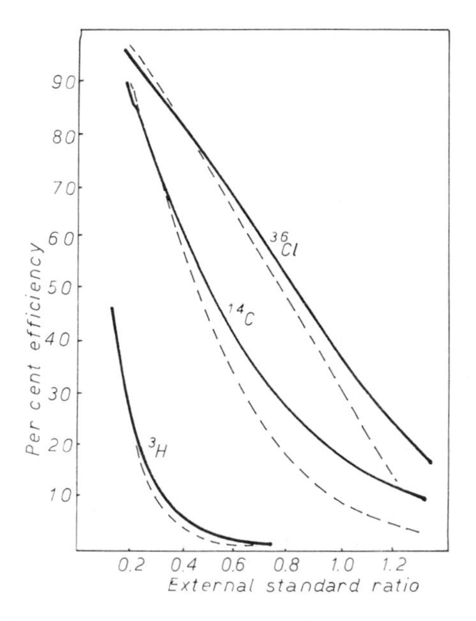

Fig. 11.6. A plot of counting efficiency against a ^{137}Cs external standard source (placed outside the freezer cabinet, approx. 2 mCi), using chloroform ——— and methyl red – – – – – – – as quenching agents. In both cases a second channel was set at 100–250 at 10% amplification and a third channel at 50–1000 at 3% amplification (Takahashi and Blanchard 1970).

Subject index p. 309

the measurement of the degree of quenching using thin-layer scrapings or paper discs, ESR may be used provided the sample is completely eluted in the vial from the absorbant. It is usually convenient to leave the absorbant at the bottom of the vial, but an external standard which is sited at the base may lose some of the gamma emission by absorption in the silica gel layer. Values from a side-sited external source would be less affected by factors of this type (Houx 1969).

Assuming that no contamination of the vials occur, background values can often be more accurately determined by the use of the external standard ratio (Takahashi and Blanchard 1969). A single set of accurate measurements is required at different levels of quenching, and a curve of external standard ratio against background cpm is drawn, and the background levels are then taken from the external standard value. This is a very useful procedure provided that no vial contamination had occurred; it is probable that a quick run through the vials containing the scintillant ready for addition of samples would pick up any gross contamination, and the more accurate background count is then obtained from the external standards ratio which can be more accurately counted. This is a particularly suitable method where there are coloured impurities in the sample and where the window size is large enough to allow a relatively high background counting rate.

An interesting report by Rauschenbach and Simon (1971) suggested that ^{137}Cs or ^{133}Ba external sources gave external standard ratio values for scintillant mixtures in polyethylene vials which decreased over the first 24 hr, whereas the actual cpm and channels ratio values remained the same during the same period. Glass vials or the use of a ^{226}Ra/^{241}Am source was said not to produce such a decrease. This observation needs further investigation, under different instrument and sample preparation conditions.

11.2. Multiple isotope counting

One of the most important aspects of the modern scintillation spectrometer is its ability to measure pulse heights within discriminator settings which can be varied over a wide range of values. This enables suitable setting to be made to separate overlapping pulse-height spectra. This ability to assay a number of isotopes simultaneously allows concurrent or consecutive biochemical processes to be studied. Two chemically different precursors labelled with different isotopes (e.g. tritium and carbon-14) are sometimes used. These may be administered together or separately in time. Alternatively the same precursor at one time labelled with tritium and at another time with carbon-14 may be used to study temporal relationships in biochemical systems. It is fortunate that many of the isotopes that are used in this way are relatively easily separable on the spectrometer. The only serious problem is encountered with carbon-14 and sulphur-35 mixtures where the spectra are very closely similar. In this case it is necessary to degrade the material to its oxidation products and chemically separate the two isotopes before assaying.

A homogenous counting system should be aimed for as often as possible. Serious problems of differential absorption can arise in heterogenous systems. It is a considerable advantage to be able to use the external standard ratio in counting, and the homogenous systems alone allow this to be done.

The pulse height spectrum of tritium decreases as the temperature of the counting system increases. This relationship is essentially linear between +23°C and −25°C (Kaczmarczyk and Ruge 1969). However as the quencher concentration increases, the increase in counting efficiency on cooling is greater. Thus for mixed isotopes, especially those mixtures containing tritium, cooling the system does assist in the separation.

11.2.1. Two beta-emitters (e.g. ^{14}C and ^{3}H)

It is of some importance that one should consider at the outset of a mixed isotope experiment, the technique intended for the preparation of samples for liquid scintillation counting. So often, in the excitement of the biochemical experiment itself, one ends up with a product for counting which destroys the advantage of using two isotopes, simply because one cannot separate them in the final assay.

In all cases of mixed isotope assessment, it is essential to use a homogenous counting system for accurate isotope separation. A blended toluene or dioxane-naphthalene scintillant should wherever possible be used. Heterogenous systems, although useful to provide qualitative and rough data, are not suitable for accurate assays.

Isotope exclusion method

The pulse height spectra of the two isotopes, tritium and carbon-14 are normally easily separable, and there are several ways in which their separate determination has been simplified. The most obvious method as can be seen from Fig. 11.7 is to place the lower discriminator of the upper window at a point which excludes most of the tritium counts (D_2). In an amplification situation where the carbon-14 spectrum fills the instrument full window setting, the tritium spectrum will occupy only a relatively small proportion of the whole. By careful choice of the discriminator level, i.e. by moving it down to a position

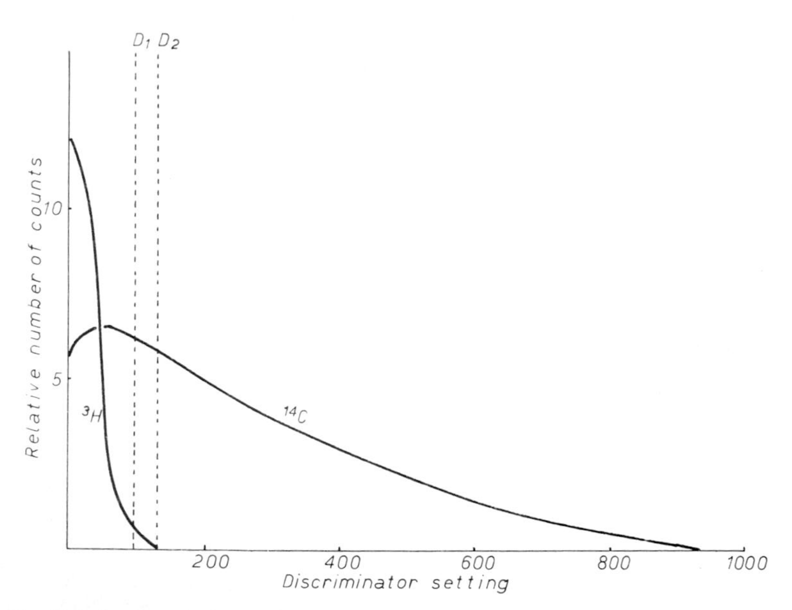

Fig. 11.7. Placing the lower discriminator of the upper window to exclude the tritium spectrum is the simplest method of separating the tritium from the ^{14}C counts (see text).

D_1, one can allow small losses of tritium counts in the lower window, but at the same time allow for much greater numbers of carbon counts to be measured in the upper window. Table 11.2 (from Kobayashi and Maudsley 1970) shows the effect.

TABLE 11.2

Effect of increasing tritium spectrum spillover into ^{14}C counting channel on ^{14}C counting efficiency.

Tritium counting efficiency lost from the lower window (%)	Carbon-14 counting efficiency gained in the upper window (%)
0	50.9
0.01	57.5
0.05	62.5
0.1	65.0
0.5	70.3
1.0	72.5

The procedure is as follows (refer to Fig. 11.1)

1) Set the amplification and discriminators D_1 and D_3 to cover the whole tritium pulse height spectrum.
2) Starting with D_2 at a point about 10 divisions (in a 1000 division scale) below D_3, measure the counts due to tritium and those due to carbon-14 (or sulphur-35) in both channels $D_2 - D_1$ and $D_3 - D_2$, using separate standard samples.
3) Gradually increase $D_3 - D_2$ by steps of 10 or more divisions until $D_3 - D_2$ is equal to $D_3 - D_1$. The channel $D_3 - D_2$ will represent the carbon spill over.
4) Plot the efficiency of tritium in $D_3 - D_2$ against the carbon-14 (or sulphur-35) efficiency in the same channel and obtain the curve illustrated in Fig. 11.8. The tangent drawn from the origin will depart from the curve at the best setting (i.e. 26% for tritium and 5% for carbon-14 in the illustrated example).

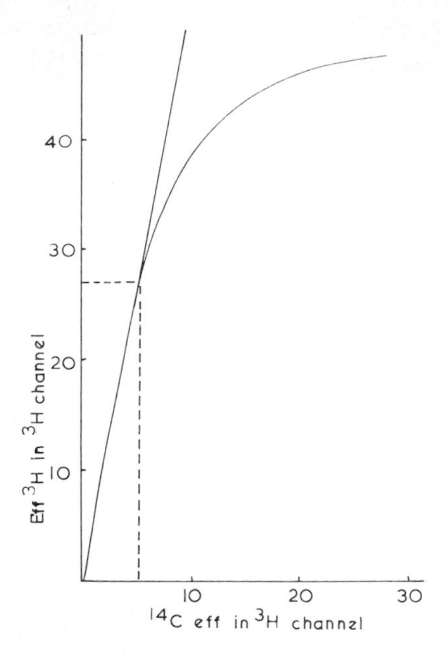

Fig. 11.8. A method for determining the optimal conditions for separation of counts from ^{14}C and tritium. A tangent drawn from the origin on to the curve, leaves the curve at a point where the relative difference in counting efficiency of tritium in the tritium channel from the overlapped ^{14}C counts in the same channel is greatest.

Simultaneous equations. The calculation of the relative contribution of two or more isotopes may be simply made for any window settings by the use of simultaneous equations. The procedure was outlined for gamma-emitters by Gilbert (1960), but has been adopted by many authors for applications to liquid scintillation counting.

Let E_c^A and E_c^B be the efficiencies of the carbon-14 source of N_c total dpm in channels A and B respectively, and E_h^A and E_h^B, those corresponding to tritium (N_h total dpm) in the channels same. The count rate observed in each channel will be n^A and n^B respectively.

$$\text{where } n^A = N_c \times E_c^A + N_h \times E_h^A \qquad (11.4)$$

$$n^B = N_c \times E_c^B + N_h \times E_h^B \qquad (11.5)$$

Thus the total carbon-14 counts N_c

$$N_c = \frac{n^A \quad n^B\left(\frac{E_h^A}{E_h^B}\right)}{E_c^A - E_c^B\left(\frac{E_h^A}{E_c^A}\right)}$$

$$N_h = \frac{n^B - n^A\left(\frac{E_c^B}{E_c^A}\right)}{E_h^B - E_h^A\left(\frac{E_c^B}{E_c^A}\right)} \tag{11.6}$$

It is usually possible to place the discriminator D_2 (Fig. 11.1) in a position which excludes all the tritium counts from the upper window (B) i.e. $D_3 - D_2$ and in this case

$$N_h E_h^B = 0 \tag{11.7}$$

thus $n^B = N_c E_c^B$

$$\text{or } N_c = \frac{n^B}{E_c^B} \tag{11.8}$$

By solving eqs. (11.2) and (11.6)

$$N_h = \frac{n^A - n^B\left(\frac{E_c^A}{E_c^B}\right)}{E_h^A} \tag{11.9}$$

It is obvious however, that this procedure may only be adopted for a number of mixed isotope samples where the level of quenching is similar.

A method for the simultaneous assay of tritium and sulphur-35 under conditions of variable quenching was described by Weltman and Talmage (1963). Essentially similar methods of isotope exclusion were used, but in this case the level of quenching was determined by employing two channels within the upper window to measure the sample channels ratio (i.e. B_1/B_2, Fig. 11.9). There is no fundamental

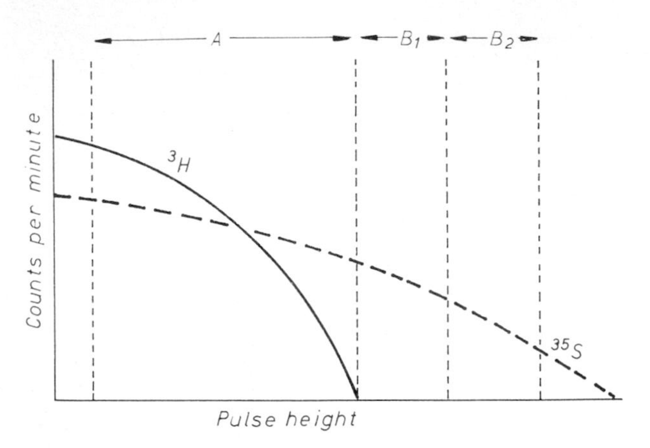

Fig. 11.9. Simultaneous assay of tritium and ^{35}S under conditions of variable quenching. The upper discriminator of channel A is placed at the top of the tritium spectrum. The upper channel $B_1 + B_2$ is split in the way shown in order to determine the degree of quenching using the ratio of the ^{35}S counts in each of the upper two windows (Weltman and Talmage 1963).

reason why this procedure should not be adopted for any pair of isotopes showing sufficient pulses above the lower exclusion window (A) to allow the region to be split and still give meaningful channel ratios. However, a number of different methods have been devised based on this general principle, but designed to give greater statistical accuracy to the assessment of the degree of quenching. The method adopted by Hendler (1964) was again to exclude the lower isotope in window (A) see Fig. 11.10, but to use an overlapping pair of upper windows (B and C) as shown. The C/B ratio was used to determine the degree of quenching and in this case, a Fortran IV programme was designed to handle the necessary mathematics. This method was later criticised by Nimomiya (1966) on the grounds that this channel ratio was relatively insensitive to large changes in A/B ratios, required to determine the degree of overlap of the carbon-14 into window A. In order partially to overcome this problem, Nimomiya suggested that the C/B ratio should be kept as small as possible but his suggestion that the removal of water from each sample and addition of a small

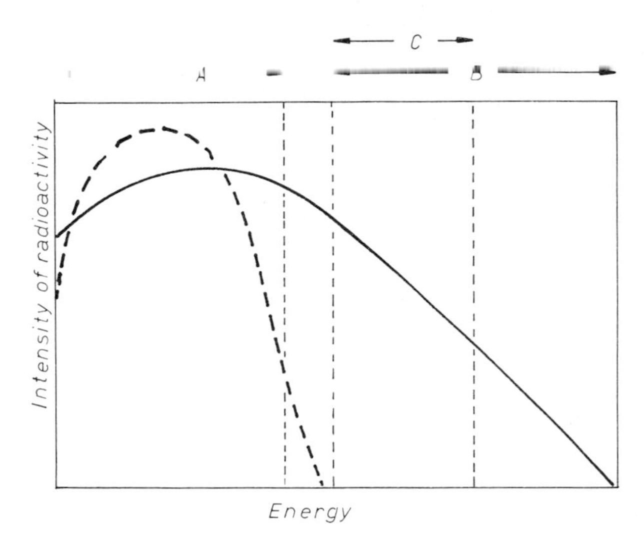

Fig. 11.10. The use of overlapping windows to separate tritium and ^{14}C by the method of Handler (1964) (see text).

fixed quantity of water is not practicable for large numbers of samples.

An alternative simplification of the procedure was suggested by Hetenyi and Reynolds (1967) who employed a graphical procedure, coupled with an assessment of the degree of quenching using the external channels ratio. In view of the fact that these calculations do not require computer assistance, this method will be described in greater detail. Only two channels are used (A, B-see Fig. 11.10), together with the external standards ratio. The lower channel 'A' is set for optimum value of tritium. In channel 'B', the lower discriminator is raised to exclude 99.9% of the counts arising from tritium. The external standard ratio between channels A and B (A/B Ext st) is assessed after each count.

Both tritium and carbon-14 water-quenched standards are used and the counts in channel A for both isotopes and channel B for carbon-14 are determined together with the external standards ratio for all samples. The results are plotted as shown in Fig. 11.11 and between the limits of efficiency shown, the curves are reasonably linear. In order

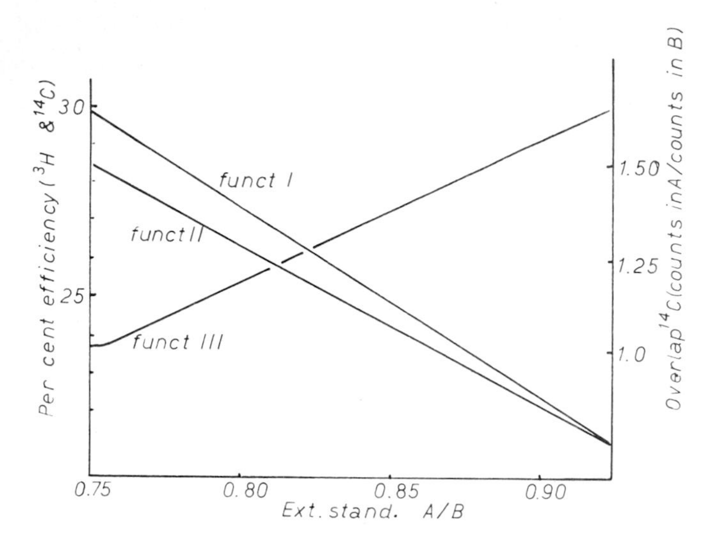

Fig. 11.11. A graphical procedure for the separation of counts from tritium and ^{14}C (Hetenyi and Reynolds 1967). Funct I is a plot of efficiency of tritium counting in channel A against the external standard ratio. Funct II is a similar plot for ^{14}C in channel B. Funct III is a plot of the ratio ^{14}C in channel A to ^{14}C in channel B against the external standard ratio. The latter is used to estimate the degree of ^{14}C overlap into channel A from the counts present in channel B.

to compute the counts in a sample, the external standard is assessed and from curve III, the proportion of overspill into channel A expected can be calculated from channel B.

$$\text{channel B (cpm)} \times (\text{A/B}) = \text{channel A overspill (cpm)} \quad (11.10)$$

The dpm due to carbon-14 can be obtained from the channel B cpm, applying the efficiency data from line II. The dpm due to tritium is obtained by subtracting the overspill carbon-14 cpm (from eq. 11.10) from the total cpm in channel A and applying the efficiency data from line I. A similar procedure for three isotopes (e.g. ^{32}P, ^{14}C and ^{3}H) using three channels is described by Davies and Hall (1969).

A major effect of quenching is to cause the pulse height spectra of the two isotopes to overlap to a greater extent as the quenching is increased. This leads to an increasing difficulty in effectively separating the spectra contributed by each isotope; the choice of discriminator position will therefore become more and more critical with quenching and the choice of the correct setting becomes more difficult. For a more detailed treatment for dual isotope counting under such quenched conditions, readers should consult Bush (1964) and Kobayashi and Maudsley (1970).

An alternative procedure is to use the so-called 'Engberg' plot, originally described by H. Engberg (Nuclear Chicago), which is as follows. Using a fixed window, the instrument gain is increased stepwise from 0 to the maximum. This is done for two isotopes under consideration and their relative efficiencies at each gain step are plotted against each other on log-log paper. The result for tritium and carbon-14 is shown in Fig. 11.12 as an example. The arrow direction indicates increasing gain. It can be seen that little change in the carbon-14 efficiency occurs from 0.01 to 1.0% efficiency in tritium, but significant losses of both isotopes occur at high gains. The best ratio between tritium and carbon-14 can be obtained by drawing a 45° line from the base of the curve (as shown in Fig. 11.12) and noting the point of departure from the line (i.e. point A).

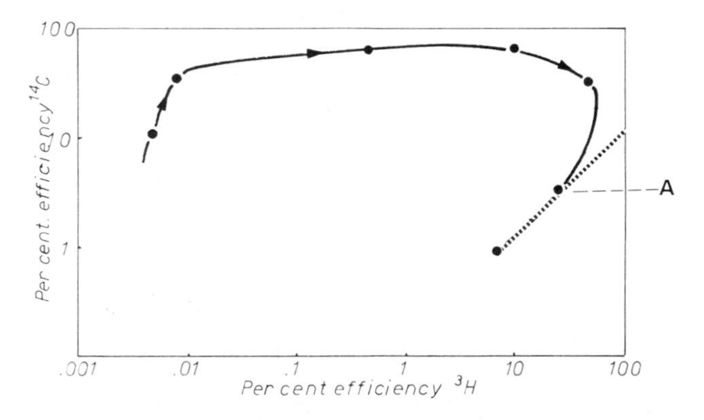

Fig. 11.12. The 'Engberg' plot (see text) is designed to determine the conditions under which two isotopes can be most efficiently separated.

11.2.2. Gamma- and beta-emitters together

There are several situations where two isotopes may be used together, one of which is primarily a gamma-emitter, such as ^{131}I or ^{59}Fe and the second a beta-emitter. In many cases, the gamma-emitter may also be a beta-emitter and will be measurable by the liquid scintillation technique. However, where the level of isotopes are low, especially with respect to the gamma-emitter, it may be more efficient to measure the mixture on a gamma counter, since the whole sample may be used without contamination by fluors and then an aliquot can be assayed by the liquid scintillation counting (LSC) technique. Any contribution by the gamma-emitter to the liquid scintillation assay may be determined by a series of quenched standards. It should be noted however, that the amount of 'overspill' into the liquid scintillation counting by the gamma-emitter will be liable to quenching in the LSC but not the gamma counting system. The level of LSC 'overspill' will therefore be related to both the level of gamma counts in the gamma counter corrected for quenching by means of the external standard ratio. A relationship between the external standard ratio and the gamma: LSC counter ratio should first of all be obtained for the counting conditions used. This relationship will depend on the amount of aliquot which must be accurately taken for LSC counting.

The separation of ^{131}I from ^{3}H in biological material before counting was achieved by Brown and Reith (1965) who combusted the material in oxygen and then dissolved the products in methanol. By reaction of this solution overnight with methanolic silver nitrate, and removal of the silver iodide by centrifugation and washing, the supernatant was used for the assay of tritium. The ^{131}I could then by assayed by planchette counting, using a Geiger Muller end-window system. There is however, little contribution to liquid scintillation counting from iodine-131, and this fact was used by Roche et al. (1962) who assessed dual labelled (^{3}H and ^{131}I) tri- and tetra-iodo thyronines on paper strips following paper chromatography. Either the same or duplicate samples are assayed in both LSC and Geiger Muller systems to separate the isotope.

11.2.3. Automatic quench correction (AQC)

As the pulse height of a beta-emitter decreases with increased quenching, the total area beneath the pulse height spectrum within a full window decreases, and hence the efficiency decreases. If two isotopes are present, such as ^{14}C and ^{3}H, then the spectrum of the more energetic beta-emitter will more and more overlap the spectrum of the weaker, and increase the difficulties of discriminating between them.

A possible method of reducing this problem is to increase the amplification as quenching increases (Wang 1970). Since quenching may be monitored by means of an external standard ratio measurement, this value could be inversely linked in the instrument – usually by modifying the anode potential of the photomultiplier to increase amplification to a sufficient extent to attempt to restore the spectrum to the unquenched condition. This process of gain restoration, does shift the spectrum to higher voltage levels and thus allow for a better discrimination between isotopes, but does not restore the lost counts in the full window.

An additional advantage of this gain restoration device, is that it decreases the number of quenched background counts in the lower window used for tritium determination.

An interesting application of this device is to combine the gain restoration with an assessment of the isotope by the balance point quench correction technique (§ 11.1.1). This combination of procedures will allow for constant efficiency counting over a range of quench conditions. However, it is hoped that as these more sophisticated correction procedures are applied it will still be possible to obtain data which are independent of the external standard ratio, to avoid errors arising in heterogenous counting systems. Basic data from fixed gain measurements should accompany all modified data on readout information from the instrument.

11.3. Data evaluation

Within a biochemical experiment, the analysis of liquid scintillation counting data from one or more isotopes should have the least influence on the accuracy of the experiment as a whole. There are now adequate techniques for preparing good samples and for refining the output data from the instrument so that this stage of the experiment should not be regarded as the most error prone.

Only broad outlines of the principles involved that specifically apply to the handling of liquid scintillation counting data are considered here. The purpose of the statistical analysis of data is usually a) to determine the accuracy of counting a low sample count in relation to the background count rate, or b) to analyse the quality of the data obtained to ensure that it follows an expected Poisson distribution, and that no instrument fault or environmental interference is modifying the count rate.

In order to carry out these calculations repetitively for large numbers of routine samples computer technology has been applied to some extent. Only a brief general survey can be given of this aspect, since considerable local variation of instrumentation and programme requirements occur between operators.

11.3.1. Some statistical considerations

The disintegration of a single atom of an isotope is an unpredictable event. In a block of such atoms of isotope however, it is possible to predict with accuracy when one half of the total number will have disintegrated (isotopic half life, τ). Furthermore, this predictable half-life will be the same irrespective of the stage of decay already reached so that the full lifetime of the block of isotope, asymptotically extends to infinity, i.e. it decays exponentially. Thus the number of isotope atoms at any given time (t) will be related to the number of isotope atoms (N_0) at zero time according to

$$N = N_0 \exp^{-\lambda t} \qquad (11.11)$$

where λ is the transformation or disintegration constant, i.e. the number of disintegrations per unit time from a source assumed to contain 1 atom. If N becomes $N_0/2$, i.e. when 0.5 of the atoms initially present have disintegrated (in time $t = \tau$, the isotope half life)

$$\text{then } 0.5 = \exp^{-\lambda\tau}$$

$$\text{or } \tau = \frac{0.693}{\lambda} \tag{11.12}$$

The extent of decay of any isotope can be readily determined by plotting on semi-log paper as shown in Fig. 11.13. When the information is routinely required for internal standardization it is usually more convenient to plot the curve for tritium decay with absolute dpm

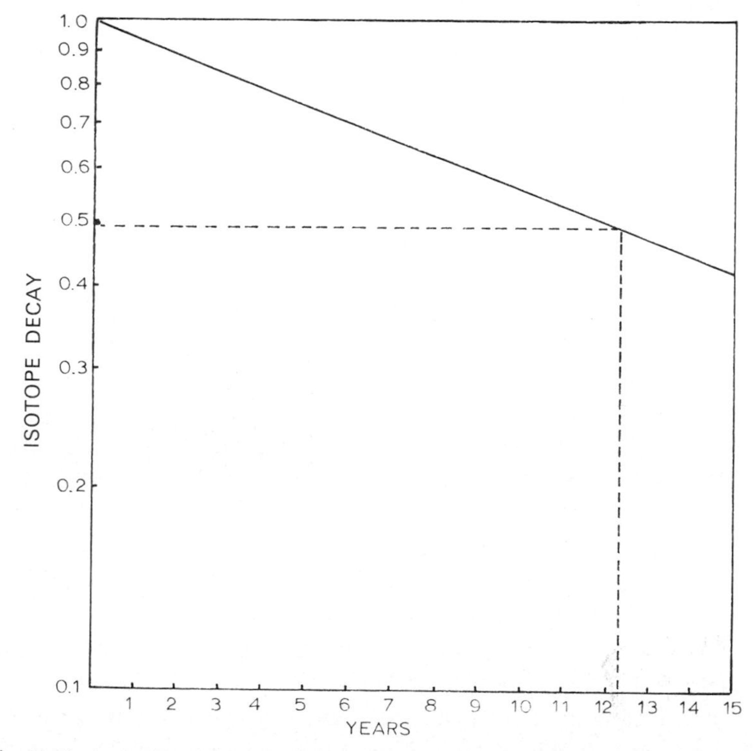

Fig. 11.13. A simple semi-log-linear plotting technique for rapidly assessing the correction factor required to determine the decay of an isotope.

Subject index p. 309

as ordinate and calendar month as absicca. Although this is a shallow curve, it is useful to draw it for a few years from the date of standardization as shown in Fig. 11.14. The actual dpm for any given date in

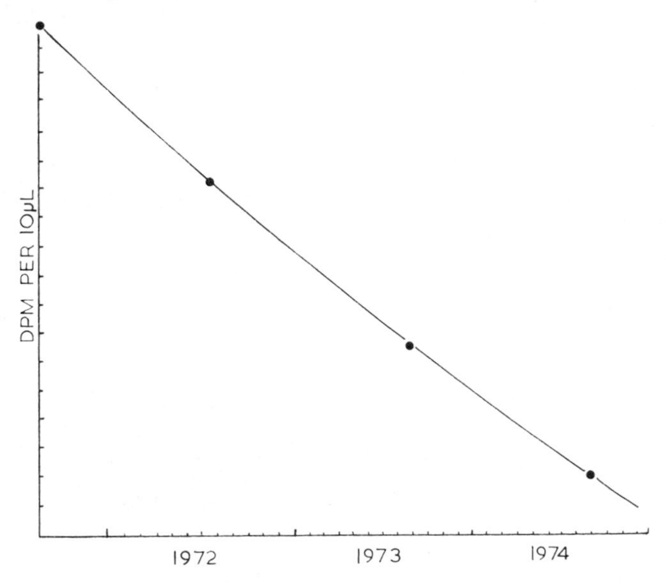

Fig. 11.14. A tritium standard plotted directly on linear paper with an expanded ordinate and the present and future calender as absissa.

the future or past can then be read off directly, which avoids the application of a division calculation each time. Alternatively, decay tables (Tables 2–8; Appx. I) may be used where the tabulation interval automatically maintains a useful accuracy level.

Herberg (1963) suggested that for use as internal standard, the addition of 50,000 dpm (22.5 mCi) of carbon-14 and 150,000 dpm (68 mCi) of tritium is adequate to give an internal standard which only requires a one minute count with an acceptable statistical accuracy of 1 % at a 95 % confidence level.

The disintegration of an isotope at any given time follows a Poisson distribution. At higher counting rates the Poisson distribution approaches the normal distribution.

For a normal distribution, the standard deviation of a count σ is equal to $\sqrt{N}$. The standard deviation (σ) around the average value of the count ($\overline{N}$) i.e. $\overline{N} \pm \sqrt{N}$, represents the spread of 68.3% of the total probability of the counts appearing within these limits. Twice the standard deviation (2σ), i.e. $\overline{N} \pm 2\sqrt{N}$ will represent the range of cpm values in which there is a 95.5% probability of finding the actual measurement observed. A number of instruments use this 2σ value as a percentage of the total counts accumulated; the so-called 'percentage error'. As these increase with longer and longer times of measurement, this 'percentage error' will decrease. Thus the 'percentage error' can also be regarded as inversely related to the total counts accumulated in any particular count and is often used in place of the total counts. A list of 'percentage error' values and the absolute counts they represent is given in Table 11.3.

TABLE 11.3

The percentage error (2σ) corresponding to observed total counts per minute.

Percentage error*	Total counts	Percentage error	Total counts
0.09	5×10^6	2.00	10,000
0.10	4×10^6	2.83	5,000
0.20	1×10^6	3.00	4,400
0.28	500,000	4.00	2,500
0.30	444,000	5.00	1,600
0.40	250,000	6.00	1,100
0.45	200,000	6.31	1,000
0.50	160,000	7.00	820
0.60	111,000	8.00	630
0.63	100,000	8.93	500
0.70	81,600	9.00	490
0.80	62,500	10.00	400
0.89	50,000	11.53	300
0.90	49,400	14.13	200
1.00	40,000	20.00	100

* Percentage errors are derived from 95.5% confidence level.

If it is necessary to estimate the standard deviation of the counts of a sample, when the error in counting the background is of a different order, the following relationship holds. If the counts per minute (S) of the sample (including the background count rate) and background count (B) are N_s/t_s and N_b/t_b respectively, the standard deviation of the sample count is $\sqrt{(N_s/t_s)}$ and that of the background is $\sqrt{(N_b/t_b)}$.

Thus the (2σ) standard deviation of the sample less the background count rate $2(\sigma_{S-B})$ will be

$$2\sigma_{S-B} = 2\sqrt{\sigma_S^2 + \sigma_B^2} \tag{11.13}$$

$$\text{i.e. } 2\sigma_{S-B} = 2\sqrt{\frac{N_s}{t_S^2} + \frac{N_b}{t_B^2}} \tag{11.14}$$

$$= 2\sqrt{\frac{S}{t_S} + \frac{B}{t_B}} \tag{11.15}$$

or the 'percentage error' of the sample counts alone

$$= \frac{\sqrt{\dfrac{S}{t_S} + \dfrac{B}{t_B}}}{S - B} \times 200 \tag{11.16}$$

It is often necessary to decide how long the background count should be if the sample count is also very low. The maximum precision in a given time is given by

$$\frac{t_S}{t_B} = \sqrt{\frac{S}{B}} \tag{11.17}$$

Thus if the sample + background is 4 times the background count, then the sample should be assayed $\sqrt{4/1}$ i.e. twice as long as the background. Since the accuracy of the counting will largely be governed by the size of the sample plus background count rate, then shorter background counts will be required as the sample count rate increases. The background count rate on most modern scintillation counters employing coincidence counting is usually very low, i.e. between 15 to 30 cpm, depending on the channel settings used. There is usually very little point in assessing an accurate background count unless the sample count itself is close to the background count.

A useful nomogram (Fig. 11.15) was published by Davidon (1953) which relates the time required to achieve 5% error corresponding to a total count rate and background rate. If a different percentage error is required from the 5% as indicated, then the time to be used is inversely proportional to the square of the accuracy desired. Thus for 1% standard error, then the counting time should be multiplied by 25.

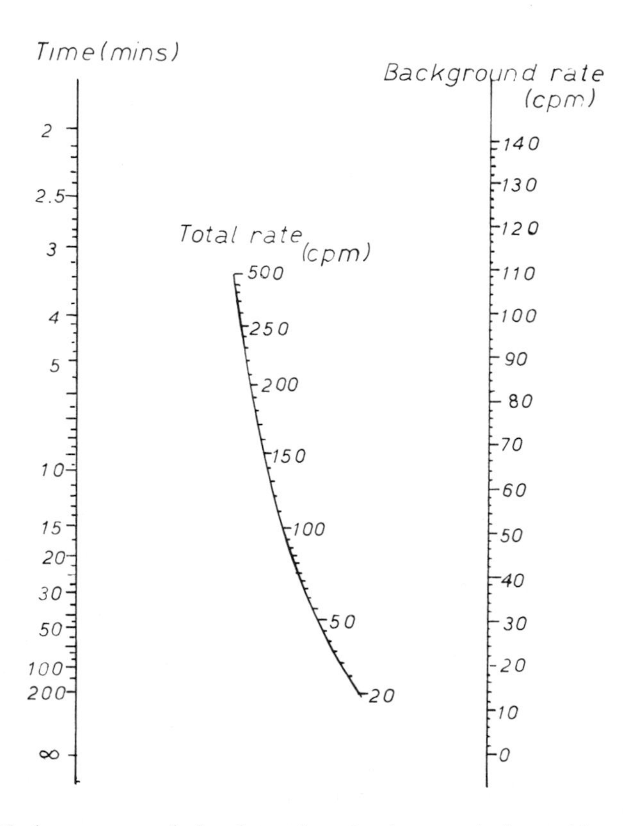

Fig. 11.15. A nomogram designed to relate the time required to achieve 5% error corresponding to a total count rate and background rate (Davidon 1953). (Reproduced by kind permission of McGraw Hill Publ. Co.)

Subject index p. 309

The background count rate arises from a variety of different causes, such as thermal noise on the photocathode, Cerenkov light, cosmic rays and ^{40}K in the glass of the vial. Some of these pulses are quenched like beta particles, but others are not. Thus the over-all background count will be non-linearly related to the actual quenching or in turn, the external standard ratio. A non-linear calibration curve may thus be drawn which relates the background count rate to the external standard ratio (Takahashi and Blanchard 1969; Scales 1963) and provided there is no contamination of the vial, will give an accurate background count in as much time as is necessary to carry out the external standard ratio measurement. A standard quenched series of backgrounds must first of all be assessed by an accurate reading for the use of the calibration curve. A similar series of coloured samples can also be prepared and the absorbance, say at 400 nm may be used to determine the background expected.

If a variation is observed in different sets of counts of the same sample, it may be due to either the presence of a rapidly decaying chemiluminescence, or varying external sources near the measuring instrument, or faults in the instrument itself or some other non-random series of events. In order to test whether or not one set of counts differ significantly form another set, taken on the same sample, a Student's 't' test may be used.

Alternatively, in a series of measurements, from which we are deriving a mean value, we may wish to know whether the variation experienced amongst these counts are within the limits of variation expected of a random distribution, or whether there has been some factor which has increased the similarity of each count or increased the variability of the counts outside what we should have expected from disintegrations from the isotope alone. Such variations are likely to be experienced due to the considerable sensitivity of the instrument detecting system itself, which will tend to amplify such variations. A convenient test to determine this is the chi-squared test.

Student's 't' test. If the mean of two different sets of readings (total number, $n = n_A + n_B$) are $\bar{A}$ and $\bar{B}$ respectively and the standard deviation of each is σ_A and σ_B respectively, then

$$`t' = \frac{|\bar{A} - \bar{B}|}{\sigma_{(A-B)}} \tag{11.18}$$

An estimate of the standard error of the difference $\sigma_{(A-B)}$ can be obtained from

$$\sigma_{(A-B)} = \sqrt{\frac{\sigma_A^2}{n_A} + \frac{\sigma_B^2}{n_B}} \tag{11.19}$$

The numerator $|\bar{A} - \bar{B}|$ is always regarded as positive, irrespective of the sign.

The number of degrees of freedom is calculated and will be equivalent to $n - 1$ for those conditions in which Poisson distribution is being considered, whereas it is equivalent to $n - 2$ for the normal distribution. For large numbers of estimates, this difference is not significant. From the Student '*t*' test tables, the probability that the difference observed is due to chance can then be assessed. A less than 5% value of this probability should be suspected, and a fault looked for.

Chi squared test (χ^2). In this case a number of readings are taken and the mean estimated. The test then examines whether the deviation of each count from the mean is more or less than would be expected from strictly random distribution. To assist us, if the mean of n sets of cpm estimates is $\bar{A}$ of which A is an example, then

$$\chi^2 = \sum_1^n \frac{(A - \bar{A})^2}{\bar{A}} \tag{11.20}$$

The number of degrees of freedom are compared with the χ^2 value in either the standard table or a graphical presentation and the probability level is estimated. Values of P greater than 95% or less than 2% should be regarded as suspicious and a fault suspected.

Chauvenet's criterion. A quick estimate of whether a particular cpm estimation should be included in a list of cpm estimates required to calculate an accurate mean value, can be made by means of a graphical analysis known as Chauvenet's criterion. First, the deviation from the mean of the suspicious cpm value is divided by the standard deviation.

(this cpm value could have been used in the estimation of the mean value and thus unfairly biased it). If the number of cpm estimates are *n*, then from Fig. 11.16, if the value obtained falls within the hatched area, it should be rejected from the estimate of the mean.

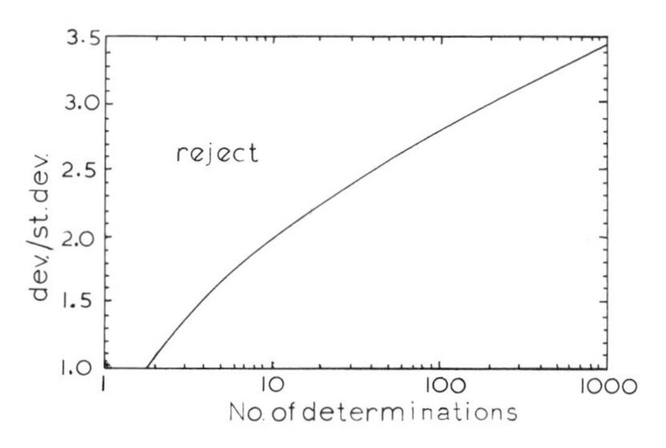

Fig. 11.16. A graphical procedure to determine whether a particular cpm estimate from a group of repeated cpm estimates should be included in an estimate (Chauvenet's criterion).

11.3.2. Computer-assisted data handling

Computers can carry out a series of repetitive and tedious calculations which are often required in the conversion of raw data derived from the liquid scintillation spectrometer into a form which is readily utilizable by the operator. The facility is particularly valuable in those situations where routine analysis of very similar samples are being undertaken and especially where statistical analysis may be required frequently throughout the assay.

The type of instrumentation used will be determined by the nature of the work to be done, the routine nature of the assay and local financial considerations. A detailed survey of the procedures currently employed is thus not possible within the scope of this book.

Basically there are two broad groups of facilities, on-line and off-line. The on-line computer, which is part of the scintillation spectrometer instrument itself, can derive its information directly from the pulses within the scintillation counter. However, it suffers the disadvantage that it is usually limited in its use to the scintillation counter of which it forms a part. The use of such computing as a shared facility, which gives priority to the scintillation counter or which can be used with several counters, has, however, been considered. For those laboratories employing liquid scintillation counting as a routine facility, such as in beta-emitter applications of radioimmunoassay, this on-line facility can be an important consideration.

In multiuser laboratories however, off-line computers can provide the operator with a facility for storing basic data in the form of paper tape, punched cards or even magnetic tape, which will allow the operator to draw up different programme requirements for use on the same data if necessary.

Several programmes have been published for use with different computers ranging from the simple desktop systems to the larger capacity instruments. They are capable of dealing with both single and multi isotope experiments under varying quench conditions. For further details, readers should consult the works of Spratt (1965), Mathijssen (1966), Plotka et al. (1966), Blanchard et al. (1968), Walkenstein et al. (1968), Stanley (1972), and Boeckx et al. (1973) and a review by Grower and Bransome (1969).

Where both long and short lived isotopes are being measured simultaneously, the additional problem of calculating decay can give rise to further computational problems (Cramer et al. 1971). In situations where detailed statistical analysis is also required, readers should consult the work of Carrol and Houser (1970).

Subject index p. 309

CHAPTER 12

Geophysics and archaeology

The methods developed within these fields achieve high sensitivity and have considerable application in biology and medicine by enabling a much lower level of isotope to be used as tracer. Furthermore, many of the fall-out surveys require the sensitive assay of a variety of isotopes in biological material to determine isotope concentration and transfer processes within the environment. However, due to the high specialization of many of these fields, their publications do not reach many biochemical specialists who could usefully apply such information in their own fields if they were aware of them.

One of the chief requirements when radioisotopes are measured or used as tracers in geophysics and archaeology, is a very high sensitivity for the assay of beta-emitters. There are many problems in which the level of activity being measured is only a small fraction of the background rate itself and thus it becomes necessary to a) concentrate the isotope to be measured as far as possible and b) employ the method of assay which shows the highest sensitivity for a large volume, i.e. in effect has a high merit value.

The most sensitive means of measuring these low levels of radioactivity has been by means of proportional counting, but with the advent of the newer colloidal systems with high merit values, the two methods have become comparable in their relative sensitivity. For groundwater monitoring following tracer application, the proportional counter is most suitably adapted as a probe or flow monitoring device.

One of the most useful applications of low-level tritium measurement has been in hydrological surveys of groundwater, surface water movement in rivers, lakes and estuaries and in determining the age of water deposits in desert wells, at different water layers in lakes, different snow depths and similar situations using tritium and in some cases carbon dating methods. Intimately linked with these investigations are the studies of the distribution and extent of fall-out from thermonuclear explosions as well as the associated meteorological studies. One interesting result from the study of tritiated water from nuclear explosions in the upper atmosphere is the better understanding of moisture movement in the troposphere.

The study of the natural beta-emitters in the soil is a very complex process due to the presence of thorium and uranium families of isotopes with their wide variety of beta particle energy spectra. Gamma spectroscopy is of far greater assistance in this field. However, the study of added ^{32}P phosphates in plant utilization studies has been widely used. Different soil types show different absorption patterns for different isotope species and chromatographic concentration of certain isotopes can also be achieved in certain soils. Such movements of isotopes, especially in relation to oil-water transfer through oil shales is an important aspect in understanding the formation and movement of oil deposits.

The dating of water samples by the measurement of natural tritium content as well as the well known radiocarbon dating techniques are also now employing liquid scintillation counting technology. In each case, considerable concentration of the isotope is necessary prior to measurement. As more sophisticated measurements of other isotopes become available, a choice of half lives will no doubt become available which will enable other historical events, e.g. volcanic activity or movement in geological plates, to be dated with accuracy.

For a review of the radioecological concentration processes, readers are advised to consult Agnedal (1967) and other specialised works.

12.1. Hydrology applications and tritium dating

Liquid scintillation counting techniques have found several applications in situations where the movement of large volumes of water requires to be known. An example is the study of fresh water and sea water mixing in estuaries, and in particular the effect of tidal movements on deposited sewage and reactor waste products. Other applications include the plotting of the movement of water phase in oil shales. These methods include the injection of tritiated water into a suitable site and tracing the course of dilution of the tritium in water and brine samples taken at different points.

In the above and other examples, the essential requirement is the measurement of very low levels of isotope, usually (but not always) tritium or chlorine-36 in a large dilution of water. It is convenient to employ a unit of measurement which can be readily used in comparative studies and Tamers (1964) has proposed the use of Tritium Unit (TU). This is defined as

$$1 \text{ TU} \equiv \text{one tritium atom per } 10^{18} \text{ hydrogen atoms.} \qquad (12.1)$$

100 TU ≡ 6.5 dpm tritium per g hydrogen *or* 9 g water.

Some examples of the level of tritium activites are shown in Table 12.1.

TABLE 12.1

The change of tritium level (Tritium Units) in some naturally occurring water from 1951 to 1963. (From Tamers 1964.)

	1951	1963
Rain	6	100–1000
Rivers and shallow lakes	1–6	10–1000
Underground waters	0–5	0–1000

An alternative method for the measurement of low levels of isotope such as would be derived from these types of measurements, is the *Y* value (the minimum detectable activity) of Moghissi et al. (1969).

This is defined as the minimum limit (in pCi) of the detection at a 1σ confidence level and one minute counting time. The value takes into account the background (B), counting efficiency E (cpm/dpm) and the amount of the sample in the scintillation mixture (M) in grams.

$$Y(\text{pCi per gram isotope}) = \frac{\sqrt{B}}{2.22 \times E \times M} \qquad (12.2)$$

This figure is of considerable value in comparative studies, but there is often insufficient data in the older literature to be able to convert the information given into this parameter. Natural tritium is known to be produced in cosmic processes and in view of its relatively short half life (12.3 yr), provides a means of studying the dynamics of mixing and exchange in water of lakes and underground water supplies, as well as in many meteorological problems.

The advent of thermonuclear explosions has added another source of tritium and thus an accurate continuous assay of low levels need to be made. By measuring the tritium units present at any given time, an assessment may be made of the 'age' of a water mass. Tritium dating has been used to confirm the regular mixing in some newly cut desert wells, and it has been necessary to employ carbon dating in certain Arabian desert wells and which have confirmed that the water present may be as much as 20,000 years old (Thatcher et al. 1961).

In order to achieve sensitivity in the measurement of low levels of tritiated water in water, it is necessary to concentrate the tritiated water. Electrolysis of water preferentially liberates hydrogen rather than tritium from water and thereby concentrates the tritiated water. Several reports on the procedure involved have appeared and readers are recommended to read the survey of methods of electrolytic enrichment by Cameron (1965) for further details.

Early methods of electrolytic enrichment used iron and nickel electrodes, but were later superseded by stainless steel and later still by a phosphate treated-stainless steel (Zutschi and Sas-Hubicki 1966). In the latter case, the mild steel electrodes, after degreasing and cleaning in dilute HCl for 15 min, are washed in very hot water and

Subject index p. 309

then immersed in pure concentrated phosphoric acid for one hour at 70°C. After washing with very hot water and drying, they present a dull grey appearance, but become uniformly shining black after the first electrolytic run.

Electrolysis is usually carried out at 0°C with a current of 9 A through the cells with a total of 700 A over 4 days. An 18-fold enrichment is possible in a one stage electrolysis with a reproducibility of about 3%. The tritiated water is then distilled from the electrolyte with a recovery of about 80%.

The analysis of the enriched water was initially carried out in a water-accepting dioxane-based scintillant mixture. However Sauzay and Schell (1972) have employed a colloid scintillation counting system, in this case Instagel (Packard), to measure the tritium content of the enriched sample. Eight ml of tritium-enriched water is added to 12 ml of Instagel in a polyethylene vial and assayed over 300 min at approx. 23% efficiency. The sensitivity of this method is 5.0 ± 1.2 TU.

If these high values of sensitivity are not required, such as in tracer studies in groundwater surveys, direct measurement of the tritiated water may be undertaken. This is usually the procedure undertaken in oil field investigation, and often requires the assessment of relatively low levels of tritium in the presence of brine. Van der Laarse (1967) used Toluene:Triton X-100:water (55:25:20) or Toluene:Triton X-100:10% NaCl (38:22:40) mixtures for the direct measurement of tritium in fresh and brackish water. In this case the sensitivity of measurement was 0.002 μCi/l (i.e. approx. 40 TUs). Workers concerned with the injection of the tritiated water into the geological system may be subject accidentally to ingestion of tritiated water, and thus the monitoring of distilled urine is carried out in the same way. The detection limit by this procedure was well below the maximum permitted level of tritium in the urine (approx. 10 μCi/l, i.e. 3×10^6 TUs).

The distillation of the urine for these estimations is most accurately carried out using a procedure described by Moghissi et al. (1969) and adapted from Osburn and Werkman (1932). A total of 30 ml of urine is treated with 1 to 2 g potassium persulphate at 75 to 95°C in an all-glass vessel for 1 hr. The temperature is then raised and 25 ml is distilled over. The distillate is then extracted with approx. 15 ml toluene to remove any distillable impurities and the resulting purified water used in a colloid counting system as described above.

The alternative method of converting tritium into acetylene by treating tritiated water with carbide, and then converting into benzene as in the Radiocarbon estimation (see § 12.3), is also a sensitive procedure, the lowest detectable level being 24 TU. By using tritiated water to hydrogenate isopropenyl acetylene over Mg at 600°C, followed by palladised charcoal catalysis the lowest detectable level is 36 TUs (Perschke and Florkowski 1970).

Phosphorus-32 in water may be assayed by employing the high merit values obtainable by counting Cerenkov emission. Total sample volumes of 22 ml may easily be measured in 25 ml polyethylene vials, at efficiencies of 44 %. To avoid artifacts arising from wall absorption, a stopping solution is added, which consists of EDTA to chelate any calcium ions present, and phosphate ion to reduce any precipitation which may occur on the walls. The procedure suggested by Curtis and Toms (1972) is to add 21 ml of sample and 1 ml of a 'stopping solution' (disodium EDTA, 2.7 g/l, KH_2PO_4, 12.1 mg/l and formalin, 1 %) before counting. Although a number of other isotopes have been used in hydrological studies, chloride-36 appears to be the least selectively absorbed by different clays. For more detailed information, see more specific works such as the review by Gaspar and Oncesgu (1972).

12.2. Fall-out and meteorology

Following an atmospheric nuclear test, rainwater will contain the highest concentration of radioactive fission products. Later, passage of this water into streams and rivers will result in considerable dilution. It is from this diluted phase that natural concentration and fixation processes will occur. There are thus two ecosystems which are most useful for determining the level of fall-out at positions remote from the site of explosion, a) rainwater and b) concentrating biosystems.

Subject index p. 309

Germai (1964) has described the method for the estimation of ^{137}Cs, ^{90}Sr and ^{131}I in rainwater. The method relies on a suitable efficient precipitation technique and assay as a suspension in Cab-O-Sil (see § 2.2.2). ^{137}Cs and ^{90}Sr are precipitated as the perchlorate and sulphate respectively. Iodide-131 is best precipitated as the silver salt and assayed using a sodium iodide well-system. The ^{137}Cs and ^{90}Sr in fish-laden rivers, enter into fish muscle and bone respectively and are actively concentrated in these tissues. Levels of these from a number of localities in Colorado in the period 1965 to 1970 were reported by Whicker et al. (1972) although in this case, well counting was used throughout. Repeated samples throughout the years under study showed a definite fall in levels throughout the period. Levels of ^{137}Cs ranging from 116 to 5802 pCi/kg wet-weight muscle and of ^{90}Sr from 6 to 238 pCi/g were recorded in trout tissue from various sites.

Plutonium-241 and plutonium-239 are important ingredients of nuclear reactors, and with the increasing scale of activities in this field, the possible accidental contamination of the environment with these isotopes also will increase. The estimation of these isotopes in biological tissues has been the subject of a number of detailed studies. Early methods employed electrode deposition of the actinides on steel electrodes, and by subsequent removal of this film and assaying it in a liquid scintillation counter, considerable sensitivities were achieved. However a very convenient system arose from the studies of Eakins and Brown (1966) during the development of precipitation methods for the assay of ^{55}Fe and ^{59}Fe in blood samples. The method (Eakins and Lally 1972) depends on the formation of a frerriphosphate complex with the actinides and the subsequent measurement of the precipitate by a suspension counting method described in detail in § 10.2.

The maximum permissable body burden of these two isotopes are ^{241}Pu (0.9 Ci) and ^{239}Pu (0.04 Ci).

12.3. Radiocarbon dating

The formation of carbon-14 in the upper atmosphere by external high energy particles has been continuing at an apparantly constant rate for many millenia. Living organisms, which utilize carbon either directly from carbon dioxide or indirectly from ingestion of plant tissue will incorporate both the stable carbon-12 and the carbon-14 present at the time of their life. This fixed carbon will then cease to equilibrate with external carbon dioxide and the carbon-14 will decay with a half life of 5730 $\pm$ 40 years. An estimation of the relative proportion of the two isotopes present in the sample would then give the age of the sample.

One of the earliest pioneers in the field of radiocarbon assaying was W.F. Libby whose book 'Radiocarbon Dating' (1952) represented a significant landmark in this field of study. The half life of carbon-14 was considered at this time to be 5568 years and this is often referred to as the 'Libby half life' in the literature regarding these dating techniques. This appellation was recognised by the 5th Radiocarbon Dating Congress held in Cambridge in 1962 (see Godwin 1962a).

The earliest methods employed proportional gas counting techniques and in order to achieve accuracy in order to measure low levels of carbon-14 produced by combustion of the sample, very large volume counters were required. Some reduction of volume was achieved by converting the natural carbon to acetylene rather than carbon dioxide (Barker 1953) and instrumentation for the measurement of acetylene was described by Crathorn (1953).

In 1954, however, Arnold recognised the possibility of using a liquid scintillation counter and used 100 ml vial volume in a single photomultiplier system and achieved considerable success in comparison with the proportional and screen-wall counting techniques originally described by Libby. The main problem however was the large volume of gas necessary to achieve a sufficiently high count rate to be statistically meaningful in relation to the background level. However, Tamers et al. (1961) and Noakes et al. (1961) realised that by conversion of the carbon dioxide or acetylene into benzene, a very considerable concentration of the carbon could be achieved, at the same time producing a primary solvent which would also effectively concentrate the isotope within the scintillant mixture without quenching. Benzene in fact, contains 92.26 % of its weight as carbon.

Subject index p. 309

The method of synthesis of benzene from carbon dioxide was originally designed by Shapiro and Weiss (1957) and Weiss and Shapiro (1958) who used diborane-activated silica-alumina catalyst to assist the conversion. Noakes et al. (1963) described in detail an apparatus for undertaking this conversion in 50 to 60% yield in a form suitable for liquid scintillation counting. In order to undertake this process however, cylinder-stored diborane was employed, a somewhat costly and hazardous requirement. McDowell and Ryan (1966) described a laboratory production of diborane, which decreased these hazards.

Diborane is produced by reacting sodium borohydride with boron trifluoride ether adduct in diethyleneglycol dimethyl ether (Diglyme) in a helium atmosphere $3NaBH_4 + 4BF_3(C_2H_5)_2O \rightarrow 2B_2H_6 + 3NaBF_4 + 4(C_2H_5)_2O$.

The synthesis of benzene is undertaken by combusting the sample in oxygen to carbon dioxide which is trapped in ammonium hydroxide as the carbonate. By addition of strontium chloride, strontium carbonate is obtained which is dried and heated with magnesium to give strontium carbide. Treating the latter with water produces acetylene. Losses up to this stage range from 1.5 to 3.5% and correspond to an error of 80 years in the age determination of the sample (Suess 1954).

Using diborane activated silica catalyst the final conversion of acetylene to benzene was stated to be about 80% efficient and hence a small amount of fractionation of the carbon-14 from the carbon-12 can be anticipated due to an isotope effect. By using samples of accurately determined age (i.e. available from other archaeological evidence), a reproducible fractionation can be achieved and a correction for this effect can be applied. More recently vanadium-activated silica-alumina catalyst has been used with higher yields.

A count rate of 7.37 cpm/g was obtained for carbon laid down in 1966 (McDowell and Ryan). In order to achieve some form of standardization of half-life data, zero dates etc., several International Congresses have been held. An important decision was made at the 5th Radiocarbon Dating Conference (1962) when it was agreed that from the radiocarbon dating point of view, the zero age for radiocarbon should be 1950 AD and that in assessments of age, the half-life of carbon should be taken as 5568 years, the so-called Libby half-life. A more accurate assessment is 5730 ± 40 years (Godwin 1962a, b). Furthermore, from tree ring data, the amount of ^{14}C was found to actually decrease from 1900 to 1950 due to rapid increase in the formation of CO_2 from fossil fuels.

In order to improve the counting system for such low count measurements as are required in this type of work, an efficient shielding system is required. Mercury has been suggested as the shielding material of choice (Arnold 1958). For a useful review on these methods see Gibbs (1962).

In addition to the formation of carbon-14 by external high-energy sources, 1954 also marked the date when artificially produced carbon-14 was introduced into the atmosphere from nuclear bomb trials; by March 1958, 4.8×10^{27} artificially produced carbon-14 atoms had been introduced. Thus, wood samples or organic material that was laid down earlier than this date should be used as standards. Standard carbon-14 oxalic acid is obtainable from the National Bureau of Standards and several 'check' samples of ancient wood have been used to test the methods described by McDowell and Ryan (1966).

Samples for assay are first cleaned free of contemporary materials by treatments with cold dilute HCl and or alkali. Up to 12 g of pretreated sample is used and ignited in a 3 l stainless steel bomb containing pure oxygen at 10 atmos pressure. About 9 l of carbon dioxide is formed and this is led directly over purified lithium metal to form the carbide, with subsequent production of acetylene by addition of distilled water (see Burleigh 1972 for details).

The upper detectable age limit in radiocarbon dating at present is around 45,000 years for which counting times up to week in duration are required to accumulate sufficient counts to be statistically meaningful.

The results of application of this particular technique to the study of archaeological samples are reported in the journal 'Radiocarbon' as well as various archaeological journals.

CHAPTER 13

Miscellaneous applications and future prospects

From certain specific applications of liquid scintillation counting that have been made, there is potentially a much wider possible use of the technique in unsuspected fields.

The principles involved in heterogenous scintillation counting provide a number of possible ways in which to exploit the system as a whole. Apart from the liquid-liquid systems used in colloidal counting there is also the solid-liquid system in which the scintillant may be in either phase. By fixing the scintillant in the solid phase, it is possible to measure the radioactivity of flowing liquids from chromatographic eluates or gases from gas chromatography. By incorporating an isotope into this solid scintillant phase, a convenient internal light source may be obtained, which can be used to determine the level of colour quenching brought about by a series of coloured solutions.

A potentially most rewarding result of the fundamental study of those parameters which determine quenching, is the possibility of learning much more about the structural aspects of biochemical systems, especially those involving changing organelle structures. By locating soft beta-emitters within these structures by suitable selection of precursor molecules, much information could be obtained about the changing environment from the observed changes in quenching.

There is also considerable scope for further developments in the field of controlled chemiluminescent reactions, especially in the ultra-sensitive analysis of both trace metals and organic materials. Further bioluminescent systems employing different substrates could also provide powerful tools for ultra-sensitive assay of important biochemical substrates.

13.1. Flow cells

The monitoring of a flowing stream of liquid or gas which contains a beta-emitter may be undertaken in two basic ways. The stream can be continuously monitored by passing it over an insoluble scintillant (heterogenous method) or by sampling the eluate either continuously or at intervals and assaying by conventional liquid scintillation methods (homogenous method). The maximum efficiency for the heterogenous method for tritium is about 1.0% using anthracene crystals and less than 2.0% for diphenyl oxazole (PPO). However, for the more energetic isotopes, there is a significant advantage in being able to recover the sample easily with negligible running costs in scintillant. Thus for carbon-14, sulphur-35 and similarly energetic beta-emitters there is an advantage in using the suspended scintillator method.

Until relatively recently, the homogenous method was mechanically not very efficient but it is evident that the sensitivity of the method, particularly for tritium, is likely to be much superior to that of the heterogenous technique.

The early aspects of flow monitoring have been reviewed by Rapkin (1963). A useful review of the methods and problems associated with the monitoring of aqueous solutions by means of the heterogenous flow methods is given by Schram (1970) and some aspects of the homogenous method by Hunt (1968). The two methods have been critically compared by Schutte (1972), and readers should consult these reviews for a more detailed appraisal of the different methods.

Heterogenous flow methods. The use of plastic scintillator beads, europeum-activated fluorspar, cerium-activated lithium glass and similar scintillator beads have not proved as successful as anthracene for the monitoring of tritium-containing aqueous eluates. The use of diphenyloxazole (PPO) and butyl PBD have proved to be better solid scintillators from the point of view of counting efficiency for both ^{14}C

Subject index p. 309

and tritium, but they are slightly soluble in water, especially in acid pH conditions and could lead to artifacts and clogging of the cell.

One of the problems is the fact that nucleotides appear to absorb on to most of the solid scintillants examined. From this point of view, cerium-activated lithium glass appears to be the least troublesome, and could be advantageously employed in these circumstances in spite of the relatively low efficiencies experienced. Anthracene crystals on the other hand seem to be useful in the monitoring of amino acid eluates from automatic analysers and this is still the heterogenous method of choice for this purpose.

It is often inconvenient to use the usual counting instrument for flow-monitoring alone, and single sample coincidence counters are available commercially to undertake this. However, a cheaper single

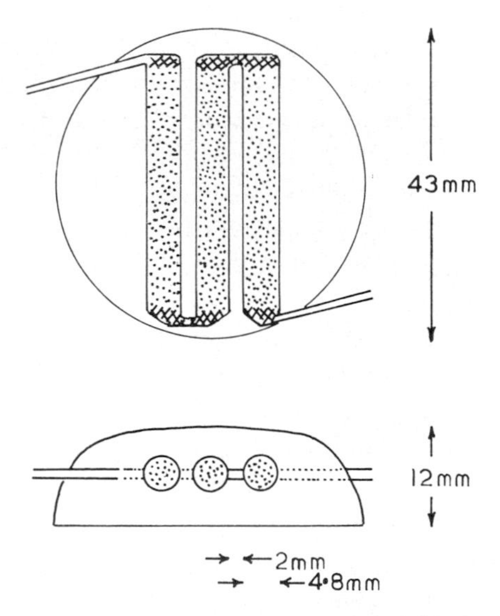

Fig. 13.1. A flow cell constructed from a 2 1/4 × 2 1/4 inch block of Lucite by drilling three closely and evenly spaced holes, 0.48 cm in diameter and 3.7 cm deep. The interconnections are constructed of 20 guage stainless steel hypodermic needles fixed by means of an epoxy resin. The packing consists of glass wool plugs and the plastic scintillator NE 102 (Nuclear Enterprises) ground to 100 to 200 mesh powder (Tkachuk 1962). (Reproduced by kind permission of the National Research Council of Canada.)

photomultiplier system can be used to obtain a qualitative record of the position of labelled peaks, and a simple device constructed from a suitably light-shielded perspex block and is often all that is required. Such devices have been available commercially (Nuclear Enterprises), but a simple apparatus (Fig. 13.1) which could probably be constructed in any reasonably equipped workshop was described by Tkachuk (1962). A useful pressure-resistant flow-cell (Fig. 13.2) for accelerated amino acid analysis (White and Mencken 1970) was intended to be used with the conventional coincidence counter.

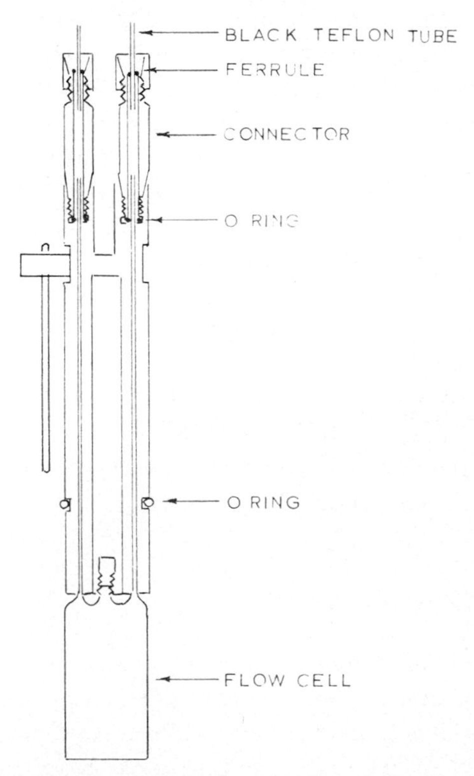

Fig. 13.2. A flow cell constructed of Perspex and packed with detergent coated-anthracene is designed to fit into the well of a standard liquid scintillation counter. It is intended for use with accelerated amino acid analysis procedures (White and Mencken 1970). (Reproduced by kind permission of Academic Press.)

Homogenous systems. The monitoring of tritium-containing eluates and low levels of carbon-14, requires a homogenous system where the sample is either continuously or intermittently monitored in the more conventional liquid scintillation counting system. A typical flow diagram was described by Schutte (1972) where a proportional pump was used to divert the flowing stream and to ensure mixing with a water-accepting liquid scintillation composition. A typical flow diagram is illustrated in Fig. 13.3a with a coiled tube inserted into a scintillation vial as shown in Fig. 13.3b, inserted into a scintillation counting well. The mixing spiral is a tube of internal diameter 2 mm and 70 cm long. Since mixing is assisted by a stream of air bubbles which periodically interrupts the flow, the resolution obtainable is maintained. The coil inside the vial has a total volume of 1.4 ml and the net flow rate is approx. 2.3 ml/min. In this system, a counting efficiency for tritium of 30% and for carbon-14 of 80% is recorded.

With this system therefore, if only a sensitive record of label is required, the splitter can be dispensed with and a sensitivity of 5 nCi per peak can be obtained. In addition, for dual isotope experiments, good resolution of the isotopes can be made as pointed out by Hunt (1968).

For those isotopes with a greater energy max than 0.5 MeV, the Cerenkov emission may be usefully employed since a simple coil of plastic immersed in a vial is all that is necessary for the monitoring of the label (Colomer et al. 1972).

13.2. *Gas chromatography*

In the gas chromatographic technique, organic materials are heated and chromatographed over hot columns in the vapour state. On many of these absorbant columns, there is also a stationary liquid phase, such as silicone oil. The gases are monitored in a variety of different ways, but it is often useful to be able to follow the level of radio isotope that may be used as a label in one of the components. However, to do so, the hot gases need first of all be concentrated either by

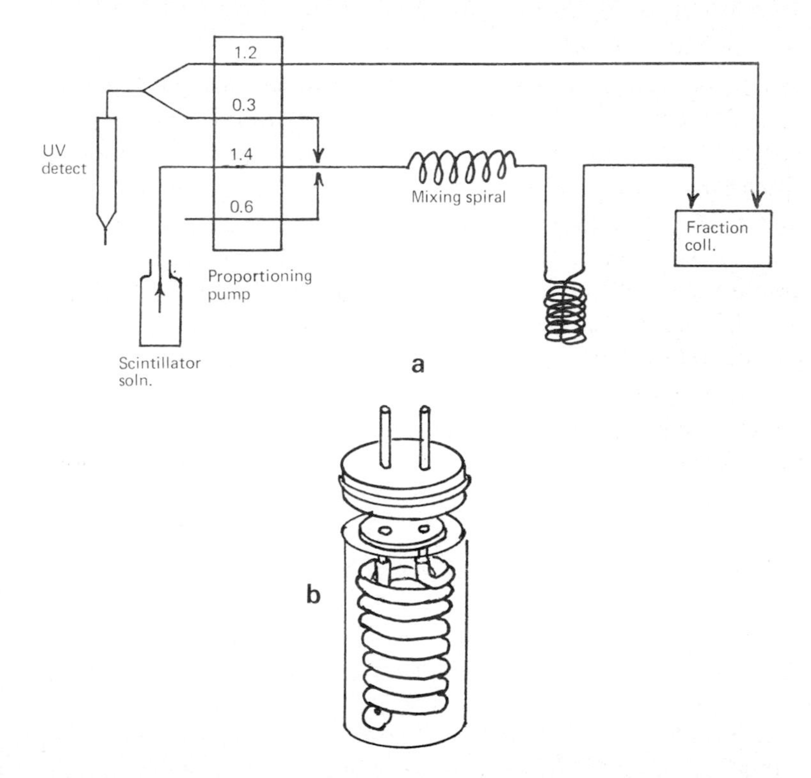

Fig. 13.3. A flow diagram (a) for the homogenous assay of chromatographic eluates using a Technicon Analyser proportionating pump by diverting a constant fraction of the stream with scintillant via a mixing spiral into a coil (b) placed in the scintillation vial. Adapted from Hunt (1968). The tubing and coil are made of a toluene resistant plastic (e.g. Acidflor, Technicon) and an interspacing of air bubbles prevents tailing of the peaks (For a more detailed description, see Schutte 1972.)

condensation on cooling or by absorption into some system which can be readily adapted for liquid scintillation counting.

One of the most straight-forward, though not the most convenient ways to do this, is to lead the gases into an evacuated chamber. They are then passed from this chamber into a catalytic tube, where the carbon compounds are oxidised to carbon dioxide, which is then absorbed in a suitable absorbant scintillant, such as one based on ethanolamine (Bosshart and Young 1972). By using a weighed ethanolamine trap, the specific activity of the carbon dioxide collected can be calculated and together with the known mole fraction values normally determined in the gas chromatographic technique, the specific activity can be assayed for each of the peaks from the chromatogram. A detection limit of 0.04 dpm/ml can usually be obtained.

An alternative procedure is to have a short section of gas chromatography tubing (2″ long and 1/4 ins int dia.) at the exit port of the column, containing anthracene, coated with the liquid stationary phase, e.g. silicone oil (5 % w/w, oil/anthracene). The first two inches will trap the majority of the effluent if this tube is maintained at room temperature. It can then be assayed by counting inside a vial in a conventional liquid scintillation spectrometer (Karmen and Trich 1960).

13.3. Analytical applications

There are a number of instances where the liquid scintillation spectrometer has been employed in an unconventional manner to assay light photons from sources other than a scintillator-isotope system. Mention has already been made of the extreme sensitivity of the bioluminescent system of the luciferin-luciferase (§ 2.4). If the origin of chemiluminescence was better understood, there is clearly a potentially important field for the analysis of many inorganic and organic species. A useful chemiluminescent reaction which has already been exploited for a number of analyses (in most cases not using a liquid scintillation counter), is afforded by the peroxide-induction of light with luminol (5-amino-2,3-dihydro-1,4-phthalazine dione). This reaction is catalysed by several inorganic ions, such as cadmium, cobalt, copper, iron, lead and vanadium, and the rate of catalysis is related to the concentration of the ion. Other organic molecules either catalyse or inhibit the reaction; in some instances, such as with the organophosphorus compounds, the chemiluminescent reaction has been used as an alternative to the cholinesterase reaction to determine the extent of anticholinesterase activity possessed by these molecules. Many other chemiluminescent reactions are now known, and it is likely that in this field the use of the liquid scintillation counter may develop.

An alternative method of using the counter is to create a constant source of photons and measure the degree of quenching by changing colour reactions, particle density etc. An example of this application has already been described (§ 8.3 and Fig. 1.1f) for the assay of lipid mass. The method has also been modified for the assay of sulphydryl groups (Snyder and Moehl 1971).

Sulphydryl group assay. The light source is immaterial, but the liquid source consists of a solution of ^{14}C-tripalmitin in toluene containing approximately 98,000 cpm in 900 μl mixed with Tpp scintillant in the central tube of a glass vial of the type illustrated in Fig. 1.1f. The SH compound is dissolved in 0.1 m phosphate buffer (4 ml) pH 8.0 (reduced glutathione is usually employed as a reference in these determinations). An aliquot (200 μl) of a solution of 39.6 mg 5,5′-dithio-bis-2-nitrobenzoic acid in 10 ml 0.1 M potassium phosphate buffer (pH 7.0) is added in a standard vial. Either a liquid scintillator insert (e.g. Fig. 1.1f) or scintillator beads are then added. A good linear correlation (Fig. 13.4) of the glutathione concentration with the degree of quenching is achieved. The method is a useful alternative to the spectrophotometric method, since, if several light sources of known light emission properties are made, the system may be automated. A plastic source may be more conveniently used as an alternative to a liquid one and these are compared in Fig. 13.4.

Another use of an internal light source is that of counting particles in a suspension by addition of a Cerenkov emitter to the suspension. This procedure was described by Ashcroft (1969) for the measurement of *Escherichia coli* suspensions as well as 0.5 nm diameter polystyrene latex beads. In this case, 0.5 μCi of ^{36}Cl is employed in 0.9 % saline per vial as the source of Cerenkov light.

An observation of Paus (1972) made during the assay of ^{14}C labelled RNA, appears to be capable of considerable exploitation. The level of quenching increases as the molecule of labelled RNA associates with water molecules. These surround the beta-emitter and appear to reduce the degree of penetration of the beta particle into the scintillant.

Subject index p. 309

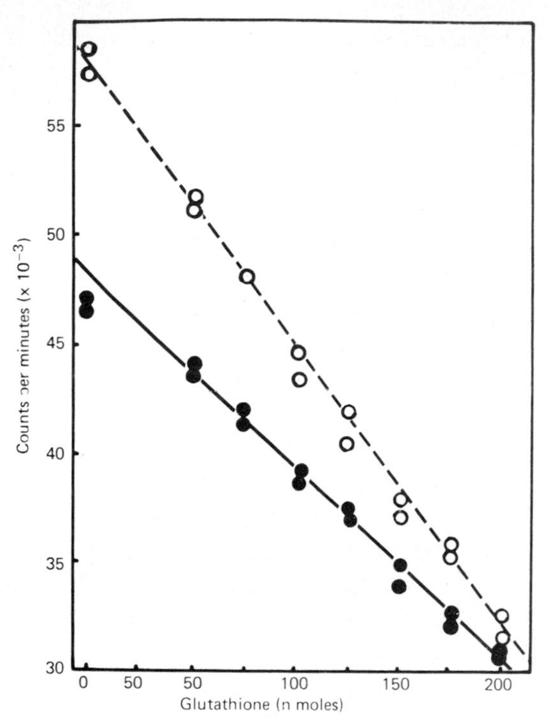

Fig. 13.4. The effect of two different scintillation sources on the estimation of glutathione by colour quenching. Solid plastic light source ○— — —○, PPO-POPOP liquid source ●———● (Snyder and Moehl 1971). (Reproduced by kind permission of Academic Press Inc.)

This association of water molecules appears to be a relatively loose one, since the quenching temporarily decreases on shaking the mixture. It would appear that further elaboration of the phenomenon could lead to a powerful method of assaying the rate and extent of water absorption to macromolecules of this type in relation to structure changes which may have important biological implications.

The use of multi-isotope analyses have already been referred to but there are several ways in which use of such mixtures has been made, apart from the more obvious applications. Heidelberger et al. (1953) required to establish the identity of a metabolite of a labelled carcinogenic hydrocarbon following skin painting on an animal. Suspected metabolites were prepared unlabelled, and added in known amounts to the tissue digest prior to paper chromatography. Label associated with one of the added metabolite spots on the subsequent chromatogram was shown to be associated through repeated chromatographic runs and it was concluded that this was a true metabolite. The basic principle of employing isotope dilution is, of course, not new, but a combination of two isotopes may sometimes be employed to advantage. Prior to the widespread use of scintillation counting, Keston et al. (1950) described a procedure for the assay of amino acids on paper chromatograms by using labelled pipsyl chloride (in this case labelled with ^{131}I). The ^{131}I-pipsyl amino acids were then run with known quantities of ^{35}S-pipsyl amino acid markers, and a constant $^{131}I/^{35}S$ ratio could then be used to confirm both radiochemical purity and to determine the level of the original amino acid present. There appears to be no obvious reason why ^{14}C or ^{3}H compounds should not be used in a similar manner.

Multiple isotope analysis has also been employed for a somewhat complex, but in certain cases a very convenient, way of assaying intracellular pH in whole animal tissue, in this case dogs (Schloerb and Grantham 1965). The method consists of the intravenous injection of a mixture of 100 μCi tritiated water, 10 μCi ^{14}C, 5,5-dimethyl-2,4-oxazolidine dione and 10 μCi ^{36}Cl in 10 ml of NaCl (0.9 %). After an equilibration time of not less than 3 hr, biopsy or tissue specimens, together with plasma samples are taken and the relative amounts of each isotope present are measured. The principles involved had previously been described by Waddel and Butler (1959). The resulting intracellular pH_i is then related to tissue (t) and plasma (p) radioactivity and to the extracellular pH_e according to equation

$$pH_i = 6\cdot 13 + \log\left\{\left[\frac{^{14}C_t\,^3H_p}{^{14}C_p\,^3H_t}\left(1 + \frac{^{36}Cl_t\,^3H_p}{1\cdot 05\,^{36}Cl_p\,^3H_t - {}^{36}Cl_t\,^3H_p}\right) - \left(\frac{^{36}Cl_t\,^3H_p}{1\cdot 05\,^{36}Cl_p\,^3H_t - {}^{36}Cl_t\,^3H_p}\right)\right] \times [1 + 10^{(pH_e - 6.13)}] - 1\right\}$$

Appendices

APPENDIX I

Isotope tables and decay charts

During the course of liquid scintillation counting, it is often desirable to have ready access to data which relate to different aspects of the liquid scintillation process. A list of those isotopes which may be measured by means of this technique is given in Table A.I.1, together with an indication of those isotopes which may also be assayed by means of their Cerenkov emission (see also § 2.3).

The final efficiency of counting of the isotope cannot be given in detail since it will depend on the pulse height spectrum and the efficiency of the scintillant used, together with the level of quenching, etc. as outlined in the earlier chapters. The Cerenkov emission will also depend on the photocathode and solvents used, as well as on the presence of colour.

A number of decay charts are also given (Tables A.I.2 to A.I.8) for some of the more commonly used isotopes in a form which may be readily employed for calculation of isotope decay. In the case of tritium, either the decay chart (Table A.I.8) or the decay curve based on the standard used (Fig. 11.14) may be consulted. The latter method has found most application in the author's laboratory.

TABLE A.I.1

Isotopes useful in liquid scintillation counting in biomedicine.

Isotope	Atomic No.	Half-life	Particle energy (KeV)	%[1]	Cerenkov
Barium	140	12.8 days	β^-480	25	+
			β^-590	10	
			β^-900	5	
			β^-1010	60	
			IC^+	75	
Caesium	137	30 years	β^-510	95	2.1 (S11)
			β^-1170	5	
Calcium	45	165 days	β^-254	100	
Calcium	47	4.7 days	β^-690	82	7.5 (S11)
			β^-2000	18	
Carbon	14	5760 years	β^-159	100	
Cerium-Pr	144	285 days	β^-190	19.5	54 (S11)
			β^-240	4.5	
			β^-320	76	
Chlorine	36	3×10^5 years	β^-714	98.3	2.3 (S11)
			EC	1.7	
Chromium	51	27.8 days	EC	100	
Copper	66	5.2 months	β^-760	0.2	+
			β^-1590	9	
			β^-2630	91	
Gold	198	2.7 days	β^-290	1.2	5.4 (S11)
			β^-960	98.8	
			β^-1370	0.025	
			IC	4.0	
Hydrogen (Tritium)	3	12.26 years	β^-18	100	
Iodine	125	60 days	EC, 35	7	
			IC	93	

(cont.)

TABLE A.I.1 (cont.)

Isotope	Atomic No.	Half-life	Particle energy (KeV)	%[1]	Cerenkov
Iodine	131	8.04 days	β^-250	2.8	+
			β^-330	9.3	
			β^-610	87.2	
			β^-810	0.7	
			IC	4.5	
				0.4	
				1.5	
Iron	55	2.7 years	EC	100	
Iron	59	45 days	β^-130	1	
			β^-270	46	
			β^-460	53	
			β^-1560	0.3	
Krypton	85	10.6 years	β^-150	0.7	+
			β^-670	99.3	
Magnesium	27	9.5 months	β^-1590	30	+
			β^-1750	70	
Magnesium	28	21.4 hours	β^-420	100	
			IC	4	
Manganese	54	314 days	EC	100	
Manganese	56	2.58 hours	β^-330	1	+
			β^-750	15	
			β^-1050	24	
			β^-2860	60	
Mercury	203	47 days	β^-210	100	
Nickel	63	120 years	β^-67	100	
Phosphorus	32	14.3 days	β^-1710	100	25 (S11)

(cont.)

TABLE A.I.1 (cont.)

Isotope	Atomic No.	Half-life	Particle energy (KeV)	%[1]	Cerenkov
Phosphorus	33	25 days	β^-250	100	
Potassium	40	1.3 × 10^9 years	β^-1320	89	14 (S11)
			EC	11	
Potassium	42	12.4 hours	β^-2(3.6)	18(82)	60 (S11)
Promethium	147	2.6 years	β^-225	100	
Promethium	149	53 hours	β^-780	2	+
			β^-1070	98	
Rubidium	86	18.7 days	β^-680	8.5	23 (S11)
			β^-1770	91.5	
Rubidium	87	4.7 × 10^{10} years	β^-275	100	
Ruthenium-Rh	106	1.0 years	β^-39	100	62 (S11)
			(β^-1500–3600)		
Silver	108	2.4 month	β^-1020	2	
			β^-1650	94	
			EC	4	
Sodium	24	15 hours	β^-1390	100	18 (S11)
Strontium	85	65 days	EC	100	
Strontium	89	51 days	β^-1460	100	+
Strontium-Y	90	28 years	β^-540	100	+
Sulphur	35	87.2 days	β^-167	100	
Yttrium	91	59 days	β^-330	0.3	+
			β^-1540	99.7	

[1] Proportion of the total electron emission characterised by the energy of the particle.

Subject index p. 309

TABLE A.I.2

Calcium-45 (half-life 165 days).

Days		Days		Days		Days	
0	1.000	50	0.811	100	0.657	150	0.533
5	0.979	55	0.794	105	0.643	155	0.522
10	0.959	60	0.777	110	0.630	160	0.511
15	0.939	65	0.761	115	0.617	165	0.500
20	0.919	70	0.745	120	0.604	170	0.490
25	0.900	75	0.730	125	0.592	175	0.480
30	0.882	80	0.715	130	0.579	180	0.470
35	0.863	85	0.700	135	0.568	185	0.460
40	0.845	90	0.685	140	0.555	190	0.450
45	0.828	95	0.671	145	0.544	195	0.441

TABLE A.I.3

Iron-59 (half-life 45 days).

Days	0	2	4	6
Weeks				
0	1.000	0.970	0.941	0.912
1	0.898	0.871	0.844	0.819
2	0.806	0.782	0.758	0.735
3	0.724	0.702	0.681	0.660
4	0.650	0.631	0.611	0.592
5	0.583	0.566	0.549	0.532
6	0.524	0.508	0.493	0.478
7	0.470	0.456	0.442	0.429
8	0.422	0.410	0.397	0.385
9	0.379	0.368	0.357	0.345
10	0.340	0.329	0.320	0.311

TABLE A.I.4

Phosphorus-32 (half-life 14.29 days).

Days	0	1	2	3	4	5	6
Weeks							
0	1.000	0.953	0.908	0.865	0.824	0.786	0.748
1	0.712	0.678	0.646	0.616	0.587	0.559	0.532
2	0.507	0.483	0.460	0.438	0.418	0.398	0.379
3	0.361	0.344	0.328	0.312	0.297	0.283	0.270
4	0.257	0.245	0.233	0.222	0.212	0.202	0.192
5	0.183	0.174	0.166	0.158	0.151	0.144	0.137
6	0.130	0.124	0.118	0.113	0.107	0.102	0.098
7	0.093	0.088	0.084	0.080	0.077	0.073	0.069
8	0.066	0.063	0.060	0.057	0.055	0.052	0.049
9	0.047	0.045	0.043	0.041	0.039	0.037	0.035
10	0.034	0.032	0.030	0.029	0.028	0.026	0.025

TABLE A.I.5

Rubidium-86 (half-life 18.7 days).

Days	0	1	2	3	4	5	6
Weeks							
0	1.000	0.964	0.929	0.895	0.862	0.831	0.800
1	0.771	0.743	0.716	0.690	0.665	0.641	0.618
2	0.595	0.574	0.553	0.533	0.513	0.495	0.477
3	0.459	0.443	0.427	0.411	0.396	0.382	0.368
4	0.354	0.342	0.329	0.317	0.306	0.294	0.284
5	0.273	0.265	0.254	0.245	0.236	0.227	0.219
6	0.211	0.203	0.196	0.189	0.182	0.175	0.169
7	0.163	0.157	0.151	0.146	0.140	0.135	0.130
8	0.126	0.121	0.117	0.112	0.108	0.104	0.101
9	0.097	0.093	0.090	0.087	0.083	0.080	0.078
10	0.075	0.072	0.069	0.067	0.064	0.062	0.060

TABLE A.I.6

Sodium-24 (half life 15 hr).

Day	0	1	2
Hour			
0	1.000	0.330	0.109
2	0.912	0.301	0.099
4	0.831	0.274	0.091
6	0.758	0.250	0.083
8	0.691	0.228	0.075
10	0.630	0.208	0.069
12	0.575	0.189	0.063
14	0.524	0.173	0.057
16	0.476	0.158	0.052
18	0.435	0.144	0.047
20	0.397	0.131	0.043
22	0.362	0.119	0.039

TABLE A.I.7

Sulphur-35 (half-live 87.2 days).

Days		Days		Days		Days	
0	1.000	50	0.672	100	0.452	150	0.304
5	0.961	55	0.646	105	0.432	155	0.292
10	0.923	60	0.621	110	0.417	160	0.281
15	0.888	65	0.597	115	0.400	165	0.270
20	0.853	70	0.572	120	0.385	170	0.259
25	0.820	75	0.551	125	0.370	175	0.249
30	0.788	80	0.529	130	0.356	180	0.239
35	0.757	85	0.509	135	0.342	185	0.230
40	0.728	90	0.489	140	0.329	190	0.221
45	0.699	95	0.470	145	0.316	195	0.212

TABLE A.I.8

Tritium (half-life 12.3 years).

Years	0	1	2	3	4	5	6	7	8	9	10
Months											
0	—	0.945	0.893	0.844	0.798	0.754	0.713	0.674	0.637	0.602	0.569
1	0.995	0.941	0.889	0.841	0.794	0.751	0.710	0.671	0.634	0.599	0.567
2	0.991	0.936	0.885	0.837	0.791	0.747	0.706	0.668	0.631	0.597	0.564
3	0.986	0.932	0.881	0.833	0.787	0.744	0.703	0.665	0.628	0.594	0.561
4	0.981	0.928	0.877	0.829	0.783	0.740	0.700	0.661	0.625	0.591	0.559
5	0.977	0.923	0.873	0.825	0.780	0.737	0.697	0.658	0.622	0.588	0.556
6	0.972	0.919	0.869	0.821	0.776	0.733	0.693	0.655	0.619	0.585	0.553
7	0.968	0.915	0.865	0.817	0.772	0.730	0.690	0.652	0.616	0.583	0.551
8	0.963	0.910	0.860	0.813	0.769	0.727	0.687	0.649	0.614	0.580	0.548
9	0.959	0.906	0.856	0.810	0.765	0.723	0.684	0.646	0.611	0.577	0.546
10	0.954	0.902	0.852	0.806	0.762	0.720	0.680	0.643	0.608	0.575	0.543
11	0.950	0.898	0.848	0.802	0.758	0.716	0.677	0.640	0.605	0.572	0.541

APPENDIX II

Properties of solutes, solvents and scintillation mixtures

The physical properties of most of the commonly used primary and secondary solutes used in liquid scintillation counting are given in Table A.II.1. The emission spectral data are approximate only, since the spectrum is usually complex and dependent on the solvent used for the measurement as well as the concentration of the solute in the solvent. For a more detailed account of the fluorescent spectra of these types of molecules, readers should consult Berlman (1965).

The physical properties of some primary solvents and blenders are given in Table A.II.2. The relative pulse height given is usually related to a 4 g/l solution of PPO in the appropriate solvent and relative to that of PPO in toluene.

A selected number of scintillation mixtures are given in Tables A.II.3 to A.II.5. In these tables and throughout the book, a simple code designation for these mixtures has been adopted. The first one or more capitals refer to the primary solvent or solvents used as the electron-transfer system within the mixture. The first lower-case letter refers to the primary solute and the second lower case letter refers to the secondary solute or waveshifter, where present. The final capital letter refers to any additional blender solvent or antifreeze components that are added. In the case of some of the non-blended mixtures (Table A.II.3) it has been found useful to add a numeral which

TABLE

Physical properties of solutes

	Solute	Abbrev.	Mol. Wt.	Melt Pt.
1.	Oxazole, 2,5-diphenyl-	PPO	221.3	72–73
2.	Oxazole, 2-(1-naphthyl)-5-phenyl	αNPO	271	104–106
3.	Oxazole, 2,5-di(4-biphenylyl)-	BBO	373	237–239
4.	Oxazole, 2(4′-biphenylyl)-6-phenyl benz-	PBBO	347.4	189–190
5.	Oxazole, 2,5 di(2-naphthyl)-	β-NNO		
6.	Oxazole, 5-phenyl-2(4-biphenylyl)-	PBO		
7.	1,3,4 Oxadiazole, 2-phenyl-5(4″-biphenylyl)-	PBD	298.3	167–169
8.	1,3,4 Oxadiazole, 2(4′t-butylphenyl-5(4″-biphenylyl)-butyl)-	butyl PBD	354.5	138–139
9.	1,3,4 Oxadiazole, 2-(1-naphthyl)-5-phenyl-	αNPD	272	370
10.	1,3,4 Oxadiazole, 2,5-diphenyl-	PPD	222	
11.	1,3,4 Oxadiazole, 2,5-di(4-biphenylyl)-	BBD	374	
12.	Benzene, 1,4-di-2(5-phenyl oxazolyl)-	POPOP	364.4	245–246
13.	Benzene, 1,4-di-2(4-methyl-5-phenyl oxazolyl)-	DMPOPOP	392.4	231–234
14.	Benzene, p-bis (o-methyl styryl)-	bisMSB	310.4	180–181
15.	Benzene, 1,4-di-2(5-α-naphthyl)oxazolyl-	α-NOPON	424.5	215–217
16.	Stilbene, 4,4′-diphenyl	DPS	332	308–310
17.	Anthracene		178.2	215–217
18.	Naphthalene		128.2	79–80
19.	Diphenyl	DP	154.2	69–71
20.	p-Terphenyl	TP	230.3	212–213
21.	p-Quaterphenyl	QP	308	
22.	p-Quaterphenyl, 4,4-bis(2-butyl octyloxy)-	BIBUQ	675.06	
23.	Anthracene, 9,10-diphenyl		330	
24.	Thiophene,2,5-bis-(5′-t-butyl benzyloxazolyl)-	BBOT	430.6	201–202
25.	1,6-diphenyl hexa-1,3,5-triene	DPH		

1) Relative Pulse Height (RPH) measured in toluene.

2) Figs in bracjets are conc (g/l).

3) As a 0.1 g/l solution with TP (4 g/l).

4) Fluorescence decay half-life.

A.II.1

used in liquid scintillation counting.

	Solubility (g/l) 20°C		Emission spec.		Relative pulse height[1]		τ
	toluene	xylene	max	mean	primary[2]	secondary[3]	(ns)[4]
1.	414	395	365	394	1.01(4)	0.97	1.6
2.	108	83	399	432		1.21(1.82)	2.0
3.	3.1	1.3	410			1.44	1.4
4.	5.0(25°)						
5.					1.08(5)	1.33	
6.					1.52(7.5)		
7.	21	18	365	388	1.28(10)		1.2
8.	119	77	365		1.53(12)		1.2
9.	49		370				2.0
10.	70	48	355				1.5
11.	2.5	1.0	380				1.4
12.	2.2	1.4	419	444		1.45, 2.29	1.5
13.	3.9	2.6	430				1.5
14.			416				
15.			442	469		2.12	
16.	1.5	0.9	409				1.2
17.			402		2.00		4.9
18.	265	210	335				110
19.	440		316				
20.	8.6	6.0	344	360	1.00(8)		1.2
21.	0.2		369		0.2(sat)		
22.			385		1.60(24)		
23.			428				9.4
24.	53(25°C)		435		1.23(8)		
25.			430	468		1.55	

Subject index p. 309

TABLE

Physical properties of

Solvent	Synonyms	Mol. Wt.	Boiling pt. °C
Toluene	Toluol, methyl benzene	92.13	110.8
p-Xylene	Xylol, 1,4 dimethylbenzene	106.16	138.5
m-Xylene	Xylol, 1,3 dimethylbenzene	106.16	138.8
o-Xylene	Xylol, 1,2 dimethylbenzene	106.16	144
Benzene	Benzole	78.11	80.1
pseudo-Cumene	1,2,4 Trimethylbenzene	120.19	169.8
s-Triethylbenzene	1,3,5 Triethylbenzene	162.27	218
Mesitylene	1,3,5 Trimethylbenzene	120.19	169.8
Dioxane	1,4 Dioxan, glycol ethylene ether	88.10	101.5
Anisole	Methoxybenzene	108.13	155(42.8/ 10 mm)
n-Butylbenzene	1-Phenyl butane	134.21	180
p-Cymene	Cymene, p-isopropyl toluene	134.21	176
Ethyl alcohol	Ethanol, methyl carbinol	46.07	78.5
Methyl alcohol	Methanol, carbinol	32.04	64.65
Ethyl cellosolve	2-Ethoxy ethanol	90.12	135.1
Methyl cellosolve	2-Methoxy ethanol	76.09	124.3

[1] Density values are at 20°C, referred to water at 4°C.

[2] Relative pulse height in cyclohexane.

[3] Fluorescence decay half-life.

A.II.2

primary solvents and blenders.

Melt pt. °C	Refractive index (°C)	Solubility in water (g/l) (°C)	Density[1] (g/ml)	RPH[2]	τ (ns)[3]
−95	1.49728(16)	0.47(16)	0.866	1.00	34
13.2	1.4942(23.4)	insol	0.8611	1.12	30
−53.6	1.49962(14.85)	insol	0.8641	1.09	
−29	1.50777(15.5)	insol	0.8745	0.98	
5.51	1.50142(20)	0.82(22)	0.8794		29
−57.4	1.50672(15.3)	insol	0.876		
	1.4939	insol	0.863	0.96	
−57.4	1.50672(15.3)	insol	0.876		
11.7	1.4232	∞	1.0353		
−37.3	1.51791(20)	insol	0.9954	0.83	
−81.2	1.494(13)	insol	0.862	0.88	
−73.5	1.49474(15)	insol	0.8570	0.80	
−117.3	1.36242(18.35)	∞	0.7893		
−97.8	1.33118(14.50)	∞	0.7928		
			0.9311		
			0.9660		

Table A.II.3

Composition of some non-blended scintillation mixtures.

Component[1]	Abn[2]	Mixtures[3]									
		Tpp4	To4	Ttm4	Tb7	DNpp (Butler)	Xd10	DNpp7	DNb10 (Pfeffer)	Td12	Tpd4
Toluene (ml)	T	1000	1000	1000	1000					1000	1000
Xylene (ml)	X						1000				
Dioxane (ml)	D					1000		900	1000		
Naphthalene (g)	N					120		100	60		
PPO (g)	p	4				4		7			4
PBD (g)	d						10			12	
Butyl PBD (g)	b				7				10		
p-terphenyl (g)	t			4							
BBOT (g)	o		4								
POPOP (g)	p	0.05				0.075		0.2			0.05
DMPOPOP (g)	d										
bis MSB	m			1							
Cost (p)/10 ml		1.68	1.78	1.98	2.01	3.98	4.13	4.29	4.46	4.69	1.68

[1] See Table A.II.1 for abbreviations of solutes, solvent, etc.

[2] Abbreviations used in this Table.

[3] The first capital letter is the primary solvent, second capital letter (if present) is the secondary solvent (e.g. naphthalene), first small letter of primary solute, second small letter (if present) is secondary solute or waveshifter. Any large letters following the small letters indicate any blending solvents present and the figures after all the letters refer to the concentration of the primary solute in g/l. The concentration of the primary solute may be varied under different conditions and can be indicated on the code.

indicates, in g/l, the concentration of the primary solute. In some cases, the mixtures are well known by the author's name, such as 'Bray's', and these, where appropriate have also been given.

An additional line indicating an assessment of the cost (in UK pence, 1973) per 10 ml of scintillant has been given, based on the price (retail) of the components purchased in 'large lab' quantities of a kg or a few litres as appropriate. It should be em-

TABLE A.II.4

Composition of some blended toluene-based scintillation mixtures.

Component[1]	Abn[2]	Mixtures[3]					
		TpEGN (Sarnat and Jeffay)	TppE (Phosphor C)	TpdA	TNppX (Phosphor A)	TpC (Mahin and Lofberg)	TpdD
Toluene (ml)	T	1000	1000	1000	1000	1000	750
Naphthalene (g)	N				150		
PPO (g)	p	11.4	7	4	7	6	5
DMPOPOP (g)	d			0.05			0.3
POPOP (g)	p		0.6		0.6		
Ethyl alcohol (ml)	E	1000	500				
Ethylene glycol (ml)	G	250					
Methoxy ethanol (ml)	C					600	
Ethoxy ethanol (ml)	X				300		
Nitric acid (ml)	N	12.4					
Dibutyl phosph. (ml)	D						250
Gl. Ac. Acid (ml)	A			1			
Cost (p)/10 ml		1.38	1.58	1.69	2.06	2.49	3.27

[1] See Table A.II.1 for abbreviations of solutes, solvents, etc.

[2] Abbreviations used in this Table.

[3] As in footnote to Table A.II.3.

phasised that the mixtures given in each Table are selected primarily from published recipes used most frequently, and which in many cases, especially in the case of blended mixtures, are the result of an empirical choice of ingredients, often aptly referred to as a 'cocktail'. A more systematic approach to the construction of mixtures, having in mind maximum blending ability with minimum quenching, is very much needed in many of these mixtures. The type of approach adopted by Little and Neary (1971) would appear to be the most useful and logical. In general, the blended mixtures given in Tables A.II.4 and A.II.5 are subject to considerable variation especially with regard to the level of primary and secondary solutes present. In some instances, such as

TABLE A.II.5

Composition of some blended dioxane-based scintillation mixtures.

Component[1]	Abn[2]	Mixtures[3]					
		DNXppE (Phosphor 2)	DXpdEA (Paus)	TDNpdM	DNXpnE (Kinard)	DNppGM (Bray)	DAppD (Polyether 611)
Toluene (ml)	T			500			
Xylene (ml)	X	772	385		500		
Dioxane (ml)	D	772	385	500	500	1000	750
Naphthalene (g)	N	160		100	105	60	
Anisole (ml)	A						125
PPO (g)	p	10	5	6	6.5	4	12
POPOP (g)	p	0.1				0.2	0.4
DMPOPOP (g)	d		0.075	0.15			
α-NPO (g)	n				0.065		
Ethanol (ml)	E	682	230		300		
Ethylene glycol (ml)	G					20	
Methanol (ml)	M			300		100	
Dimethoxyethane (ml)	D						125
Acetic acid (ml)	A		1				
BHT[4] (g)							1
Cost p/10 ml		2.43	2.44	2.65	2.66	3.62	4.61

[1] See Table A.II.1 for abbreviations of solutes, solvents, etc. [2] Abbreviations used in this Table. [3] Coding as described in legend to Table A.II.3. [4] BHT is di-t-butyl-4-hydroxytoluene and is referred to in chapter 2, as an antioxidant.

Some commercial mixtures used in scintillation counting.

Supplier[1]	Toluene-based		Xylene-based	Dioxane-based		Solubilizer	Gel scintillant	Colloidal system
	Conc.[2]	Drymix[3]	Conc.[2]	Conc.[2]	Drymix[3]			
Nuclear Enterprises	NE 233		NE 216	NE 220 NE 240 NE 250			NE 221	NE 520
Nuclear Chicago						NCS		PCS
Beckmann	Fluor- alloy TL2 alloy TLY alloy TLX	Fluor- alloy TLA alloy TLB			Fluor- alloy D × A alloy D × B	BIOSOLV BBS 1		BIOSOLV BBS 3
Packard	Perma- fluor I fluor II fluor III fluor IV	Perma- blend I blend II blend III				Soluene 100		Instagel
New England Nuclear	Liquifluor	Omni-fluor		Aqua- fluor	Omni-fluor	Protosol		Aquasol
Koch-Light	Scintol 1 Scintol 2 Scintol 3	Scintimix Scintipak	KL 359 KL 360	KL 371 KL 372			KL 368	Unisolv I Unisolv II
ICN Tracer	Liqui-Mix	Soli-Mix 1 Soli-Mix 2		Aqua- scint I scint II				

[1] Addresses in Appx. III, Tables 1 and 2. [2] Concentrate, dilute with primary solvent before use. [3] Powder formulation, add to mixture before use.

Subject index p. 309

TpEGN and TpdD (Table A.II.4), the mixtures have been designed for specific uses and are thus not to be regarded as alternatives to the other mixtures in the Table.

Many commercial mixtures are available and a number of these are listed in Table A.II.6. This list is not comprehensive as some organizations do not release the composition of their mixtures. This practice should be deprecated since should incompatibility between solute or solvent and sample occur, there is no means of investigating the cause. The possibility of counting heterogeneity always exists where detergents are used in the counting mixtures, and all such mixtures should at least indicate that a detergent is present, so that appropriate, unequivocal quench correction procedures may be employed. The safest procedure is to employ only those mixtures whose exact composition is clearly indicated.

A list of manufacturers of liquid scintillation counting instruments as at December 1973 is given in Table A.II.7, but a more useful review is given by Rapkin (1972).

TABLE A.II.7

List of scintillation spectrometer manufacturers*.

Manufacturer⁺	Country	Date commenced
Aloka	Japan	1965
Beckmann Instruments Inc.	U.S.A. & U.K.	1965
Friesecke & Hoepfner	Germany	1966
Intertechnique	France	1968
Nuclear Chicago	U.S.A.	1961
Nuclear Enterprises	U.K.	1960
Packard	U.S.A.	1954
Phillips	Holland	1963
Tracerlab	U.S.A. & Belgium	1954
Wallac (LKB)	Finland	1968

* Data as at 1973, adapted from Rapkin (1972).

⁺ See Appx. III, Table 1 for addresses.

APPENDIX III

Analytical key to sample preparation

Where only casual use of the liquid scintillation counting technique is envisaged, it is often necessary to try and decide from a very large and confusing literature the best approach to the assay of any particular sample. Often, the experimenter simply follows the procedure adopted by a fellow worker or an earlier experimenter in the field, without an examination of the literature to determine if a better method exists. In most cases, the procedure adopted by this method is probably the best, but this attitude may also lead to a blind faith in the truth of the numbers which are released by the instrument. Unless the existence of chemiluminescence, quenching and counting heterogeneity are recognised, then unexplained fluctuations in the results may be attributed to the biology rather than the assay procedure itself.

The following key is intended only as a guideline to a suitable procedure, and cannot of course be definitive. It is based largely on local experience, conversation and broad overview of the literature used in the present work.

The operation of the key is similar to those keys employed in taxonomic literature in biology. Starting from 1, the choice of a, b or c is made for the operator's particular problem. Having made this choice, the right hand figure is noted and the choice of questions further along the key is then considered. The choice of either a or b is made and the right hand figure is noted and the questions designated by the same figure further along the key are again considered, and so on. This process is continued until a double letter, e.g. AA, BB, etc. is arrived at; this refers to a suggested method or methods of counting, which are listed at the end of the Key. The ultimate method chosen will, of course, depend on specific requirements of the operator, but it is hoped that the key will assist in arriving at the most suitable method.

KEY

1	a	Solid	2
	b	Liquid	3
	c	Gas	4
2	a	Tissue origin	5
	b	Non-tissue origin	6
3	a	Aqueous, i.e. polar	7
	b	Non-aqueous i.e. non-polar	8
4	a	Acidic (CO_2, SO_2, H_2S, etc.)	9
	b	Basic (NH_3, CH_3NH_2, etc.)	10
	c	Neutral (HCHO, CH_4, etc.)	11
5	a	Wet	12
	b	Dry or pellet	13
6	a	Organic	14
	b	Inorganic	15
7	a	Labelled macromolecule only	16
	b	Label in macromolecule and/or low mol wt substances	17
8	a	Solute labelled (e.g. lipid, hydrocarbon, etc.)	18
	b	Solvent labelled	AA
9	a	From respiration and enzyme degradation studies	FF, EE
	b	From combustion techniques	CC, FF

Subject index p. 309

10	a	From enzyme and degradative studies	GG
	b	From reduction processes	GG
11	a	Volatile aldehydes	HH
	b	Aliphatic hydrocarbons	AA
12	a	Soft (lung, liver, faeces, etc.)	19
	b	Hard (bone, collagen, teeth, etc.)	20
13	a	Bone, dry solid, etc.	20
	b	Pellet after centrifuging, residue on paper, TLC scrapings etc.	21
14	a	Label soluble in toluene	AA
	b	Label insoluble in toluene	22
15	a	Label present in macroscopic quantities	23
	b	Label present in trace amounts	24
16	a	Solution $<$ 1000 cpm/0.2 ml	25
	b	Solution $>$ 1000 cpm/0.2 ml	26
17	a	Solution $<$ 1000 cpm/0.2 ml	27
	b	Solution $>$ 1000 cpm/0.2 ml	28
18	a	Solution $<$ 1000 cpm/0.2 ml	AA
	b	Solution $>$ 1000 cpm/0.2 ml	JJ, AA
19	a	Single or very few samples	CC, FF, DD
	b	Many samples, routine assay system	29
20	a	High spec. act. e.g. $>$ 500 cpm/20 mg	DD
	b	Low spec. act. e.g. $<$ 500 cpm/20 mg	30
21	a	Absolute activity required	31
	b	Relative activity between samples only required	32
22	a	Soluble in water	33
	b	Insoluble in water	CC
23	a	Mixed of inorganic ions containing labelled ion	34
	b	Pure inorganic ion	35
24	a	Forms complex soluble in di- or tri-butyl phosphate	LL
	b	Forms specific ppt with organic acid	JJ, MM
	c	Does not form either (a) or (b)	MM
25	a	Salts or sucrose absent	II
	b	Salts, sucrose etc. present	36
26	a	Salts, sucrose etc. absent	OO
	b	Salts, sucrose etc. also present. Take aliquot	NN
27	a	Total activity required	25
	b	Separate labelled species required to be assayed	NN + II
28	a	Macromolecule label low, small mol has high spec. act.	PP + II or BB
	b	Small mol high, macromol. low spec. act.	NN + II
29	a	High spec. act. $>$ 2000 cpm/20 mg	DD
	b	Low spec. act. $<$ 2000 cpm/20 mg	FF, CC

30	a	Calcium, strontium, etc.	KK
	b	Tritium, ^{14}carbon, ^{35}sulphur etc.	DD
31	a	Soluble in toluene	AA
	b	Insoluble in toluene	FF, DD
32	a	Visible pellet, TLC scrapings, etc.	FF, DD
	b	Very thin layer on solid support, disc etc.	JJ
33	a	High spec. act. $>$ 1000 cpm/0.5 ml	BB
	b	Low spec. act. $<$ 1000 cpm/0.5 ml	II
34	a	Inorganic ion forms dibutyl phosphate soluble complex	LL
	b	Does not form such a complex	37
35	a	Soluble in water	38
	b	Insoluble in water and toluene	MM
36	a	Fraction from CsCl, sucrose gradient etc. i.e. in small vol $<$ 0.5 ml	II
	b	Fract from chromatog vol usually $>$ 0.5 ml	39
37	a	Isotope extractable with acid or alkali	II
	b	Isotope not extractable with acid or alkali	MW
38	a	High spec. act. $>$ 500 cpm/10 mg	soln then BB or MM
	b	Low spec. act. $<$ 500 cpm/10 mg	MM
39	a	High spec. act. $>$ 1000 cpm/0.5 ml	BB
	b	Low spec. act. $<$ 1000 cpm/0.5 ml	II

SUGGESTED METHODS

AA Use a toluene-based scintillant such as Tpp4 (see Table A.II.3).

BB Use a dioxane-based scintillant (see Table A.II.5).

CC Combustion technique (see § 3.1).

DD Wet oxidation technique, e.g. the perchloric-peroxide digestion method of Mahin and Lofberg (see § 3.3.1).

EE Wet oxidation technique with liberation of gas (see § 3.3.2).

FF Quaternary ammonium salt solubilization method (§ 3.2).

GG Collect in dilute acid and use a colloid system (§ 2.2.3).

HH pass into a dimedone solution, decolourise by reduction, and use suspension counting (§ 2.2.2 and § 3.4).

II Colloid system (§ 2.2.3).

JJ Put on glass fibre disc, dry and assay in Tpp4 (§ 2.2.1).

KK Use the ^{45}calcium procedure of Carr and Parsons (1962) (§ 4.1.1).

LL Use the dibutyl phosphate extraction method (see § 10.1).

MM Make a suspension with Cab-O-Sil or other gelling agent as in § 2.2.2.

NN Add 5% perchloric acid and assay the precipitate on a disc see § 2.2.1.

OO Take 0.2 ml aliquot and add an appropriate enzyme as in § 3.3.3.

PP By selective absorption on charcoal, bentonite etc. (e.g. as in radioimmunoassay procedure) see § 4.3.3.

Subject index p. 309

APPENDIX IV

Glossary

The many terms employed in liquid scintillation counting literature have been derived from a number of different disciplines and several of these terms are given the same name but mean something quite different in different contexts. No attempt has been made to suggest a standardization of these terms in the following glossary.

Auger electrons Those electrons emitted from an atom due to the filling of a vacancy in an inner electron shell.

Balance point The gain and discriminator settings which result in a *channels ratio* of unity.

Balance point quench correction The use of a selected counting window centred around a threshhold discrimination level so that positive and negative deviations from the unquenched counts are equally probable.

Blender A solvent which will reduce a two-phase, liquid: liquid system into a single, homogenous phase. These solvents, usually aliphatic alcohol derivatives, will hence extend the electron transfer system (e.g. toluene).

Bremsstrahlung The electromagnetic radiation resulting from the retardation of charged particles.

Channel see Discriminator.

Channels ratio The ratio of counts observed in two *channels* or *window settings* of the *discriminators*.

Curie (*Ci*) A measure of radioactivity, named after Madame Curie, the discoverer of Radium. It is equivalent to 3.7×10^{10} disintegrations per sec.

Detection efficiency (*D*) This term applies to a scintillation counter and is defined for *single energy particles*, of energy (E) as the fraction of particles which give detectable pulses.

Diffusional quenching Excited solvent or solute molecules combine with the quencher molecules to form encounter complexes which then undergo reactions to yield deactivated products.

Discriminator The cut-off point of the gain setting of the instrument used. The purpose of the discriminator is to count the number of pulses above (*lower discriminator*) or below (*upper discriminator*) specific energy levels. When these two energy levels are spaced at different levels, the region in between the two settings is referred to as a *window* or *channel*.

Dynamic quenching see Diffusional quenching.

Efficiency The percentage of the disintegrations occurring which can be measured as counts within the discriminator window of the instrument used.

Electron capture (*EC*) A nuclear transformation whereby a nucleus captures one of its orbital electrons.

Electron transferring system This term refers to the hydrocarbon solvent or dissolved hydrocarbon system (e.g. napthalene) which is capable of absorbing energy derived from a beta-emitter and transferring this energy via excited states to the primary solute. The energy transferred to the primary solute in a toluene solution is usually about 5% of the energy leaving the beta-emitter.

Energy transfer quantum efficiency (f) The efficiency by which energy is transferred by the *solvent excitation migration* process.

Excimers A dimeric association of an excited solvent or solute molecule with an equivalent unexcited molecule resulting in an excited complex whose combined first singlet state is lower as a result of vibrational relaxation but which can be reactivated by thermal activation to the first singlet energy state of the excited monomer.

External standard The positioning of a standard gamma source near to a liquid scintillation counting vial causing Compton and recoil electrons to activate photon emission within the scintillant. This is normally executed in an automated fashion in many Instruments (AES). The gamma source can be single, e.g. ^{133}Ba or multiple, e.g. ^{226}Ra + ^{241}Am.

External standard ratio (ESR) The ratio of the counts due to the external standard in two selected windows.

Figure of merit This term has been used for several different purposes in scintillation counting. The following are some of the many definitions.

1) The overall efficiency of the scintillator and the photomultiplier tube (P) (Horrocks and Studier 1961), where $P = n_E/(EF_E)$, (n_E = average number of electrons at the first stage of the photomultiplier; E = electron energy in the scintillator; F_E = correction for energy losses into vial walls).

2) Scintillator figure of merit. The component of the pulse height which is due to the scintillator (F) (Birks 1969), where $F = sfqm$, (s = solvent conversion energy; f = solvent-solute energy transfer quantum efficiency; q = solute fluorescence quantum efficiency; m = spectral matching factor).

3) A criterion of performance in a liquid scintillation counter. The value sets the minimum activity which can be distinguished from background in a given measuring time (F'), where $F' = (E_f)^2/B$, (E_f = counting efficiency for an isotope when the background is B). For a 1 min counting time, the minimum dpm that can be detected with a probability of 95% is $2/\sqrt{F'}$.

4) Pulse height discrimination figure of merit (M). The value is a property of the scintillator and determines the degree of discrimination that can be achieved. The value also gives a theoretical estimate of the particle rejection ratio. $M = t/(\tau_\gamma + \tau_n)$, ($t$ = time or voltage difference between peaks; τ_γ = full width half max. of γ peak; τ_n = full width half max. of neutron peak).

5) In radiocarbon dating, as the ratio of the contemporary net carbon count rate to the square root of the background (Aitken 1961) = $S_0/\sqrt{B}$.

Fluor The combination of primary and secondary solutes used in converting the energy derived from the electron transferring systems into photons.

Subject index p. 309

Fluorescence quantum efficiency (q) The ratio of the number of fluorescence photons emitted to the number of molecules originally excited. For toluene, $q = 0.1$. For the efficient primary solutes, q approaches unity.

G value Number of events per 100 electron volts expended.

Gain factor (photocathode) (M) The number of anode pulses resulting from the emission of one electron from the photocathode passing through the dynode system. Thus the anode pulse amplitude $= MN$ where N = no of electrons leaving photocathode, and entering dynode system.

Gel Applies to those suspending agents which result in the formation of a semi-solid transparent mass supporting the aqueous solution of the beta-emitter.

Internal conversion (IC) A transition between two energy states of a nucleus in which the energy difference is not emitted as a photon (gamma ray) but is transferred to an orbital electron which is ejected from the atom.

Isotope effect A difference in the chemical or physical property of a compound due to the presence of an isotope, rather than its non-radioactive equivalent.

Libby half-life Half-life of carbon-14 used by Libby (1952) equivalent to 5568 years.

Maximum photoelectric quantum efficiency (K) The quantum efficiency at the peak of the photocathode spectral response (varies from 15% to 30%).

Merit number or *Merit value* (M) The product of the percentage by volume of radioactive aqueous solution in the counting vial and the counting efficiency (Van der Laarse 1967).

$M = \%$ aqueous sample $\times$ % efficiency.

The instrument corrected merit value (MIV) is corrected for the efficiency (instrument reference) of the isotope (E_f) in a wide window setting (to avoid artifactual changes due to loss from a narrow window during quenching).

The instrument corrected merit value stability quotient (MISQ) has been used by the author to relate colloid systems whose stability with time after mixing vary (Fox 1974)

($\%\sigma$ = percentage coefficient of variation of cpm over a 48 hr counting period).

Primary solute The organic compound which initially converts the transferred electron energy within the electron-transferring system into photons.

Pulse height A measured anode pulse related to the photon yield of an ionizing particle. The amplitude or height of the pulse is controllable by the *amplification* of the instrument. Within a given amplification setting, the pulse-height is a product of the energy by which the electron is emitted (beta spectrum); the detection efficiency spectrum of the instrument and thus the spectrum of pulse heights *pulse spectrum distribution* (PSD) will give an indication of the range of energies by which particles are emitted from an isotope.

Pulse shape discrimination (PSD) The study of all light pulses emitted from a liquid scintillation system as a means of particle discrimination.

PSD figure of merit (M) This factor is a property of the fluor. It is the ratio of the time or voltage difference between peaks in a zero crossing PSD to the sum of the full width half maxima of the two peaks (for details see McBeth, Winyard and Lutkin 1970). The larger the value of M, the greater the degree of PSD that is achieved. The value also gives a theoretical estimate of particle rejection ratios.

Phosphor A term often loosely and inaccurately applied to the scintillant mixture.

Quenching The process of reduction of pulse height distribution by interference with the conversion of beta-emitter energy to its final assessment as light energy. There are many different causes of quenching, and current evidence suggests that there may be both qualitative as well as quantitative differences between the effects of different quenching agents. The presence of coloured materials in the final scintillant mixture also reduces the efficiency of the light transfer from the fluor system to the photomultiplier tubes and is known specifically as *colour* or *optical quenching*. The many different processes which inhibit the transfer of energy to the fluor system is referred to in general terms as *chemical* or *impurity quenching*.

Recoil The motion acquired by a particle due to the ejection of another particle or photon.

Relative pulse height (*RPH*) The anode pulse related to the photon yield of a scintillant formulation relative to a standard formulation, e.g. a 4 g/l solution of PPO in toluene. The standard used has varied between authors and this should therefore be stated.

Roentgen A unit of exposure of X or gamma radiation.

Scintillation attenuation factor (*SAF*) This term was first described by Jaffee and Ford (1966) as the attenuation of scintillation efficiency produced by 1 mMole of quenching agent.

Scintillation collection efficiency (*G*) The fraction of the photons (P) emitted from the scintillant that actually impinge on the photocathode.

Scintillation efficiency (*S*) The fraction of the energy (E) of the ionizing particle which is converted into (P) fluorescent photons of mean energy (U). Thus $S = PU/E =$ (Total photon energy)/(Energy of ionizing particle).

Secondary solute The organic substance which absorbs energy emitted by the primary solute and converts it into photons of a higher wavelength, more suitable for the photomultiplier system used in the instrument.

Solubilizer A substance which will bring into solution an otherwise insoluble material, so that it may be blended into a toluene-rich fluor solution. This term should not be applied to those systems containing detergents which form gels and colloids.

Solvent excitation migration The process by which the energy of pi bond excitation moves throughout the solvent molecules, by exciting adjacent unexcited molecules forming intermediate structures known as *excimers*. Energy transfer quantum efficiency describes the extent of this process.

Solvent-solute energy transfer quantum efficiency (f) The number of excited solute molecules produced per excited solvent molecule used. $f = B/A = B/sE$, (B = no of S_{1Y} excited solute molecules created; A = no of S_{1X} excited solvent molecules used; s = solvent conversion factor; E = energy of beta particle.

Depends on *nature* of *solvent* and concentration of *solute* in *solvent*, but almost independent on nature of *solute*.

Specific activity The amount of radioactivity present (in Curies or a fraction of a Curie or as dpms) per unit of weight (either as grams or moles). The specific activity of a solution is often described as the amount of radioactivity per unit of volume. The maximum specific activity attainable by an isotope is given by the following equation

$$S_{max} = \frac{6.02 \times 10^{23}}{3.7 \times 10^{10}} \times \frac{0.693}{M\tau} = \frac{1.123 \times 10^{13}}{E\tau} \text{Ci/g}$$

where M is the atomic mass of the isotope and τ is the half-life in seconds.

Spectral matching factor (m) The relationship (normalised integral product) between the solute fluorescence spectrum and the photocathode spectral response spectrum.

Static quenching described by Weller (1959). A form of chemical quenching in which the non-fluorescent molecules combine with the solvent or solute molecules to form unexcited encounter molecules. This effectively decreases the excited fluor molecule concentration.

Thermal diffusion The Brownian movement of molecules of solvent and solute, which increases as the temperature of the solution increases.

Tritium unit (*TU*) One unit corresponds to a ratio of tritium atoms to hydrogen atoms of 10^{18} i.e. 100 TU is equivalent to 6.5 dpm ^{3}H/g hydrogen (Tamers 1964).

Wavelength shifter Any substance which absorbs light energy at a low wavelength and re-emits light at a higher wavelength. Often used as an alternative term to secondary solute, but can also be applied to the conversion of Cerenkov light to photon energies more appropriate to the photomultiplier system used. It is therefore recommended that this term be applied specifically to this use.

Window see Discriminator.

Y value Defined by Moghissi et al. (1969) as the minimum limit of detection at a 1–0 confidence level and a one minute counting time.

$$Y = \frac{\dfrac{\text{pCi/g}}{\sqrt{B}}}{2.22 \times E_f \times M}$$

(B = background, cpm; E_f = efficiency, cpm/dpm; M = amount of sample, g, in mixture).

Appendix V

Table A.V.1

Instrument manufacturers.

Beckmann Instruments Inc.,* Scientific Instruments Div., 2500 Harbor Blvd. Fullerton, California, 92634, U.S.A. (U.K. address: Beckmann RIIC, Ltd., Eastfield Industrial Estate, Glenrothes, Scotland, KY7 4NG).

Berthold/Frieseke, GmbH, Vertriebsgesellschaft für Messtechnik, 75 Karlsruhe-Durlach, Bergwaldstr. 30, Germany (U.K. agent: 232 Addington Rd., S. Croyden, Surrey, CR2 8YD).

ICN Tracerlab, 1601 Trapelo Rd., Waltham, Mass, U.S.A. (U.K. address: Ship Yard, Weybridge, Surrey).

Intertechnique, 78 Plaisir-France (U.K. address: 5, Victoria Rd., Portslade, Sussex).

LKB Wallac, LKB-Produkter AB, S-161 25 Bromma 1, Sweden (U.K. address: LKB Instruments Ltd., LKB House, 232, Addington Rd., S. Croydon, Surrey, CR2 8YD).

Nuclear Chicago Corporation*, 333 East Howard Av., Des Plaines, Illinois 60018, U.S.A. (U.K.: G.D. Searle & Co, Ltd., Lane End Rd, High Wycombe, Bucks).

Nuclear Enterprises, Ltd.,* Bankhead Crossway, Edinburgh, EH 11 ONU, Scotland, U.K.

Packard Instrument Co., Inc.,* 2200 Warrenville Rd., Downers Grove, 60515, Chicago, Illinois, U.S.A. (U.K. address: Caversham Bridge House, 13/17 Church Rd., Caversham, Reading RG4 7BR, Berks).

Phillips N.V. Gloeilampen Fabrieken, Anal. Equipt. Dept., TQ111-2, Eindhoven, The Netherlands (U.K. Pye Unicam Ltd., York St., Cambridge, CB1 2PX).

* also supply scintillation chemicals.

Table A.V.2

Commercial firms that supply scintillation chemicals.

British Drug Houses Chemicals Ltd., Poole, Dorset, BH12 4NN, U.K.

Calbiochem, 10933 N. Torrey Pines Rd., La Jolla, California, 92037, U.S.A.

Eastman-Kodak Ltd., Acornfield Rd, Kirkby, Liverpool, L33 7UF, U.K.

Emmanuel, Ralph N., 264, Water Rd., Wembley, Middlesex, HAO 1PY, U.K.

Fisons Scientific Apparatus, Bishop Meadow Rd., Loughborough, Leics LE11 ORG, U.K.

Fluka AG, Chemische Fabrik, CH-9470, Buchs, Schweiz.

Hopkins and Williams Ltd., P.O. Box 1, Romford, RM1 1HA, U.K.

ICN Pharmaceuticals Inc., Life Sciences Gp., 26201, Miles Rd., Cleveland Ohio, U.S.A.

Subject index p. 309

Koch-Light Laboratories Ltd., Colnbrook, SL30BZ, Bucks, U.K.
Merck, E., D61, Darmstadt, Germany.
New England Nuclear, 575 Albany St., Boston Mass, 02118, U.S.A.
Pfaltz & Bauer, Inc. (Aceto Chemical Co., Ltd.) 126-04 Northern Blvd, Flushing, N.Y. 11368, U.S.A.
Schuchardt Chemische Fabrik, 8 Munchen 80, Gausbergstr. 1-3, Germany.
Schwartz/Mann (Becton, Dickinson & Co), Orangeburg, New York 10962, U.S.A.
Serva Feinbiochemica, D-6900 Hsidelberg 1, P.O. Box 1505, Fed. Rep of Germany.
Sigma London Chemical Co., Ltd., Norbiton Station Yard, Kingston upon Thames, Surrey, KT2 7BH, U.K.
Rohm and Haas, Inc., Lennig House, 2 Mason's Av., Croyden CR9 3NB, U.K.

References

AGNEDAL, P.O. (1967) Radioecological Concentration Processes. *In*: Åberg, E. and F.P. Hungate, eds. (Pergamon Press, Oxford) p. 879.

AITKEN, M.J. (1961) Physics and Archaeology (Interscience, New York) p. 115.

ALFRED, J.B. (1967) Anal. Chem. *39*, 547.

ALPERS, D.H. and R. GLICKMAN (1970) Anal. Biochem. *35*, 314.

ANKER, H.S. (1970) Febs. Letters *7*, 293.

ANON (1961) Chem. Eng. News (Apr. 17), *39*, 63.

ARNOLD, J.R. (1954) Science *119*, 155.

ARNOLD, J.R. (1958) Archaeology and chemistry. *In:* Bell, C.B. and F.N. Hayes, eds., Liquid Scintillation Counting, Proceedings of Conference, North western Univ. 1958. (Pergamon Press, Oxford) pp. 129–134.

ASHCROFT, J. (1969) Int. J. Appl. Radn. Isot. *20*, 555.

ASHCROFT, J. (1970) Anal. Biochem. *37*, 268.

ATKINSON, M.R. and A.W. MURRAY (1965) Biochem. J. *94*, 64.

AURES, D. and W.G. CLARK (1964) Anal. Biochem. *9*, 35.

AVERBUCH, T. and Y. AVNIMELACH (1970) Plant Soil *33*, 260.

BAKAY, B. (1971) Anal. Biochem. *40*, 429.

BAKER, N., H. FEINBERG and R. HILL (1954) Anal. Chem. *26*, 1504.

BAGGETT, B., T.L. PRESSON, J.B. PRESSON and J.C. COFFEY (1965) Anal. Biochem. *10*, 367.

BAILLIE, L.A. (1960) Int. J. Appl. Radn. Isot. *8*, 1.

BALHARRY, G.J.E. and D.J.D. NICHOLAS (1971) Anal. Biochem. *40*, 1.

BALL, C.R., R.W. POYNTER and H.W. VANDENBERG (1972) Anal. Biochem. *46*, 101.

BARKER, H. (1953) Nature *172*, 631.

BATCHELOR, R., W.B. GILBOY, J.B. PARKER and J.H. TOWLE (1961) Nucl. Inst. Method *13*, 70.

BAXTER, C.F. and I. SENONER (1964) Anal. Biochem. *7*, 55.

BELCHER, E.H. (1960) Phys. Med. Biol. *5*, 49.

BELCHER, E.H. (1963) Proc. Roy. Soc. (A) *216*, 90.

BELL, C.B. and F.N. HAYES, eds. (1958) Liquid Scintillation Counting. Proceedings of a Conference, Northwestern Univ., Aug. 20–22, 1957 (Pergamon Press, Oxford).

BENAKIS, A. (1971) A new gelifying agent in liquid scintillation counting. *In:* Horrocks, D.L. and C.-T. Peng, eds., Organic Scintillators and Liquid Scintillation Counting (Academic Press, New York and London) pp. 735–745.

BENSON, R.H. (1966) Anal. Chem. *38*, 1353.

BERKOWITZ, E.H. (1969) Nucl. Inst. Methods *73*, 225.

BERLMAN, I.B. (1965) Handbook of Fluorescence Spectra of Aromatic Molecules (Academic Press, New York).

BERSAQUES, J. DE (1963) Int. J. Appl. Radn. Isot. *14*, 173.

BIRKS, J.B. (1964) The Theory and Practice of Scintillation Counting (Pergamon Press, Oxford).

BIRKS, J.B. (1969) Solutes and solvents for Liquid Scintillation Counting. Koch Light Laboratories, Ltd. booklet. 39 pps.

BIRKS, J.B. and G.C. POULLIS (1972) Liquid scintillators. *In:* Crook, M.A., P. Johnson and B. Scales, eds., Liquid Scintillation Counting. Proceedings of International Conference, Brighton, 1971, vol. 1 (Heyden & Son, Ltd., London, N. York, Germany) pp. 1–21.

BIRNBOIM, H.C. (1970) Anal. Biochem. *37*, 178.

BLAIR, A. and SEGAL (1960) J. Labs. Clin. Med. *55*, 959.

BLANCHARD, F.A. and I.T. TAKAHASHI (1961) Anal. Biochem. *33*, 975.

BLANCHARD, F.A., M.R. WAGNER and I.T. TAKAHASHI (1968) Adv. Tracer Methodol. *4*, 133.

BLOOM, B. (1963) Anal. Biochem. *6*, 359.

BLUMENKRANTZ, N. and D.J. PROCKOP (1969) Anal. Biochem. *30*, 377.

BOECKX, R.L., D.J. PROTTI and K. DAKSHINAMURTI (1973) Anal. Biochem. *53*, 491.

BOLLINGER, J.N., W.A. MALLOW, J.W. REGISTER, Jr. and D.E. JOHNSON (1967) ANAL. CHEM. *39*, 1508.

BOLLUM, F.J. (1959) J. Biol. Chem. *234*, 2733.

BOLLUM, F.J. (1963) Cold Spring Harbor. Symp. Quant. Biol. *28*, 21.

BOLLUM, F.J. (1966) Filter paper disc techniques for assaying radioactive macromolecules. *In:* Cantoni, G.L. and D.R. Davies eds., Proc. Nucleic Acid Res. (Harper and Row, New York) pp. 296–300.

BOSSHART, R.E. and R.K. YOUNG (1972) Anal. Chem. *44*, 1117.

BOUSEQUET, W.F. and J.F. CHRISTIAN (1960) Anal. Chem. *32*, 722.

BOWMAN, M.S. and R. ROHRINGER (1970) Can. J. Bot. *48*, 803.

BOYCE, I.S. and J.F. CAMERON (1962) A low background liquid scintillation counter for the assay of low-specific-activity tritiated water. *In:* Tritium in The Physical and Biological Sciences (Inst. Atomic Energy publ., Vienna) Vol. 1, p. 231.

BOYD, J.B. and H.K. MITCHELL (1966) Anal. Biochem. *14*, 441.

BRANSOME, E.D., Jr. (1970) ed. The Current Status of Liquid Scintillation Counting (Grune and Stratton, Inc., New York and London).

BRANSOME, E.D. Jr. and M.F. GROWER (1970) Anal. Biochem. *38*, 401.

BRANSOME, E.D. Jr. and M.F. GROWER (1971) Local absorption of low energy betas by solid supports: A problem in heterogenous counting. *In:* Horrocks, D.L. and C.T. Peng, eds., Organic Scintillators and Liquid Scintillation Counting (Academic Press, New York and London) pp. 683–686.

BRAUNSBERG, H. and A. GUYVER (1965) Anal. Biochem. *10*, 86.

BRAY, G.A. (1970) Determination of radioactivity in aqueous samples. *In:* Bransome, E.D. Jr., ed., The current status of Liquid Scintillation Counting (Grune and Stratton, Inc., New York and London) pp. 170–180.

BREITMAN, T.R. (1963) Biochem. Biophys. Acta *67*, 153.

BROWN, B.L. and W.S. REITH (1965) Biochem. Biophys. Acta *97*, 378.

BROWNELL, J.R. and A. LAUCHLI (1969) Int. J. Appl. Radn. Isot. *20*, 797.

BUCKLEY, J.P. (1971) Int. J. Appl. Radn. Isot. *22*, 41.

BURLEIGH, R. (1972) Liquid scintillation counting of low levels of carbon-14 for radiocarbon dating. *In:* Crook, M., P. Johnson and B. Scales, eds., Liquid Scintillation Counting, Vol. 2, Proceedings of the International Liquid Scintillation Counting Conference, Brighton, 1971 (Heydon and Sons, Ltd.) pp. 139–146.

BURNS, H.G. and H.I. GLASS (1963) Int. J. Appl. Radn. Isot. *14*, 627.

BUSH, E.T. (1963) Anal. Chem. *35*, 1024.

BUSH, E.T. (1964) Anal. Chem. *36*, 1082.

BUTLER, F. (1961) Anal. Chem. *33*, 409.

BUTTERFIELD, D. and R.J. MCDONALD (1972) Int. J. Appl. Radn. Isot. *23*, 249.

CADDOCK, B.D., P.T. DAVIES and J.H. DETERDING (1967) Int. J. Appl. Radn. Isot. *18*, 209.

CAIN, D.F. and R.E. PITNEY (1968) Anal. Biochem. *22*, 11.

CALF, G.E. (1969) Int. J. Appl. Radn. Isot. *20*, 611.

CAMERON, J.F. (1965) A survey of systems for concentration and low-background counting of tritium in water. *In:* Proceedings of Radiocarbon and Tritium Dating Conference, Pullman, Washington.

CAMPBELL, C.B. and L.W. POWELL (1970) J. Clin. Path. *23*, 304.

CARR, T.E.F. and J. NOLAN (1967) Liquid scintillation counting of Ca^{45} in the presence of Sr^{85}. *In:* Proceedings of the Beckmann Summer School on Techniques for Liquid Scintillation Counting, pp. 137–157.

CARR, T.E.F. and B.J. PARSONS (1962) Int. J. Appl. Radn. Isot. *13*, 57.

CARRIER, W.L. and R.B. SETLOW (1971) Anal. Biochem. *43*, 427.

CARROLL, C.O. and T.J. HOUSER (1970) Int. J. Appl. Radn. Isot. *21*, 261.

CARTER, J.G. and L.G. CHRISTOPHOROU (1967) J. Chem. Phys. *46*, 1883.

CAYAN, M.N. and P.A. ANASTASSIADIS (1966) Anal. Biochem. *15*, 84.

CERENKOV, P.A. (1934) Dokl. Akad. Nauk, S.S.S.R. *2*, 451.

CHAKRAVARTI, A. and J.W. THANASSI (1971) Anal. Biochem. *40*, 484.

CHAPMAN, D.I. and J. MARCROFT (1971) Int. J. Appl. Radn. Isot. *22*, 371.

CHIRIBOGA, J. (1962) Anal. Chem. *34*, 1843.

CHLECK, D.J. and C.A. ZIEGLER (1957) Rev. Sci. Instr. *28*, 466.

CHOULES, G.L. and B.H. ZIMM (1965) Anal. Biochem. *13*, 336.

CLULEY, H.J. (1962) Analyst *87*, 170.

COLOMER, J., M. COUSIGNE and G. METZGER (1972) Application of Cerenkov Technique to continuous measurement of radioactive isotopes isolated by an automatic analytical process. *In:* Crook, M., P. Johnson and B. Scales, eds., Liquid Scintillation Counting, Vol. 2 (Heyden & Sons, Ltd) pp. 181–187.

CONWAY, W.D., A.J. GRACE and J.E. ROGERS (1966) Anal. Biochem. *14*, 491.

COSOLITO, F.J., N. COHEN and H.G. PETROW (1968) Anal. Chem. *40*, 213.

CRAMER, C.F. and S.G. ARNOTT (1972) Int. J. Appl. Radn. Isot. *23*, 339.

CRAMER, C.F., M. NICHOLSON, C. MOORE and K. TENG (1971) Int. J. Appl. Radn. Isot. *22*, 17.

CRAMER, C.F. and B.H. ROSS (1970) Int. J. Appl. Radn. Isot. *21*, 237.

CRATHORN, A.R. (1953) Nature *172*, 632.

CREGER, C.R., M.N.A. ANSARI, J.R. COUCH and L.B. COLVIN (1967) Int. J. App. Radn. Isot. *18*, 71.

CROOK, M., P. JOHNSON and B. SCALES, eds. (1972) Liquid Scintillation Counting, Vol. 2 (Heyden & Sons, Ltd., London, N. York, Germany).

CURTIS, E.J.C. and I.P. TOMS (1972) Techniques for counting carbon-14 and phosphorous-32 labelled samples of polluted natural waters. *In:* Crook, M., P. Johnson and B. Scales, eds., Liquid Scintillation Counting, Vol. 2. (Heyden & Sons, Ltd., London, New York, Germany) 1972, pp. 167–180.

CURTIS, M.L., S.L. NESS and L.L. BENTZ (1966) Anal. Chem. *38*, 636.

DAVIDON, W.C. (1953) Nucleonics *11* (a), 62.

DAVIDSON, E.A. (1962) Packard Technical Bulletin No. 4, Revised July, 1962. Packard Instruments Ltd.

DAVIDSON, E.A. and J.G. RILEY (1960) J. Biol. Chem. *235*, 3367.

DAVIDSON, E.A. and J.G. RILEY (1960) Biochim. Biophys. Acta *42*, 566.

DAVIDSON, J.D. (1958) Round Table on Homogenous Counting Systems. *In:* Bell, C.B. and F.N. Hayes, eds., Liquid Scintillation Counting Proceedings of a Conference, Northwestern Univ., Aug. 20–22 (Pergamon Press, Oxford) pp. 88–95.

DAVIDSON, J.D. (1961) Some recent developments in liquid scintillation counting of biochemical samples. *In:* A.E.C. document, T1D 7612, Proc. Univ. New Mexico. Conf. on Organic Scintillation Detectors pp. 232–238.

DAVIDSON, J.D. and V.T. OLIVERIO (1965) Methods for pharmacological studies of methyl glyoxal-bis (guanylhydrazone-C^{14}) in man and animals. *In:* Roth, L.J., ed., Isotopes in Pharmacology (Univ. Chicago Press) pp. 343–352.

DAVIDSON, J.D. and V.T. OLIVERIO (1968) Adv. Tracer Methodol. *4*, 67.

DAVIDSON, J.D., V.T. OLIVERIO znd J.I. PETERSON (1970) Combustion of samples for liquid scintillation counting. *In:* Bransome, E.D. Jr., ed., The Current Status of Liquid Scintillation Counting (Grune and Stratton, Inc., New York and London) pp. 222–235.

DAVIES, J.W. and E.C. COCKING (1966) Biochim. Biophys. Acta *115*, 511.

DAVIES, J.W. and T.C. HALL (1969) Analyt. Biochem. *27*, 77.

DERN, R.J. and W.L. HART (1961a) J. Lab. Clin. Med. *57*, 322.

DERN, R.J. and W.L. HART (1961b) J. Lab. Clin. Med. *57*, 460.

DOBBS, H.E. (1963a) Anal. Chem. *35*, 783.

DOBBS, H.E. (1963b) Nature *200*, 1283.

DOBBS, H.E. (1966) Int. J. App. Radn. Isot. *17*, 363.

DOBBS, H.E. (1971) The role of liquid scintillation counting in the investigation of M99 (Reckitt)-a revolutionary veterinary drug. *In:* Horrocks, D.L. and C.T. Peng, eds., Organic Scintillators and Liquid Scintillation Counting (Academic Press, New York and London) pp. 669–682.

DOWNES, A.M. (1971) A study of the radioassay of some β-emitting isotopes in wool. *In:* Horrocks, D.L. and C.T. Peng, eds., Organic Scintillators and Liquid Scintillation Counting (Academic Press, New York and London) pp. 1031–1054.

DOWNES, A.M. and A.R. TILL (1963) Nature *197*, 449.

DROSDOWSKY, M. and N. EGOROFF (1966) Anal. Biochem. *17*, 365.

DULCINO, J., R. BOSCO, W.G. VERLY and J.R. MAISIN (1963) Clin. Chim. Acta *8*, 58.

DUNSCOMBE, W.G. and T.J. RISING (1969) Anal. Biochem. *30*, 275.

DUNN, A. (1971) Int. J. App. Radn. Isot. *22*, 212.

DYER, A. (1971) ed., Liquid Scintillation Counting, Vol. 1, (Heyden and Sons Ltd., London, New York and Germany) Proc. Symp. Liquid Scintillation Counting Univ. Salford, Sept. 21–22, 1970.

DYER, A. (1972) Methods of sample preparation of inorganic materials including Cerenkov counting. *In:* Crook, M., P. Johnson and B. Scales, eds., Liquid Scintillation Counting, Vol. 2 (Heyden and Sons Ltd., London, N. York and Germany) pp. 121–137.

EAKINS, J.D. and D.A. BROWN (1966) Int. J. App. Radn. Isot. *17*, 391.

EAKINS, J.D. and A.E. LALLY (1972) The simultaneous determination of plutonium alpha activity and plutonium-241 in biological materials by gel scintillation counting. *In:* Crook, M., P. Johnson and B. Scales, eds., Liquid Scintillation Counting, Vol. 2 (Heyden and Sons Ltd., London, New York and Germany) pp. 155–165.

EGAN, M.L., J.T. LAUTENSCHLEGER, J.E. COLIGEN and C.W. TODD (1972) Immunochemistry (Apr) *9*, 289.

ELKINS, R.P. (1961) Vth Annual Symp. on Advances in Tracer Methodology, Washington DC., Oct. 1961.

ELLIS, M.K., S.N. WAMPLER and R.H. YAGER (1966) Anal. Chim. Acta *34*, 169.

ERDTMANN, G. and G. HERRMANN (1963) Radiochim. Acta *1*, 98.

FALES, H.M. (1963) Atomlight (Jan) *28*, 8. (New England Nuclear publication).

FALLOT, P., A. GIRGIS, M. LAINE-BOESZOERMENYI and J. VIEUCHANGE (1965) Int. J. App. Radn. Isot. *16*, 349.

FARMER, E.C. and I.A. BERSTEIN (1952) Science *115*, 460.

FARMER, E.C. and I.A. BERSTEIN (1953) Science *117*, 279.

FLYNN, K.F. and L.E. GLENDENIN (1959) Phys. Rev. *116*, 744.

FLYNN, K.F., L.E. GLENDENIN, E.P. STEINBERG and P.M. WRIGHT (1964) Nucl. Instr. Methods *27*, 13.

FOX, B.W. (1968) Int. J. App. Radn. Isot. *19*, 717.

FOX, B.W. (1972) Sample preparation techniques in biochemistry with particular reference to heterogenous systems. *In:* Crook, M., P. Johnson and B. Scales, eds., Liquid Scintillation Counting. Vol. 2. (Heyden and Sons, Ltd., London, New York and Germany) pp. 189–204.

FOX, B.W. (1974) Int. J. App. Radn. Isot. *25*, 209.

FRANC, Z., V. SVOBODOVA, M.L. FRANCOVA and C. HORESOVSKY (1965) Coll. Czech. Chem. Comm. *30*, 2874.

FRANCIS, G.E. and J.D. HAWKINS (1967) Int. J. App. Radn. Isot. *18*, 223.

FRANCOIS, B. (1967) Int. J. App. Radn. Isot. *18*, 525.

FRANK, I. and I. TAMM (1937) Dokl. Akad. Nauk. S.S.S.R. *14* (3), 109.

FUCHS, A. and F.W. DE VRIES (1972) Int. J. App. Radn. Isot. *23*, 361.

FUNT, B.L. (1956) Nucleonics *14* (8), 83.

FUNT, B.L. (1961) Can. J. Chem. *39*, 711.

FUNT, B.L. and A. HETHERINGTON (1960) Science *131*, 1608.

FURLONG, N.B. (1963) Anal. Biochem. *5*, 515.

FURLONG, N.B. (1970) Liquid scintillation counting of samples on solid supports. *In:* Bransome, E.D. Jr., ed., The Current Status of Liquid scintillation Counting (Grune and Stratton, Inc., New York and London) pp. 201–206.

FURLONG, N.B., N.L. WILLIAMS and D.P. WILLIS (1965) Biochem. Biophys. Acta *103*, 341.

FURST, M. and H. KALLMAN (1955) Phys. Rev. *97*, 583.

FURST, M., H. KALLMAN and F.H. BROWN (1955) Nucleonics *13* (4), 58.

GABRIEL, O. (1965) Anal. Biochem. *10*, 143.

GARFINKEL, S.B., W.B. MANN, R.W. MEDLOCK and O. YURA (1965) Int. J. App. Radn. Isot. *16*, 21.

GASPAR, E. and M. ONESCU (1972) Radioactive tracers in hydrology. *In:* Acad, Bucharesti, Romania, eds. Developments in Hydrology No. 1. (Elsevier Publ. Co., Amsterdam, London, New York).

GEIGER, J.W. and L.B. WRIGHT (1960) Biochem. Biophys. Res. Comm. *2*, 282.

GERMAI, G. (1963) Bull. Soc. Roy. Sci. Liège *32*, 863.

GERMAI, G. (1964) Bull. Soc. Roy. Sci. Liège *33*, 672.

GERMAI, G. (1970) Int. J. App. Radn. Isot. *21*, 587.

GERWECK, L.E., W.R. HENDEE and C.A. PATERSON (1972) Int. J. App. Radn. Isot. *23*, 203.

GHOSH, H.P. and J. PREISS (1966) J. Biol. Chem. *241*, 4491.

GIBBS, J.A. (1962) Packard Technical Bulletin No. 8 pp. 1–13.

GILBERT, C.W. (1960) Int. J. App. Radn. Isot. *8*, 230.

GILL, D.M. (1964) Nature *202*, 626.

GILL, D.M. (1967) Int. J. App. Radn. Isot. *18*, 393.
GLEIT, C.E. and J. JUMOT (1961) Int. J. App. Radn. Isot. *12*, 145.
GODFREY, P. and F. SNYDER (1962) Anal. Biochem. *4*, 310.
GODWIN, H. (1962a) Nature *195*, 943.
GODWIN, H. (1962b) Nature *195*, 984.
GOLDSTEIN, G. and W.S. LYON (1964) Int. J. App. Radn. Isot. *15*, 133.
GOMEZ, E., J. FREER and J. CASTRILLON (1971) Int. J. App. Radn. Isot. *22*, 243.
GOODMAN, D. and H. MATZURA (1971) Anal. Biochem. *42*, 481.
GORDON, C.F. and A.L. WOLFE (1960) Anal. Chem. *32*, 574.
GOTTSCHALF, R.G., H. HOCH and G.H. STIDWORTHY (1962) Clin. Chem. *8*, 318.
GRABER, S.E., L.C. MCKEE and R.M. HEYSSEL (1967) J. Lab. Clin. Med. *69*, 170.
GRAY, R.H. and D.M. STEFFANSEN (1968) Anal. Biochem. *24*, 44.
GREEN, A.A. and W.D. MCELROY (1956) Biochim. Biophys. Acta *20*, 170.
GROWER, M.F. and E.D. BRANSOME, E.D. Jr. (1969) Anal. Biochem. *31*, 159.
GRUNDON, N.J. and C.J. ASHER (1972) J. Agric. Food. Chem. *20*, 794.
GULBINSKI, J.S. and W.W. CLELAND (1968) Biochemistry *7*, 566.
GUNDERMANN, K.D. (1968) Chemilumineszenz Organisher Verbindungen (Springer Verlag, Berlin).
GUPTA, G.W. (1966) Anal. Chem. *38*, 1356.
GUPTA, G.W. (1968) Microchem. J. *13*, 4.
HAAF ten, F.E.L. (1972) Colour quenching in liquid scintillation coincidence counters. *In:* Crook, M., P. Johnson and B. Scales, eds., Liquid Scintillation Counting, Vol. 2 (Heyden and Sons, Ltd., London, New York and Germany) pp. 39–48.
HABERER, K. (1965) Atom wirtschaft *10*, 36.
HABERER, K. (1966) Packard Technical Bulletin, No. 16.
HAGENFELDT, L. (1967) Clin. Chim. Acta *18*, 320.
HALLBERG, L. and H. BRISE (1960) Int. J. App. Radiat. Isot. *9*, 100.
HANDLER, J.A. (1963) Analyst *88*, 47.
HANSEN, D.L. and E.T. BUSH (1967) Anal. Biochem. *18*, 320.
HARDCASTLE, J.E., R.J. HANNAPEL and W.H. FULLER (1967) Int. J. App. Radn. Isot. *18*, 193.
HARLAN, J.W. (1963) Adv. Tracer Methodol. *1*, 115.
HARRIS, J.E. and L. FRIEDMAN (1969) Anal. Biochem. *30*, 199.
HARVEY, B.R. and G.A. SUTTON (1970) Int. J. App. Radn. Isot. *21*, 519.
HATTORI, T., H. AOKI, I. MATSUZAKI, B. MARUO and H. TAKAHASHI (1965) Anal. Biochem. *37*, 159.
HAVILAND, R.T. and L.L. BIEBER (1970) Anal. Biochem. *33*, 323.
HAYES, F.N. (1958) I.R.E. Trans. on Nuclear Sci. Vol. NS-5, 166.
HAYES, F.N., B.S. ROGERS and W.H. LANGHAM (1956) Nucleonics *14*, (3) 48.
HAYES, F.N., B.S. ROGERS and P.C. SANDERS (1955) Nucleonics *13* (1) 46.
HEIDELBERGER, C., H.I. HADLER and G. WOLF (1953) J. Amer. Chem. Soc. *75*, 1303.

HELF, S. (1958) Suspension counting. *In:* Bell, C.B. and F.N. Hayes, eds., Liquid Scintillation Counting (Pergamon Press, Oxford) pp. 96–100.

HELF, S. and C. WHITE (1957) Anal. Chem. *29*, 13.

HELF, S., C.G. WHITE and R.N. SHELLEY (1960) Anal. Chem. *32*, 238.

HENDLER, R.W. (1964) Anal. Biochem. *7*, 110.

HENNINGS, H. and K. ELGJO (1970) Cell Tissue Kinet. *3*, 243.

HERBERG, R.J. (1958) Science *128*, 199.

HERBERG, R.J. (1960) Anal. Chem. *32*, 1468.

HERBERG, R.J. (1963) Anal. Chem. *35*, 786.

HERCULES, D.M. (1970) Physical basis of chemiluminescence. *In:* Bransome, E.D. Jr., ed., The Current Status of Liquid Scintillation Counting (Grune and Stratton, Inc., New York and London) pp. 315–336.

HETENYI, G. Jr. and J. REYNOLDS (1967) Int. J. App. Radn. Isot. *18*, 331.

HIGASHIMURA, T., O. YAMADA, N. NOHARA and T. SHIDEI (1962) Int. J. App. Radn. Res. *13*, 308.

HILTON, H.W., N.S. NOMURA and S.S. KAMEDA (1972) Anal. Biochem. *49*, 285.

HIMMS-HAGEN, J. (1968) Can. J. Biochem. *46*, 1107.

HO, R.J. (1970) Anal. Biochem. *36*, 105.

HOCH, F.L., R.A. KURAS and J.D. JONES (1971) Anal. Biochem. *40*, 86.

HOFFMANN, W. (1963) Radiochim. Acta *1*, 216.

HOFFMANN, W. (1965) Radiochim. Acta *4*, 222.

HOLM-HANSEN, O. and C.R. BOOTH (1966) Limnol. Ocenog. *11*, 510.

HORROCKS, D.L. (1963) Rev. Sci. Instr. *34*, 1035.

HORROCKS, D.L. (1964) Rev. Sci. Instr. *35*, 334.

HORROCKS, D.L. (1964) Nucl. Inst. Methods *30*, 157.

HORROCKS, D.L. (1971) Techniques for the study of excimers. *In:* Horrocks, D.L. and C.T. Peng, eds., Organic Scintillators and Liquid Scintillation Counting (Academic Press, New York and London) pp. 75–90.

HORROCKS, D.L. and C.T. PENG (1971) eds. Organic Scintillators and Liquid Scintillation Counting (Academic Press, New York and London).

HORROCKS, D.L. and M.H. STUDIER (1961) ANAL. CHEM. *33*, 615.

HORROCKS, D.L. and M.H. STUDIER (1964) Anal. Chem. *36*, 2077.

HOUTMAN, A.C. (1965) Int. J. App. Radn. Isot. *16*, 65.

HOUX, N.W.H. (1969) Anal. Biochem. *30*, 302.

HUMPHRIES, E.R. (1965) Int. J. App. Radn. Isot. *16*, 345.

HUNT, J.A. (1968) Anal. Biochem. *23*, 289.

HUNT, L.M. and B.N. GILBERT (1972) Int. J. App. Radn. Isot. *23*, 246.

HUTCHINSON, F. (1967) Int. J. App. Radn. Isot. *18*, 136.

IANDOLO, J.J. (1970) Anal. Biochem. *36*, 6.

JACOBSON, H.I., G.N. GUPTA, C. FERNANDEZ, S. HENNIX and E.V. JENSEN (1960) Arch. Biochem. Biophys. *86*, 89.

JAFFEE, M. and L.A. FORD (1966) Trans. Am. Nucl. Soc. *8*, 318.

JEFFAY, H. (1962) Packard Technical Bulletin No. 10, Oct.

JEFFAY, H., F.O. OLUBAJO and W.R. JEWELL (1960) Anal. Chem. *32*, 306.

JELLEY, J.V. (1958) Cerenkov Radiation and its Applications (Pergamon Press, London).

JENNER, H. and K.J. OBRINK (1962) SCAND. J. CLIN. LAB. INVEST. *14*, 466.

JOHNSON, M.K. (1969) Anal. Biochem. *29*, 348.

JOHNSON, D.R. and J.W. SMITH (1963) Anal. Chem. *35*, 1991.

JOHNSON, P. (1972) discussion comment. *In:* Crook, M., P. Johnson and B. Scales, eds., Liquid Scintillation Counting (Heyden and Sons Ltd., London, New York and Germany) pp. 241–242.

JONES, G.B. (1965) Anal. Biochem. *12*, 249.

JONES, G.B. and N.F. HENSCHKE (1963) Int. J. App. Radn. Isot. *14*, 618.

KAARTINEN, N. (1969) Packard Technical Bulletin No. 18, Apr.

KACZMARCZYK, N. (1971) Relation between the Counting efficiency of beta emitters and the quencher concentration. *In:* Horrocks, D.L. and C.T. Peng, eds., Organic Scintillators and Liquid Scintillation Counting (Academic Press, New York and London) pp. 977–990.

KACZMARCZYK, N. and I. RUGE (1969) Int. J. App. Radn. Isot. *20*, 653.

KALBHEN, D.A. (1967) Int. J. App. Radn. Isot. *18*, 655.

KAMP, A.J. and E.A. BLANCHARD (1971) Anal. Biochem. *44*, 369.

KANDELL, M. and A.G. GORNALL (1964) Can. J. Biochem. *42*, 1833.

KARMEN, A. and H.R. TRITCH (1960) Nature *186*, 150.

KATZ, J., S. ABRAHAM and N. BAKER (1954) Anal. Chem. *26*, 1503.

KATZ, J.H., M. ZOUKIS, W.L. HART and R.J. DERN (1964) J. Lab. Clin. Med. *63*, 885.

KEARNS, D.S. (1969) Int. J. App. Radn. Isot. *20*, 821.

KELLY, R.G., E.A. PEATS, S. GORDON and D.A. BUYSKE (1961) Anal. Biochem. *2*, 267.

KERR, V.N., F.N. HAYES and D.G. OTT (1957) Int. J. App. Radn. Isot. *1*, 284.

KESTON, A.S., S. UNDENFRIEND and M. LEVY (1950) J. Amer. Chem. Soc. *72*, 748.

KHAN, A.U. and M. KASHA (1966) J. Amer. Chem. Soc. *88*, 1574.

KIM, T.K. and M.B. MCINNIS (1972) Combination of liquid scintillation counting and solvent extraction. *In:* Horrocks, D.L. and C.T. Peng, eds., Organic Scintillators and Liquid Scintillation Counting (Academic Press, New York and London) pp. 925–933.

KIMM, W.K. and R. ROHRINGER (1969) Can. J. Botany *47*, 1425.

KINARD, F.E. (1957) Rev. Sci. Instr. *28*, 293.

KINNORY, D.S., E.L. KANABROCKI, J. GREGO, R.L. VEATCH, E. KAPLAN and Y.T. OESTER (1958) A liquid scintillation method for measurement of radio-activity in animal tissue and tissue fractions. *In:* Bell, C.G. and F.N. Hayes, eds., Liquid Scintillation Counting (Pergamon Press, Oxford) pp. 223–229.

KIRKHAM, K.E. and W.M. HUNTER (1971) eds., Radioimmunoassay Methods (Churchill Livingstone, Edinburgh).

KLEIN, P.D. (1970) Isotope effects and their consequences in liquid scintillation counting. *In:* Bransome, E.D. Jr., ed., The current status of Liquid Scintillation Counting (Grune and Stratton, Inc., New York and London) pp. 142–155.

KMENT, V. and A. KUHN (1963) Technik des Messens radioaktiver Strahlung. (Leipzig) p. 254.

KNOCHE, H.W. and R.M. BELL (1965) Anal. Biochem. *12*, 42.

KOBAYASHI, Y. (1963) Anal. Biochem. *5*, 284.

KOBAYASHI, Y. and D.V. MAUDSLEY (1970) Practical aspects of double isotope counting. *In:* Bransome, E.D. Jr., ed., The current status of Liquid Scintillation Counting (Grune and Stratton, Inc. New York and London) pp. 76–85.

KRICHEVSKY, D. and S. MALHOTRA (1970) J. Chromatogr. *52*, 498.

KUSHINSKY, S. and W. PAUL (1969) Anal. Biochem. *30*, 465.

LAARSE, J.D. vander (1967) Int. J. App. Radn. Isot. *18*, 485.

LAGERSTEDT, H.B. and R. LANGSTON (1966) Anal. Biochem. *15*, 448.

LAHR, T.N., R. OLSEN, G.I. GLEASON and D.L. TABERN (1955) J. Lab. Clin. Med. *45*, 66.

LAUCHLI, A. (1969) Int. J. App. Radn. Isot. *20*, 265.

LAURENCOT, H.J. and J.L. HEMPSTEAD (1971) Liquid scintillation counting of biological materials in solubilized whole blood. *In:* Horrocks, D.L. and C.T. Peng, eds., Organic Scintillators and Liquid Scintillation Counting (Academic Press, New York and London) pp. 635–657.

LEON, S.A. and A.T. BOHRER (1971) Anal. Biochem. *42*, 54.

LERCH, P. and M. COSANDEY (1966) Atomlight No. 52, New England Nuclear Corporation, Boston, Mass. p. 1.

LEWIS, J.D. (1972) Int. J. App. Radn. Isot. *23*, 39.

LIBBY, W.F. (1952) Radiocarbon Dating (Univ. Chicago Press, Chicago).

LIEBERMAN, R. and A.A. MOGHISSI (1970) Int. J. App. Radn. Isot. *21*, 319.

LINDSAY, P.A. and N.B. KURNICK (1969) Int. J. App. Radn. Isot. *20*, 97.

LITLE, R.L. and M.P. NEARY (1971) Solute optimization. *In:* Horrocks, D.L. and C.T. Peng, eds., Organic Scintillators and Liquid Scintillation Counting (Academic Press, New York and London) pp. 431–439.

LITT, G.J. and H. CARTER (1970) Sample absorption problems in liquid scintillation counting. *In:* Bransome, E.D. Jr., ed., The Current Status of Liquid Scintillation Counting (Grune and Stratton, Inc. New York and London) pp. 156–163.

LLOYD, R.A., S.C. ELLIS and K.H. HALLOWS (1962). *In:* Tritium in the Physical and Biological Sciences, Int. Atomic Energy Agency, Vienna. *1*, 263.

LOCKNER, D. (1965) Scand. J. Clin. Lab. Invest. *17*, 247.

LOFTFIELD, R.B. (1960) Liquid scintillation counting of carbon-14 labelled paper chromatograms. *In:* Atomlight *13*, June (New England Nuclear publication).

LOFTFIELD, R.B. and E.H. EIGNER (1963) Biochim. Biophys. Acta. *72*, 372.

LOWRY, O.H., N.J. ROSEBROUGH, A.L. FARR and R.J. RANDALL (1951) J. Biol. Chem. *193*, 265.

LUDWICK, J.D. (1960) Anal. Chem. *32*, 607.

LUDWICK, J.D. (1961) Health Physics *6*, 63.

LUDWICK, J.D. (1964) Anal. Chem. *36*, 1104.

LUPICA, S.B. (1970) Int. J. App. Radn. Isot. *21*, 487.
LUTWAK, L. (1959) Anal. Chem. *31*, 340.
MADSEN, N.P. (1969) Anal. Biochem. *29*, 542.
MAHIN, D.T. and R.T. LOFBERG (1966) Anal. Biochem. *16*, 500.
MAIZEL, J.V. (1966) Science *151*, 988.
MALT, R.A. and W.L. MILLER (1967) Anal. Biochem. *18*, 388.
MANS, R.J. and G.D. NOVELLI (1961) Arch. Biochem. Biophys. *94*, 48.
MARLOW, W.F. and R.W. MEDLOCK (1960) J. Res. NBS *64A*, 143.
MARSHALL, J. (1952) Phys. Rev. *86*, 685.
MARTIN, L.E. and C. HARRISON (1962) Biochem. J. *82*, 18P.
MATHIJSSEN, C. (1966) Anal. Biochem. *15*, 382.
MCBETH, G.W., R.A. WINYARD and J.E. LUTKIN (1970) Pulse height discrimination with Organic Scintillators. Koch Light Booklet, 1971, Kock Light Laboratories Ltd., Colnbrook, Bucks, England.
MCCLENDON, D., M.P. NEARY, M. GALASSI and W. STEPHENS (1971) Study of the use of Bio-solve^{TM} solubilizer with biologically signficant samples. *In:* Horrocks, D.L. and C.T. Peng, eds., Organic Scintillators and Liquid Scintillation Counting (Academic Press) pp. 587–598.
MCDOWELL, W.J. (1972) Liquid scintillation counting techniques for the higher actinides. *ibid* pp. 937–950.
MCDOWELL, L.L. and M.E. RYAN (1966) Int. J. App. Radn. Isot. *17*, 175.
MCGRATH, R.A. and R.W. WILLIAMS (1966) Nature *212*, 534.
MCTAGGERT, W.G. and D. CARDUS (1971) Tritium oxide movement in body water of healthy and paralytic men. *In:* Horrocks, D.L. and C.T. Peng, eds., Organic Scintillators and Liquid Scintillation Counting (Academic Press) pp. 621–634.
MEADE, R.C. and R. STIGLITZ (1962) Int. J. App. Radn. Isot. *13*, 11.
MILLER, M., J.G. DEREIAKES and B.I. FRIEDMAN (1969) Int. J. App. Radn. Isot. *20*, 133.
MOGHISSI, A.A. (1970) Low-level liquid scintillation counting of X and X emitting nuclides. *In:* Bransome, E.D. Jr., ed., The Current Status of Liquid Scintillation Counting (Grune and Stratton, Inc., New York and London) pp. 86–94.
MOGHISSI, A.A., H.L. KELLEY, J.E. REGNIER and M.W. CARTER (1961) Int. J. App. Radn. Isot. *20*, 145.
MOIR, A.T.B. (1971) Int. J. App. Radn. Isot. *22*, 213.
MOORHEAD, J.F. and W. MCFARLAND (1966) Nature *211*, 1157.
MORIARTY, C.M. (1972) Int. J. App. Radn. Isot. *23*, 238.
MORRISON, J.F. and W.W. CLELAND (1966) J. biol. Chem. *241*, 673.
MOSS, G. (1961) Int. J. App. Radn. Isot. *11*, 47.
MOSS, G. (1964) J. Lab. Clin. Med. *63*, 315.
MURPHY, B.P. (1967) J. Clin. Endocrinol. *27*, 973.
MURPHY, B.P., A.B. HOOD and C.J. PATTEE (1964) Canad. Med. Assoc. J. *90*, 775.
MURTY, H.S. and P.P. NAIR (1969) Anal. Biochem. *31*, 1.
MYERS, L.S. Jr. and A.H. BRUSH (1962) Anal. Biochem. *34*, 342.

NADARAJAH, A., B. LEESE and G.F. JOPLIN (1969) Int. J. App. Radn. Isot. *20*, 733.

NARROG, J. (1965) Detection of beta-emitting nuclides of energy 1 MeV in urine after an accident by means of measurement of the Cerenkov effect in a liquid scintillation counting system. *In:* Personnel Dosimetry for Radiation Accidents, Int. Atomic Energy Agency, Vienna, pp. 427–434.

NATHAN, D.G., J.G. DAVIDSON, J.G. WAGGONER and N.J. BERLIN (1958) J. Lab. Clin. Med. *52*, 915.

NATHAN, D.G., T.G. GABUZDA, F.H. GARDNER, Y.B. LENURE and A.L. LIMAURO (1963) J. Lab. Clin. Med. *62*, 511.

NEARY, M.P. and A.L. BUDD (1970) Color and Chemical Quench. *In:* Bransome, E.D. Jr., ed., The Current Status of Liquid Scintillation Counting (Grune and Stratton, Inc., New York and London) pp. 273–282.

NEUJAHR, H.Y. and B. EWALDSSON (1964) Anal. Biochem. *8*, 487.

NEWSHOLME, E.A., J. ROBINSON and K. TAYLOR (1967) Biochim. Biophys. Acta *132*, 338.

NICHOLIS, P.B. and A.W. MURRAY (1968) Plant Physiol. *43*, 645.

NIMOMIYA, R. (1966) Int. J. Appl. Radn. Isot. *17*, 355.

NOAKES J.E., A.S. ISBELL and D.W. HOOD (1961) Am. Geophys. Union. Trans. *42*, 226.

NOAKES, J.E., A.S. ISBELL, J.J. STIPP and D.W. HOOD (1963) Geochim. Cosmochim. Acta *27*, 797.

NOLAN, J. (1972) Liquid scintillation counting of calcium-45 in biological samples containing environmental strontium-90. *In:* Crook, M., P. Johnson and B. Scales, eds., Liquid Scintillation Counting, Vol. 2. (Heyden & Son, Ltd., London, New York and Germany) pp. 147–154.

NUGENT, C.A. and D.M. MAYER (1967) J. Clin. Endocrinol. Metabol. *27*, 10.

NUNEZ, J. and C. JACQUEMIN (1961) J. Chromatog. *5*, 271.

OBER, R.E., A.R. HANSEN, D. MOURER, J. BAUKOMA and G.W. GWYNN (1969) Int. J. App. Radn. Isot. *20*, 703.

O'BRIEN, R.D. (1964) Anal. Biochem. *7*, 251.

OLDHAM, K.G. (1968) Radiochemical Methods of Enzyme Assay. Review, No. 9, The Radiochemical Centre, Amersham.

OLIVERIO, V.T., C. DENHAM and J.D. DAVIDSON (1962) Anal. Biochem. *4*, 188.

OLSON, A.C., L.M. WHITE and A.T. NOMA (1968) Anal. Biochem. *24*, 120.

OSBURN, O.L. and L.H. WERKMAN (1932) Ind. Eng. Chem. Anal. Ed. *4*, 421.

OTT, D.G., F.N. HAYES, J.E. HAMMEL and J.F. KEPHART (1955) Nucleonics *13* (5), 62.

OTT, D.G., C.R. RICHMOND, T.T. TRUJILLO and H. FOREMAN (1959) Nucleonics *17*, (9), 106.

OXBY, C.B. and P.A. KIRBY (1968) Int. J. App. Rad. Isot. *19*, 151.

PAGE, A.L., J.R. SIMS and F.T. BINGHAM (1964) Soil. Sci. Soc. Am. Proc. *28*, 359.

PALMER, H.E. and T.M. BEASLEY (1965) Science *149*, 431.

PARKER, R.P. and R.H. ELRICK (1966) Int. J. App. Radn. Isot. *17*, 361.

PARKER, R.P. and R.H. ELRICK (1970) Cerenkov counting as a means of assaying

X-emitting radionuclides. *In:* Bransome, E.D. Jr., Liquid Scintillation Counting (Grune and Stratton, Inc., New York and London) pp. 110–122.

PARMENTIER, J.H. and F.E.L. ten HAAF (1969) Int. J. App. Radn. Isot. *20*, 305.

PASSMANN, J.M., N.S. RADIN and J.A.D. COOPER (1956) Anal. Chem. *28*, 484.

PATTERSON, M.S. and R.C. GREENE (1965) Anal. Chem. *37*, 854.

PAUS, P.N. (1970) Anal. Biochem. *38*, 364.

PAUS, P.N. (1971) Anal. Biochem. *42*, 372.

PAUS, P.N. (1972) Liquid scintillation counting of biological macromolecules. Extraction from aqueous solution and from glass fibre filters. *In:* Crook, M., P. Johnson and B. Scales, eds., Liquid Scintillation Counting (Heyden & Sons, Ltd., London, New York and Germany) pp. 205–212.

PEACOCK, W.C., R.D. EVANS, J.W. IRVINE, Jr., W.M. GOOD, A.F. KIP, S. WEISS and J.G. GIBSON, II (1946) J. Clin. Invest. *25*, 605.

PEARCE, E.M., F. de VENUTO, W.M. FITCH, H.E. FIRSCHEN and U. WESTPHAL (1956) Anal. Chem. *28*, 1762.

PENG, C.T. (1970) A review of methods of quench correction in liquid scintillation counting. *In:* Bransome, E.D. Jr., ed., The current status of Liquid Scintillation Counting (Grune and Stratton, New York) pp. 283–292.

PENMAN, S., Y. BECKER and J.E. DARNELL (1964) J. Mol. Biol. *8*, 541.

PERRY, S.W. and G.T. WARNER (1963) Int. J. App. Radn. Isot. *14*, 397.

PERSCHKE, H. and T. FLORKOWSKI (1970) Int. J. App. Radn. Isot. *21*, 747.

PETERSON, J.I., F. WAGNER, S. SIEGEL and W. NIXON (1969) Anal. Biochem. *31*, 189.

PFEFFER, M., S. WEINSTEIN, J. GAYLORD and L. INDINDOLI (1971) Anal. Biochem. *39*, 46.

PHILLIPS, A.P. and C.A. SAMBROOK (1972) Sample preparation for tritium counting in the application of the digoxin radioimmunoassay technique to lysed blood. *In:* Crook, M., P. Johnson and B. Scales, eds., Liquid Scintillation Counting (Heyden and Sons Ltd., London, New York and Germany) pp. 217–222.

PICKERING, D.E., H.L. REED and R.L. MORRIS (1960) Anal. Chem. *32*, 1214.

PILLARISETY, R.J., D.G.T. WELLS, L.S. NARITOMI, G.S. UYESEGI and R.M. WOOD (1971) The use of liquid scintillation counting methods for study of protein syntheses by *Treponema pallidum*. *In:* Horrocks, D.L. and C.T. Peng, eds., Organic Scintillators and Liquid Scintillation Counting (Academic Press, New York) pp. 659–667.

PINTER, K.G., J.G. HAMILTON and O.N. MILLER (1963) Anal. Biochem. *5*, 458.

PITCHER, C.S., H.S. WILLIAMS, A. PARSONSON and R. WILLIAMS (1965) Brit. J. Haematol. *11*, 633.

PLOTKA, E.D., E.G. STANT, Jr., F.A. WALTZ, V.A. GARWOOD and R.E. ERB (1966) Int. J. App. Radn. Isot. *17*, 637.

PREGEL-ROTH, F. (1958) Quantitative organische Mikroanalyse (Springer, Wien).

PRINGLE, R.W., L.D. BLACK, B.L. FUNT and S. SOBERING (1953) Phys. Rev. *92*, 1582.

RADIN, N.S. (1973) Anal. Biochem. *55*, 637.

RAPKIN, E. (1961) Packard Technical Bulletin, No. 3, June (Packard Instruments, La Grange, Illinois).

RAPKIN, E. (1963) Packard Technical Bulletin, No. 11, Feb. (Packard Instruments, La Grange, Illinois).

RAPKIN, E. (1972) A history of the development of the modern liquid scintillation counter. *In:* Crook, M., P. Johnson and B. Scales, eds., Liquid Scintillation Counting, Vol. 2 (Heyden and Sons, Ltd., London, New York and Germany) pp. 61–100.

RAPKIN, E. and J.A. GIBBS (1963) Int. J. App. Radn. Isot. *14*, 71.

RAPKIN, E. and I.E. PACKARD (1960) *In:* Proc. Univ. New Mexico Conf. on Organic Scintillation Detectors, TID-7612, p. 216.

RAUSCHENBACH, P. and H. SIMON (1971) Z. Anal. Chem. *256*, 119.

REED, D.J. (1968) Adv. Tracer Methodol. *4*, 145.

RETIEF, W.L., J. DEIST and F.J. HAASBROEK (1972) Agrochemophysica *4*, 11.

REYNOSA, G., T.N. CHU, D. HOLYOKE, E. COHEN, T. NEMOTO, J.J. WANG, J. CHUANG, P. GUINAN and G.P. MURPHY (1972) J. Amer. Med. Assoc. *220*, 361.

RIPPON, S.E.H. (1963) Nucl. Inst. Methods *21*, 185.

ROBERTS, E., D.G. SIMONSEN and B. SISKEN (1963) A convenient method for the determination of metabolically liberated $C^{14}O_2$. *In:* Adv. Tracer Methodology, 7th Symp. on Adv. in Tracer Methodol., Vol. II (Plenum Press, New York) 1965, p. 93.

ROBERTS, E.M. and K.C. TOVEY (1970) Anal. Biochem. *34*, 582.

ROCHE, J., J. NUNEZ and C.E. JACQUEMIN (1962) *In:* Tritium in the Physical and Biological Sciences, Vol. 2, Int. Atomic Energy Agency, Vienna, Vol. 2, p. 395.

ROGERS A.W. and J.F. MORAN (1966) Anal. Biochem. *16*, 206.

RONCUCCI, R., G. LAMBELIN, M.J. SIMON and W. SOUDYN (1968) Anal. Biochem. *26*, 118.

RONZANI, C. and M.A. TAMERS (1966) Radiochimica Acta *6*, 206.

ROSS, H.H. (1964) Int. J. App. Radn. Isot. *15*, 273.

ROSS, H.H. (1970) Cerenkov radiation: Photon yield application to (^{14}C) assay. *In:* Bransome, E.D. Jr., eds., The current status of liquid scintillation counting (Grune and Stratton, Inc., New York and London) pp. 123–126.

ROUCAYROL, J., E. OVERHAUSER and R. SCHUSSLER (1957) Nucleonics *15*, 104.

RUNYON, W.S., M.A. FARDY and R.P. GEYER (1967) Biochim. Biophys. Acta *141*, 421.

SARNAT, B.G. and H. JEFFAY (1962) Anal. Chem. *34*, 643.

SATYASWAROOP, P.G. (1972) Ind. J. Biochem. Biophys. *9*, 101.

SAUZAY, G. and W.R. SCHELL (1972) Int. J. App. Radn. Isot. *23*, 25.

SAXENA, B.N., S. REFETOFF, K. EMERSON, Jr. and H.A. SELENKOV (1968) Am. J. Obs. and Gynecol. *101*, 874.

SCALES, B. (1963) Anal. Biochem. *5*, 489.

SCHILLING, R.F. (1964) Packard Technical Bulletin, No. 13, 1.

SCHINKE, R.T. and DOYLE, D. (1970) Ann. Rev. Biochem. *7*, 929.

SCHINDLER, D.W. (1966) Nature *211*, 844.

SCHLOERB, P.R. and J.J. GRANTHAM (1965) J. Lab. Clin. Med. *65*, 669.

SCHMUCKLER, M. and M.J. YIENGST (1968) Anal. Biochem. *25*, 406.

SCHNEIDER, P.B. (1971) J. Nucl. Med. *12*, 14.

SCHRAM, E. (1963) Organic Scintillation Detectors: Counting of Low energy Beta Emitters (Elsevier, Amsterdam, London, New York).

SCHRAM, E. (1967) Arch. Int. Physiol. Biochim. *75*, 894.

SCHRAM, E. (1970) Flow monitoring of aqueous solutions containing weak X-emitters. *In:* Bransome, E.D. Jr. The Current Status of Liquid Scintillation Counting (Grune and Stratton, Inc., New York and London) pp. 95–109.

SCHRAM, E. (1970) Use of scintillation counters for bioluminescence assay of adenosine triphosphate (ATP). *In:* Bransome, E.D. Jr., The Current Status of Liquid Scintilation Counting (Grune and Stratton, Inc., New York and London) pp. 129–133.

SCHRAM, E. and LOMBAERT, R. (1962) Anal. Biochem. *3*, 68.

SCHRAM, E. and ROOSENS, H. (1972) Semi-automatic microtransferator and cell for the bioluminescence assay of ATP and reduced NAD with scintillation counters. *In:* Crook, M., P. Johnson and B. Scales, eds., Liquid Scintillation Counting, Vol. 2 (Heyden and Son, Ltd., London, New York and Germany) pp. 115–120.

SCHUTTE, L. (1972) J. Chromatog. *72*, 303.

SCHWERDTEL, E. (1966) Int. J. App. Radn. Isot. *17*, 479.

SEBRING, E.D. and N.P. SALZMAN (1964) Anal. Biochem. *8*, 126.

SEGAL, S. and A. BLAIR (1962) Anal. Biochem. *3*, 221.

SELIGER, H.H. and B.W. AGRANOFF (1959) Anal. Chem. *31*, 1605.

SETO, S., K. OGURA and Y. NISHIYAMA (1963) Bull. Chem. Soc. Jap. *36*, 331.

SHAKHIDZHANIAN, L.G., D.G. FLEISHMAN, V.V. GLAZUNOV, V.G. LEONT'EV and V.P. NESTEROV (1959) (Chem. Abstr. 53: 18999i), 1959, Dokl. Akaud. Nauk. S.S.S.R. *125*, 1.

SHAPIRA, J. and W.H. PERKINS (1960) Science *131*, 414.

SHAPIRO, I. and H.G. WEISS (1957) J. Am. Chem. Soc. *79*, 3294.

SHAW, W.A., W.R. HARLAN and A. BENNETT (1971) Anal. Biochem. *43*, 119.

SHEPPARD, H. and W. RODEGHER (1962) Combustion of tissues for C^{14} and H^{8} analysis, Parr Bomb. *In:* Atomlight, Feb. *22* (New England, Nuclear).

SHERMAN, J.R. (1965) Anal. Biochem. *5*, 548.

SIBATINI, A. (1970) Anal. Biochem. *33*, 279.

SIGGERS, D.C., C. SALTER and P.A. TOSELAND (1970) Clin. Chim. Acta *30*, 373.

SIMPSON, J.D. and J.R. GREENING (1960) Nature *186*, 467.

SLATER, G.G., E. GELLER and A. YUWILER (1964) Anal. Chem. *36*, 1888.

SMITH, L.W. (1969) Anal. Biochem. *29*, 223.

SMITH, T.W., V.P. BUTLER and E. HABER (1969) New England J. Med. *281*, 1212.

SMITH, R.W. and S.H. PHILLIPS (1969) Int. J. App. Radn. Isot. *20*, 553.

SMITH, D.L., R.G. POLKE and T.G. MILLER (1968) Nucl. Inst. Meth. *64*, 157.

SMITH, A.L., J.W. THOMAS and H. WALLMAN (1970) Int. J. App. Radn. Isot. *21*, 171.

SNYDER, F. (1968) Advances in Tracer Methodology (Plenum, New York) pp. 81–104.

SNYDER, F. (1970) Liquid scintillation radioassay of thin layer chromatograms. *In:* Bransome, E.D. Jr., ed., The Current status of Liquid Scintillation Counting (Grune and Stratton, Inc., New York and London) pp. 248–256.

SNYDER, F. and GODFREY P. (1961) J. Lipid Res. *2*, 195.

SNYDER, F. and A. MOEHL (1969) Anal. Biochem. *28*, 503.

SNYDER, F. and A. MOEHL (1971) Mass measurements in a liquid scintillation spectrometer: Quantitation of sulhydryl moities by color quenching. *In:* Horrocks, D.L. and C.-T. Peng, eds., Organic Scintillators and Liquid Scintillation Counting (Academic Press New York and London) pp. 419–424.

SPENCER, R.P. and M.A. BANERJI (1970) Int. J. App. Rad. Isot. *21*, 431.

SPRATT, J.L. (1965) Int. J. App. Radn. Isot. *16*, 439.

STANLEY, P.E. (1970) Proc. Aust. Biochem. Soc. *3*, 51.

STANLEY, P.E. (1971) The use of the liquid scintillation spectrometer for measuring NADH and FMN by the photobacterium luciferase and ATP by the Firefly luciferase. *In:* Horrocks, D.L. and C.-T. Peng, eds., Organic Scintillators and Liquid Scintillation Counting (Academic Press, New York and London) pp. 607–620.

STANLEY P.E. (1972) Determination of absolute radioactivity in multi-labelled samples using external standardization or channels ratio: A Fortran IV programme. *In:* Crook, M., P. Johnson and B. Scales, eds., Liquid Scintillation Counting. Vol. 2 (Heyden and Sons, Ltd., London, New York and Germany) pp. 285–291.

STANLEY, P.E. and S.G. WILLIAMS (1969) Anal. Biochem. *29*, 381.

STEELE, R., W. BERNSTEIN and C. BJERKNESS (1957) J. App. Physiol. *10*, 319.

STEINBERG, D. (1958) Nature *182*, 740.

STEINBERG, D. (1959) Nature *183*, 1253.

STEINBERG, D. (1960) Anal. Biochem. *1*, 23.

STEYN, J. (1966) Proc. symp. standardization of radionuclides, Vienna, Oct. 10–14, Paper SM-79/16.

STUBBS, R.D. and A. JACKSON (1967) Int. J. App. Radn. Isot. *18*, 857.

SUESS, H.E. (1954) Science *120*, 5.

TALIBUDEEN, O. and Y. YAMADA (1966) J. Soil. Sci. *17*, 107.

TAKAHASHI, I.T. and F.A. BLANCHARD (1969) Anal. Biochem. *29*, 154.

TAKAHASHI, I.T. and F.A. BLANCHARD (1970) Anal. Biochem. *35*, 411.

TAMERS, H.A. (1964) Packard Technical Bulletin, No. 12, Jan.

TAMERS, M., R. BIBRON and G. DELIBRIAS (1962) Tritium in the Physical and Biol. Sci., Int. Atomic Energy Agency, Vienna, *1*, 303.

TAMERS, M.A., J.J. STIPP and J. COLLIER (1961) Geochim. Cosmochim. Acta *24*, 266.

THATCHER, L., M. RUBIN and G. BROWN (1961) Science *134*, 3472.

THOMAS, R.C., R.W. JUDY and H. HARPOOTLIAN (1965) Anal. Biochem. *13*, 358.

TISHLER, P.V. and C.J. EPSTEIN (1968) Anal. Biochem. *22*, 89.

TKACHUK, R. (1962) Can. J. Chem. *40*, 2348.

TOPOREK, M. (1960) Int. J. App. Radn. Isot. *8*, 229.

TREWAVAS, A. (1967) Anal. Biochem. *21*, 324.

TURNER, J.C. (1969) Int. J. App. Radn. Isot. *20*, 499.

TURNER, J.C. (1968) Int. J. App. Radn. Isot. *19*, 557.

TURPIN, R.A. and J.E. BETHUNE (1967) Anal. Chem. *39*, 362.

VAN SLYKE, D.D. and J. FOLCH (1940) J. Biol. Chem. *136*, 509.
VAUGHAN M., D. STEINBERG and J. LOCAN (1957) Science *126*, 446.
VEEN, H. (1972) Planta *103*, 35.
VEMMER, H. and J.O. GUTTE (1964) Atompraxis *10*, 475.
VOLPI, A. de and K.G. PORGES (1965) Int. J. App. Radn. Isot. *16*, 496.
WACHTER, R. de and W. FIERS (1966) Arch. Int. Physiol. Biochim. *74*, 915.
WADDEL, W.J. and T.C. BUTLER (1959) J. Clin. Invest. *38*, 720.
WALKENSTEIN, S.S., C.M. GOSNELL, E.G. HENDERSON and J. PARK (1968) Anal. Biochem. *23*, 345.
WALKER, L.A. and R. LONGHEAD (1962) Int. J. App. Radn. Isot. *13*, 95.
WANG, C.H. (1970) Quench compensation by means of gain restoration. *In:* Bransome, E.D. Jr., ed., The current status of Liquid Scintillation Counting (Grune and Stratton, Inc., New York and London) pp. 305–312.
WANG, C.H. and D.E. JONES (1959) Biochem. Biophys. Res. Comm. *1*, 203.
WARD, S., D.L. WILSON and J.J. GILLIAM (1970) Anal. Biochem. *38*, 1.
WEBB, R.A. and D.F. METTRICK (1972) J. Chromatog. *67*, 75.
WEG, M.W. (1962) Nature *194*, 180.
WEISS, H.G. and I. SHAPIRO (1958) J. Am. Chem. Soc. *80*, 3195.
WELLER, A. (1959) Disc. Farad. Soc. *27*, 28.
WELTMAN, J.K. and D.W. TALMAGE (1963) Int. J. Appl. Radn. Isot. *14*, 541.
WENZEL, M. and W. STOHR (1970) Anal. Biochem. *37*, 282.
WHICKER, F.W., W.C. NELSON and A.F. GALLEGOS (1972) Health Physics *23*, 519.
WHITE, D.R. (1967) The evaluation of liquid scintillation mixtures of aqueous samples. *In:* Proc. 1967 Beckmann Summer School., pp. 72–125.
WHITE, C.G. and S. HELF (1956) Nucleonics *14*, 46.
WHITE, F.H. Jr. and C.R. MENCKEN (1970) Anal. Biochem. *34*, 560.
WHITE, E.H., E. RAPAPORT, H.H. SELIGER and T.A. HOPKINS (1971) Bio-organic Chem. *1*, 92.
WHYMAN, A.E. (1970) Int. J. App. Radn. Isot. *21*, 81.
WIDE, L. and A. KILLANDER (1971) Scan. J. Clin. Lab. Invest. *27*, 151.
WIEBE, L.I., A.A. NOUJAIM and C. EDISS (1971) Int. J. App. Radn. Isot. *22*, 463.
WOODS, J.F. and G. NICHOLS (1963) Science *142*, 386.
WRIGHT, P.M., E.P. STEINBERG and L.E. GLENDENIN (1961) Phys. Rev. *123*, 205.
YALOW, R.S. (1973) Pharmacol. Revs. *25*, 161.
YAMAZAKI, M., H. ISHIHAMA and Y. KASIDA (1966) Int. J. App. Radn. Isot. *17*, 134.
YARBROUGH, J.D., A.F. FINDEIS and J.C. O'KELLEY (1966) Int. J. App. Radn. Isot. *17*, 453.
ZUTSCHI, P.K. and J. SAS-HUBICKI (1966) Int. J. App. Radn. Isot. *17*, 670.

Subject index

Absorption
on to solid supports 174, 222
on vial 146
self 23
self by cells 150, 151
self in plastic spiral 110
acetabularia periodicity assay 84
acetic acid chemiluminescence suppression 54, 119, 144
acetone
dimethyl acetal complex (^{45}Ca) 111
quenching by *170*, 213
radioassays in 49
solvent for lipids 179
acetonitrile as primary solvent 47
acetylene radioassay for tritium 103, 249
acid solutions
instrument damage by 166
radioassay of 164
actinides trivalent plutonium radioassay 201
adenine radioassay 59
adenosine triphosphate
luciferin-luciferase assay 83, 84, 150
sulphurase assay 84
adrenaline radioassay in blood 131
aerosil 63
aerosol 61
agar blocks radioassay in 189
agarose
radioassay of 188, 189
radioassay of nucleic acids in 177
alanine-^3H radioassay of 55
alcohol, aliphatic as blender 43
aldehyde, radioassay of 106
aldoses, radioassay of 182
algae
mean size data 153
radioassay of 63, 142, *146*, 147
alkaline earth metals, *see* Calcium, barium etc.
alkaline solution
caesium chloride 191
chemiluminescence in 167
polyacrylamide gel 187
radioassay of *166*
radioassay of macromolecules 176
radioassay of whole animal 107
trapping acid gases 88, 141
alkaline phosphatase assay 59
alkyl phenyl polyethylene glycol in colloid counting 74
alpha emitters, radioassay 47, 199
aluminium-2-ethyl hexanoate in gel counting 63
aluminium stearate
in gel counting 61
problems with 63

americium-241, radioassay 200, 201
amino acid
  activation studies 84
  radioassay of ^{14}C 55, 58, 60, 100, 116, 143, 174
  solubilization of ^{14}C 96
amino (2) napthalene 6,8 disulphonic acid, Cerenkov assay 10, 79, 80, 146
amino (1) naphthalene-4-sulphonic acid, with Cerenkov 79
ammonium
  formate, colloid counting of 163, 169
  formate, in nucleic acid assay 58
  formate, gradient assay 172
  hydroxide, radioassay in 162
amphibia, egg development, firefly syst. 84
amplification 286
anaemia, vitamin B12 assay in 47
analytical key 280
ANDA, *see* amino(2) naphthalene-6,8-disulphonic acid
anthracene
  blue violet 48, 49
  detecting high energy electrons 47
  flow cell 255
  physical properties 272
  with Triton GR-5 49
antibody, radioimmunoassay 132
anticholinesterase, chemiluminescent assay 260
antigen, radioimmunoassay 132
anti-oxidant
  BHT 10, 35, 46
  use in LSC 46
aqueous solution
  radioassay *162*
  radioassay by Cerenkov 82
archaeology, applications 244, 250
argon
  to displace oxygen 10
  ^{21}A solubility in toluene 90
Armeen L-11, gel counting with 64
ascorbic acid
  radioassay of iron 126
  subdue chemiluminescence 26, 97, 98, 119
astatine, ^{211}At, ^{217}At 199
auger
  definition 284
  emission 199
automatic amino acid analyser, effluent radioassay 176, 256, 257
automatic histology processor, radioassay of gels 187
automatic fractionator
  polyacrylamide gels 185, 186
  thin layer scrapings 194
automatic quench correction (AQC) 233
autoradiography 151, 180

Background count rate
  after vial cleaning 40
  due to vial material 27
  ESR to determine 222, 240
  in colloid counting 69
  instrumental 238
  nomogram to optimize 239
  origin 240
  statistics of 237, 238, 239
bacteria
  amino acid activation 84
  assay by quench analysis 261
  radioassay 60, 149, *150*
  size distribution 153
balance point
  definition 284
  quench correction method 219, 284
barium
  ^{133}Ba radioassay 62, 221
  ^{140}Ba data 265
  carbonate-^{14}C radioassay 60, 61, 62, 63, 102, 147, 160
  chloride solution for $^{14}CO_2$ 147

hydroxide as $^{14}CO_2$ trap 160
sulphate-^{35}S radioassay on discs 54
barley, ^{86}Rb in root 146
basophil, size 153
BBOT
see Tables A.II.1,3
as primary solute 113
physical properties 272
with solubilizers 97
BBS 2 & 3 *see* Biosolve
benzene
as extracting solvent 169
^{14}C radioassay via acetylene 140, 251, 252
half quencher molar conc. 22
physical properties 274, 275
radiocarbon dating 251, 252
tritium assay via acetylene 103, 249, 251
benzidinium sulphate radioassay as ^{35}S compd. 60
benzonitrile, as primary solvent 47
benzoyl peroxide, as bleaching agent 81, 123
betaine hydrochloride, radioassay as ^{14}C compd. 144
beta particle, *see* electron
BHT, *see* antioxidant
BIBUQ
see Tables A.II.1
as primary solute 18
half molar quench concentration 22
physical properties 272
bile pigments, colour quenching by 118
bio-luminescence
see luciferin-luciferase
assay of 150
Biosolve
polyacrylamide gel assay 187
solubilizer 97, 100, 114, 124, 279
biuret, det. protein in solubilizer 98
bleaching agents
benzoyl peroxide 81
in Cerenkov assay 81
in radioimmunoassay 133
of haemin 123, 133
radioassay of plant material 144
sodium bisulphite 81
sodium borohydride 123
blender
definition 284
use of 21, 43, 45
blood
colour interference 118
iron turnover studies 100, 126
radioassay 97, 121, *123*, 124, 126
solubilization 97
wet oxidation 24, 100
body water, assay of 134, 137
bone
^{45}Ca specific activity *111*
collagenolytic activity 116
fission products in fish 250
radioassay of *109*, 111, 121
^{89}Sr in 63, *109*
boric acid
glucotriazole deriv 182
scintillant with 182
bottle, non-spillable and Mason jar 103
bovine serum albumin as carrier 171, 191
brain, rat, radioassay 100
bremsstrahlung definition 284
brine, colloid counting of 248
bromonaphthalene, Cerenkov assay in 82
bromophenacyl-formate (^{14}C), glucose radioassay 125, 183
broth, culture, radioassay 154, 155
buffer solution
colloid counting of 169
radioassay of nucleic acids in 177
radioassay of proteins in 174
butoxy ethyl alcohol, blending action of 125
butyl acetate, Cerenkov assay in 81

butyl PBD
see Table A.II.1,3
artifacts with solubilizers 97
physical properties 272
scintillant mixture with 115

Cab-O-Sil
description 63
radioassay $^{14}CO_2$ in KOH 167
radioassay environmental fission products 250
radioassay iron 127
radioassay milk 138
radioassay plant tissue 146
radioassay plasme 129, 131
suspending calcium salts 110, 147
suspending glass fibre 21
suspending inorganic ppts. 203
suspending lithium carbonate 141
suspending thin layer lipids 180
cadmium
chemiluminescent assay of 260
^{109}Cd-^{109}Ag radioassay 199
caesium
centrifugation studies 171, 191
^{137}Cs as external standard 221
^{137}Cs in rainwater 250
^{137}Cs radioassay 64, 80, 199, 205
^{137}Cs radioassay as perchlorate 63
half life, energy data ^{137}Cs 265
radioassay nucleic acids 177
calcein, in calcium assay 112
calcium
^{45}Ca chloride radioassay 110
^{45}Ca, faeces 123
^{45}Ca, half life, energy data 265, 268
^{47}Ca, half life, energy data 265
^{45}Ca hydroxide radioassay 109
^{45}Ca, milk 138
^{45}Ca, nitrate radioassay 110
^{45}Ca, oxalate radioassay 110, 112
^{45}Ca perchlorate radioassay 112
^{45}Ca phosphate radioassay 189
^{45}Ca, plant tissue 144, 146
^{45}Ca, plasma 131
^{45}Ca radioassay 49, 109, 110, 131
^{45}Ca suspension assay 122
^{45}Ca total body assay 108
^{45}Ca, urine 136
^{45}Ca, wet oxidation 101, 114, 121
^{45}Ca + ^{85}Sr, radioassay 112, 113
^{45}Ca + ^{90}Sr, radioassay 112, 113
carbide to assay ^{3}H-H_2O 103
^{14}C-carbonate pellet 189
^{14}C-carbonate radioassay 60, 110
chloride as carrier 145
calibration curve, efficiency-solute conc. 23
californium, ^{252}Cf radioassay 199
caproate, inorganic salt radioassay 200, 205
carbamates, radioassay 42
carbohydrates, wet oxidation 101
carbon
^{14}C, carbonate, radioassay 102, 157
^{14}C, colloid counting 65
^{14}C, combustion 90, 94
^{14}C, Cerenkov assay 82
^{14}C, emission energy 148
^{14}C, flow cell 255
^{14}C, half life 148, 253, 265
^{14}C, half value thickness 148
^{14}C, nucleic acids 177
^{14}C, proteins, wool etc. 115, 175
^{14}C, quench correction 221
^{14}C, radioassay 48, 49, 51, 109
^{14}C, soil 147, 148
^{14}C, wet oxidation 101, 114, 121
tetrachloride, half molar quench 170
tetrachloride, lipid solvent 179
tetrachloride, quenching agent 213
carbon dioxide
^{14}C from proteins 174
^{14}C from wet oxidation 101

^{14}C radioassay 62, 125, 157
^{14}C vial for uptake 30, 125, 157
^{14}C respired 140, 146
^{14}C trapping methods 88, 89, 103, 140, 141, 156, 158, 159, 260
carbon-hydrogen bonds 26
carcinoembryonic antigen (CEA), radioimmunoassay of 133
carotenoids, effect on Cerenkov 142
cartilage 190
cation exchange paper, Amberlite SA-1 58
cells
mammalian 61, 149, 151
in culture 148
cellulose
as absorbant in RIA 133
as bulk in combustion 147
DEAE 10, 56, 58, 171
esters, pore sizes 154
esters, radioassay on 50, 56, *59*, 60, 191
filter paper 152
Celotate, pore sizes 154
centrifugation
pellets from 189
radioassay after 174, 189
Cerenkov
acid solutions 81, 166
and background count rate 240
channel setting for 80
chloroform soln. 179
chlorophyll on 142
$^{144}Ce + ^{144}Pr$ 77
colour quenching of 218
directional aspect 77, 82
in solvents 81
of ^{36}Cl 145, 146
^{32}P nucleic acids 179
^{32}P on TLC 195, 196
^{32}P phospholipids 179
radioassay by 23, *76*, 80, 82, 150, 199
relation 76, 78
waveshifters 78
with plant tissue 112, 115
cerium
$^{144}Ce + ^{144}Pr$, Cerenkov 77
$^{144}Ce + ^{144}Pr$, half life, etc. 165
cetyl trimethyl ammonium, precipitate 106, 178
channels ratio
correction procedure, ^{14}C *216*
correction procedure, ^{35}S *216*
correction procedure, ^{3}H *215*
sample 72, 73, 74, 80, *214*, 216, 217, 284
charcoal
decolourizing agent 135
radioimmunoassay 132
Chauvenet's criterion 241
chemiluminescence
acid solutions 164, 165, 166
alkaline solutions 167
artifacts by 165
assay of metal ions by 47, 260
colloid counting 69
effect of temperature on 96
future prospects with 254, 260
mechanism of 24, 25, 26, 33
prevention on discs 54
solubilizers 96, 97, 98, 124, 144
chi-squared test 240, *241*
chlorine
^{36}Cl, assay of cell populations 261
^{36}Cl, by Cerenkov 145, 146
^{36}Cl half life, energy data 265
^{36}Cl, hydrology 246
^{36}Cl, quench correction 221
^{36}Cl, radioassay 62
peroxide, hazard in wet oxid. 101
photon yield, ^{36}Cl 77, 80
sodium chloride-^{36}Cl assay 204
chloroform
phospholipid solvent, Cerenkov 179
quenching action of *170*, 213
solvent for lipids 179, 195

chlorophyll
  colour quenching by 146
  effect on Cerenkov 142
chromatography
  cpm correction for gradient 171
  eluate, colloid counting 75, 168, 169
  eluate, hydroxylapatite 169
  eluate, radioassay 82, 161, 162, 178
  filter paper *196*
  isotope concentration 120, 203
  thin layer *194*
chromic acid, wet oxidation with 101
chromium, ^{51}Cr half life, energy data 265
chromophore 23
clays, chromatographic effect of 249
cobalt
  chemiluminescent assay of 260
  ^{60}Co environmental 200
  ^{57}Co radioassay 47
  ^{60}Co radioassay 47, 199
coefficient of variation, colloid counting 69
coincidence
  in LSC 24
  delay time 24
collagen, radioassay 109, *114*, 116, 117
colloid counting
  caesium chloride fract. *191*
  cell pool assay 193
  chemiluminescence in 166
  chromatography 171
  commercial products 279
  counting systems 23, 55, *64*, 131, 160, 162, 163
  in RIA 133
  instability 67
  milk 138
  nucleic acid soln. 177
  of acid soln. 108, 163, 169
  plant assay 144
  plasma 130, 131, 133
  polyacrylamide gel 185, 187
  protein solutions 174, 176
  quench correction 212
  solubilizers and 97, 114
  starch and agarose gels 189
  strong acids 165
  structure 19, 39
  sucrose gradient *192*
  temperature 67
  tritiated water 248, 249
colour quenching 96, 121, 123, 135, 197, 218
combustion
  animal tissue 114, 115, 117, 151
  flask electrodes 91, 92
  milk 138
  paper chromatogram 197
  Parr bomb 90, 123, 147
  plant tissue 143
  plastic bag 94, 95
  polysaccharides 182
  proteins 174
  ^{35}S in plants 145
  scintillants for 45
  use in sample preparation 21, 24, *87*, 93
competitive binding, cortisol radioassay 133
complex formation
  ^{45}Ca radioassay 111
  inorganic application 200, *203*
Compton electrons, external standard 16, 208, 220
computer
  on and off line 242, 243
  shared facility 243
  use of 31, 35, 242
conversion electrons 220
Coomassie blue
  protein stain, polyacrylamide gel 188
  colorimetric assay of 188
copper
  chemiluminescent assay of 47
  combustion catalyst 90

^{66}Cu half life, energy data 265
corticosteroids, radioimmunoassay 133
cortisol, radioimmunoassay 133
cosmic rays, on background count rate 240
cost aspects 9
counter current, radioassay of fractions 170
cover slips, radioassay on 151
curium-242 199
cyclic AMP 59

DAppD scintillant 157, *278*
data evaluation 234
DEAE cellulose discs 10, 56, 58, 171
decay curve
 ^{35}Ca 268
 ^{59}Fe 268
 ^{3}H 271
 kinetics 25, 234
 ^{24}Na 270
 ^{32}P 269
 ^{86}Rb 269
 ^{35}S 270
 tables 236
decontamination methods 36, 38
degradation
 enzymatic 99, *104*, 164, 177, 190
 methods 98
 nucleic acids *104*, 178
 polysaccharides 182
 with volatilization of isotope 99, *101*, 143, 174
 without volatilization of isotope *99*, 100, 114, 115, 118, 138, 143, 146, 151, 182
dehydration
 plants before assay 146
 solvents for LSC 170
dehydroluciferin 83
deoxyribonuclease, DNA degradation 164, 177, 190
desert wells, tritium and ^{14}C dating of 247
detection limits
 efficiency definition (D) 284
 ^{55}Fe 128
 Y value 247
detergent
 colloid counting 26, 65, 74
 washing 36
deuterons, detection of 47
diallyl tartardiamide
 electrophoresis gel 187
 synthesis 188
diborane activated catalyst, preparation and use 252
dibutyl phosphate, use in LSC 110, 200
di-2-ethyl hexyl phosphoric acid in inorganic radioassay 201, *202*
diffusional quenching 284
digitalis, radioimmunoassay 132, 133
diglyme in diborane catalyst prep. 252
diluter *see* blender
dimedone complex radioassay 106, 116, 182
dimethyl formamide, wet oxidation with 100, 120
dimethyl oxazolidone dione, intracellular pH assay 263
dimethyl POPOP as secondary solute 42, 43, 44, 89, 145, 152, 196, 201
dimethyl sulphoxide solvent for plasma 129
Dimilume 26, 30, 166
dioxane
 as scintillant solvent 22, 25, 35, 42, 43, 44, *45*, 129, 134, 144, 168, 203
 chemiluminescence in 96
 physical properties 274, 275
diphenyl stilbene 48
discs
 assay macromolecule on 171, 192
 counting of protein on 174
 drying of 55, 56

extraction of quencher from 53
of salt solution 169
ppt. radioassay 105, 106, 163
protein lysis on 163
punching out 56, 57
quenching artifacts 59
radioassay of 21, 36, 49, 50, 108, 112, 124, 126, 153, 163, 196
discriminator definition 284
disintegration constant 234
dispensers
for scintillants 35, 36
for standards 209
distillation, tritiated water from urine 134
DNA
adsorption on discs 60, 211
as carrier 171
assay of synthesis in cells 152
enzymic assay of 104
from caesium chloride 191
from sucrose gradients 192, 193
loss of MW on storage 119
radioassay of pellet 190
tritiated 59
DNb scintillant 99, 100, *276*
DNpp scintillant 55, *276*
DNppGM scintillant 45, 168, *278*
DNXpnE scintillant 45, 120, 138, *278*
DNXppE
scintillant 104, *278*
assay of enzyme digests 190
quenching of acids in 165
quenching of solvents in 170
double isotope
^{14}C and ^{3}H 88, *223*
cooling 223
effect of quenching 231
Engberg plot 231
gamma and beta 232
graphical procedure *229*
^{3}H spill-over 225
optimal channel setting *226*
^{35}S and ^{3}H 88
^{35}S and ^{14}C 223
simultaneous equation method 226
three isotopes 230
drug distribution study
general 107
radioimmunoassay 134
Duralon filter, pore size 154
DXpdEA scintillant 193, *278*
dynamic quenching 284

EDTA
$BaCO_3$ solubilization 160
^{45}Ca specific act. estimate 112
chelating agent 249
promethium assay 204
radioassay of nucleic acids 106
efficiency definition 284
elapsed time 34
electrolysis
electrode, preparation 248
enrichment by 247
Fe radioassay in blood 126
electron
auger 199
beta particle 14
capture 126, 199, 284
Compton 16, 208, 220
conversion 220, 286
internal conversion 286
pi 13
positive 47, 199
sigma 13
transfer process 46, 49, 285
electrophoresis
agarose *188*
as preprocessing method 131, *184*
polyacrylamide gel *184*
starch *188*
energy
excitation 26
internal conversion 14

non radiative transfer 17
transfer quantum efficiency 15, 285
transfer solute-solute 15
transfer solvent-solute 15, 287
transfer solvent-solvent 14, 21
vibrational 13
vibrational relaxation 15
Engberg plot 231
environmental studies 61, 138, 142, 147, 150, 200, 245, 246, 249, 250
enzyme
collagenase 116
degradation with 99, *104*, 190
kinetic studies 53, *58*, 157
pancreatic DNAase for DNA *104*, 190
pronase B for protein 105, 190
RNAase for RNA 105, 190
use on discs 175
eosinophil, size 153
erythroblast, size 153
esters, radioassay 42
estuaries, water movement in 245, 246
ethanolamine, trapping acid gases 88, 89, 102, 125, 126, 141, 157, 260
ethers
radioassay in 169
radioassay of 169
ethoxy ethanol
after wet oxidation 101, 102, 190
peroxides in 46
physical properties 274, 275
quenching by 213
use as a blender 21, 35, 42, 43
ethyl acetate, Cerenkov assay in 81
ethyl alcohol
blending 45, 96
blood iron radioassay 127
fixation by 152
physical properties 274, 275
precipitation by 145, 203
quenching by 170, 213
radioassay in 49
ethylene diacrylate radioassay of ^{45}Ca 109
ethyl naphthalene 46
excimer 14, 285
excitation energy 26
excited states 26
excretion rate 121
extraction
advantages 156, 161
radioassay following 161
solvent 119, *200*
external standard
artifacts 156, 177
background count rate with 222, 240
colloid counting 72, 73, 74, 75, 124
definition 285
lipid radioassays 179
ratio (ESR) 10, 16, 156, 208, 220, *221*, 285
source dependence 222
eye
solubilization 117
wet oxidation 100

Faeces
radioassay in 61, *121*, 122
synthetic faecal ash 122, *123*
fall out
^{55}Fe in 128
in milk 138
surveys 244, 245, *249*, 250, 253
fat, wet oxidation 100
ferriphosphate, plutonium assay *203*, 250
ferritin, granule size 153
ferrous *see* iron
figure of merit
colloid counting 64
definitions 285
instrumental 49, 285
scintillator 285
solubilization 97
filamentous organisms, radioassay 142

filter paper
elution from *198*
radioassay on 50, *51*, 111, 126, 152, 157, 176, 192, *197*, 211
^{35}S assay on 52
self absorption artifacts 59
filtration, characteristics 153
firefly enzyme 83
Fiske–Subbarow reagent 180
flask
electrodes for 91, 92
Kjeldahl, wet oxidation 103
Shöniger combustion 24
Warburg, wet oxidation 103
flavin mononucleotide (FMN), luciferin assay 83, 85
flow monitoring
Cerenkov 82
chromatographic eluates 162, 165, 255
gas chromatography 258, 260
heterogenous system *255*, 257
homogenous system *256*, 258
polyacrylamide gel 185
fluorescence
quantum efficiency (q) 15, 286
spectrum 16, 42, 45
yield 47
food, radioassay 138, *139*
formaldehyde, ^{14}C radioassay 106, 116, 182, 183
formamide, solubilization 96, 150, 189
formic acid
^{14}C radioassay 115, 125, 144, 183
Cerenkov radioassay in 81
colloid counting of 166
merit values (MIV) in 166
fractionator
thin layer chromatography 194
polyacrylamide gel 185, 186
sucrose gradient centrifug. 192
freeze drying
artifacts 156, 174
plant tissue 143, 144
plasma 131
freezing
animal tissue 119
plant material 144

G value definition 286
gain factor 286
galactokinase 58
gamma
detection 47, 132, 199
radioimmunoassay 133
gas
chromatography 258
liberated *156*
respired *140*, 149
gastrointestinal tract
CEA in tumours 133
iron isotopes 127, 128
Geiger Muller 62
gel
assay 49, *60*, 62, 127
definition 286
fractionator 185, *186*
gelatine vials 94
gelling agents 54
geometry
Cerenkov 82
disc orientation 51, 197
effect of 17, 19, 125, 220
geophysical *244*, 245
glass
fibre, centrifugation assay 192, 193
fibre, multiple 52, 53
fibre, nucleic acid assay 177, 178, 191
fibre, polyacrylamide gel 185
fibre, radioassay on 50, 52, *53*, 111, 152, 162, 174, 177, 178, 182, 191
fibre shedding 111
vial material 27, 28, 78, 80
wool, $^{14}CO_2$ absorbant 125
gluconate, ^{14}C radioassay 52

glucose
^{14}C assay in blood 106, 124
^{14}C colloid counting 66
^{14}C loss in plant assay 143
penta-acetate 124
phosphate 125
triazole deriv. 106, 182
glutathione, quench assay *261*
glycerol
in tissue assay 120
kinase assay 59
glycine, radioassay 100
glycogen
granule size 153
radioassay 153
gold, half life, enery data ^{198}Au 265
gradients
colloid counting of 169
correction for *171*, 192
nucleic acids on 193
ground water
colloid counting of 67, 244, 248
Survey 245

Haemin
colour problem 23, 123
^{14}C radioassay 125
hair
diameter, quench assay 214
radioassay 109, *114*
half value quench molar concentration
acid solutions *165*
definition 22
organic solvents *170*
half value thickness, data 148, 150
hapten 132
health physics 75, 134
HeLa cell 151
heptane, Cerenkov in 81
herbicides
radioassay in plants 142
radioassay in soil 147
heterogenous LSC
counting 17, 20, 42, *48*, 62, 125, 208, 234
counting organelles 153
polyacrylamide gel assay 185
precipitates 123
quench correction 211
solubilizer system 97
hexadecane, labelled standard 66, 209
hexamethylene tetramine, combustion artifact with 62
hexane
Cerenkov plant solvent 146
lipid extractant 194
hexokinase assay 59
histidine decarboxylase, assay 157, 158
homogenization, tissue and faeces 118, 122
homogenous LSC
blood iron 127
cells 151
definition 41
eluates 171, 172, 178
gradient correction 172
macromolecule assay 104, 163, 178, 190
quench correction 207
salt solutions 169
solvent radioassay 170
subcellular organelles 153
horn, radioassay 109
Hyamine 10X
for gels 187
plant tissue 143
quenching 213
solubilizing action 96, 117, 118, 123, 125, 129, 146, 167
trapping acid gases 25, 62, 65, 88, 91, 103, 140, 157, 159
hydrocarbon assay 42
hydrochloric acid
chemiluminescence prevention 98, 119

colloid counting of 166
faecal ash solubilizer 122
merit value (MIV) 166
hydrofluoric acid, silica gel solubilization 196
hydrogen peroxide
bleaching agent 123, 125
polyacrylamide gels 184, 187
wet oxidation with 24, 101, 143, 175, 190
hydrogen sulphide, ^{35}S, trapping of 156
hydrology 245, *246*, 249
hydroxamate, ^{14}C radioassay 58
hydroxy apatite 16
hydroxy lysine, ^{14}C radioassay 116
Hylene TM 65, *see* toluene di-isocyanate
hypochlorite, bleaching action 133

Igepal CA 720, in colloid counting 74
immunoglobulin, ^{131}I antigoat, in RIA 133
inert gases, radioassay 42
infra-red, in combustion 95
inorganic 199
insect cuticle, radioassay 99
insecticides, in plants, radioassay 142
Instagel
colloid counting 166, 279
^{3}H-H_2O radioassay 248
polyacrylamide gel 185
instrument
choice 31
manufacturers, list 280, 289
internal, conversion electrons 286
internal standard
accuracy 236
^{14}C, $BaCO_3$ radioassay 212
^{14}C-benzoic acid 209
^{14}C-oxalic acid 210
^{14}C-sodium bicarbonate 210
colloid counting 189
enzyme digests 190
^{3}H-water 210
homogenous counting *209*, *211*
labelled hexadecane 209
labelled toluene 209
volume temperature data 210
with Cerenkov 80, 96
intestine
cleaning 119
epithelium 120
wet oxidation 100
intracellular pH, assay 263
intrinsic factor, in RIA 133
iodide peroxide detection 46
iodine
$Ag^{131}I$ in rain 63, 250
^{127}I by neutron activation 79, 80
^{125}I half life, energy data 265
^{125}I in RIA 132, 133
^{131}I in RIA 133
^{131}I with beta emitter 52, 232
^{131}I half life, energy data 266
ion exchange
chemiluminescence 165
concentration on resin 156
radioassay on paper 50, 53, *56*
iron
chemiluminescent assay 260
deficiency 126, 127, 128
di-(2-ethyl hexylphosphate) 128
^{59}Fe decay table 268
^{55}Fe in environment 128
^{55}Fe, ^{59}Fe half lives, energy data 266
ferric benzene phosphonate 127
^{59}ferric citrate 127
ferriphosphate 126
ferrous perchlorate 126
^{59}Fe wet oxidation 100
^{59}Fe with beta emitter 232
hydroxide in radioassay 126
isotope, channel settings 128
isotope, clinical 127
isotopes in blood 250

isotopes in food 139
isotopes in plasma *129*
isotope radioassay 126, 127, 128, 199
stable 128
iso-enzymes electrophoresis 188
isoprenyl acetylene hydrogenation with ^3H 249
iso propyl alcohol, Cerenkov assay in 81
isopropyl PBD, incompatibility with solubilizer 97
isotope
exclusion method 224
half life 234, 235
isotope effect
definition 286
steroids 181

Kits radioimmunoassay 132, 133
krypton
^{85}K half life, energy data 266
^{85}K in air 206
^{85}K radioassay 125, 206
^{85}K solubility 206

Lactate
CO_2 assay 125
radioassay 106
lakes
TU in 246
water movement in 245
lead
chemiluminescent assay of 260
tetra-butyl in LSC 133
leaves radioassay 142, 144, 146
lens radioassay 144, 117
lens cleaning tissue radioassay on 50, 51
Libby half life definition 251, *253*, 286
light collecting factor 17
lipids
extraction for radioassay 169, 180, 194
radioassay 42, 86, 179
removal 152
lipophyllic salts, use in organic 205
liquid scintillation counting comparative study 152
Liquifluor use 55, 125, 279
lithium
carbonate 141
hydroxide to trap $^{14}CO_2$ 141, 167
liver
assay ^{14}C 60, 64
solubilization 97, 98
low sample reject 33
luciferin-luciferase assay system 82, 150, 160
luminol chemiluminescent metal assay 260
lung, recoveries from tissue 118
lymphoblast cell size 153
lymphocytes cell sizes 153
lysosome sizes 153

Macromolecules
centrifugation fractions 171, 190, 193
in cells 149
in column eluates 162
quench correction 208
radioassay 105, *173*, 174, 177
magnesium
^{27}Mg, ^{28}Mg half life, energy data 266
nitrate in wet oxid. of ^{35}S 100, 120
perchlorate, drying 170
sulphate, drying 170
magnetic tape in data processing 243
mammalian cells
DNA strand assay 192
radioassay 61, 149, *151*
manganese
^{56}Mn, Cerenkov 79
^{54}Mn, ^{56}Mn half life, energy data 266
matching factor (m) 16, 17, 19
maximum
permissable body burden 250
photoelectric quantum eff. 17

medium cell culture assay 75, 117, 149, 154, 155, 159
megakaryocyte cell size 153
megaloblast nucleus size 153
membrane
  pore sizes 150, *154*
  radioassay 150
  radioassay of nucleic acid 17
mercury
  ^{203}Hg, E max 115, 266
  ^{203}Hg, half life, energy data 266
  ^{203}Hg in wool 115, 116
  radiation shield 253
merit value
  blood, plasma, radioassay 129, 130
  colloid counting 64, 65, 67, 69, 70, 75, 155, 164
  definition 62, 286
  gradient systems 172
  instrument corrected (MIV) 9, 136, 139, 166, 286
  instrument and stability corrected (MISQ) 9, 286
  solubilization 176
mesitylene
  physical properties 274, 275
  solvent 22
metabolic cage design 121, *122*
metaperiodate sodium, as oxidant 116
meteorological 245, *249*
methacrylate as gel 63
methoxy ethanol
  as blender 35, 97, 106, 144, 178
  peroxides in 46
  physical properties 274, 275
methyl alcohol
  blender 45, 88, 94, 96, 106, 168, 178
  in lipid assays 180
  physical properties 274, 275
  quenching 170, 213
  radioassay in 49
methylene-bis-acrylamide in gel 184
methyl naphthalene 46
methyl red quenching agent 213
methyl umbelliferone, Cerenkov assay with 78
micelles colloid counting 73
microsyringe standard dispensing 209
microtransferator 84, 85
Microweb pore size 154
milk
  colloid counting 75
  radioassay 138
Millipore cellulose ester, pore size 154
mineral oil primary solvent 47
Mitef, pore sizes 154
mitochondria sizes 153
molecular sieve isotope concentration 156
molybdate phospholipid assay 180
monocyte mean cell size 153
MSB-bis
  colloid counting 74, 135
  physical properties 272
multi-isotope 223, 263
multi-sample 99, 118, 212, 222
muscle
  fish, environmental studies 250
  recoveries of isotopes from 118
  solubilization of 97
  water balance in 136
  wet oxidation 100
myeloblast mean cell size 153
myelocyte mean cell size 153

Naphthalene
  colloid counting 74
  in homogenous LSC 25, 42, 43, 44, *45*, 97, 144
  physical properties 272
  purity 45
NCS (Nuclear Chicago)
  solubilizer 97, 98, 116, 117, 123, 124, 178, 279
  use with dicsc 54, 55

with polyacrylamide gels 185, 187, 188
NE 213 (Nuclear Enterprises) use 129
neutron
activation of ^{127}I 79, 80
capture 200
detection of 47
neutrophil mean cell size 153
nickel
caproate 205
^{63}Ni 199, 204, 266
tetrapyridine dithiocyanate 205
nicotinamide adenine dinucleotide (NADH), luciferin assay of 83, 85
ninhydrin radioassay of amino acids 176
nitric acid
oxidising agent 99, 100, 103, 126, 144, 146
scintillant component 110
solubilization 109, 120
nitromethane quenching agent 43, 44
nitrogen oxygen replacement 89, 94
noble gases 206
non radiative energy transfer 17
nonyl decylamine sulphate, inorganic ion extractant 201
nonyl phenyl polyglycol ether 65
nor adrenaline radioassay in blood 131
normoblast mean cell size 153
NPO (α) 21
nuclear power station
Cerenkov assay of effluent 82
radioassay of effluent 246, 250
nucleic acid
and precursor assay 174
enzyme degradation of 104
in polyacrylamide gels 184
in-vial assay of synthesis *151*, 152
radioassay of *177*, 178, 179
wet oxidation of 101
nucleii
mean sizes 153
radioassay 152
nucleotides, radioassay 177

Octanoates inorganic ion radioassay 200, 205
octyl alcohol blender 129
oil shales ^{3}H-H_2O surveys in 245, 246
oligonucleotides
radioassay 177
radioassay in urea solutions 164
organic solvents radioassay in *169*
optimization
colloid counting 67, 72, 75
scintillator solute 44
osazone radioassay of 182
oven drying plant tissue 143
oxalate
^{45}Ca 110
co-precipitation 203
oxidation, wet 24, *99*, 100, 101, 102, 114, 115, 118, 120, 121, 138, 143, 146, 151, 182
oxygen
atmospheric 21
dimers 26, 90
peroxides from 46
quenching by 90
solubility in toluene 90
use of 18, 21, 24, 26

Paper tape 243
paraterphenyl
primary solute 18, 22, 45, 63, 201
physical properties 272
partition 201
PBBO
solute 114
physical properties 272
PBD
primary solute 18, 22, 41
physical properties 272
pellet radioassay 150, 174, 175, 189, 190
pentyl alcohol RNA extraction from sucrose 193

peptide
  radioassay 123
  radioimmunoassay of hormone 134
peptone quenching by 155
percentage error 237
perchloric acid
  colloid counting of 166
  lipids 180
  oxidizing agent 99, 101, 103, 120, 126, 144, 146
  pellet from 189, 190
  precipitant 105, 110, 152, 162, 191
  quenching by 164
  radioassay of 175, 177, 188
  salt removal by 174
periodate
  solubilization as acid 188
  wet oxidation 101
peroxides
  see also hydrogen peroxide
  chemiluminescence 96, 119
  detection 46
  formation 26, 35, 45, 169
persulphate wet oxidation with 102
pesticides
  radioassay in plants 142
  radioassay in soils 147
petri dish use in urine radioassay 134
pH
  gradient correction 171
  solid scintillator efficiency 49
phase diagram
  colloid counting *67*, 69, 74, 75, 127
  construction *67*
  gradient studies 172
  milk 139
  optimization 144
  urine 136
phenyl alanine ^{14}C radioassay 60, 143
phenyl ethylamine
  radioassay of proteins in 175
  trapping acid gases in 41, 88, 89, 94
phenyl glucosazone radioassay 106
phosphate buffer 146
phospholipids
  Cerenkov assay 79, 179
  radioassay 179, *180*
  TLC radioassay 195
  phosphomolybdate
  ^{32}P assay 139
  ^{32}P phospholipid 180
phosphorescence
  dioxane based scintillants 46
  origin 24, 25, 27
  solubilization 118
phosphorus
  Cerenkov ^{32}P 79, 81, 145, 166
  nucleic acids ^{32}P 177, 179
  ^{32}P, decay table 269
  ^{32}P, in bone 111, 112
  ^{32}P in cells 150
  ^{32}P in faeces 122
  ^{32}P in food 139
  ^{32}P in plants 142, 245
  ^{32}P in soil 142, 245
  ^{32}P in tissue 109, 121
  ^{32}P on TLC 195
  ^{32}P in water 249
  ^{32}P, ^{33}P, half life, energy data 266
  photon yield ^{32}P 77
  radioassay ^{32}P 199
phosphotransferase ATP-creatine assay 59
phosphotungstic acid polyacrylamide gel 187
Photinus pyralis 83
Photobacterium fischeri 83
photocathodes
  description 16
  thermal noise 240
photomultipliers
  description 16
  radioassay of discs on 50
photon

number 16, 24
trapping 19
photo
oxidation 89
synthesis, $^{14}CO_2$ in 147
Pilot B beads 48
piperidine solubilization of gels 187
pipsyl (^{131}I) chloride amino acid assay 263
Pirie's reagent, recipe and use 120
placenta lactogen, radioimmunoassay 134
plant
assay growth regulators 189
Cerenkov assay in 79
ion transport studies 145
radioassay *142*, 143
wet oxidation 101, 182
plasma
colloid counting 75
iron isotopes in 128, 129
radioassay 63, *123*, 128, 129, 131
radioimmunoassay 132
solubilization 97
plastic scintillator
beads 255
Pilot B 48
shavings 206
spiral 109
platelets, mean size 153
plutonium
^{239}Pu M.P.B.B. 250
^{241}Pu M.P.B.B. 250
^{236}Pu radioassay 199
^{239}Pu radioassay 199, 200, 203, 205, 250
^{241}Pu radioassay 199, 200, 203, 250
Poisson distribution 234, 236
polio mean size of virus 153
pollen mean size 153
polyacrylamide gel
fractionation 185, 186
maceration 185
protein estimation in 188
radioassay 50, 98, 177, *184*
staining 188
Poly-Gel B gelling agent 64
polymer suspension 61
polynitro compds, radioassay of 62
poly-olefins gelling agents 64
polysaccharides
enzyme degradation 104
presence during radioassay 142
radioassay 177, 181
radioimmunoassay 133
polystyrene
gel counting 63
vial material 78
Polyvic membrane pore size 154
pool
labelling 173
radioassay in cells 193
POPOP
in LSC 17, 42, 45, 60, 62, 63, 74, 86, 89, 94, 115, 144, 175
physical properties 272
unsuitable for 101
positrons
detection of 47
emitters 199
post mortem plasma radioimmunoassay 132
potassium
alcoholic hydroxide solubilizer 123
hydroxide radioassay 166
hydroxide to trap gases 88, 147
^{40}K Cerenkov 77
^{40}K emission energies 148
^{40}K half life 148
^{40}K hal thickness data 148
^{40}K in KOH 167
^{40}K in soil 147, 148
^{40}K in vial glass 240
^{40}K, ^{42}K, half life, energies etc. 267

^{40}K radioassay 27, 64
pouter, TLC collection 194, *195*
PPO
colloid counting 130, 135, 152
in flow cells 255
in LSC 15, 17, 18, 22, 40, 41, 42, 44, 45, 60, 63, 74, 86, 89, 94, 144, 159, 196
in solubilizers 175
physical properties 272
praseodymium co-precipitation with ^{147}Pr 203
precipitation
isotope concentration *105*, *203*
artifacts from 123
nucleic acid 177
Pregl-Roth combustion system 87
Primene 81R trapping acid gases 88
print out 32
promethium
^{147}Pr, ^{149}Pr half life, etc. 267
^{147}Pr radioassay, urine 201, 203
proportional counting 142, 144
protein
electrophoretic gels 184, 188
enzyme degradation 104, 105, 163
estimation in solubilizer 98
in blood 123
on discs 60
pellet 189
quenching by 155
radioassay 25, 42, 75, 123, 150, 174, 175
radioimmunoassay 132
removal of 131, 162, 163
solvent for 96, *98*
staining 188
synthesis on discs 175
wet oxidation 101
protons, detection of 47
pulse height
discrimination 286
relative (RPH) 17, 18, 21, 286
spectrum 215, 229
spectrum $^{45}Ca + ^{90}Sr$ 113
spectrum $^{236}Pu + ^{238}Pu$ 204
spectrum $^{85}Sr + ^{45}Ca$ 113
pulse shape
discrimination (PSD) 286
figure of merit (M) 256
punched cards 243
pyridoxal phosphate enzyme assay 157
pyrophosphorylase ADP-glucose assay 59

Quantum efficiency
solute fluorescence (q) 15
photoelectric (K) 17, 286
quartz vial material 27
quench
automatic correction 233
balance point correction 219
by solvents 199
colour 19, 20, 23, 197, 218, 287
comparative data 47
correction based on protein 131, 175
correction in colloid counting 72, 124, 131
correction methods *207*
correction precision 217
curve construction 2
definition 287
diffusional 284
double isotope correction 231
dynamic 284
external standard ratio (ESR) 220
faecal ash curve 122
future prospects 254
impurity 19, 20, 287
in inorganic application 200
isotope exclusion 224
occurrence 19, 20, 21, 23, 127, 155, 164, *165*, 194, 197, 218
on lipid extraction 179
sample channels ratio (SCR) 212

source 20, 21
use of carriers 171

Radiocarbon
bomb method 253
dating 245, *250*, 251
LSC method 251
proportional counting 251
radioimmunoassay
absorbants 132
CEA 133
digitalis 132, 133
double isotope 133
drug distribution studies 134
general *131*, 132, 133, 134
^{131}I in 133
post mortem plasma 132
steroids 181
radium
Cerenkov 76
use as external source 222
rainwater
^{137}Cs, ^{96}Sr and ^{131}I in 63, 250
fission products in 249
tritium in 246
recoil atoms
detection of 47
definition 287
red blood cell, mean size 153
refractive
dispersion on Cerenkov 82
index, on Cerenkov 76, 77, 78, 82
relative pulse height (RPH) 17, 18, 21, 287
reproducibility 8
resin
as strip in RIA 133
use in RIA 132, 133
rhodium, ^{220}Rh radioassay 199
ribonuclease in RNA degradation 164, 177, 190
ribonucleotides radioassay by Cerenkov 79
ribosomes radioassay 152
rivers
fission products in 249, 250
tritium units in 246
water movement assay 245
RNA
enzyme assay of 104
pellet radioassay 190
radioassay on discs 54, 60
sucrose gradients 193
synthesis in cells 152
water occlusion on 178, 262
roots 146
rubidium
nucleic acids in solution 177
^{86}Rb Cerenkov assay 79, 145, 146
^{86}Rb decay table 269
^{87}Rb emission energy 148
^{86}Rb half life, energy data 267
^{87}Rb half life, energy data 148, 267
^{87}Rb half thickness 148
^{87}Rb in soil 147, 148
^{87}Rb radioassay 205
ruthenium
^{106}Ru half life, energy data 267
^{106}Ru, radioassay 199

Salt solution
colloid counting 64, 143
effect on Bray's 168
effect on LSC 123
proteins, radioassay 174
radioassay 168
samarium
^{147}Sm radioassay 205
^{151}Sm 199
sample channels ratio (SCR)
^{14}C, ^{35}S 216
curve construction *214*
^{3}H 215

method 208, 212
sample multiplicity 7, 208
Sartorius membrane pore size 154
scanning procedure comparison with LSC 152, 176
scintillant
anthracene 255, 256, 260
boric acid containing 182
Bray's 60, 66, 72, 156, 167, 168, 177, 181, 205
butyl PBD 255
cerium activated lithium glass 255, 256
code definition 271, 276
commercial mixtures 2, 279
commercial suppliers 289
europium activated fluorspar 255
in solid phase 48, 49, 255, 256, 260
Kinard's 88, 94, 187
large volume mixture 47
lead loaded mixture 30
mixtures, blended 277, 278
mixtures, non-blended 276
Pilot B beads 48
plastic beads 255
plastic shavings 206
silicone oil coated 260
tin loaded 30, 48
scintillation
attenuation factor (SAF) 287
collection efficiency 287
efficiency 287
self absorption
by cells 150, 151
in silica gel 222
occurrence 23
plastic scintillator 110
self cleaning ovens 38
Sephadex as absorbant in RIA 133
serum
electrophoretic radioassay 188
fatty acids, using ^{63}Ni 205
iron isotopes in 129
radioassay in 25, 98
sewerage 246
Shöniger flask combustion 24
silica
as Cab-O-Sil 63
TLC support 179, 194
silver
^{108}Ag, half life, energy data 267
nitrate, catalyst 102
skin
radioassay 109, 114, 115
wet oxidation 100
sodium
bisulphite bleach 81
^{3}H acetate, radioassay 60
hydroxide, radioassay in 166
^{24}Na, decay table 270
^{24}Na, half life, energy data 267
^{22}Na radioassay 62, 199
^{24}Na radioassay 199
sulphate as drying agent 170
soil
bioluminescent assay, bacteria 150
^{32}P in 142
radioassay 61, 147, 245
solubilizer
alcoholic KOH 123
blood 124
^{45}Ca, skunk toe nails 114
colloid nature of 74
commercial list 279
definition 287
disc treatment 54, 60, 152, 163, 178
for cells on cover slips 151
of precipitates 105, 150
of protein 175
polyacrylamide 184
quaternary ammonium 25, 41, 42, 43, 62, 95, 117, 118
skin 115
trapping acid gases 88
water requirement 178

Soluene-100
solubilizer 97, 98, 279
extraction of RNA 193
polyacrylamide gel 185
solute
data list 272
fluorescence life time 15
fluorescence quantum eff. (q) 15
primary 17, 18, 42, 43, 272, 273
secondary 18, 42, 43, 272, 273, 287
secondary, with Cerenkov 78
solute-solute energy transfer 15
solvent
excitation migration 285, 287
primary 13, 274
secondary 47, 274
physical properties 274, 275
solvent-solute energy transfer 15, 287
solvent-solvent
conversion factor 14
energy transfer 14, 21
Somogi's reagent, blood radioassay 124
specific activity definition 288
spectrum
see pulse height
Cerenkov 78
matching factor 16, 288
sensitivity 16
spectrometer use as 31
spleen
poor isotope recovery from 118
wet oxidation 100
spores fungal, mean size 153
squames adhesive tape radioassay 115
stability
colloid counting 67, 69, 70
plot 72
stainless steel reactors, radioassay 204
stain effect on paper, radioassay 196
standard
^{14}C count rate (archaeol) 253
external 124, 156, 177, 222, 240
deviation 237, 238
internal 189, *209*, 210, *211*, 212
internal accuracy 236
starch electrophoresis, radioassay 188
static quenching 288
statistics 81, 234
stems radioassay 142
Stern-Volmar, use of equation in quench 21, 23
steroids
Cerenkov assay 79
conjugates 181
radioassay 42, 179, 181
radioimmunoassay 181
solvents used, radioassay 169
Sterox DJ, colloid counting with 74
stilbene, secondary solvent 47
storage
labelled material 144
scintillation mixtures 35, 36
strontium
carbide 252
^{85}Sr + ^{45}Ca radioassay 112
^{90}Sr + ^{45}Ca radioassay 112
^{90}Sr Cerenkov 79
^{90}Sr in milk 138
^{89}Sr in plants 146
^{90}Sr in rainwater 63, 250
^{89}Sr on discs 54
^{89}Sr radioassay 109, 112
^{90}Sr radioassay 62, 64, 109
^{89}Sr suspension 63
^{85}Sr, ^{89}Sr, ^{90}Sr, half life, energy data 267
Student's *t* test, use *240*
subcellular
components radioassay 149, 152
organelles sizes 153
sucrose solution
Cerenkov assay 77
colloid counting in 163
gradient radioassay 171, 192

nucleic acids 177, 192, 193
Sudan IV colour quench 123
Sudan red colour quench 23, 213
sugar radioassay of 181
sulphur
^{35}S combustion 94
^{35}S decay table *270*
^{35}S disc counting 52
^{35}S flow cell assay 255
$^{35}SO_2$ from combustion 144, 145
$^{35}SO_2$ wet oxidation 101
^{35}S half life, energy data 267
^{35}S in plants 144, 145
^{35}S in soft tissue 120
^{35}S in wool 115
^{35}S radioassay 199
$^{35}SO_2$ radioassay 88, 94
$^{35}SO_2$ trapping 156
^{35}S wet oxidation 100, 101, 114
sulphuric acid
in $^{14}CO_2$ assay 125
to precipitate ^{35}sulphate 145
wet oxidation with 101, 102, 126
sulphydryl assay 261
sunlight photoinduction by 25
suspension
barium carbonate radioassay 101
cell radioassay 151
counting 49, *60*
faeces, radioassay 122
inorganic 200
iron radioassay 127
milk radioassay 138
nucleic acid radioassay 152
phosphomolybdate radioassay 139
plant radioassay 146
polymer 61
sel absorption 60
TLC assay 179

Tachyarrhythmia, supraventricular, digitalis assay in 132
tape, adhesive in skin radioassay 115
Tb scintillant 194, *276*
TDNpdM scintillant 117, 137, *278*
teeth, radioassay of *109*
temperature effects 25, 33
Tesla coil, combustion igniter 91, 92
testes, wet oxidation of 100
tetrabutyl, tin, gamma assay 133
tetra decanal, in firefly assay 83
tetramethyl ammonium hydroxide 98
Texafor FN 11, colloid counting with 131
thallium, ^{204}Tl, Cerenkov 77
thermal
diffusion 288
movement 14, 288
noise 240
thin layer chromatography (TLC)
lipids, radioassay 179, *180*
radioassay 63, *194*
removal of support for assay 194, *195*
thiocyanates, ^{60}Co radioassay 200
thixcin, gel counting with 61, 62, 65
thixotropy, gel counting *61*
thorium
assay in yttrium oxide 201, *202*
^{232}Th emission energy 148
^{232}Th half life 148
^{232}Th half thickness data 148
^{232}Th in soil 147, 148
^{232}Th radioassay 199
thymidine
DNA synthesis radioassay 115, 151
kinase assay 58
removal during assay 58
thyroid
hormone radioimmunoassay 133
$^{131}I + {}^{3}H$ radioassay 232
tibia, collagen synthesis in 116
tin
loaded scintillant 30, 48
$^{113}Sn + {}^{113m}In$ radioassay 199
tetrabutyl 48, 133

tissue
Cerenkov assay 79
combustion of 90
plutonium isotopes in 203
radioassay 61, 86, 96, *107*, *109*, *117*, *119*
wet oxidation 100, *101*, 103
TNppX scintillant 43, 179, *277*
toe nails, skunk, radioassay 114
toluene
as extraction solvent 169
di-isocyanate (Hylene TM 65) 64
^{3}H labelled as standard 66, *209*
hazards 39
physical properties 274, 275
primary solvent 22, *42*, 97
temperature volume changes *210*
tomato plants, radioassay during storage 144
Tp scintillant 101
TpC scintillant 190, 277
Tpd scintillant 145, 146, 201, 202, 276
TpdA scintillant 193, *277*
TpdB scintillant 110
TpdD scintillant 112, *277*
TpEGN scintillant 110, *277*
Tpp scintillant 48, 51, 56, 58, 86, 91, 103, 108, 110, 111, 115, 117, 120, 123, 125, 129, 140, 141, 147, 157, 158, 162, 168, 170, 175, 180, 185, 187, 188, 189, 192, 194, 261, *276*
extraction with 161
assay of ^{85}Kr 206
TppE scintillant 9, 121, 167, 179, 190, *277*
transferase, purine phosphoribosyl assay 58
transformation, constant 234
transparancy in colloid counting 70, 71
tributyl phosphate scintillant mixture 110
trichloracetic acid
colloid counting of 166
disc washing with 175
merit values (MIV) 166
nucleic acid extraction 177
polyacrylamide gel 188
precipitating agent 105, 117, 162, 175
protein pellet from 189
removal of salts with 174
quenching by 164, 213
trinitrobenzene, 1-sulphonic acid (TBNS) 197
tripalmitin, ^{14}C radioassay 261
triplet-triplet, annihilation reaction 26
tritiated water
body water in injury 136
colloid counting of 66
electrolytic concentration 247, 248
from combustion 91, 120
from nuclear explosions 245
in lakes 245
internal standard 66, 189
plant radioassay 142
plasma radioassay 131
troposphere 245
urine radioassay 134, 136, 248
wet oxidation 103
tritium
colloid counting 65
combustion 90, 94
contamination 40
counting efficiency 45, 101, 162
dating 246
$^{3}H + ^{131}I$ 52
half life, energy data *271*
in blood, merit value 129
in cells 151
in wool 115
proteins 175
quenching of 208
radioassay of 48, 49, 117
radioassay, low level 245
self absorption effects 51
TU units definition *246*, 288

wet oxidation 114
Triton GR-5, with anthracene 49
Triton N 101, colloid counting with 65, 74, 135
Triton X 100
colloid counting with 26, 55, 65, 66, 69, 70, 71, 74, 108, 115, 125, 127, 133, 144, 152, 160, 163, 164, 169, 189, 193, 194, 196
caesium chloride 191
media, broths etc. 155
merit values 129, 130
milk 138, 139
plasma 129, *130*
protein solutions 176
radioimmunoassay 133
urine 135, 136
with solubilizer 124, 129
Triton X 114 colloid counting with 65, 74
troposphere tritiated water in 245

Ultrasonication
suspensions 61, 127
degassing 21, 90
solubilization 97
increased efficiency with 160
ultra-violet
absorbance, acid solns. 164
avoidance 40
uranium
^{238}U emission energy 148
^{238}U half life 148
^{238}U half value thickness 148
^{238}U in soil 147, 148
^{233}U radioassay 199
^{236}U radioassay 199
urea colloid counting of 75, 163, 164
uridine use in RNA synthesis assay 151, 152
urine
bacteria, bioluminescence in 150
^{45}Ca in 110
Cerenkov 79
colloid counting 75, 135
inorganic isotopes in *201*
radioassay 121, *134*, 135
tritiated water in 134, 135, 248

Vanadium
chemiluminescent assay of 260
silica-alumina activator 252
variability count rate, origins 240
vial
chipping 39, 40
combustion 28, 30
culture in 151
effect on Cerenkov 78, 81, 82
ESR artifacts 29
gamma counting 28, 48, 133
gas absorption 157
gelatine, combustion 94
geometry effects 220
glass 27, 28, 78, 80
^{40}K in glass 240
minivial 28
phosphorescence in 27, 39
polyethylene 27, 29
polyfluohydrocarbon 29
polypropylene 80, 196
polystyrene 78
preprocessing in 151
quartz 27
roughening on efficiency 27
wall absorption 63
washing 36, 37
virus
mean size 153
radioassay 149, *150*
viscosity
chemiluminescence and 25
colloid counting and 70, 71
high, gel counting 63
vitamin
B12 radioassay 47

B12 radioimmunoassay 133
volatilization 99
volcanic activity, isotope assay in 245

Warburg apparatus *156*
washing
methods 36, 37
ultrasonics 40
water
age, radioassay 245
body, assay *134*
Cerenkov assay of 81
colloid counting 69, 163
fresh, radioassay 246
in sample 43, 121
sea, bioluminescence in 150
sea, radioassay of 246
waveshifters use of 17, 79, 80, 150, 288
well counting 133
wet oxidation
blood 24, 100
^{45}Ca 101, 114, 121
carbohydrates 101
ethoxyethanol after 101, 102, 190
eye 100
fat 100
^{59}Fe 100
hydrogen peroxide in 24, 101, 143, 175, 190
intestine 100
muscle 100
nucleic acids 101
plant material 101, 182
protein 101
^{35}S 100, 101, 114
$^{35}SO_2$ from 101
tissue 100, *101*, 103
tritiated water 103, 114
with chromic acid 101
with dimethylformamide 100, 120
with periodate 101
with persulphate 102
wheat stem radioassay *107*
wool
fibre diameter assay *115*, 116
tritiated 61, 114, *115*

Xylene
colloid counting 69, 74
extracting solvent 169
physical properties 274, 275
scintillation mixture 22, 67, 201

Y value
^{36}Cl 204
definition 247, 288
^{85}Kr 206
yeast
mean size 153
radioassay 142, *146*
yttrium
assay of Th in 202
^{90}Y, radioassay 199
^{91}Y, half life, energy data 267

Zero age definition (archaeol) 253
zinc granulated 35, 46, 169
zirconium ^{95}Zr radioassay 199

ISOELECTRIC FOCUSING

P.G. Righetti

Department of Biochemistry,
University of Milano,
Milano 20133, Italy

and

J.W. Drysdale

Tufts University School of Medicine,
Boston, Mass., U.S.A.

To Professor H. Rilbe
father of the 'fine art' of isoelectric focusing

Acknowledgements

Many people have contributed invaluable advice and assistance to the making of this book. Prof. T.S. Work has carefully and competently corrected the manuscript. Prof. H. Rilbe has provided much of the mathematical equations pertaining to isoelectric focusing. H. Davies, H. Haglund, Å. Johansson and C. Karlsson, all from LKB Produkter AB, have given invaluable criticism and suggestions and have provided several of the drawings in the book. T. Wadström and O. Vesterberg have critically read and commented individual chapters of the book. We are particularly grateful to the many friends and colleagues who made published and unpublished material available to us. We thank particularly P. O' Farrel and R. Rüchel for their stimulating work well before publication. We owe a last thank to Miss Mimma Musotto who melted away in a hot summer in Milano typing the final manuscript.

Contents

Chapter 1. Theory and fundamental aspects of IEF *341*

1.1. Introduction 341
1.2. Background 345
1.3. Theory of IEF 347
1.3.1. Artificial pH gradients 347
1.3.2. Natural pH gradients 347
1.3.3. Conductivity 348
1.3.4. Buffering capacity 349
1.3.5. Law of pH monotony 351
1.3.6. Resolving power 353
1.3.7. Peak capacity 354
1.3.8. Mass content of a protein zone 355
1.4. Synthesis of carrier ampholytes 357
1.5. Laboratory synthesis of carrier ampholytes 358
1.5.1. Reagent distillation 358
1.5.2. Coupling process 360
1.5.3. Detection of carrier ampholytes 362
1.6. Fractionation of carrier ampholytes 366
1.7. Properties of Ampholine 367
1.7.1. Molecular size 368
1.7.2. Optical properties 369
1.7.3. Buffering capacity and conductivity 372
1.7.4. Chelating properties 372
1.7.5. Biological toxicity 373
1.8. pH gradients generated by other means 373
1.8.1. Bacto peptone carrier ampholytes 373
1.8.2. Thermal pH gradients 374
1.8.3. Dielectric constant pH gradients 374

1.8.4. pH gradients generated by buffer diffusion 375
1.8.5. pH gradients obtained with mixtures of acids 375

Chapter 2. Preparative IEF 377

2.1. Preparative IEF in liquid media 378
2.1.1. The LKB columns 378
2.1.2. The ISCO columns 389
2.1.3. Rilbe's columns 392
2.1.4. Zone convection IEF according to Valmet 393
2.1.5. Zone convection IEF according to Talbot 397
2.1.6. Multi-compartment electrolyzers 400
2.1.7. Free-flow, high voltage IEF 400
2.2. Preparative IEF in gels 404
2.2.1. Fawcett's continuous-flow apparatus 404
2.2.2. IEF in granulated gels 405
2.2.3. IEF in multiphasic columns 411
2.2.4. IEF in polyacrylamide gel cylinders 412
2.2.5. IEF in polyacrylamide gel slabs 415
2.2.6. Concluding remarks – load capacity 417

Chapter 3. Analytical IEF 419

3.1. IEF in small density gradient columns 420
3.2. Analytical IEF in gel media 424
3.2.1. IEF in granulated gels and in agarose matrices 424
3.2.2. Properties and structure of polyacrylamide gels 425
3.2.3. Highly cross-linked polyacrylamide gels 428
3.2.4. IEF in polyacrylamide gels 430
3.2.5. Apparatus 431
3.2.6. Gel composition 433
3.2.7. IEF in gel cylinders 437
3.2.8. Methodology 439
3.2.9. Gel preparation 440
3.2.10. Sample application 445
3.2.11. Electrolysis conditions 447
3.2.12. Micro-isoelectric focusing 448
3.2.13. IEF at sub-zero temperatures 449
3.2.14. Thin-layer slab technique 450
3.2.15. IEF in thin layers of granulated gels 463

3.3. Detection methods in gels 465
3.3.1. Staining procedures 465
3.3.2. Densitometry of focused bands 469
3.3.3. Specific stains and zymograms 472
3.3.4. Autoradiography 483
3.3.5. Immunotechniques 487
3.4. Two-dimensional procedures 488
3.4.1. IEF – immunoelectrofocusing 488
3.4.2. IEF – gel electrophoresis 491
3.4.3. IEF – electrophoresis in gel gradients 492
3.4.4. IEF – SDS gel electrophoresis 493
3.4.5. IEF in urea-gradient gels 497
3.5. Transient state IEF (TRANSIF) 498

Chapter 4. General experimental aspects *501*

4.1. Isoelectric precipitation 501
4.2. Additives 506
4.3. Sample application 508
4.4. Choice of pH gradient – production of narrow pH gradients 509
4.5. Measurement of pH gradients 512
4.6. Removal of carrier ampholytes after IEF 517
4.7. Possible modification of proteins during IEF. 520
4.8. Power requirements 523
4.9. Instability of pH gradients 525

Chapter 5. Applications of IEF *527*

5.1. IEF of cells, subcellular particles, bacteria and viruses 527
5.2. IEF of hormones 530
5.3. IEF of peptides 534
5.4. IEF of glycoproteins 536
5.5. IEF of lipoproteins 539
5.6. IEF of membranes 542
5.7. IEF of metallo-proteins 543
5.8. IEF of immunoglobulins 549
5.9. IEF of tissue extracts 551
5.10. IEF as a probe of interacting protein systems 552
5.11. Clinical applications 558
5.12. Curiosities 562

5.13. On heterogeneity – facts and artifacts 565
5.14. Avoidable artifacts in polyacrylamide gels 568
5.15. General artifacts 569
5.16. Conclusion 570

References *573*

Subject index *587*

Theory and fundamental aspects of isoelectric focusing

1.1. Introduction

Electrophoretic mobility has long been used as basis for separating and characterising proteins. The resolving power of conventional electrophoresis is limited, however, and characterisation by mobility is not unique because of its dependence on the composition, pH and ionic strength of the medium. The development of isoelectric focusing (IEF) represents a major advance in the field of electrophoretic separations of proteins and other amphoteric substances. Isoelectric focusing is essentially an equilibrium electrophoretic method for segregating amphoteric macromolecules according to their isoelectric points in stable pH gradients. The method offers unique advantages for both analytical and preparative procedures.

Isoelectric focusing has now been refined to the point where one can display all components whose isoelectric points differ by as little as 0.01 pH unit. Such exquisite resolution is not normally obtainable by other procedures based on charge separations such as electrophoresis or ion exchange chromatography. With these latter methods, specially tailored conditions usually have to be devised for particular separations. By contrast, the built-in resolution of IEF allows the separation in a single experiment of all components with measurably different isoelectric points. IEF is, therefore, a more definitive technique for examining charge heterogeneity and is particularly suitable for differentiating closely related molecules. By the same token, IEF

Subject index p. 587

is also a rigorous test of homogeneity. In addition, because molecules are concentrated during separation by IEF, the technique lends itself to preparative as well as analytical purposes. Finally, IEF defines an important parameter, the isoelectric point (pI), which gives information about the composition of macromolecules and also allows a rational approach to other experimental manipulations.

The principle of IEF is outlined in Fig. 1.1. A stable pH gradient increasing progressively from anode to cathode is established by electrolysis of carrier ampholytes in a suitable anticonvective liquid medium. When introduced into this system, a protein or other amphoteric molecule will migrate according to its surface charge in

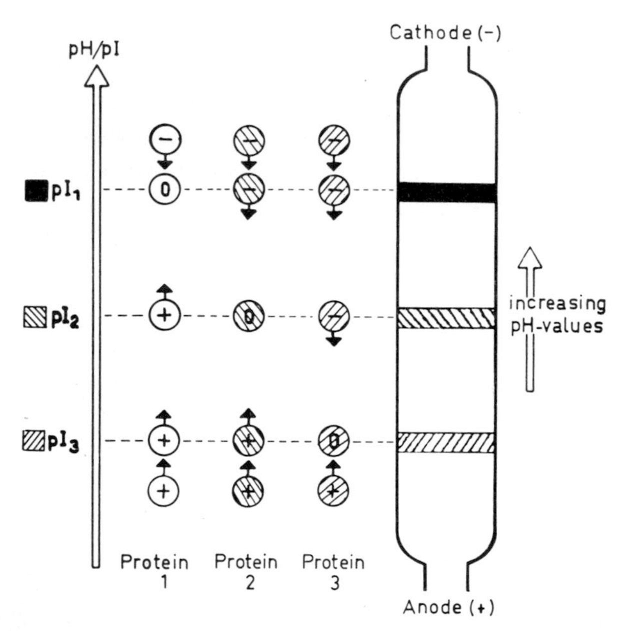

Fig. 1.1 Principle of IEF. Three amphoteric molecules, represented as pI_1, pI_2 and pI_3, migrate to their respective isoelectric points in a stable pH gradient generated by electrolysis of carrier ampholytes in an anticonvective medium. Each species will eventually reach an equilibrium zone where its net electrical charge is zero, i.e. its pI (Courtesy of LKB Produkter AB).

the electric field. Should its initial charge be positive, it will migrate towards the cathode into regions of higher pH. As it does so, the molecule will gradually loose positive charges and gain negative charges, e.g. through deprotonation of carboxyl or amino functions. Eventually, it will reach a zone where its net electrical charge is zero, i.e. its pI. Should the molecule migrate or diffuse away from its pI, it will develop a net charge and be repelled back to its pI. Thus by countering back diffusion with an appropriate electrical field, a protein or other amphoteric macromolecule will reach an equilibrium position where it may be concentrated into an extremely sharp band. As might be expected, the degree of separation of two ampholytes is a function, among other parameters, of the slope of the pH gradient in which they are focused. The shallower the pH gradient, the better is the separation (Fig. 1.2). This principle is discussed more fully later (§1.3.6).

IEF was originally developed as a preparative technique. Fractionations were conducted in sucrose density gradients which served as an anticonvective medium to stabilize the pH gradient and focused protein zones. Experiments were usually performed in column volumes of 110 or 440 ml and usually required 2–4 days to reach equilibrium. These systems were, however, rather expensive in time and required careful standardization. They also suffered from some practical problems arising from isoelectric precipitation and excessive diffusion during elution of the column in the absence of the electrical field. Finally, sample detection was laborious and often difficult. Consequently, these early systems were not convenient for routine analytical procedures. Fortunately, many of the problems inherent in IEF in sucrose density gradients have now been overcome with the use of more suitable anti-convective media such as polyacrylamide gel or Sephadex beds. With these media many of the potential advantages of IEF for high resolution separations of amphoteric substances can be realised. The methodology for IEF in gels has now been fairly well standardized for both analytical and preparative procedures. Lastly, IEF in gels is a remarkably forgiving technique in which a minimum of technical skill and effort is amply rewarded in the quality of the fractionations achieved.

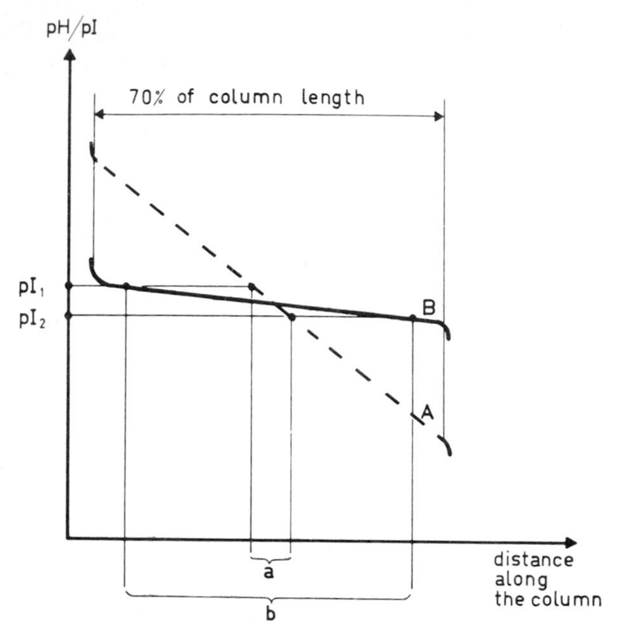

Fig. 1.2. Separation of ampholytes as a function of the slope of the pH gradient. Two amphoteric molecules, represented by pI_1 and pI_2, are separated only by the distance 'a' in a steep pH gradient (broken line A) but by a greater distance (b) in a shallower pH gradient (solid line B). (Courtesy of LKB Produkter AB.)

This book deals with theoretical and practical developments of IEF and offers detailed methodology for many of the commonly used procedures, such as IEF in gels. It is intended both as a reference guide and a practical manual. The first section of the book deals with the theory and development of the technique of isoelectric focusing. The second section discusses tactical and practical considerations of analytical and preparative aspects and also details current experimental methodology. The third section reviews some applications of IEF in biological and biomedical research which indicate the wide applicability of this exciting new technique.

Needless to say, this monograph cannot possibly cover all aspects of IEF in detail. Since the advent of present day techniques, more than

1500 papers have appeared on the use of IEF and three International Symposia have been held. For additional information, the reader is directed to the publications of these meetings (Catsimpoolas 1973a; Arbuthnott and Beeley 1975; Righetti 1975a) and to other review articles on specialised topics (Vesterberg 1968, 1971a; Catsimpoolas 1970a, 1973b, 1975; Haglund 1971; Wrigley 1971; Williamson 1973; Righetti and Drysdale 1974; Drysdale 1975; Righetti 1975b). A very useful literature reference list, called Acta Ampholinae, is published by LKB Produkter AB. A volume covering the years 1960–1974 is now available. Supplements are published yearly covering the fields of IEF and isotachophoresis.

1.2. Background

The history of IEF dates to 1912 when Ikeda and Suzuki achieved a coarse separation of amino acids from plant protein hydrolysates in a three-chambered electrolysis cell. They noticed that the amino acids tended to arrange themselves according to increasing pI between the anode and cathode. In so doing, the separated amino acids formed a pH gradient between the electrodes. In 1929, Williams and Waterman extended this work by designing a multichamber compartment which gave better resolution by reducing diffusion and convective disturbances. Practical applications were limited, however, because variable field strengths between the electrodes prevented formation of stable pH gradients for true equilibrium focusing conditions. Nevertheless these and other applications of the principle of IEF provided useful separation of peptides and proteins (du Vigneaud et al. 1938).

One of the major obstacles in these pioneer experiments was the lack of suitable ampholytes for developing smooth pH gradients that were also sufficiently stable to allow true equilibrium focusing. Happily, these problems have been solved, largely through the efforts of three chemists, Kolin, Svensson (now called Rilbe) and Vesterberg. Kolin pointed out the importance of using electrolytes with a high buffering capacity and of stabilizing the pH gradient against convective mixing (Kolin 1954, 1955). He described the separation and concentration of proteins by electrical transport in a pH gradient generated through a sucrose density gradient in a Tiselius-like apparatus. The substance to be separated was placed at the interface of an acidic and a basic buffer which were allowed to diffuse against one another while subjected to an electric field. The separation force in these experiments was, therefore, the resultant of several factors: a pH gradient, a density gradient, an electrical conductivity gradient and a vertical temperature gradient. Unfortunately, the pH gradients were unstable because of the rapid migration of the buffers during electrolysis, and the separated components could not be easily recovered.

The rapid practical application of IEF followed the work of Svensson who laid the theoretical foundations for present-day systems. In a series of articles entitled 'Isoelectric fractionation, analysis and characterisation of ampholytes in natural pH gradients' (Svensson 1961, 1962a,b; Rilbe 1973a), he introduced the law of pH monotony and advanced the idea of developing a 'natural' pH gradient by electrolysis of amphoteric molecules. From theoretical considerations he recommended that ampholytes should have good buffering capacity, good conductivity and solubility at their pI and be easily distinguishable and separable from proteins. He studied the protolytic properties of ampholytes, their conductivity and buffering capacity and their titration curves. He also calculated the resolving power of IEF in the presence of ampholytes capable of generating stable pH gradients.

Initial experiments using protein hydrolysates and synthetic peptides as ampholyte (Vesterberg and Svensson 1966) confirmed many of Svensson's calculations, but these ampholytes were useful over only a small pH range. Svensson's concepts were, however, soon realised in practice with the synthesis by Vesterberg (1969) of carrier ampholytes with many of the properties prescribed by Svensson. These ampholytes formed smooth pH gradients between pH 3.5 and 10 and thus encompassed the pI range of most proteins. They have been commercially available for several years in a variety of pH ranges covering either a wide pH range (3.5–10) or several narrower ranges of 0.5 to 3 pH units. Recently, the pH range of Vesterberg's ampholytes has been extended to cover the pH range 2.5–11. These ampholytes or others prepared by the same principle but with different reagents may be readily synthesised in the laboratory (Vinogradov et al. 1973; Pogacar and Jarecki 1974; Righetti et al. 1975a). IEF with synthetic ampholytes has now been adapted for a wide range of analytical and preparative procedures either separately or in combination with other techniques. Their usefulness for high-resolution separations of proteins and other amphoteric macromolecules has been amply demonstrated by the vast amount of new information currently becoming available.

1.3. Theory of IEF

1.3.1. Artificial pH gradients

The pH gradients obtained by diffusion of non-amphoteric buffers of different pH under an electric field (Kolin 1954, 1955, 1958), were termed 'artificial' pH gradients by Svensson (1961), since the gradient is affected by changes in electric migration and diffusion of the buffer ions. The most favourable conditions are obtained with pairs of buffers of the same ionic species, e.g. two acetate buffers whose initial pH's encompass the pI's of substances of interest. Such gradients can never be expected to give more than quasi-equilibrium positions of amphoteric molecules.

1.3.2. Natural pH gradients

In the early 60's Rilbe (Svensson 1961) introduced the concept of 'natural' pH gradients and established a theoretical basis for IEF in its present form. If the natural pH gradient were created by the current itself it should be stable over long periods if protected from convective disturbances. IEF should be, therefore, a true equilibrium method. This concept has several parallels with equilibrium (isopycnic) density gradient centrifugation and, indeed, Giddings and Dahlgren (1971) have derived similar equations for the resolving power and peak capacity of both methods. Svensson predicted that stable pH gradients would be obtained by isoelectric stacking of a large series of carrier ampholytes, arranged under the electric current in order of increasing pI from anode to cathode. The pH in every part of the gradient would

Subject index p. 587

then be defined by the buffering capacity and conductivity of the specific isoelectric carrier ampholyte located in that particular region. The stability of such natural pH gradients is, however, contingent on other factors such as convection and diffusion. The electrical load should not exceed the cooling capacity of the column to prevent convective mixing. Ideally, the conductivity should be uniform throughout the gradient to prevent local overheating in regions containing low levels of ampholyte or ampholytes exhibiting poor conductivity and buffering capacity. Instability of the pH gradient could also arise by anodic oxidation or cathodic reduction of carrier ampholyte. Consequently, strong acids and bases were recommended for use at the electrodes to repel the ampholytes from the electrode compartments.

1.3.3. Conductivity

An important prerequisite for a good carrier ampholyte is that it has a good conductivity at its pI. Regions of low conductivity will not only cause local overheating, but will also absorb much of the applied voltage. This reduces the field strength and hence the potential resolu- in other parts of the gradient.

The degree of ionization, α, of a diprotic ampholyte can be written:

$$\alpha = \frac{C_{\text{amph}_+} + C_{\text{amph}_-}}{C_{\text{amph}}} \tag{1}$$

Were C_{amph_+} is the molar concentration of positive ions, C_{amph_-} is the molar concentration of negative species and C_{amph} is the total molar concentration of ionic and undissociated ampholyte. Neglecting activity coefficients, the following equations follow from the law of mass action:

$$C_{\text{amph}_+} = C_{\text{amph}_0}\,(10^{\text{p}K_1 - \text{pH}}) \tag{2}$$

$$C_{\text{amph}_-} = C_{\text{amph}_0}\,(10^{\text{pH} - \text{p}K_2}) \tag{3}$$

where C_{amph_0} is the molarity of zwitterionic and undissociated forms.

The total molar concentration, C_{amph}, can be written:

$$C_{amph} = C_{amph_+} + C_{amph_-} + C_{amph_0} \tag{4}$$

Elimination of C_{amph_0} gives, after rearrangement:

$$C_{amph}/C_{amph_+} = 1 + 10^{pH - pK_1} + 10^{2(pH - pI)} \tag{5}$$

$$C_{amph}/C_{amph_-} = 1 + 10^{pK_2 - pH} + 10^{2(pI - pH)} \tag{6}$$

The relationship between the pK's of a diprotic ampholyte and the isoionic point pI is:

$$pI = \frac{pK_1 + pK_2}{2} \tag{7}$$

where pK_1 is defined to be smaller than pK_2.

By substituting the proper values of eqs. (5), (6) and (7) into (1), the degree of protolysis at the isoelectric state (pH = pI) will be given by:

$$\alpha = \frac{2}{2 + 10^{(pI - pK_1)}} \tag{8}$$

The highest possible value for α at maximum conductivity is 1/2, since $pI - pK_1$ has a lower limit of log 2 (Svensson 1962a; Rilbe 1971). Eq. (8) shows that good conductivity is associated with small values of $pI - pK_1$. This is also true for the buffering capacity of an ampholyte. Thus the parameter $pI - pK_1$ becomes the most important factor in selecting carrier ampholytes exhibiting both good conductivity and buffering capacity. As shown in Fig. 1.3, glycine, which is isoelectric between pH 4 and 8 ($\Delta pK = 7.4$, corresponding to a degree of ionization of 0.00038 in the isoprotic state) is a very poor carrier ampholyte, while histidine, lysine and glutamic acid, with sharp isoelectric points (low $pI - pK_1$ values) are quite useful ampholytes.

1.3.4. Buffering capacity

Svensson (1962a) has demonstrated that ampholytes with good conductivity also have a good buffering capacity, and vice-versa. If the charge Q is plotted against pH, the buffering capacity can be expressed as:

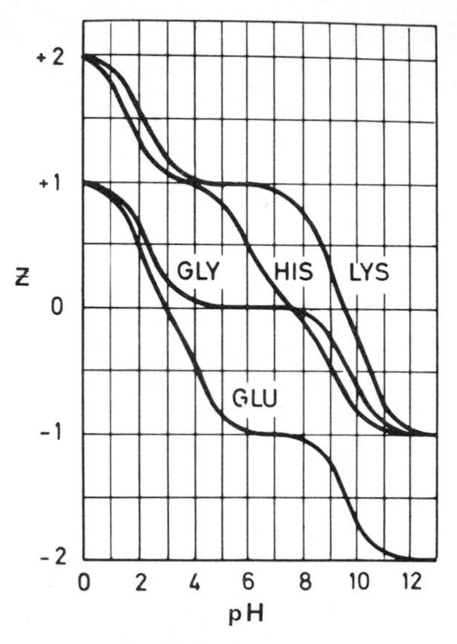

Fig. 1.3. Titration curves of glutamic acid, glycine, histidine and lysine. The three amino acids with sharp isoionic points (low $\mathrm{pI} - \mathrm{p}K_1$ values) are useful as carrier ampholytes, while glycine, which is isoelectric over a wide pH range, is a useless ampholyte. (From Svensson, 1962a, by permission of Munksgaard.)

$$\frac{-\mathrm{dQ}}{\mathrm{d(pH)}} = \frac{\mathrm{d}(C_{\mathrm{amph}-}/C_{\mathrm{amph}})}{\mathrm{d(pH)}} - \frac{-\mathrm{d}(C_{\mathrm{amph}+}/C_{\mathrm{amph}})}{\mathrm{d(pH)}} \tag{9}$$

when pH = pI, eqs. (9), (5) and (6) give:

$$\frac{-\mathrm{dQ}}{\mathrm{d(pH)}} = \frac{2\ \ln 10}{2 + 10^{(\mathrm{pI} - \mathrm{p}K_1)}} \tag{10}$$

It is therefore evident that the buffering capacity is also a function of $\mathrm{pI} - \mathrm{p}K_1$. Its numerical value is largest for small differences between $\mathrm{p}K_1$ and pI or between $\mathrm{p}K_2$ and $\mathrm{p}K_1$, since $\mathrm{p}K_2 - \mathrm{p}K_1 = 2(\mathrm{pI} - \mathrm{p}K_1)$. The buffering capacity reaches a maximum value of $0.25\ \ln 10$ (=0.575), corresponding to an $\alpha = 0.5$ for a weak acid. As the difference between $\mathrm{p}K_2$ and $\mathrm{p}K_1$ increases, the buffering capacity diminishes, linearly at first, and then exponentially. For

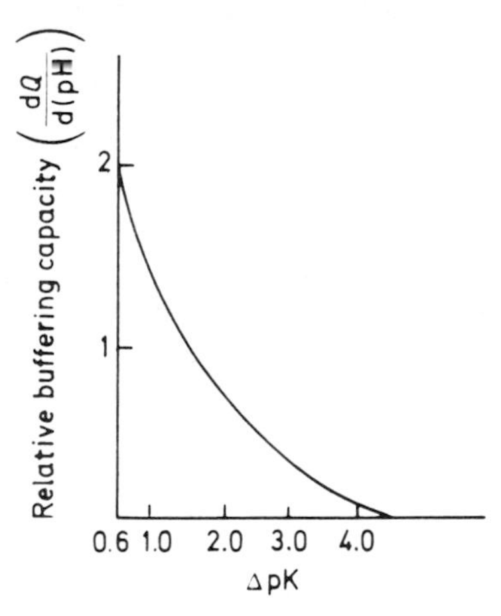

Fig. 1.4. Relative buffering capacity, in units of the maximum capacity of a monovalent weak protolyte, of an ampholyte in the isoionic state as a function of the pK differences between the dissociation steps on either side of the isoionic point. The buffering capacity is at a maximum when ΔpK is at a minimum (log. 2). (From Rilbe 1971, by permission of Munksgaard.)

example, the illustration in Fig. 1.4 shows that when pK_2 and pK_1 lie 1.5 pH units apart, 50% of the limiting buffering capacity is still retained. However, when pK_2 − pK_1 = 3.2 pH units, only 1/10 of the maximum buffering capacity remains and the compound can be considered a poor carrier. Thus, good carrier ampholytes should have a pair of pK values which lie as close as possible to the pI of the ampholyte.

1.3.5. Law of pH monotony

Formulated by Svensson (1967), this law states that a natural pH gradient, developed by the current itself, is positive throughout the gradient as the pH increases steadily and monotonically from anode to cathode. Reversal of the pH gradient at any position between the electrodes is thus incompatible with the steady state. Thus in stationary electrolysis, two ampholytes can never be completely separated from one another, unless a third ampholyte, with intermediate pI, is present in the system. For instance, in the case of Asp (pI 2.8) and Glu (pI 3.2) a complete separation could only be achieved if a zone of pure water (pI 7) were to develop between them. This situation is excluded by the law of pH monotony, therefore the gaussian distributions of two amino acids must overlap in the steady state if no intermediate ampholyte is present.

Ideally the ampholytes should be evenly distributed over the entire pH range. It is particularly important to have a suitable ampholyte in the neutral pH range to prevent formation of a zone of pure water. Development of a region of pure water within the electrofocusing column will have undesirable effects, notably in creating local heating effects. The theoretical conductance of pure water (at 0°C) is:

$$X_w = 1.0 \times 10^{-8}\ \text{ohm}^{-1}\ \text{cm}^{-1}$$

The field strength is given by the equation:

$$E = I/x$$

and the Joule heat per unit volume by the equation:

$$q = I^2/x$$

where I is the current density and x the distance along the separation column. Even with I as low as 10^{-4} A/cm^2, the field strength in pure water would be 10,000 V/cm, and the Joule heat 1W/cm^3. Any region of pure water will absorb almost all of the available voltage, thereby decreasing the focusing effect and allowing neighbouring electrolytes to diffuse with the developing region of pure water. Theoretically, and compatible with the law of pH monotony, a zone of pure water could develop at a neutral pH and, in fact, water is an ampholyte, isoelectric at pH 7. Cationic water (H_3O^+) is repelled from the anode and anionic water (OH^-) from the cathode, so that water as a whole tends to concentrate at pH 7. However, it is an extremely poor carrier ampholyte, with little, if any, buffering capacity and conductivity at its pI, and it is, therefore, important to have other ampholytes in the neutral pH region.

1.3.6. Resolving power

Svensson (1962a) has derived an equation for the resolving power of IEF which expresses the balance between electrical and diffusional mass transport. With the latter given in units of mass per unit time and cross-sectional area, one has:

$$CuE = D\left(\frac{\mathrm{d}c}{\mathrm{d}x}\right) \tag{11}$$

where C is the protein concentration, u its mobility, D its diffusion coefficient, E the field strength and x the distance moved. For narrow zones, the mobility u can be regarded as a linear function of x within it. By introducing the proportionality factor p, defined as:

$$u = -px \tag{12}$$

eq. (11) can be written:

$$\frac{\mathrm{d}c}{C} = \frac{-Ep}{D} x\mathrm{d}x \tag{13}$$

In a given experiment, for a given protein, E, p and D can be treated as constants, and eq. (13) can be integrated to give:

$$C = C_{(0)} \exp\left(-pEx^2/2D\right) \tag{14}$$

where $C_{(0)}$, the integration constant, physically means the local concentration maximum in the region where the protein is isoelectric. Eq. (14) expresses a gaussian concentration distribution with inflection points lying at:

$$\sigma = \pm\sqrt{D/pE} \tag{15}$$

If the proportionality factor p is written as a derivative:

$$p = -(\mathrm{d}u/\mathrm{d}x) = [-\mathrm{d}u/\mathrm{d(pH)}]\cdot\left[\frac{\mathrm{d(pH)}}{\mathrm{d}x}\right] \tag{16}$$

eq. (15) can be given in the form:

$$\sigma = \pm\sqrt{DE^{-1}(-\mathrm{d}u/\mathrm{d(pH)})^{-1}(\mathrm{d(pH)}/\mathrm{d}x)^{-1}} \tag{17}$$

If two adjacent zones of equal mass have a peak to peak distance three times larger than the distance from peak to inflection point, there will be a concentration minimum approximating the two outer inflection points. Taking this criterion for resolved adjacent proteins, Rilbe (1973a) has derived the following equation for minimally but definitely resolved zones:

$$\Delta\mathrm{pH} = \Delta x[\mathrm{d(pH)/d}x] = 3\sigma[\mathrm{d(pH)/d(pH)/d}x] \quad (18)$$

By inserting eq. (17) in (18) one obtains, at the isoelectric point:

$$\Delta(\mathrm{pI}) = 3\sqrt{\frac{D[\mathrm{d(pH)d}x]}{E[-\mathrm{d}u/\mathrm{d(pH)}]}} \quad (19)$$

Eq. (19) shows that good resolution should be obtained with substances with a low diffusion coefficient and a high mobility slope ($\mathrm{d}u/\mathrm{d(pH)}$) at the isoelectric point-conditions that are satisfied by all proteins. Good resolution is also favoured by a high field strength and a shallow pH gradient. If the pH gradient ($\mathrm{d(pH)/d}x$) is known and the value of σ is obtained from the zone breadth at theoretical ordinates $e^{-1/2} = 0.61$ of the peak height, the numerical value of ΔpH can be estimated. Such calculations have shown that the resolving power of IEF allows separation of molecules which differ in pI by as little as 0.01 pH unit (Vesterberg and Svensson 1966).

A similar equation has been derived by Giddings and Dahlgren (1971) for the resolving power (R_s) in IEF:

$$R_s = \frac{\Delta\mathrm{pH}}{4}\sqrt{\frac{-FE[\mathrm{d}q/\mathrm{d(pH)}]}{RT[\mathrm{d(pH)/d}x]}} \quad (20)$$

where F is the faraday (96500 coulombs) E the electric field strength, R the gas constant, T the absolute temperature, q the effective charge of the particle and x the distance moved.

1.3.7. Peak capacity

Giddings and Dahlgren (1971) have defined the peak capacity 'n' as the maximum number of components resolvable by a given technique under specified conditions. In IEF the peak capacity is given by:

$$n = \sqrt{\frac{-FE[\mathrm{d}q/\mathrm{d(pH)}][\mathrm{d(pH)}/\mathrm{d}x]L^2}{L6\,RT}} \tag{21}$$

where L is the total path length.

This equation shows that peak capacity increases in proportion to total path length and is a function of the square root of both the slope of the pH gradient and the protein mobility slope. The alternate form is obtained by replacing [d(pH)/dx] with (pH$_L$-pHo), the total pH increment. In this case, assuming a uniform-gradient model, i.e. a constant value of (d(pH)/dx), we have:

$$n = \sqrt{\frac{-FE[\mathrm{d}q/\mathrm{d(pH)}][\mathrm{d(pH)}/\mathrm{d}x]L^2}{16RT}} \tag{22}$$

Thus the peak capacity is directly proportional to the square root of the electric field, path length and total pH increment. These three parameters may be varied simultaneously or independently according to experimental needs.

1.3.8. Mass content of a protein zone

Rilbe and Pettersson (1975) have derived a relationship for the mass load (m) of a protein zone in IEF:

$$m < 5q\sigma^2(\mathrm{d}c/\mathrm{d}x) \tag{23}$$

where q is the cross sectional area of the separation column, σ the inflection points of the gaussian protein zone, and dc/dx the sucrose density gradient along the column. We now introduce the volume V, the height H and the zone width r, taken as 2σ in units of column length:

$$\text{i.e. } r = 2\sigma/H \qquad \sigma = rH/2 \tag{24}$$

Inequality (23) can then be written in the form:

$$m < 1.25VHr^2(\mathrm{d}c/\mathrm{d}x) \tag{25}$$

If we now assume a linear sucrose concentration gradient from 0 to 0.5 g/cm^3, dc/dx will have the numerical value:

$$\mathrm{d}c/\mathrm{d}x = (1/2H)\ \mathrm{g/cm^3} \tag{26}$$

by substituting this value in eq. (25), we obtain:

$$m < 0.625\ Vr^2\ \mathrm{g/cm^3} \tag{27}$$

for a 100 cm^3 column, one thus has:

$$m < 62.5\ r^2\ \mathrm{grams} \tag{28}$$

The carrying capacity thus rises with the square of the zone width.

For a 1 mm thick zone, in a 25 cm high column, equation (28) shows that the mass content cannot be more than 1 mg. But if in the same the protein zone is allowed to extend over 1/10 of the column height, its maximum protein content will be 625 mg, whereas a zone extending over one fifth of the column can support up to 2.5 g protein.

By combining eqs. (15), (16), (19) and (23) the mass load of the protein zone can be related to the resolving power, thus obtaining the inequality:

$$m < \frac{45qD^2\,(\mathrm{d}c/\mathrm{d}x)}{E^2(\mathrm{pI})^2\,[\mathrm{d}u/\mathrm{d(pH)}]^2} \tag{29}$$

Thus, the capacity of a density gradient column rises with the square of the resolving power with which it operates for a given protein system. This means that an experiment with a pH gradient favouring the purity of protein fractions obtained also favours the quantity of pure protein that can be separated. From this standpoint, IEF has considerable advantages over other separation techniques in which a good yield axiomatically excludes a high purity and vice versa.

No equations have yet been derived for the load capacity of IEF in gel media. Radola (1975) studied this aspect experimentally in Sephadex gel layers, and found an upper load limit of approximately 10–12 mg protein/ml gel suspension. In these experiments, he defined load capacity as the weight (mg) of protein per ml gel suspension used for preparing the gel layers. He was able to fractionate 10 g Pronase E using a pH range 3–10 in a 400 × 200 mM through with a 10 mm thick gel layer (total gel volume of 800 ml). The resolution was comparable to that of analytical systems. At these remarkably high protein loads, the focused bands could be seen directly in the gel layer as translucent zones. This load limit may only apply to wide pH range gradients (3–10). From theoretical considerations (Eq. 29) Radola had predicted much higher protein loads in narrower pH ranges. These figures dramatise one of the major advantages of IEF in gel media in stabilizing protein zones. The load capacity will of course also depend on the actual number of components and their solubility at their pI.

1.4. Synthesis of carrier ampholytes

Svensson's delineation of the theory of pH monotony and specifications of suitable ampholytes led to an extensive, but essentially fruitless, search for commercial sources. Some synthetic peptides with limited applicability were found. For example, solutions of histidyl-histidine allowed excellent resolution of haemoglobins, pI ~ 7, but were useful only over a limited pH range.

Except for protein hydrolysates, which were difficult to distinguish and separate from the protein in question, no other suitable substances were found.

Efforts were then directed toward synthetic processes and in 1969, Vesterberg found the answer with a remarkably simple process. He synthesized a mixture of a large number of homologues and isomers of aliphatic polyamino polycarboxylic acids with different pK's and pI's closely spaced in the pH 3–10 range. The synthesis involves the coupling of propanoic acid residues to polyethylene polyamines. Carrier ampholytes of suitable properties are formed by reacting appropriate amounts of acrylic acid with different polyethylene polyamines in water at 70°C until all the acrylic acid has been consumed. The synthesis goes via an anti-Markovnikov addition after the mechanism of the Michael reaction and therefore β-amino acids are obtained. The general structure of these ampholytes may be depicted as follows:

$$
\begin{array}{c}
---CH_2-\underset{\underset{(CH_2)_x}{|}}{N}-(CH_2)_x-\underset{\underset{R}{|}}{N}-CH_2---
\end{array}
$$

where $x = 2$ or 3 and R = H or $-(CH_2)_xCOOH$.

Subject index p. 587

1.5. Laboratory synthesis of carrier ampholytes

Vinogradov et al. (1973) showed that it was possible to reproduce Vesterberg's synthetic procedure with simple laboratory equipment. They obtained suitable carrier ampholytes in the pH range 4–8 by coupling pentaethylenehexamine (PEHA) with acrylic acid. These preparations were satisfactory for many purposes but their conductivity profile was not uniform over the whole pH range. These experiments, nevertheless, indicated the relative ease with which suitable ampholytes could be prepared. These workers also described optical methods to identify focused ampholytes and thereby allow some much needed characterisation of ampholyte banding patterns.

On the basis of these observations, Righetti et al. (1975a) used a combination of different polyamines with acrylic acid to produce ampholytes which formed smooth and stable pH gradients over the pH range 3–9. The method involves the coupling of hexamethylenetetramine (HMTA), triethylenetetramine (TETA), tetraethylenepentamine (TEPA) and PEHA either singly, or in a mixture, to acrylic acid. All these reagents are available from several commercial sources. Experimental details for this synthesis are given below.

1.5.1. Reagent distillation

All reagents should be distilled under nitrogen and reduced pressure immediately before use. The distillation process for TETA, TEPA, and acrylic acid is quite simple, since these substances have low boiling points under reduced pressure. Perhaps the most difficult task is the distillation of PEHA which distils over the range 200–290°C under a pressure of 500 μ Hg. A suitable distillation apparatus is shown schematically in Fig. 1.5. High quality nitrogen should be used. Less pure grades should be freed from oxygen either by passage through a catalytic burner or an appropriate filter. The nitrogen is bubbled into the solution of the reagent to be distilled via a very thin capillary reaching the floor of the distillation chamber. Because of the high viscosity of PEHA at low temperatures, the condenser should only be air cooled. The whole system is maintained under reduced pressure

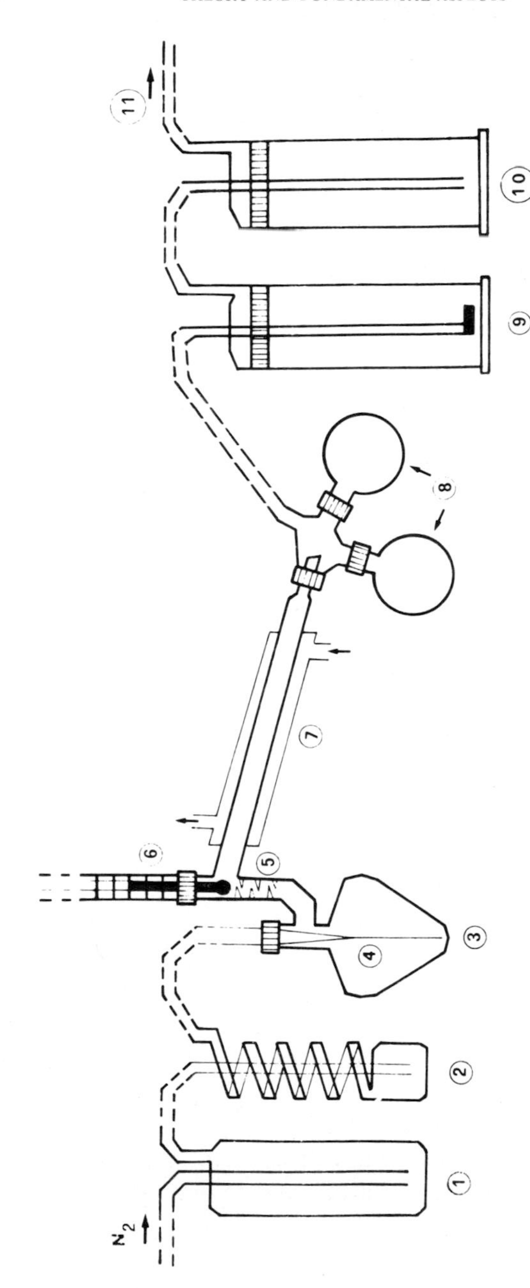

Fig. 1.5. Scheme of the distillation apparatus. 1) copper wire catalyst, kept in an oven at 450°C; 2) cooling serpentine; 3) distillation flask; 4) capillary for nitrogen flushing; 5) Vigreux column; 6) thermometer; 7) condenser; 8) collection flasks; 9) Fresenius tower with conc. H_2SO_4; 10) $CaCl_2$ trap; 11) connection to the vacuum pump. The shaded regions represent ground-glass joints. (From Righetti et al. 1975a, by permission of Elsevier.)

by a vacuum pump which should be protected by a Fresenius tower containing concentrated H_2SO_4, to absorb polyamine vapours and by a second tower containing drierite. The distillation of reagents and the subsequent coupling reaction should be performed within the same day. Acrylic acid should be distilled last since it polymerises readily in oxygen once the polymerization inhibitor is removed. Excess acrylic acid may be stored frozen under nitrogen.

1.5.2. Coupling process

The distilled amines and acrylic acid are then combined in appropriate amounts. A suitable reaction chamber (Fig. 1.6) consists of a two-necked flask, equipped with a thin capillary in one arm for nitrogen flushing, and a burette for acrylic acid addition. The burette has a built-in side arm to provide for nitrogen escape from the flask, and for an inert N_2 atmosphere in the burette itself. The mixture is stirred with a bar magnet. The acrylic acid is added dropwise to a 1.5 M solution of the polyamine in water under continuous stirring, over a period of 60 min, to a final N/COOH ratio of 2 : 1. The flask is then stoppered and held for 16–20 hr at 70°C in a Dubnoff shaker. The reaction can be summarized as follows:

$$\text{HMTA} + \text{TETA} + \text{TEPA} + \text{PEHA} + \text{acrylic acid} \xrightarrow{H_2O,\ 70^\circ C} \xrightarrow{16\text{–}20\ h} \text{carrier ampholytes.}$$

The ampholytes thus obtained are diluted to 40% (w/v) and stored frozen in dark bottles. Yellow compounds formed during the synthesis can be removed by repeated charcoal treatment, as reported by Vinogradov et al. (1973). This may not be necessary, however, since the chromophores are also ampholytes and serve a useful function as 'built-in' markers for the pH gradient. While 'chromophoric' ampholytes need not interfere with protein separation and detection in analytical system, in gel rods or slabs, they might interfere with protein detection in preparative systems, especially in sucrose density gradients, where protein is often analyzed in presence of carrier ampholytes.

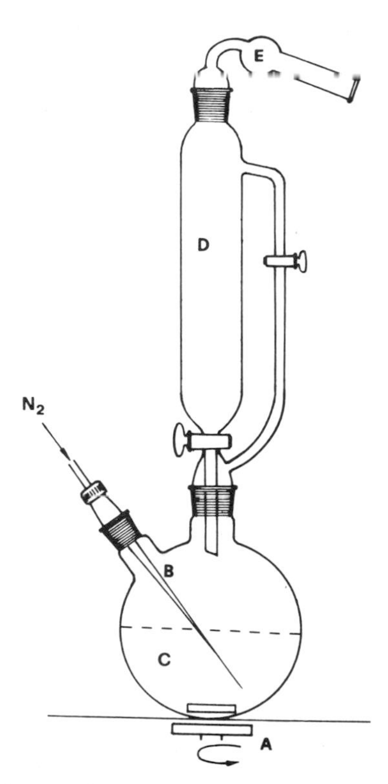

Fig. 1.6. Reaction chamber for carrier ampholyte production. A) magnetic stirrer; B) capillary for nitrogen flushing; C) polyamine solution; D) acrylic acid in a burette with a side arm; E) $CaCl_2$ trap for gas escape. (Gianazza, E. and Righetti, P.G., unpublished.)

By several criteria, the product described above is equivalent to commercially available Ampholine, the main difference relates to the strong absorption at 315 and 368 nm (Righetti et al. 1975a). The synthesis is simple, the starting reagents are readily available, and large amounts can be produced in a short time.

Other acids and amines may also be used to produce carrier ampholytes. Pogacar and Jarecki (1974) coupled TEPA and PEHA to either propansultone, vinylsulfonate, or chloromethylphosphonate.

These 'sulfonic ampholytes' were most suitable for focusing in the pH ranges 2–3.5 and 6–9.5 but did not buffer in the pH region 3.5–6. These compounds could be useful addition to 'carboxyl ampholytes' to extend their fractionation range down to pH 2. Sulphonic ampholytes covering the pH range 3–10 are now commercially available.

1.5.3. Detection of carrier ampholytes

The quality of the synthetic product can be checked by inspecting focused ampholyte patterns and by measuring conductivities in focused gradients. One simple method of visualizing ampholytes is by their different refractive indices (Rilbe 1973b). In gel, focused ampholytes can be detected by shining a light at a shallow angle across the face of a thin gel slab. Conductivity measurements may be made with a variety of small electrodes. Fig. 1.7 shows the patterns given by ampholytes prepared from coupling TETA and acrylic acid at a nitrogen:carboxyl ratio of 2:1. Several major bands of ampholytes are revealed by this method. These bands probably represent clusters of ampholytes rather than individual components. This preparation does not give smooth pH gradients between pH 3 and 10 and its conductivity is very uneven all along the pH gradient. Inspection of the focused plate revealed several rather large gaps corresponding to regions of low ampholyte concentration.

Better results were obtained with ampholytes produced with TEPA. These ampholytes gave better pH gradients than those produced with TETA. Moreover, TEPA ampholytes gave many more ampholyte species especially in the acidic regions. The ampholyte clusters were tighter and more closely spaced, with fewer regions of low ampholyte concentration. Although the basic region contained more ampholyte species than the TETA ampholytes, there appeared to be fewer basic ampholytes than acidic ampholytes (Fig. 1.8). In addition, the conductance varied considerably throughout the gel, being lowest between pH 5–7.

The best results with a single amine were given with ampholytes prepared by coupling PEHA to acrylic acid. Fig. 1.9 shows the pattern of ampholyte clusters with different diffraction indices: while acidic

Fig. 1.7. TETA ampholytes focused in a gel slab. At equilibrium, the gel was photographed against a black background with side illumination. The rope-like structures are clusters of focused ampholytes. This is a picture of a transparent, unstained gel and the ampholytes are detected on the basis of different refractive indeces. The anode (+) and the cathode (−) are marked. (From Righetti et al. 1975a, by permission of Elsevier.)

Fig. 1.8. TEPA ampholytes focused in a gel slab. The direction of the pH gradient is indicated by + (anode) and − (cathode). All other conditions as in Fig. 1.7. (From Righetti et al. 1975a, by permission of Elsevier.)

Fig. 1.9. PEHA ampholytes focused in a gel slab. The direction of the pH gradient is indicated by + (anode) and − (cathode). The arrows indicate ampholyte gaps. All other conditions as in Fig. 1.7. (From Righetti et al. 1975a, by permission of Elsevier.)

and neutral species appear to be similar to TEPA ampholytes, the basic region in PEHA seems to be markedly improved over TEPA ampholytes. PEHA ampholytes gave a smooth pH gradient between pH 3–9.5. They also had a higher and more uniform conductance throughout this pH range than had either TETA or TEPA ampholytes. However, they still presented at least three major regions of low ampholyte concentration. The UV profile at 280 nm was low and uniform throughout the pH range. However, the scan at 368 nm showed several peaks corresponding to chromophoric ampholytes which were apparent as yellow bands to the eye. The nature of these chromophores has not yet been determined (Righetti et al. 1975a).

These experiments demonstrate clearly that suitable ampholytes are readily made in the laboratory. Conceivably, the number of ampholyte species could be increased by a judicious mix of ampholytes prepared by coupling different amines to acrylic acid.

Other methods have been described to detect focused Ampholine patterns. Thus Felgenhauer and Pak (1973) reveal them by the glucose caramelization technique. After focusing in a Sephadex thin layer, by the method of Radola (1973a, b), they roll on to the Sephadex gel a sheet of filter paper soaked in 5% glucose. Upon incubation at 110°C, for 8 min, the Ampholine-glucose caramel products can be evaluated by their fluorescence at 350 nm or in daylight. This detection technique has revealed 38 distinct bands in commercial Ampholine pH range 3.5–10.

Another detection method is based on the formation of insoluble complexes between Ampholine and Coomassie Violet in half saturated picric acid containing 5% acetic acid. This method reveals acidic and neutral Ampholine but fails to detect basic ampholytes since they form soluble complexes (Frater 1970). Ampholytes can also be separated by ion-exchange chromatography. Application of this technique indicates the presence of a very large number (more than 60 major and many more minor) components in ampholytes prepared as above.

1.6. Fractionation of carrier ampholytes

Two methods may be used to prepare ampholytes that buffer in a narrow pH range. The first is to alter the relative amounts of amine and acid in the coupling reaction. Although feasible, this method is difficult to control and excess free amine or acid contaminate the product (Righetti et al. 1975a) and need to be removed. A better method is to fractionate the wide pH range ampholytes by IEF, as is done commercially. Narrow pH ranges, encompassing only one or two pH units (Gianazza et al. 1975) may be obtained by a simple method based on the continuous-flow isoelectric focusing principle of Fawcett (1973). This method has advantages over multi-compartment electrolysers, as described by Rilbe (1970) and Vesterberg (1969), since it avoids problems of osmotic pressure and polarization at membranes, and also reduces anodic oxidation of ampholytes.

Ampholine can also be separated in narrow ranges by using the zone convection electrofocusing apparatus of Valmet (1969). By this technique, this author has separated carrier ampholytes encompassing the pH range 3.5–10 into a number of fractions each covering about 0.2 of a pH unit.

Finally, it may conceivably be possible to fractionate ampholytes by ion exchange chromatography. Since the presence of salt in ampholytes is undesirable, it may be advisable to use volatile buffers e.g. ammonium acetate or ammonium bicarbonate to elute ampholytes either with a pH or salt gradient. Unwanted salt could subsequently be removed by lyophylisation and the ampholytes reconstituted in water. Further studies are required in this area to test the applicability of this approach.

For preparation of small quantities of limited pH range ampholytes it is sufficient to fractionate by IEF in Pevicon, a preparation of granular plastic beads. Recovery is achieved simply by centrifuging the suspension through a sintered glass (Otavsky, Saravis, Woodworth and Drysdale, unpublished).

1.7. Properties of Ampholine

Carrier ampholytes prepared according to Vesterberg's patent were produced commercially by LKB Produkter AB, Bromma, Sweden, in the late 1960's, under the trade name 'Ampholine'. These Ampholine were soon found to be of great practical value and led to the widespread use of IEF today. Preparations are available that cover a wide pH range, pH 3.5–10, or intermediate narrower pH range of between 1.5 and 3 pH units.

Recently, Vesterberg has extended the upper pH range of the original ampholytes to pH 11, by using polyamines with amino groups separated by more than three methylene groups rather than the previously used polyethylene polyamines (Vesterberg 1973a). More acidic ampholytes have also been synthesized by Vesterberg (1973b)

Subject index p. 587

that extend the range down to pH 2.5. Their chemical composition has not yet been described in detail but they are of the polyamine polycarboxyl type (H. Davies, personal communication). In addition to acrylic acid, Vesterberg has also suggested other monocarboxylic acids such as crotonic and methacrylic acid and dicarboxylic acids such as maleic and itaconic acids.

Ampholytes made from these substances have many of the properties suggested by Svensson (1962a) for good buffering capacity and conductivity at their pI. The pK_as of the amino groups of complex polyamines, such as TEPA and PEHA, are gradually distributed over an ample pH interval, covering the range from pH 11 to pH 1.5, whereas the pK_as of the carboxyl functions in the acids only range from pH 2–5. Presumably, many of these pK_as are substantially altered after polymerisation so that the resulting ampholytes will have many closely spaced pKs, the prerequisite for good buffering capacity and conductivity (page 348–351).

Unfortunately, despite their widespread use, relatively little is yet known of the exact composition of Ampholine. Vesterberg (1973a) theorised that more than 360 homologues and isomers might be generated in his synthesis of carrier ampholytes. However, the exact distribution of these species in the various pH ranges has still not been determined.

Vesterberg's ampholytes meet many, but not all, of Svensson's criteria for ideal ampholytes. Thus, although their chemical structure is quite different from that of proteins, they nevertheless react with ninhydrin, Folin's reagent and biuret as well as with many commonly protein stains (Davies 1970). They also have a low, but significant UV absorbancy, which varies along a focused gradient (Righetti and Drysdale 1971). Also, as will be discussed later, Ampholine chelate divalent metal ions.

1.7.1. *Molecular size*

The M.W. of Ampholine was reported to be in the range 300–1000 daltons, with the main peak centered at 600 daltons (Haglund 1971). A small fraction may, however, be as large as 5000 daltons (Gasparic and Rosengren 1975). These authors also found the peak of the gaussian distribution of Ampholine apparent M.W., as determined by gel filtration in Sephadex G-25 with polyethylene glycols as standards, to be centered around 400 instead of 600 daltons. This is consistent with a calculated M.W. of approximately 450 from the condensation of three propanoic acid residues per molecule of pentaethylenehexamine (PEHA) (provided the polyamines do not polymerise). Gasparic and Rosengren (1975) have also synthesized small M.W. ampholytes for peptide separations having a distribution curve centered around 300 daltons, only 0.2% of the total above 1000 daltons and 1.47% above 600 daltons. This 'restricted size' Ampholine should prove useful for separating peptides with apparent molecular weights of 2000 daltons and above. Baumann and Chrambach (1975a) have also found some ampholytes with apparent molecular weights as high as 20–30,000 in recent batches of Ampholine. The cause or distribution of such high molecular weight ampholytes is not known. Fortunately, they only seem to be present in trace amounts since they might otherwise seriously compromise many analyses.

1.7.2. Optical properties

The optical properties of Ampholine are also very favourable for protein fractionations, since most do not absorb appreciably around 280 nm. Consequently carrier ampholytes do not usually interfere with protein detection at this wavelength. From Fig. 1.10 it can be seen that the UV absorbance of Ampholine is generally low, and is usually confined to components of low pI. Although this is generally true for analyses of IEF experiments in sucrose density gradients where fairly large fractions are collected, it is not always true when scanning gels or columns under voltage where much sharper resolution is obtained and the absorbancy of individual ampholytes may be considerable (Righetti and Drysdale 1971; Catsimpoolas, 1973c).

UV spectra of different Ampholine ranges, by Righetti et al. (1975b) have shown that most Ampholine ranges have a peak absorbance around 365 nm (Fig. 1.11). The pH range 9–11 also exhibits a peak at 290 nm and the pH range 3.5–5 has a strong chromophore at 310 nm.

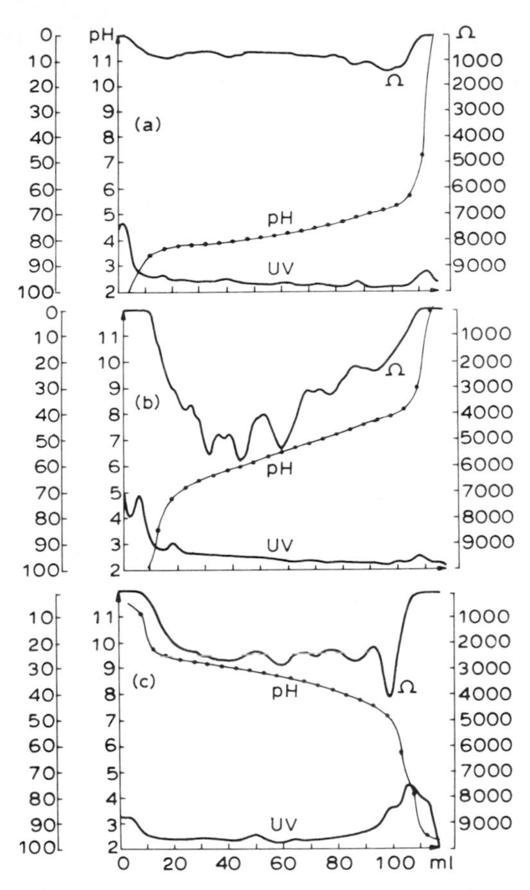

Fig. 1.10. Conductance course (Ω), pH gradient and UV profile of three different ranges of Ampholine carrier ampholytes: pH 3–6 (a); pH 5–8 (b) and pH 7–10 (c). While (a) and (c) show a rather good conductivity, pH 5–8 contains low conductivity ampholytes around pH 7 and shows a more uneven conductance course. This gradient is consequently more sensitive to thermal convection caused by Joule's heat in the low conductance region. (From Davies 1970, by permission of Pergamon Press.)

This acidic pH range has a marked UV absorbancy and is also distinctly yellow in color compared to other pH ranges. Some of these chromophores are pH-dependent, and they also exhibit pH-dependent

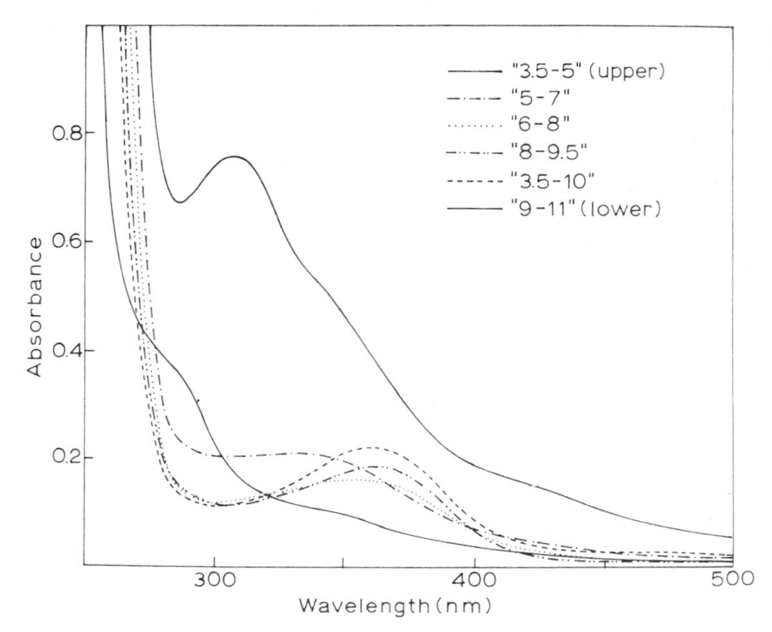

Fig. 1.11. Ultraviolet/Vis spectra of different Ampholine carrier ampholytes pH ranges. Upper solid curve: pH range 3.5–5; (– .– .– .), pH range 5–7; (. . . . . . .), pH range 6–8; (– . .– . .– . .), pH range 8–9.5, (– – – – – –), pH range 3.5–10. Lower solid curve: pH range 9–11. All spectra taken in a 4% solution, using a 3 cm light path cuvette, in a Jasco uv/ORD5 spectropolarimeter. (From Righetti et al. 1975b, by permission of Academic Press.)

fluorescence spectra (especially in the acidic pH ranges). Vinogradov et al. (1973) and Righetti and Drysdale (1974) attributed the characteristic spectra to heterocyclic nitrogen structures among the population of amphoteric molecules, but this has not been substantiated by NMR analyses (Righetti et al. 1975a).

Recent batches of Ampholine do not exhibit any ORD spectrum, and this is consistent with their known structure. However, in batches prepared before 1970, the two extreme pH ranges were found to rotate the plane of polarized light (Righetti et al. 1975b). This suggests that the extreme pH ranges were reinforced with glutamic and aspartic acid (in the acidic side) and with lysine and arginine in the basic side (as also pointed out by Davies, 1970). Similar evidence for the supplementation of ampholytes was also obtained by amino acid analyses (Earland and Ramsden 1969). Present day Ampholine, however, are said to be a mixture of polyamino polycarboxylic acids, prepared according to the original synthetic procedure by Vesterberg (1969), and do not contain amino acid or derivatives thereof.

1.7.3. Buffering capacity and conductivity

One of the great advantages of Vesterberg's synthetic ampholytes is their good conductivity and buffering capacity in the pH range 3–11. Fig. 1.10 shows the conductance, pH gradient and UV absorbance profiles of three Ampholine preparations covering different pH ranges between 3 and 10 and fractionated by IEF in a sucrose density gradient. In general, the conductance course is fairly even, except in the pH range 5–8, where it is rather variable with few regions of low conductivity. This pH range is consequently more sensitive to thermal convection caused by Joule's heating in regions of low-conductance. As might be expected, the buffering capacity of the ampholytes follows their conductivity profile (Davies 1970).

1.7.4. Chelating properties

It has been known for a long time that carrier ampholytes chelate metals (Davies 1970) but little quantitative information has been available on this subject. Recently, Galante et al. (1975) have found that metal chelation by Ampholine usually involves two adjacent nitrogen functions, giving rise to 5-membered rings. Chelation maxima occur in the pH region 8.5 to 10.5, where most of the nitrogen functions are unprotonated. Mixed type chelation on a nitrogen and carboxyl function is a possible though less likely event, since it generates less stable seven-membered rings. Little, if any, chelation occurs solely on carboxyl functions since this would lead to twelve-membered rings.

As compared to EDTA, the chelating power of Ampholine is generally weak. With the exception of divalent copper a given amount of divalent cation requires a 100 to 500 molar excess of ampholytes for chelation in contrast to EDTA, which usually binds metals in the pH range 3–11 in a 1/1 molar ratio. Carrier ampholytes are, however, as powerful chelators of divalent copper as EDTA, and consequently, many copper containing proteins may be rapidly inactivated on exposure to ampholytes in IEF.

In order to minimise metal chelation, proteins should be run to equilibrium from the anodic side. It is also advisable to apply samples to pre-focused gradients to minimise their contact time with ampholytes. Ideally the samples should be introduced as close as possible to their expected pI in prefocused gradients. Fortunately, loss of activity of many enzymes by metal chelation is seldom irreversible. Often, full enzymatic activity may be restored by incubating the focused enzyme with appropriate amounts of the sequestered cofactor. The main problem with metal chelation is the generation of multiple banding patterns, due to multiple forms containing differing amounts of metal.

1.7.5. Biological toxicity

Recent studies by Wadström et al. (1974) have also indicated that Ampholine are fairly innocuous as judged by both animal and in vitro cytotoxicity studies. Ampholine do not appear to interfere in immunodiffusion tests or in the hemagglutination inhibition test. Since Ampholine is easily removed from a purified protein, and no clear evidence of stable complexes between a protein and Ampholine exists, it should be possible to use Ampholine for preparative purposes either in isoelectric focusing or in isotachophoresis for preparation of proteins for human or animal use.

1.8. pH gradients generated by other means

1.8.1. Bacto peptone carrier ampholytes

Blanicky and Pihar (1972) have prepared ampholytes from bactopeptone by removing proteins by precipitation in 60% ethanol. After ethanol elimination, they obtained a mixture of possibly several dozen peptides, which appear to be satisfactory for IEF. The conductivity and buffering capacity in the neutral region was improved by adding histidine. This type of carrier ampholyte is similar to those described

Subject index p. 587

by Vesterberg and Svensson (1966) and may therefore suffer from the same inconveniencies. A similar approach has been used by Molnarova and Sova (1974) who used casein hydrolysates to obtain good pH gradients down to pH 2.5.

1.8.2. Thermal pH gradients

Another interesting approach is that of Luner and Kolin (1970) who perform IEF in a 'thermal pH gradient'. This system is characterized by quick focusing times (15 min for a complete fractionation). In this method, the pH determination amounts to a temperature measurement with a thermistor, which gives the pH in a zone to better than 0.01 pH unit, from the known pH versus temperature curve. Broadly speaking, this approach and the one by Troitzki et al. (1974) (see following section) are similar, since they are based on p*K* variations of buffers either in gradients of organic solvents or in temperature gradients. It is too early perhaps to assess the impact of both methods. However Lundahl and Hjertén (1973) report that they can only be of limited applicability in isoelectric focusing, because the pH of ordinary buffers and the pI of proteins do not exhibit sufficiently large differences in temperature coefficients.

1.8.3. Dielectric constant pH gradients

Troitzki et al. (1974) have formed pH gradients by using common buffers in gradients of organic solvents, such as ethanol, dioxane, glycerol, or in poly-ol gradients such as mannitol, sucrose, sorbitol. Taking advantage of the pH variations of these buffers in different concentrations of these solvents, they were able to generate gradients of approximately 1.5 pH units in different regions of the pH scale. These pH gradients were stable up to 12 days of IEF under voltages up to 1000 V. They have shown separations of rabbit haemoglobin in pH gradients 7–8.6 formed with borate buffer in a mixed gradient of 0.5% glycerol and 0–30% sucrose, and of human serum albumin in a pH gradient 4.5–5.8 formed with acetate buffer in a 0–90% glycerol gradient.

1.8.4. pH gradients generated by buffer diffusion

These gradients have already been described briefly in pages 346 and 347. They were termed 'artificial' pH gradients by Svensson (1961) since they are obtained by diffusion of buffers of different pH and rapidly decay under an electric field. The pH gradient was generated by placing the substance to be separated at the interface between an acidic and basic buffer, in a Tiselius-like apparatus, and allowing diffusion to proceed under an electrical field (Kolin 1954, 1955, 1958). In these 'artificial' pH gradients Kolin was able to obtain 'isoelectric line spectra' of dyes, proteins, cells, microorganisms and viruses on a time scale ranging from 40 sec to few minutes, a rapidity still unmatched in the field of electrophoresis.

This technique was also used by Hoch and Barr (1955) for separating serum proteins and by Tuttle (1956) and Maher et al. (1956) for separating haemoglobins. Due to poor reproducibility and instability of the pH gradients, this approach was soon abandoned.

1.8.5. pH gradients obtained with mixtures of acids

Before the synthesis of Ampholine pH 2.5–4, the lack of suitable carrier ampholytes had precluded analysis of acidic proteins, such as pepsin and glycoproteins and of acidic dyes. However, Pettersson (1969) found that acidic pH gradients between pH 1 and 3 could be formed by electrolysing a mixture of acids and ampholytes. As in the early experiments of Ikeda and Suzuki (1912), Pettersson also showed that a mixture of acids, in this case dichloroacetic acid (p*K* 1.48), phosphoric acid (p*K* 2.12), monochloroacetic acid (p*K* 2.85), citric acid (p*K* 3.08), formic acid (p*K* 3.75) and acetic acid (p*K* 4.75), mixed with two ampholytes, glutamic acid (pI 3.22) and aspartic acid (pI 2.98), would distribute themselves according to increasing p*K*'s under an electric field. The strongest acid was closest to the anode, and the weaker acid was almost completely non-ionized at its anodic side.

Although steady state conditions are never quite achieved, a quasi-equilibrium develops which changes so slowly that ampholytes have enough time for focusing. A sucrose-free histidine (pI 7.47) solution placed at the cathode on top of the density gradient provided good conductivity by convective recirculation. This system produced undesirable UV-absorbing by-products from acid hydrolysis of sucrose (anodic oxidation of fructose). This might have been avoided by substituting sorbitol or glycerol for sucrose. Nevertheless, Pettersson (1969) succeeded in focusing pepsin (pI 2.5) and separated two red components from bilberry sap, with pI's of 1.4 and 2.3 and as many as seven colored components with pI's between 1.3 and 3.1 from red beet sap. The pH gradient in this region, however, increased in a stepwise manner rather than as a smooth function.

A similar approach was used by Stenman and Gräsbeck (1972) for the fractionation of the R-type vitamin B_{12}-binding protein of human amniotic fluid. They avoid the use of aromatic acids and of halogenated organic acids, to prevent harmful effects on proteins. The pH gradient is obtained with a mixture of the following acids: maleic acid (pK_1 4.34), malonic acid (pK_1 2.18), lactic acid (pK 3.08), malic acid (pK_1 3.40), formic acid (pK 3.75), succinic acid (pK_1 4.16), glutaric acid (pK_1 4.34), acetic acid (pK 4.75) and propionic acid (pK 4.87). To this mixture two ampholytes, glutamic and aspartic acid, are added. Upon focusing for 5 days at 400 V, under quasi-equilibrium conditions, they obtained a smooth pH gradient from 1.5 to 4.5 and sharp resolution of the R-type protein into six peaks, isoelectric between pH 2.37 and pH 4.22. To obtain a reasonable field strength also in the very acid part of the gradient, the anode is always kept at the bottom of the column and a very high glycerol concentration (70%) is used in this region to diminish the conductivity of the strong acids.

Preparative IEF

Over the past few years, IEF has been shown to have many favourable attributes as a preparative method. Ideally, a good preparative method should provide high resolution separations of large amounts of material and give good recoveries, without altering the physico-chemical or biological properties of samples. IEF meets many of these criteria and also has additional advantages. For instance, unlike other methods, IEF has the almost unique advantage of affording improved resolution at increased load levels. We have already discussed how high sample load is compatible with increased resolution and how both are favoured by focusing in shallow pH gradients where a broad zone may accomodate more protein than the same pH increment in a wider pH range (see p. 355). Another advantage is that, by judicous selection of the pH range, many unwanted substances can be eliminated from the separation zone. Finally, since IEF is an equilibrium method, dilute samples are concentrated as they are separated.

Several approaches have been used to achieve the inherent potential of IEF, have confirmed theoretical predictions of load capacity and zone resolution and have given good recoveries of samples and of biological activity. Among the successful variations for preparative IEF are vertical columns with density gradient stabilisation, horizontal troughs with self-generating density gradients, zone convection systems, continuously polymerised gels such as polyacrylamide or granulated gels such as Sephadex either in flat beds or in columns. Each system has its own particular advantages. Some capitalize on

Subject index p. 587

phenomena, such as isoelectric precipitation, which are anathema to many others. This section will describe many of the presently available techniques for preparative IEF, with particular reference to capacity, resolution, sample detection and recovery. Three basic systems will be discussed: (1) IEF in liquid media, (2) IEF in Sephadex beds, (3) IEF in polyacrylamide gel. They allow protein separation in the milligram to gram scale.

2.1. Preparative IEF in liquid media

2.1.1. The LKB columns

One of the earliest methods for preparative IEF is the use of columns in which the pH gradient is stabilized against convection by a vertical sucrose density gradient. LKB adapted a design by Svensson (1962b) and produced two columns of 110 ml (LKB 8100-1) and 440 ml (LKB 8100-2) capacity. The design of these columns is shown in Fig. 2.1. Approximately 5 to 25 mg protein per zone can be focused in the smaller column and approximately 20 to 80 mg per zone in the larger. The upper load limit depends on the number of components, their solubility at their pI and the shape of the pH gradient. For most proteins, the load should not exceed 5 mg/cm^2 of cross sectional area. These columns require a considerable amount of carrier ampholytes. For most purposes a minimum concentration of 1 % (w/v) Ampholine is recommended, corresponding to about 1.1 g and 4.4 g Ampholine, respectively, for each column. However, Ampholine ranges 2.5–4 and 9–11 are used at a concentration of 0.5 % (w/v), since their conductivity is twice that of the other pH ranges.

The density gradient The density gradient is usually prepared with sucrose solutions, although other solutes may also be used. Ideally, the solute should have a high solubility but low viscosity in water. It should be non-ionic and have a high density, a minimum density gradient of 0.12 gm/cm^3. Finally, the material in the gradient should not react with the sample nor interfere with its subsequent detection. Analytical reagent grade sucrose is recommended for the preparation of the sucrose density gradient. The gradient can be formed by mixing

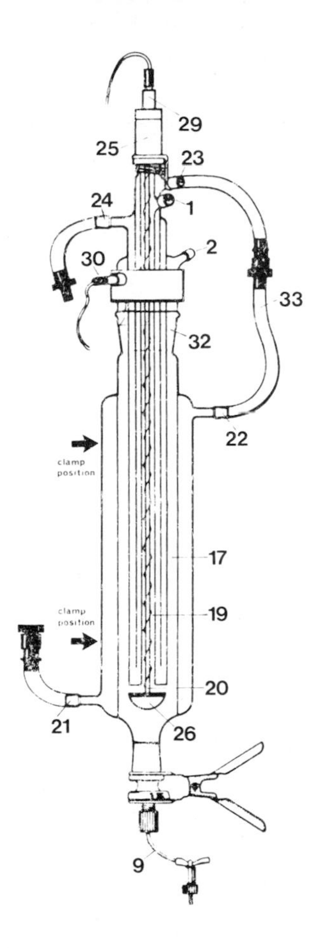

Fig. 2.1. Sketch of the LKB 8100-1, 110 ml column, (1) gas escape for the lower electrode; (2) gradient and sample inlet; (9) elution capillary; (17) electrofocusing compartment, consisting of a hollow cylinder thermostatted on both sides by cooling jackets; (19) lower electrode compartment; (20) teflon support for the lower electrode; (21) and (22) inlet and outlet, respectively, for the outer cooling jacket; (23) and (24) inlet and outlet, respectively, for the inner cooling jacket; (25) setting device for opening and closing valve (26) used to shut off the lower electrode compartment (19) from the electrofocusing chamber (17); (29) and (30) lower and upper electrode terminals, respectively, for connection with the power supply; (32) upper platinum electrode; (33) tubing connection between the two cooling jackets. (Courtesy of LKB Produkter AB.)

a light solution containing only ampholytes and sample with a dense solution containing ampholytes, sample and sucrose. These solutions may be mixed in certain proportions to produce a series of solutions of increasing densities. These are then layered sequentially on top of the dense electrolyte solution in the lower chamber of the electrofocusing column. The layers are allowed to diffuse into one another to produce a fairly smooth gradient. Alternatively, more reproducible density gradients may be prepared with a gradient mixer such as LKB 8121 and LKB 8122 Ampholine gradient mixers, for use with the 110 and 440 ml columns, respectively. These mixers provide linear density gradients from two equal volumes of light and dense solutions having a density difference of 0.2 g/cm^3. This density difference corresponds to light and dense solutions containing 5 and 50% sucrose, respectively.

Other solutes, such as sorbitol or glycerol, may be used instead of sucrose as anticonvective media, especially when performing IEF in alkaline regions where sucrose ionizes. Ethylene glycol can be used at concentrations of 60–70% in the dense solution, because of its lower density. Other useful agents include mannitol and polyglycans, such as Dextran and Ficoll (Pharmacia, Uppsala, Sweden) (Vesterberg 1971a). Recently, Frederiksson and Pettersson (1974) and Fredriksson (1975) have described the use of deuterium oxide as a medium for density gradients. D_2O gradients have so far only been used in the small (1.5 ml) spectrophotometric cell of Jonsson et al. (1973). These gradients are formed in situ by free interdiffusion of three D_2O solutions for 3 min (Rilbe and Pettersson 1968). While the density increment obtainable with D_2O is rather small (0.1 g/cm^3), its viscosity at 5°C is, however, about 50% lower than the viscosity of an aqueous sucrose solution of similar density. This means that a more favourable distribution of the field strength along the column and shorter focusing times can be obtained. However, pH readings in D_2O gradients are higher than in sucrose gradients, due to shifts in pK_a values of the protolytic groups in proteins and to shifts of the asymmetry potential of the glass electrode in D_2O compared with H_2O. It remains to be seen whether or not D_2O gradients will be of general applicability also in preparative columns.

The density gradient with ampholyte and sample is layered on top of a dense cushion containing an electrolyte at the bottom of the column. The second electrolyte is now floated on top of this gradient (see Fig. 2.2). Table 2.1 lists three types of density gradient stabilization (sucrose, glycerol or ethylene glycol) for the 110 ml and 440 ml

TABLE 2.1

Density gradient solutions for the 110 ml and 440 ml LKB columns for three types of stabilizing media (sucrose, glycerol and ethylene glycol).

Density gradient solution	110 ml Column LKB 8100-1	440 ml Column LKB 8100-2
Sucrose stabilization		
Dense gradient solution		
Sucrose	27 g	107.5 g
Volume of H_2O + Ampholine (+ sample if added)	37 ml	150 ml
Total volume	54 ml	215 ml
Concentration of sucrose	50% (w/v)	50% (w/v)
Light gradient solution		
Sucrose	2.7 g	10.75 g
Volume of H_2O + Ampholine (+ sample if added)	53 ml	207 ml
Total volume	54 ml	215 ml
Concentration of sucrose	5% (w/v)	5% (w/v)
Glycerol stabilization		
Dense gradient solution		
Glycerol – 87% (v/v)	30 ml	118 ml
Volume of H_2O + Ampholine (+ sample if added)	24 ml	97 ml
Total volume	54 ml	215 ml
Concentration of glycerol	60% (w/v)	60% (w/v)
Light gradient solution		
Total volume of H_2O + Ampholine (+ sample if added)	54 ml	215 ml
Concentration of glycerol	0%	0%

TABLE 2.1 (continued)

Density gradient solutions for the 110 ml and 440 ml LKB columns for three types of stabilizing media (sucrose, glycerol and ethylene glycol).

Ethylene glycol stabilization		
Dense gradient solution		
Ethylene glycol – 100% (v/v)	40.5 ml	161 ml
Volume of H_2O + Ampholine (+ sample if added)	13.5 ml	54 ml
Total volume	54 ml	215 ml
Concentration of ethylene glycol	75% (v/v)	75% (v/v)
Light gradient solution		
Total volume of H_2O + Ampholine (+ sample if added)	54 ml	215 ml
Concentration of ethylene glycol	0%	0%

columns. Table 2.2 gives suggestions for electrode solution to be used in both columns either when the anode or when the cathode are uppermost.

Electrolyte The anodic solution is usually 1–5% ethanolamine or ethylendiamine, or 8% NaOH. When possible, it is preferable to have the anode at the bottom of the column to prevent ionization of the dense sucrose solution by strong base at the cathode (Flatmark and Vesterberg 1966). The choice of the electrode position depends also on the pI of the components of interest. Ideally, conditions should be selected so that components of most interest focus in the lower half of the column. This will minimize diffusion during subsequent elution which is performed in the absence of the electric field. However, the choice of polarity depends also on conductance considerations. If the position of the anode is chosen so that the carrier ampholytes with the lowest conductance are at the top of the column, the low conductance of the ampholytes will be counteracted by the high conductance of the density solution. At the bottom of the column, the low conductance of the density solution will be counteracted by the high conductance of the carrier ampholytes. In this way the most uniform possible conductance will be obtained. As a consequence, it is usual to have the

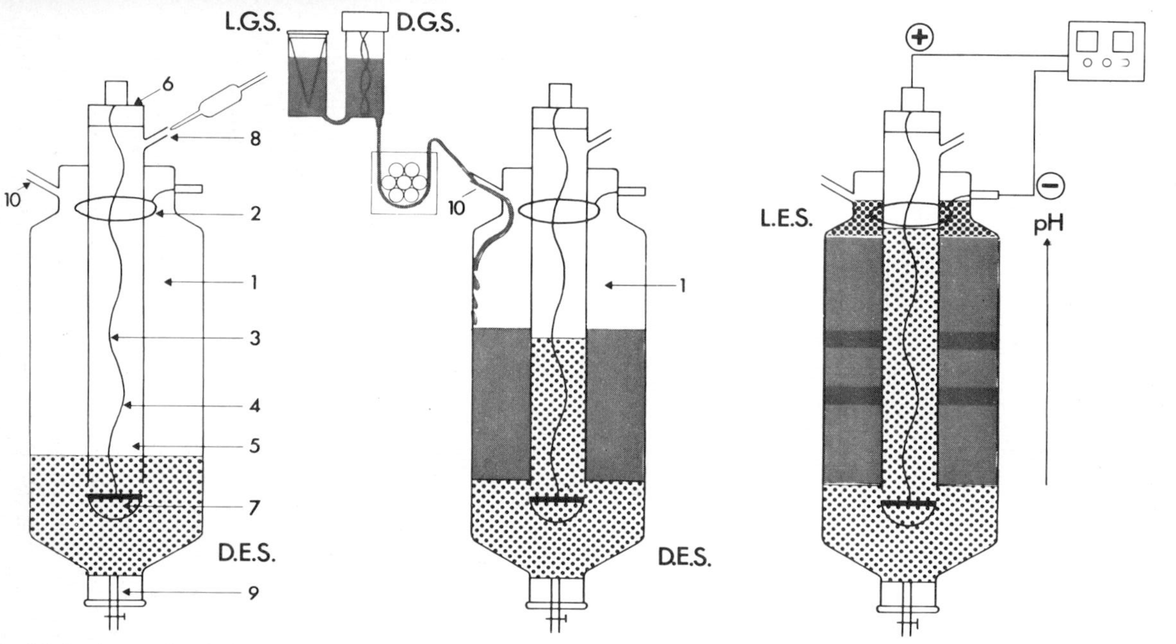

Fig. 2.2. Setting up Ampholine Electrofocusing column. Left: the dense electrode solution (D.E.S.) is layered at the bottom of the column with the valve (7) of the central electrode compartment (5) open. This solution is pipetted in via the gas escape (8), to avoid contamination of the separation chamber. Center: the density gradient is introduced in the column via the sample inlet (10) with the aid of a peristaltic pump. The density gradient is best produced with a gradient mixer, shown schematically, L.G.S. (light gradient solution) and D.G.S. (dense gradient solution). Right: the light electrode solution (L.E.S.) is now floated on top of the gradient, until the platinum loop (2) is submerged. The electrodes are then connected to a power supply. (Courtesy of LKB Produkter AB.)

TABLE 2.2

Electrode solutions for the 110 ml and 440 ml LKB columns to be used either with the cathode or with anode uppermost.

Electrode solution	110 ml Column LKB 8100-1	440 ml Column LKB 8100-2
Anode at Top of Column		
Cathode solution	15 g sucrose	48 g sucrose
(pH – 11.7 approx.)	+ 10 ml H_2O	+ 30 ml H_2O
	+ 6 ml 1 M NaOH	+ 20 ml 1 M NaOH
Total volume	25 ml	80 ml
Concentration of sucrose	60 % (w/v)	60 % (w/v)
Anode solution	1.5 ml 1 M H_3PO_4	6 ml 1 M H_3PO_4
	+ 8.5 ml H_2O	+ 34 ml H_2O
Total volume	10.0 ml	40 ml
Cathode at Top of Column		
Cathode solution	2.5 ml 1 M NaOH	10 ml 1 M NaOH
	+ 7.5 ml H_2O	+ 30 ml H_2O
Total volume	10.0 ml	40 ml
Anode solution	15 g sucrose	48 g sucrose
(pH – 1.2 approx.)	+ 12 ml H_2O	+ 38 ml H_2O
	+ 4 ml 1 M H_3PO_4	+ 12 ml 1 M H_3PO_4
Total volume	25 ml	80 ml
Concentration of sucrose	60 % (w/v)	60 % (w/v)

anode at the top of the column for pH ranges above 6 and the anode at the bottom of the column for pH ranges below 6. In particular, when using Ampholine pH 9–11, it is a must to have the anode at the top of the column. At the opposite extreme, when using Ampholine pH 2.5–4, the cathode must be at the top of the column.

Sample application Because IEF is an equilibrium method, the method of sample application is not as critical as in other forms of electrophoresis where a thin starting zone is necessary. The sample may be incorporated into either the dense or the light sucrose-ampholyte solution, or into both. Since samples are concentrated during the experiment, the volume of sample is not critical and may be as much as 80% of the final column volume. Alternatively, the sample may be introduced as a narrower zone close to the place where it is expected to focus. In this way the run will be shortened, and exposure to pH extremes avoided. As has been discussed in the previous chapter, the upper load for the sample depends on several factors. These include the geometry of the column, the number and distribution of components in the sample, their solubility at their pI, the pH range and level of carrier ampholytes and the strength of the electrical field. Since large cross-sectional areas and narrow pH ranges tolerate higher loads, the larger column which has the same height as the smaller column usually carries proportionably more sample than might be indicated by the 4-fold difference in volume. The amount of salt and buffer in the sample should be low and should not exceed 0.5 nmole for the 110 ml column and 1.5 nmoles for the 440 ml column. High salt concentrations will create high current densities at the beginning of the experiment and also generate acid at the anode and base at the cathode. A good solution is to dialyse the sample against 1% glycine or 1% Ampholine solutions at an appropriate pH. Since most proteins are generally more stable in weak alkaline solution, the pH 6–8, 7–9 or 8–10 ranges are often appropriate.

Electrolysis conditions As in IEF experiments, the total power applied to the column should not exceed its capacity to dissipate the heat generated. Since the separation column is a hollow cylinder, with cooling on all sides, heat dissipation in the LKB columns is very efficient. With a temperature of circulating coolant of 2–4°C, it is customary to run the small column (110 ml capacity) at about 2–3 W and the larger column (440 ml) at about 6–9 W. However, with the introduction of constant wattage power supplies, these upper limits have been increased and the running time considerably shortened (see §4.9). It is advisable not to thermostat the column below 2–4°C (Haglund 1971). The temperature coefficient of the viscosity and conductivity of the sucrose-Ampholine mixture is nearly constant down to around 1°C but rises steeply below this point. Consequently, small temperature differences across columns maintained below 2°C may cause wavy zones and impair resolution.

The current is highest at the beginning of the experiment when the ampholytes are randomly distributed throughout the column consequently, the initial voltage must be carefully controlled to avoid excessive heat. Under normal conditions with a level of 1% (w/v) Ampholine, an initial voltage of 400–500 volts may be appropriate. Later, when the conductivity decreases as the ampholytes migrate to their pI, the voltage may be increased up to 1600–2000 V to improve resolution. This is done automatically with constant wattage power supplies. The system reaches equilibrium when the current stabilises at a low plateau. The actual time required to achieve equilibrium will, of course, depend on the actual electrolysis conditions and the temperature of the circulating coolant, but is usually between 24 and 72 hr with both columns when run with conventional power supplies. This time may be reduced by using constant wattage power supplies (see §4.9).

Columns with narrow pH ranges or more than 1% ampholyte usually require considerably longer focusing periods. The correct electrolysis period can only be determined experimentally. It is emphasized that the final movement to equilibrium of carrier ampholytes and proteins occurs so slowly that the final drop in conductivity may not be apparent. Consequently, it is advisable to err on the safe side and continue electrolysis for a few hours after the apparent plateau.

When the system has reached equilibrium the contents of the column are either drained from the bottom or displaced upwards with a denser solution. It is customary to collect fractions of 1–4 ml and to determine the amount of protein and the pH of each fraction. This may be done manually or with the aid of flow cells which monitor UV absorbancy and pH. Ideally, the pH of the fractions should be determined at the same temperature as that prevailing during electrofocusing (see pp. 513–517).

Although these columns have proved useful for many purposes, they are subject to some technical problems which limit the amount of sample that may be focused and impair the resolution actually obtained after fractionating the column contents. Most proteins tend to be least soluble at their pI and may precipitate from solution.

Another form of precipitation can occur when the density of a focused band of a large amount of protein exceeds that of the supporting gradient. Other problems arise from diffusion during lengthy elution periods in the absence of the electric field and from remixing of zones in the lower chamber as the column contents merge into a continuous stream. Since additional mixing of zones can also occur in the parabolic flow profile of the capillary path between the column and fraction collector, this path should be kept as short as possible. If flow cells are used, the volume of their chambers should be kept to a minimum. Another source of diffusion is from peristalsis generated by pumps, especially when they operate on the eluting capillary. This situation may be avoided by emptying the column by gravity. Alternatively the pump tube may be inserted into the top of the column (inlet No. 10 in Fig. 2.3) so that the water displaces the column content downwards. The rate of outflow will thus be dictated by the speed of the pump.

Figs. 2.4 and 2.5 demonstrate the loss of resolution that may occur on elution after IEF in the LKB column. Fig. 2.4 shows a photograph of a column in which a coloured protein, horse spleen ferritin, has been resolved into multiple components, which were well separated. However, on elution by gravity displacement without pumping, much of this resolution was lost and only 2 major peaks were apparent (Fig. 2.5).

Boddin et al. (1975) have recently devised a method for preparative IEF in sucrose density gradients in which the column contents may be eluted under the electric field. They adapted a commercial polyacrylamide gel apparatus by closing the lower end of the column with a simple one hole stopper and separated the sucrose density gradient from the lower electrode with a membrane which allowed direct electrical contact. It will be interesting to see if this development solves the problem of diffusion on elution in the absence of an electrical field. For other practical aspects, the reader is referred to the new LKB manual (LKB 8100 Instruction Manual I-8100-EO4).

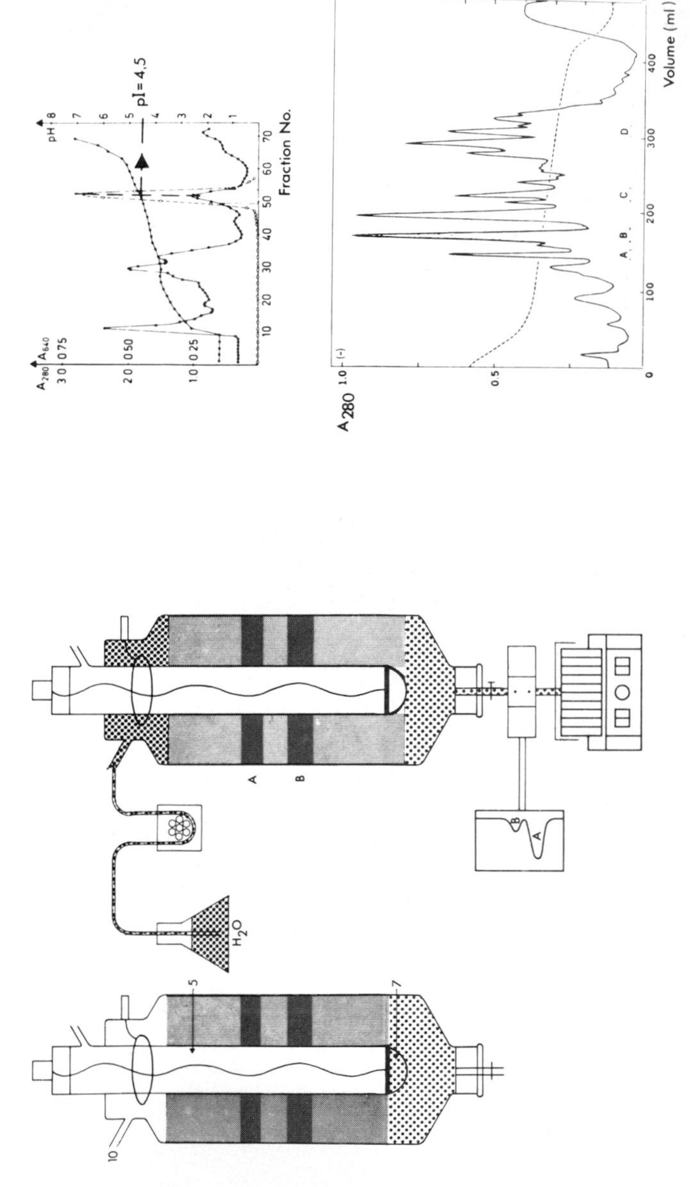

Fig. 2.3. Emptying the Ampholine Electrofocusing Column. Left: at the end of the experiment, the central valve (7) is closed and the light electrode solution removed via inlet (10). Center: a tubing connected to a peristaltic pump and to a distilled water reservoir is plugged into inlet (10). The upper part of the column is filled with H_2O and then inlet (10) is tightly stoppered. When possible, the elution capillary is connected to a UV flow cell and a recorder. Elution by a pump at the column top is superior to elution by gravity flow or by a pump connected with the elution capillary at the column bottom. Right: examples of elution profiles and pH gradients obtained by this method. The graph on the upper right side shows how to determine the pI from a UV peak. (Courtesy of LKB Produkter AB.)

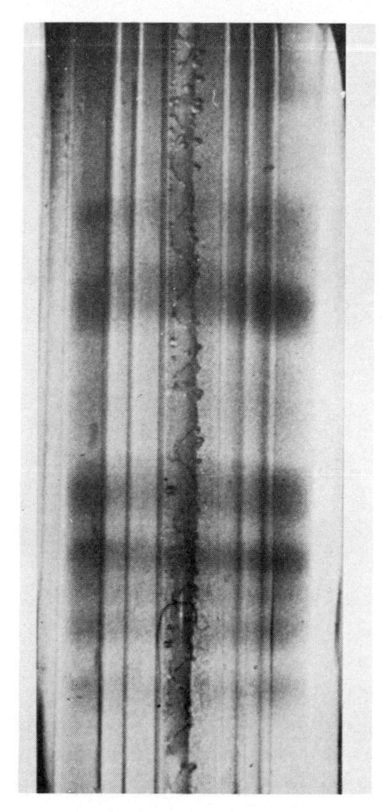

Fig. 2.4. Isoelectric fractionation of horse spleen ferritin in the LKB Ampholine Electrofocusing Column, in the pH range 3–5. At least six major, well resolved components are seen in the column. (From Niitsu, Y., unpublished.)

2.1.2. The ISCO columns

Instrumentation Specialty Company (ISCO) has produced two columns for isoelectric focusing, an analytical or small-scale preparative unit (model 212) and a preparative version (model 630). The former has a bore of 1 cm and is about 30 cm long. The design of this column is shown in Fig. 2.6. The maximum gradient volume is 23 ml and the maximum load capacity is about 10 mg of protein. When cooled at

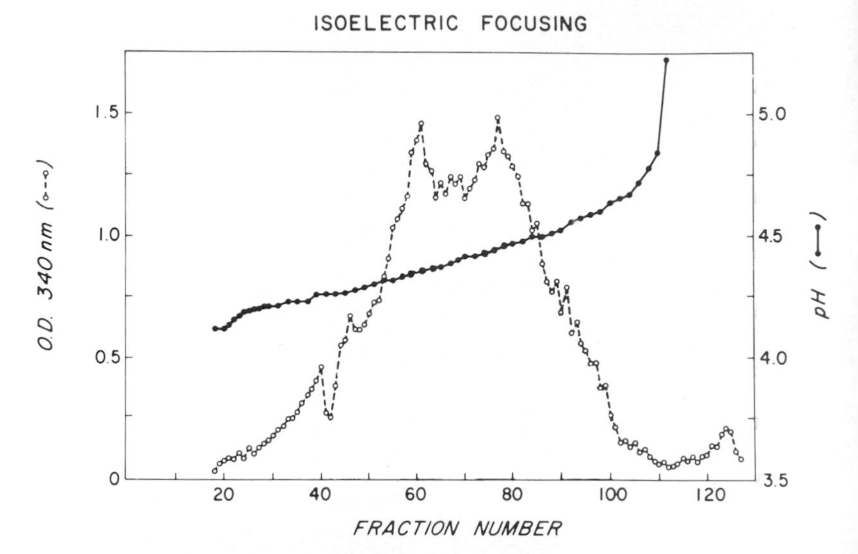

Fig. 2.5. Example of the loss of resolution obtained when the fractions of Fig. 2.4 are eluted by gravity flow. Here, only two major peaks are apparent. (From Drysdale, 1974, by permission of the Biochemical Society, London.)

2°C an electrical load of up to 15 W can be applied to the column. The larger model is built on the same principle and can accomodate gradients of up to 160 ml in volume and 32 cm in length. When cooled at 2°C it can be run at approximately 20 W. An interesting feature of both systems is that the electrode chambers are built coaxially around the separation column. Two concentric membranes separate each electrode from the gradient to inhibit the introduction of electrolysis products and pH changes into the column. This membrane also prevents net transfer of gradient components into or out of the inner column.

Another advantage is that the progress of the experiment may be monitored by scanning the column. To do this, the current is switched off and the entire gradient column is raised past a densitometer to produce an absorbance profile at one or more wavelengths. Thus progress on sample separation can be recorded at selected intervals

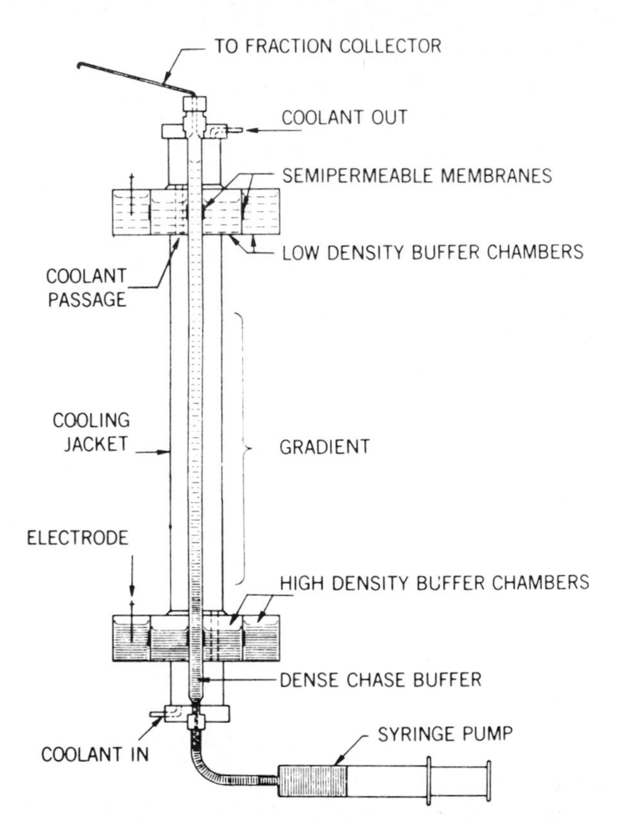

Fig. 2.6. Schematic drawing of the ISCO model 212 column. In the set up for column elution, a syringe pump containing a dense chase buffer is attached to the bottom of the column and the density gradient displaced from the column top. (Courtesy of ISCO.)

and the attainment of equilibrium confirmed. In addition, a special applicator allow the sample to be introduced into the center of the column at any desired time. Thus a preformed pH gradient can be scanned before sample application to provide a baseline for Ampholine absorbance (Righetti and Drysdale 1971). This is achieved by injecting a dense sucrose solution into the bottom of the column to force the gradient through a flow-cell and into a reservoir. Since scanning and unloading procedures are performed with pulseless syringe pumps, and the apparatus has no mixing chambers, sharp zone recoveries are possible without undue problems or convection, turbulence of laminar flow associated with other systems. The principle of these columns has been described by Brakke et al. (1968). This type of intermittent scanning is, of course, different from the method developed by Catsimpoolas (1973c) for in situ scanning of columns while still under voltage. This technique, called transient state isoelectric focusing, can provide much useful information on the kinetics of IEF (Catsimpoolas 1973c,d,e,f).

Fawcett (1975a) has pointed out that similar experiments may be conducted with much simpler apparatus. For example, he has built modified U-tubes, with a quartz column in one limb (separation arm) and a three-way tap at the bottom bend. This tap is used for filling and emptying the column. The separation limb is fitted with a cooling jacket. The column is emptied by removing the upper electrode and pumping dense sucrose solution into the base of the quartz tube to displace the gradient upwards. A Uvicord II detector head is fitted to the upper part of the quartz tube, so that a UV scan is obtained during elution. An adapter moulded of silicone rubber attached to the column top allows fractions to be collected. Fawcett (1975a) has also built multiple U-tube units (such as two and three-U-tube apparatus) to be used when precipitates occur or when high concentrations of impurities spread into focused zones. These multiple U-tube units are fitted with stop cocks in appropriate places, to facilitate removal of focused bands. These columns are usually cooled by immersion in a tank.

2.1.3. Rilbe's columns

In addition to the type of vertical column produced for LKB (Svensson 1962b), Rilbe and his colleagues have developed new vertical columns and have also improved the design of horizontal multi-compartment electrolysers. In 1973 in Glasgow, Rilbe described a system for preparative IEF in short density gradient columns with vertical cooling. Following Philpot's (1940) early suggestion of using short and thick columns with vertical heat dissipation, Rilbe and Pettersson

(1975) built two types of columns which differ essentially in the method of cooling. Both are extremely short and thick. In one, the distance between electrodes is only 1.55 cm and the cross sectional area 283.5 cm^2 with a column volume of 440 ml. The electrodes are sheets of an alloy of 75% palladium and 25% silver and are soldered to brass plates at the column extremities. Since the electrodes are also the major cooling surfaces, a vertical temperature gradient is created in the separation column. Equilibrium is achieved in a very short time, 30 min, in a final potential gradient of 75 V/cm. Because of its geometry, the column must be filled and emptied from a lateral position. Drawbacks of this type of cell include possible contamination from the metal electrodes by anodic oxidation and rapid diffusion of focused bands during the time required to turn the column through 90° prior to fractionation.

The second type of column is similar in principle and has a capacity of 110 ml and smaller electrodes of platinum wire. This column is vertically cooled through cellophane membranes by the electrolytes which are chilled and stirred (Fig. 2.7). Focusing is usually achieved in 90 min at a final potential difference of 125 V. This column must also be filled and emptied from a lateral position. In this type of cell, Rilbe and Pettersson (1975) fractionated more than 1 g of sperm whale myoglobin in a pH 7–9 gradient in 3 hr. The main band (MbI, pI 7.6 in the ferrous form or pI 8.3 in the ferric form) contained approximately 800 mg protein, an appreciable amount to be carried by a density gradient. The great advantage of this column type is therefore its high load, the capacity and the relatively short focusing time.

2.1.4. Zone convection IEF according to Valmet

In 1969, Valmet described a new method for preparative IEF based on a new electrophoretic principle, called 'zone convection isoelectric focusing'.

A unique feature of this method is that it does not require any anti-convective medium. Separations are conducted in a series of linked U-tubes.

The separation cell (shown schematically in Fig. 2.8) is made of

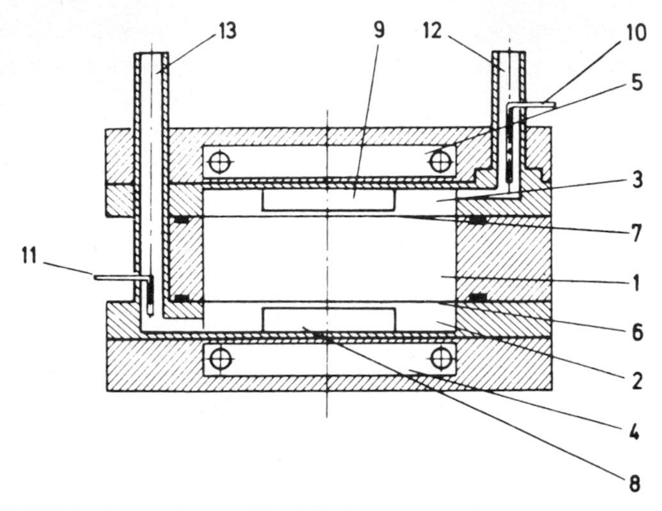

Fig. 2.7. Rilbe's short column with chilled semipermeable membranes and small platinum electrodes. (1) separation column. (2) and (3) compartments for catholyte and anolyte. (4) and (5) compartments for circulating refrigeration medium. (6) and (7) cellophane membranes. (8) and (9) rotating magnets. (10) and (11) platinum electrodes. (12) and (13) plexiglass gas-escape tube. (From Rilbe and Pettersson 1975, by permission of Butterworths.)

plastic with a wall thickness of 1 mm. It resembles a series of Tiselius cells with a cross section of 3 × 40 mm. Each 'U-tube unit' is about 10 mm high. The channels are spaced 10 mm apart. The cell consists of two separate units, the trough and the lid. The trough is a shallow rectangular box with a corrugated bottom. The lid has similar indentations so that when it is lowered on top of the trough, the interdigitating projections force the liquid in each pocket to overflow, thus closing the electric circuit on the cell. The electrodes are attached to both ends of the apparatus. Band stabilization is achieved by the Ludwig (1856) Soret (1879) effect. As a protein focuses into a pocket, it concentrates at the bottom against the cold wall, thereby generating a vertical density gradient within the channel, perpendicular to the current flow.

At the end of the run, the lid is removed thus causing the liquid to flow back into each U-tube so that the trough now acts as a fraction

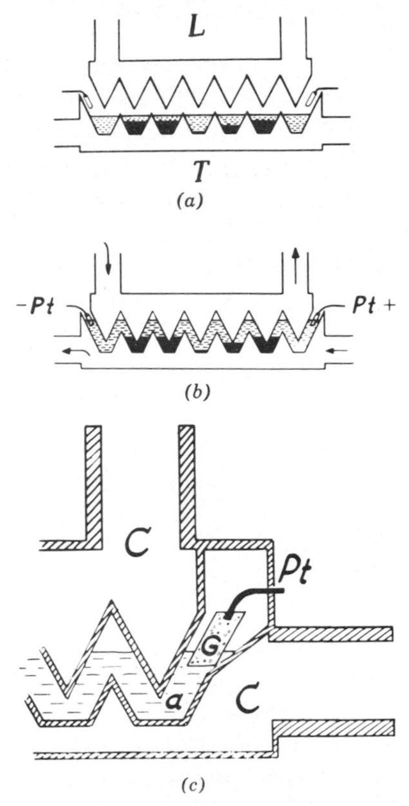

Fig. 2.8. Schematic drawing of the Valmet zone convection electrofocusing apparatus. (a) the lid (L) and the trough (T) before or after the experiment; (b) lid and trough put together, as during the experiment; (c) detail of one of the electrode compartments. (From Valmet 1969, courtesy of LKB Produkter AB.)

collector. The concentrated protein fractions can be recovered from the bottom of the U-tube with the aid of a peristaltic pump, allowing the remaining carrier ampholytes to be reused for other separations. In the case of Valmet's trough, isoelectric precipitation is an advantage rather than a disadvantage, since zone detection is simplified and samples may be recovered by centrifugation. On the other hand, non-isoelectric precipitation can result in serious contamination.

Valmet has also successfully used this apparatus for fractionating ampholytes into narrow pH ranges. By using a 30-pocket cell, he was able to separate the pH 3–10 Ampholine into a number of fractions each covering about 0.2–0.3 of a pH unit. The sample load in Valmet's horizontal trough is much higher than in conventional vertical density gradient apparatus. He has reported separation of 350 mg serum protein in 50 ml of 1% Ampholine. The high sample load, the economically favorable Ampholine protein ratio, and other unique features make this apparatus particularly attractive for large-scale IEF. Unfortunately, it has not yet become available commercially because of technical difficulties in construction.

This method has been successfully used by Bodwell and Creed (1973) for separating a wide variety of proteins. Fawcett has used an essentially similar approach by modifying an earlier design of Kalous and Vacik (1959). Although he achieved discrete fractionation of ovalbumin in a wide pH range, he did not obtain good resolution with myoglobin components in alkaline pH ranges, apparently because of poor cooling and electro-osmotic effects.

An apparatus built on the principle of the Valmet (1969) column, has been described by Macko and Stegeman (1970). IEF takes place in a coil of polyethylene tubing, 100 cm long, 4 mm inner diameter, 1 mm wall thickness. This tubing is coiled, with the aid of a double-face tape, around a copper tube of 20 mm diameter. The coiled tubing is held in place by a piece of wire screen fastened at the ends with strings. This is in fact an apparatus for zone convection IEF, where each turn of the spiral represents one compartment of the Valmet apparatus. The coil is filled with 10 ml of 10% sucrose and 1% Ampholine. The total sample load is in the order of 10–15 mg. The two tube extremities, which are allowed to stand up vertically, are filled with anolyte and catholyte. During IEF the coil is immersed in a stirred ice bath or a 2°C thermostat, with the loose ends above the water level. After IEF, the coiled tubing is cooled down quickly, preferably by immersing into liquid nitrogen, after which it is cut into segments and the frozen contents emptied into small vials.

A similar approach to free isoelectric focusing has been described by Lundahl and Hjertén (1973). Here the separation chamber is a water-cooled horizontal quartz tube with an inner diameter of 3 mm, that rotates around its longitudinal axis at 40 rpm to counteract convective disturbances. Coating of the tube with methyl cellulose eliminates electro-osmosis. The tube is scanned in the UV light and the ratio of absorbancies at 320 and 280 nm is recorded. This minimises disturbances due to irregularities in the rotating tube, dirt on its surface, dust in the cooling water, etc. At the two tube extremities, polyacrylamide beads are packed to avoid convective mixing of anolyte and catholyte with the Ampholine solution in the tube. A cellophane membrane at the tube ends prevents hydrodynamic liquid streaming.

An apparatus built on the same principle of the Valmet (1969, 1970) trough has also been described by Rose and Harboe (1970). It consists of a series of U-shaped compartments with outlets in the bottom, linked in a wavy form resembling a sea-serpent.

2.1.5. Zone convection IEF according to Talbot

Fortunately, Valmet's concept of zone convection IEF was not forgotten. Talbot (1975) devised a more practical apparatus embodying Valmet's idea. This apparatus, besides eliminating difficulties associated with non-isoelectric precipitation of material, provides easy access to the pH gradient and also incoporates a facility for autofractionation of the gradient at the end of the experiment. The separation trough is fashioned from a block of aluminium and is 1 inch high $2\frac{1}{2}$ inches wide and 24 inches long. The trough has a corrugated base and sides which form 52 pockets. The trough is coated with a varnish for electrical insulation and has cooling coils in its base for effective heat dissipation through direct metal contact. It is fitted with a lid to minimize evaporation and has electrode plates at each end. The block is held in a supporting framework, and can be pivoted about pins and locked either in a 45° position or in a horizontal position.

For operation, 30 ml of 1% (w/v) Ampholine is poured into the trough and distributed throughout the corrugations by tilting the whole apparatus from end to end. With the trough in the 45° position, the solution bridges the corrugations and forms a continuous electrical path along the length of the cell. 1% (v/v) phosphoric acid is used at the anode and 1% (v/v) N, N, N′, N′-tetramethylethylene diamine at the cathode. Fractionations are generally performed at 4°C. The pH gradient is established before insertion of sample. An initial voltage of 450 V at 1 mA is applied and stable pH gradients are usually formed in about 24 hr. After the introduction of sample, focusing is continued for a further 72 hr.

Fig. 2.9a shows the Talbot trough in the operating position and Fig. 2.9b in autofractionation position to demonstrate the principle of operation. Fig. 2.9c is an exploded diagram of the apparatus. The

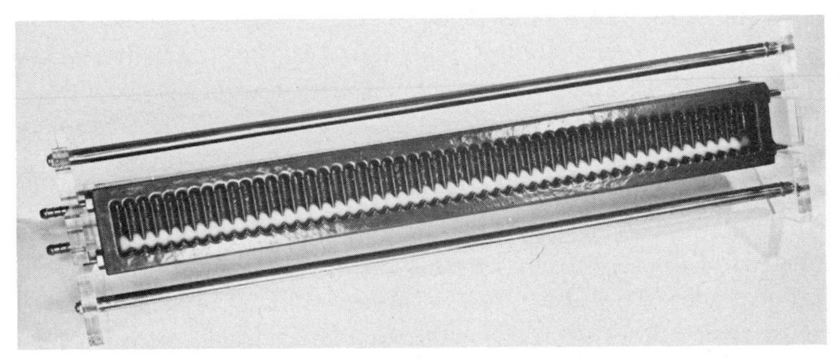

A

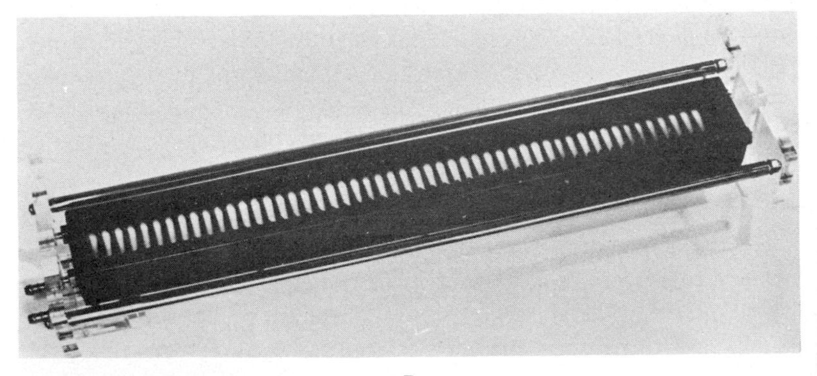

B

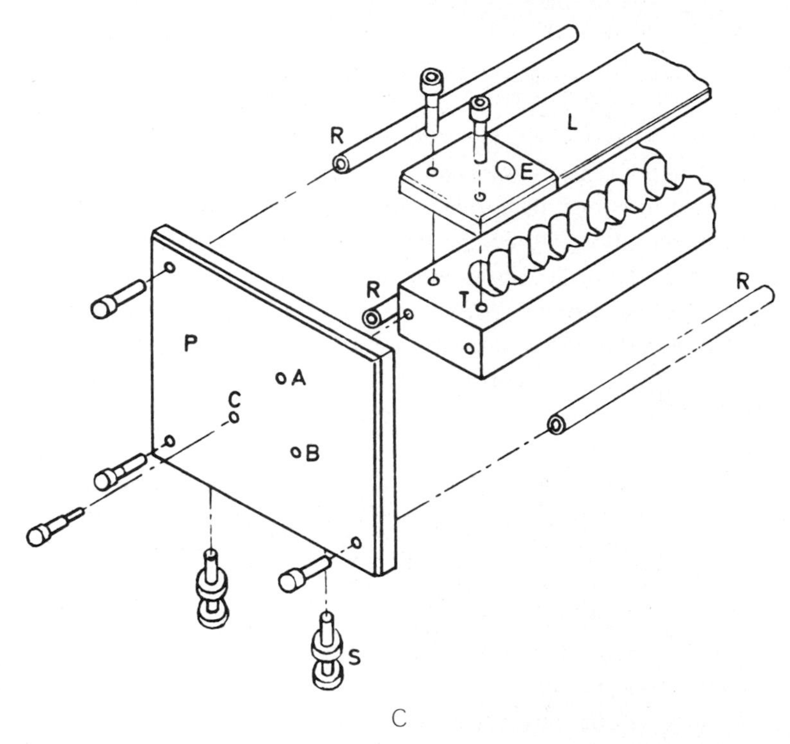

C

Fig. 2.9. A) The Talbot trough in operating position. The trough is built of aluminium and contains 52 corrugations. To the left, the inlet and outlet for cooling water are visible. During operation, the trough is tilted 45°. (Courtesy of P. Talbot, unpublished.) B) The Talbot trough in autofractionation position. This is accomplished by returning the cell to the horizontal position. The electrodes and cover lid are not shown. (Courtesy of P. Talbot, unpublished.) C) Exploded diagram of part of horizontal trough apparatus for isoelectric focusing. T) separation trough; L) lid; E) electrode – locating plate; P) end plate; R) stainless steel tie rods; S) levelling screws; C) pivot pin; A,B) locking screws. (From Talbot and Caie 1975, by permission of Butterworths.)

focused gradient is fractionated simply by returning the apparatus to the horizontal position so that the solution no longer bridges the trough corrugations and the focused fractions are separated in the bottom of each compartment. The pH of each fraction may be measured directly in the apparatus with a combination microelectrode.

Working with labelled viral preparations, Talbot and Caie (1975) have reported sample recoveries of 80 to 100%.

2.1.6. *Multi-compartment electrolyzers*

Rilbe's group has also continued along the classical line of stationary electrolysis in efforts to improve multicompartment apparatus. For reviews see Svensson (1948) and Rilbe (1970). In 1974, in Milano, Rilbe presented the latest developments on three types of electrolysers: one with open cells, internal cooling, and no stirring; a second type with closed cells, external cooling and stirring. The last model embodies most of the desired features of a good electrolyser. It is shown schematically in Fig. 2.10 (a and b). This design solves most of the technical difficulties connected with electrolysers, i.e. heat dissipation, electro-osmosis, homogeneity of solutions in each compartment, and isolation of separated components without remixing. One of the promising applications of this type of cell is in the very large-scale fractionation of proteins. Stationary electrolysis can now be completed in about 24 hr. Since the volume in these 20-compartment electrolysers is usually between 500 and 1000 ml, and since rather high concentrations of protein (about 4–5%) can be used, the load capacity is in the order of several g/day. Electrolysers of this sort should offer a major improvement over many other preparative techniques using density gradients (Rilbe et al. 1975).

2.1.7. *Free flow, high voltage IEF*

The concept of continuous flow, carrier-free IEF was first introduced by Rilbe (1969) and applied to protein separations by Seiler, Thobe and Werner in Frankfurt. They further developed the apparatus of Barrolier et al. (1958) and modified by Hanning (1961) for free-flow electrophoresis. The cell and its application in IEF have been described by Seiler et al. (1970a,b). The cell consists of a rectangular chamber, 11 cm wide, 36 cm tall and with a buffer film thickness of 0.5 mm. As many as 48 fractions are collected from the bottom of the cell by capillary tubes connected to a peristaltic pump. The separation chamber and the electrode cells are separated by semipermeable membranes. Solutions of 5% acetic acid and 1.5% ethanolamine are used at the anode and cathode respectively, and are continuously pumped through the electrode compartments during IEF. The cooling system consists of a teflon coated aluminium block, cooled by circulating liquid from a thermostat at 1°C. The temperature within the separation cell is controlled by a thermistor. Samples are usually injected into the chamber by a micro pump connected to the cooling system. A typical run is performed at a field strength of 110 V/cm and at a liquid flow rate of 1 ml/min.

This group has recently extended the method to include cells and subcellular organelles. For example, they were able to separate mixtures of human, mouse and rabbit red blood cells in the pH ranges 3–10 and 5–7, at an injection rate of 3×10^7–7×10^7 cells/min. The flow-through time for the cells was only 7 min. Solutions of 1 mM EDTA were used to prevent cell clumping. Just et al. (1975) have also reported preliminary data on the separation of the light mitochondrial fraction of rat liver into lysosomes and mitochondria. In this case, they used polyanionic substances such as heparin, chondroitin sulfate, polyvinyl sulfate, dextran sulfate and polyanetholesulfonic acid to prevent aggregation.

One obvious advantage of the technique is that separated zones remain sharply focused until their removal from the bottom of the cell. The method can also process considerable quantities of material, but at a considerable cost in ampholyte.

This technique appears to have considerable promise in the field of cell and subcellular organelle separation. The method is very mild to most cells since it avoids exposure to pH extremes and allows recovery of fully active and non-aggregated fractions. It is also rapid (a few minutes of sample flow-through), and offers an alternative to conventional techniques such as differential or density gradient centrifugation.

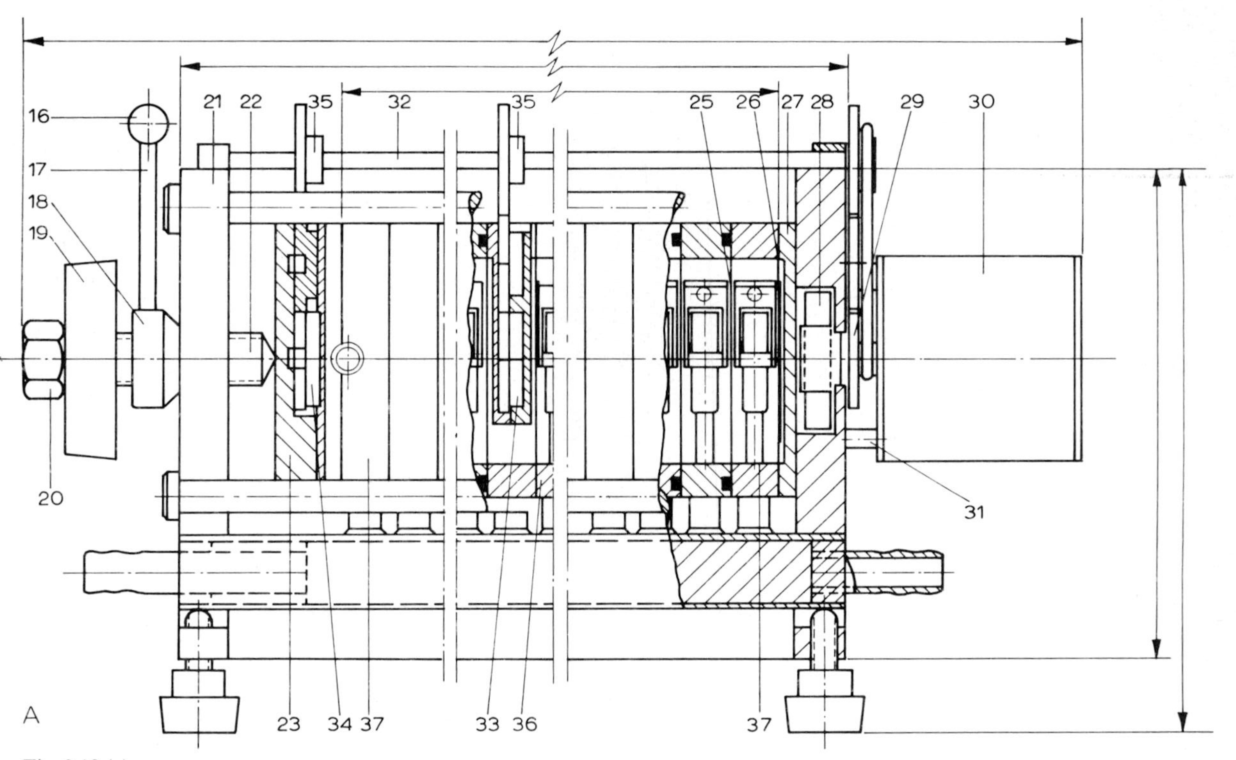

Fig. 2.10 (a).

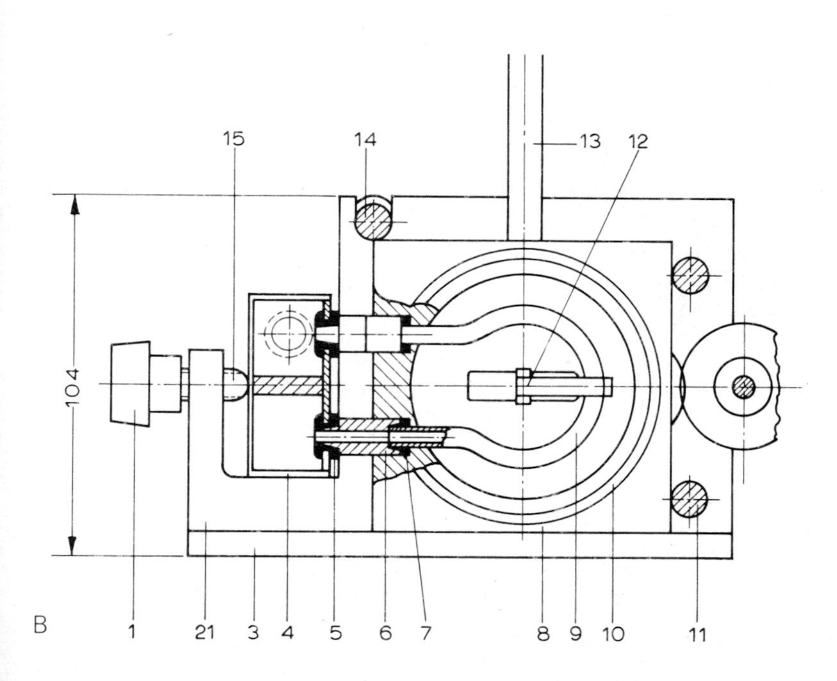

Fig. 2.10. Multicompartment electrolyser with closed cells, internal cooling and stirring. (a) front view; (b) side view; (1) two knurls, which operate two screws (15) to press the cooling water distributor (4) against the perspex tubes (6) used for the cooling loops (9) of each cell; (2) and (21) end plates; (3) base plate; (5) rubber rings fitting to the perspex tubes (6); (7) O-ring-sealed holes, used to fix the cooling loops (9) of each cell; (8) one of the 20 compartments of the electrolyser; (10) O-ring connecting each cell to the adjacent one; (11) and (14) two of the three pull rods which hold together the 20 cells; (12) bearings for the stirring magnets, affixed to the cooling glass loops (9); (13) gas-escape tubes for the hydrogen and oxygen formed at the electrodes; (16) to (20) lever and locking nut system which operates the screw (22) acting upon the brass plate (23); (26) platinum net electrodes, glued to the two end walls (27) of the two end compartments (37); (28) magnet driven directly by synchronous motor (30); (32) axle used to drive the magnet (34); (33) one of the 20 internal capsulated stirring magnets.

(From Rilbe et al. 1975, by permission of ASP Biol. Med. Press B.V.)

2.2. Preparative IEF in gels

2.2.1. Fawcett's continuous-flow apparatus

Fawcett (1973) has described two types of continuous-flow apparatus. One uses a packed Sephadex bed with vertical flow. The other uses a continuously flowing density gradient which is pumped horizontally through a vertical trough. Although good resolution is achieved with both, we will limit our discussion to the former since the latter requires rather complicated equipment. For example, as many as 54 gradient fractions are pumped in and out of the apparatus and a pump with at least 108 channels is required. By contrast, gel stabilized layers require a rather simple experimental design. In one version, Gianazza et al. (1975) have fractionated synthetic carrier ampholytes in a chamber with only 12 outlets without recirculation of anolyte and catholyte. This system can be run for weeks with very little attention by arranging a continuous sample input fed by constant hydrostatic pressure via a Mariotte flask.

Fawcett's technique is based on the continuous-flow electrophoresis method first described by Svensson and Brattsten (1949) and by Grassman (1950). The difference between the two principles is illustrated in Fig. 2.11. The separation chamber consists of two cooling plates 23 cm

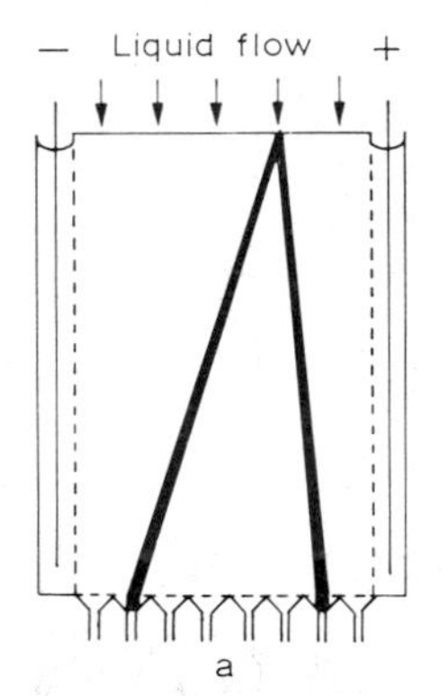

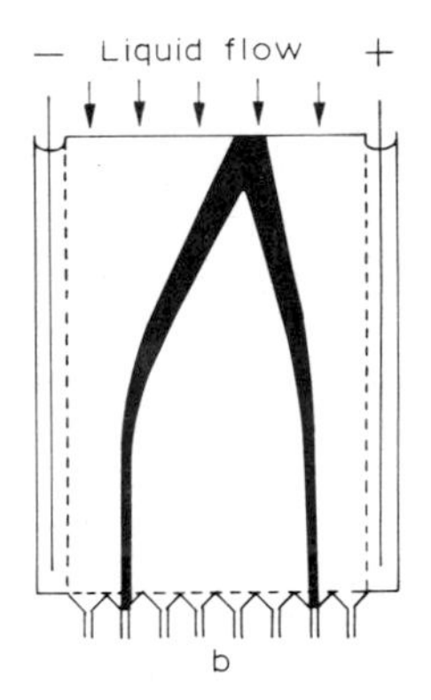

Fig. 2.11. Diagram illustrating the principle of (a) continuous-flow electrophoresis and (b) continuous-flow isoelectric focusing. (From Fawcett 1973, by permission of the New York Academy of Sciences.)

wide, 30 cm high and 0.3 cm apart. Semipermeable membranes fixed to the sides separate the trough from the electrode vessels. The membranes are porous polyethylene sheetings, impregnated with polyacrylamide gel. The cell is packed with Sephadex G-100 beads (or polyacrylamide gel particles) supported by a filter membrane. The 54 exit tubes on the chamber floor are connected to a multichannel peristaltic pump, built on the delta principle.

Continuous-flow IEF offers distinct advantages over density gradient stabilized columns. The zones remain under focus from the applied potential right until removal from the apparatus. They are, therefore, not subjected to diffusion effects or to remixing during elution as commonly occurs in density gradient columns. Large quantities of material can be fractionated by continuous-flow IEF, since high concentrations of protein can be tolerated in focused zones. Fawcett reports separations of 500 mg protein/day. Furthermore, the system can be run in a cascade form, that is, first in a wide pH range for an initial screening and then, sequentially, in a narrower pH gradient for better separation of components of interest.

2.2.2. IEF in granulated gels

Radola (1973a,b, 1975) has described an excellent method for large-scale preparative IEF in troughs coated with a suspension of granular gels, such as Sephadex G-75 'superfine' (7.5 g/100 ml) or Sephadex G-200 'superfine' (4 g/100 ml) or Bio-Gel P-60, minus 400 mesh (4 g/100 ml). The trough consists of a glass or quartz plate at the bottom of a lucite frame (Fig. 2.12). Various sizes of trough may be used; typical dimensions are 40 × 20 cm or 20 × 20 cm, with a gel layer thickness of up to 10 mm. The total gel volume in the trough varies from 300 to 800 ml. A slurry of Sephadex (containing 1% Ampholine) is poured into the trough and allowed to dry in air to the correct consistency. Achieving the correct consistency of the slurry before focusing seems to be the key to successful results. Best results seem to be given when 25% of the water in the gel has evaporated and the gel does not move when the plate is inclined at 45°. The plate is run on a cooling block maintained at 2–10°C. The electrical field is applied

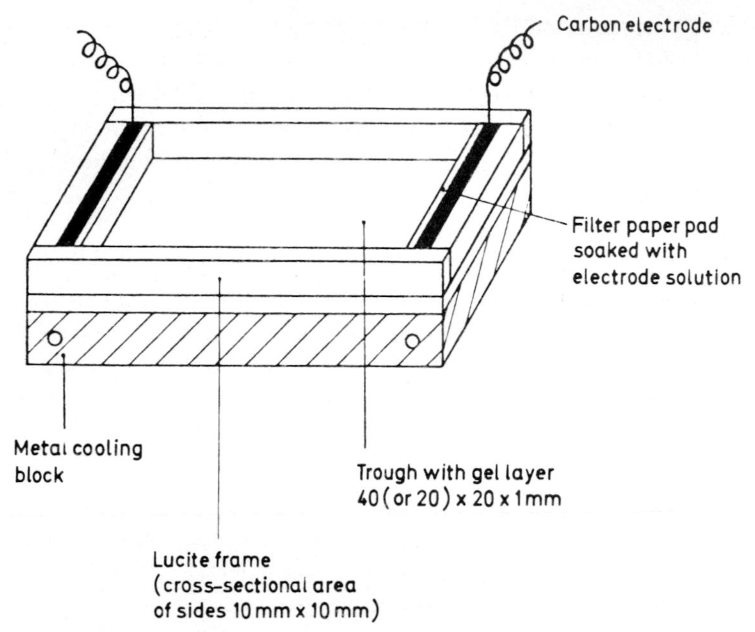

Fig. 2.12. Schematic representation of preparative isoelectric focusing in a trough of granulated gels. Notice that the carbon electrodes and filter paper pads are resting on their side. (From Radola 1975, by permission of Butterworths.)

via flat carbon electrodes or platinum bands which make contact with the gel through absorbent paper pads soaked in 1 M sulphuric acid at the anode, and 2 M ethylene diamine at the cathode. In most preparative experiments initial voltages of 10–15 V/cm and terminal voltages of 20–40 V/cm were used. As much as 0.05 W/cm^2 were well tolerated in 1 cm thick layers. Samples may be mixed with the gel suspension or added to the surface of preformed gels either as a streak or from the edge of a glass slide. Larger sample volumes can be mixed with dry gel (approx. 60 mg Sephadex G-75 or 30 mg Sephadex G-200 per ml) and poured into a slot in the gel slab. Samples may be applied at any position between the electrodes. Ideally, the gel slab should be covered with a lid to prevent it from drying out. A suitable apparatus, the 'Double Chamber' is now available from Desaga, Heidelberg, W. Germany.

After focusing, proteins are located by the paper print technique (Radola 1973a). Focused proteins in the surface film of the gel are absorbed onto a strip of filter paper which is then dried at 110°C. The proteins may then be stained directly with dyes of low sensitivity such as Light Green SF or Coomassie Violet R-150 after removing ampholytes by washing in acid. Fig. 2.13 gives examples of the resolution

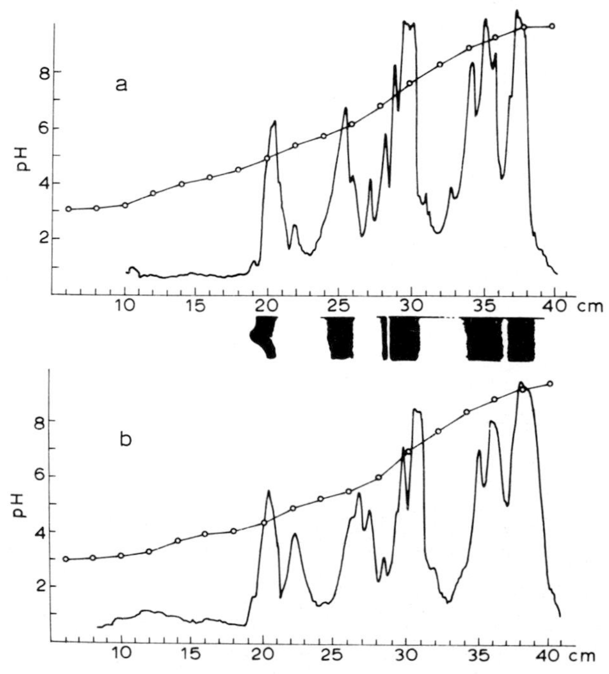

Fig. 2.13. Preparative IEF of 2 g protein in a 40 × 20 × 1 cm trough in pH 3–10 Ampholine. Gel stabilization: Sephadex G-75 'superfine', thickness of the gel layer: 1 cm. Proteins (from left to right): ovalbumin, horse myoglobin, ribonuclease and cytochrome C, 500 mg of each. Focusing (a) 400 V for 20 hr followed by 800 V for 10 hr; (b) 400 V for additional 10 hr and 800 V for 2 hr. Densitogram of a print stained with light green SF. (o–o–o–o) pH gradient. (From Radola 1973b, by permission of the New York Academy of Sciences.)

obtainable with this method. Alternatively, proteins can be detected directly in gels cast on a quartz plate by densitometry in transmission at 270–280 nm. The pH gradient in the gel can be measured in situ with a combination microelectrode sliding on a calibrated ruler.

Radola's technique offers the advantages of combining high resolu-ion, high sample load and easy recovery of focused components. As much as 5 to 10 mg protein per ml gel suspension may be fractionated in wide pH ranges. Radola (1975) fractionated 10 g pronase E in 800 ml of gel suspension and obtained excellent band resolution at this remarkable load capacity. At these high protein loads, even uncolored samples can be easily detected, since they appear in the gel as translucent zones. Proteins are easily recovered in a small volume at relatively high concentrations by elution. The absence of sucrose is a further advantage in the subsequent separation of proteins from ampholytes. As there is no seiving effect for macromolecules above the exclusion limits of the Sephadex, high molecular weight substances, such as virus particles, can be focused without steric hindrance. The system has a high flexibility, since it allows analytical, small scale and large scale preparative runs in the same trough, merely by varying the gel thickness.

When no suitable methods are available for detecting specific biological activities by the paper print technique, it is usually necessary to fractionate the gel and test for activity in eluates. This can be a rather laborious procedure. Fawcett (1975a) has described an alternative approach to IEF in granulated gels with semiautomatic sample collection. Instead of a flat bed, he used a vertical glass column such as described by Brakke et al. (1968) to support the granulated gel. After IEF, the upper electrolyte is removed and the Sephadex gel is displaced upwards by a dense sucrose solution and into a fraction-collecting device. The fractionating device removes aliquots of the emerging gel and the fractions are washed through a funnel and into a collecting tube (Fig. 2.14). Fawcett was able to obtain fractions equivalent to 3 mm sections of the gel with only moderate disturbance of focused zones.

For simplicity, flexibility, high resolution, and high load capacity,

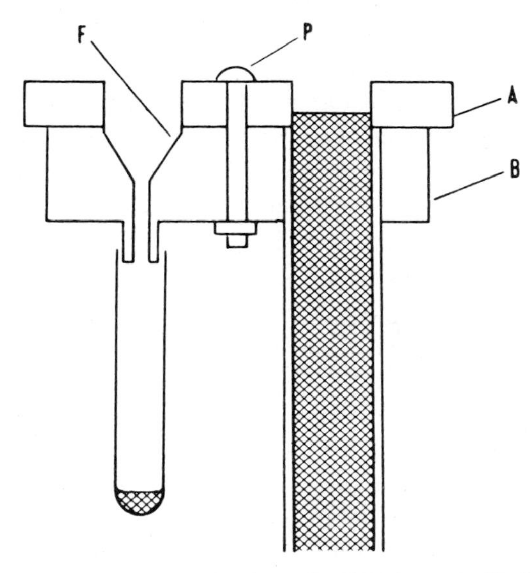

Fig. 2.14. Apparatus for collecting fractions after IEF in a tube filled with Sephadex G-200. The fraction cutter has two Teflon discs (A) and (B), the lower one (B) being fixed to the column. A dense solution pumped into the bottom of the column desplaces the Sephadex up into the top disc (A) which is then rotated about the pivot (P) and the Sephadex fraction washed through the funnel-shaped hole (F) into the collecting tube. (From Fawcett 1975a, by permission of Butterworths.)

Radola's technique has much to offer. Yet, even though this technique was described as long ago as 1969, few laboratories have adopted it. Its major defect was that pH gradients were unstable and distorted zones were often obtained. These problems were thought to be due to carboxyl groups on the Sephadex matrix which Radola (1973a) suggested might be removed by treatment with propylene oxide. This problem has been thoroughly investigated by Winters et al. (1975) who have solved most of the drawbacks of this technique. They have stressed three important points in the practical procedure: (1) the Sephadex itself has to be pre-washed; (2) the Sephadex gel should be of the superfine grade, G-75 being the best when the print technique has to be used; (3) the final water content of the gel has to be carefully controlled. If too dry, the gel bed cracks, if too wet the protein zone tend to sediment into the lower part of the gel.

Pre-washing of the Sephadex is extremely important. Most Sephadex batches contain charged low-molecular weight contaminants which interfere with the pH gradient formation. Many of these contaminants are removed by washing swollen Sephadex with distilled water on a Buchner funnel. The water within the gel is then displaced by ethanol and the gel is subsequently dried under vacuum.

Winters et al. (1975) have also developed a trough, for preparative IEF in Sephadex gels, which can be fitted onto the LKB Multiphor 2117 apparatus (see §3.2.14). This trough, equipped with a sample applicator and with a stainless steel frame divided into 30 channels, for gel fractionation, is now commercially available from LBK Produkter AB (see Fig. 2.15). This is a useful addition, since it allows analytical and preparative runs with the same equipment. When using this trough, Winters et al. (1975) suggest gel thickness up to 5 mm, since a greater thickness will result in band distortion

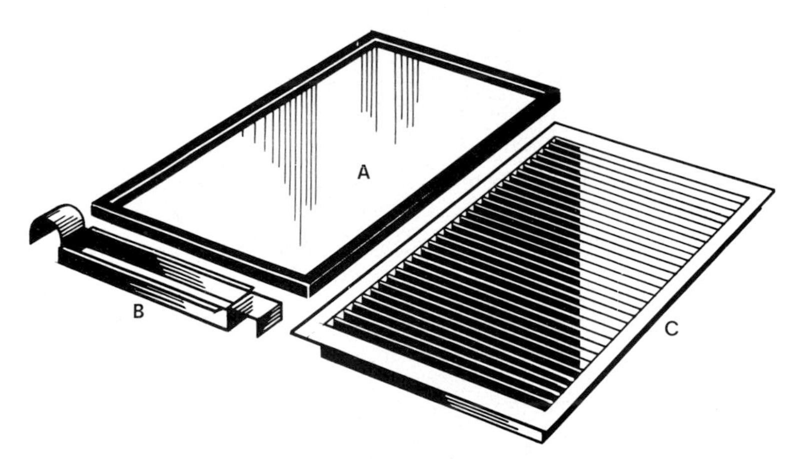

Fig. 2.15. Trough (A), sample applicator (B) and fractionation grid (C) for preparative IEF in granulated gels using the LKB 2117 Multiphor apparatus. After IEF, grid (C) is lowered on trough (A) thus fractionating the gel layer into 30 fractions. Each fraction is scooped up from each channel with a spatula. (Courtesy of LKB Produkter AB.)

due to more efficient cooling at the bottom as compared with the gel layer at the top of the gel slurry.

The sample load in this trough ranges from 200 to 400 mg protein, depending on sample heterogeneity. The gel slurry (approximately 5 g dry gel in 100 ml of 2% Ampholine) is poured into the tray, without waiting for complete gel swelling. Water is evaporated from the slurry with the help of a small, desk-type fan mounted 1 m above the tray, until 20–25% of the water in the gel has evaporated (this requires 2–3 hr). The gel is then transferred to the cooling plate of the Multiphor, the sample applied with the sample applicator (see Fig. 2.15) and the experiment run at 10°C, for about 10 hr, at 10 W, with a constant wattage power supply. The electrodes are usually placed on the short side of the trough, so that proteins are separated along the long axis (25 cm). Thirty fractions are collected by compartmentation of the gel with the aid of the fractionation grid (see Fig. 2.15) and by scraping off the gel with a spatula. The protein is eluted from the gel simply by placing the gel fraction into a syringe equipped with glass-wool as bottom filter, adding a volume of eluant equal to the gel volume, and ejecting the eluant with the syringe piston (Winters et al. 1975).

K. Saravis, W. Otavsky and J.W. Drysdale (unpublished) have recently explored other charge free anticonvective media for use in granulated gels. They have obtained highly encouraging results with flat beds made of Pevicon, an inert plastic in the form of granules 100 μ in diameter. This material has little if any electroendosmosis and forms very stable beds with high liquid retention. It may be packed by gravity or with thin layer spreaders and forms a cake which may be cut with excellent precision. Focused proteins are readily recovered by centrifugation through a coarse membrane. This material shows great promise for both analytical and preparative uses.

2.2.3. IEF in multiphasic columns

Recently, Stathakos (1975) has designed a new type of vertical column for multiphasic isoelectric focusing. The idea is somewhat similar to the principle of multicompartment electrolysers described by Rilbe et al. (1975). The column is designed on the principle of building blocks, i.e., it is composed of alternating separable gel segments and liquid interlayers extending between two electrode compartments. The column, constructed in a modular fashion, can be shortened or lengthened according to experimental needs. The scheme of the apparatus, assembled in six units of equal size, is shown in Fig. 2.16. This type of column allows great experimental flexibility and offers several interesting features. Samples can be introduced practically anywhere in the column: at the top, in any of the liquid interlayers (2.5 mm high) between adjacent blocks, or in one or more of the building blocks. The pH can be monitored at any time through the various liquid interlayers (which form small chambers with inlet and outlet nipples, as shown in Fig. 2.16) by withdrawing a few μl of liquid or by pumping out the whole solution, and returning it after pH determination. Since the porosity of each gel unit can be varied independently, unwanted compounds in the sample can be separated by making the gel 'restrictive' for them, or vice versa. The proteins collected in any gel block can be further purified by rerunning in a shallower pH gradient simply by incorporating this gel block into a second column. Finally, the proteins collected in a block can be retrieved electrophoretically, and at the same time separated from Ampholine, by a method similar to the one described by Suzuki et al. (1973).

2.2.4. IEF in polyacrylamide gel cylinders

Because of the excellent resolution and load capacity afforded by IEF in cylinders of polyacrylamide gel, we have scaled up an analytical apparatus (Righetti and Drysdale 1973) for preparative purposes by substituting an interchangeable core with larger tubes (see Fig. 2.17). The tubes are held between rubber grommets in a water-tight compartment through which coolant at 1°C is circulated. The platinum electrodes are circular and are positioned close to the extremities of the tubes to minimize loss of ampholytes from the gel and thus reduce the cathodic drift of the pH gradient. The core accomodates six gels: 3 of 50 ml capacity (2 cm I.D.), 2 of 20 ml (1.2 cm I.D.) and an indicator gel of only 2 mm in diameter and 2 ml capacity. All gels are cast in 16 cm long glass or plastic tubes. High porosity gels ($T = 4\%$;

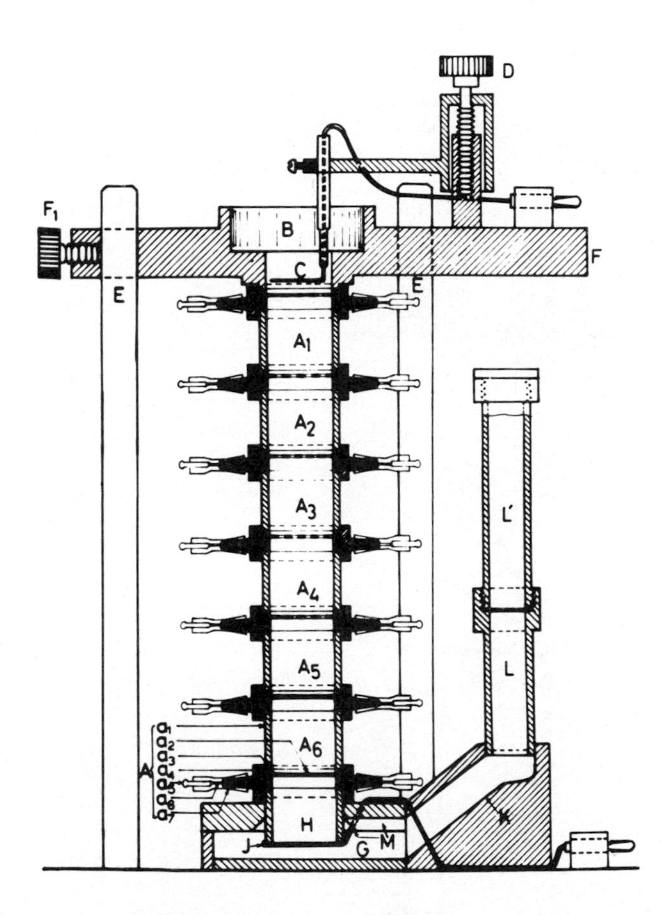

Fig. 2.16. Vertical section through a six-unit column for multiphasic isoelectric focusing. A_1–A_6, separable, independent gel units; a_1, tubular part; a_2, supporting membrane; a_3, silicone rubber O-ring; a_4, structural ring; a_5, glass plug; a_6, tygon jacket; a_7, outlet nipple; B, upper electrode chamber; C, upper electrode, movable vertically be means of screw D; E, metal rods; F, upper platform; F_1, screw for securing platform; G, lower electrode chamber; H, base cylinder; J, lower electrode; K, sloped connection to tube L; L, auxiliary counterpressure tube; L′, tube extension; M, groove for escape of electrolysis gases. (From Stathakos 1975, by permission of ASP Biol. Med. Press B.V.)

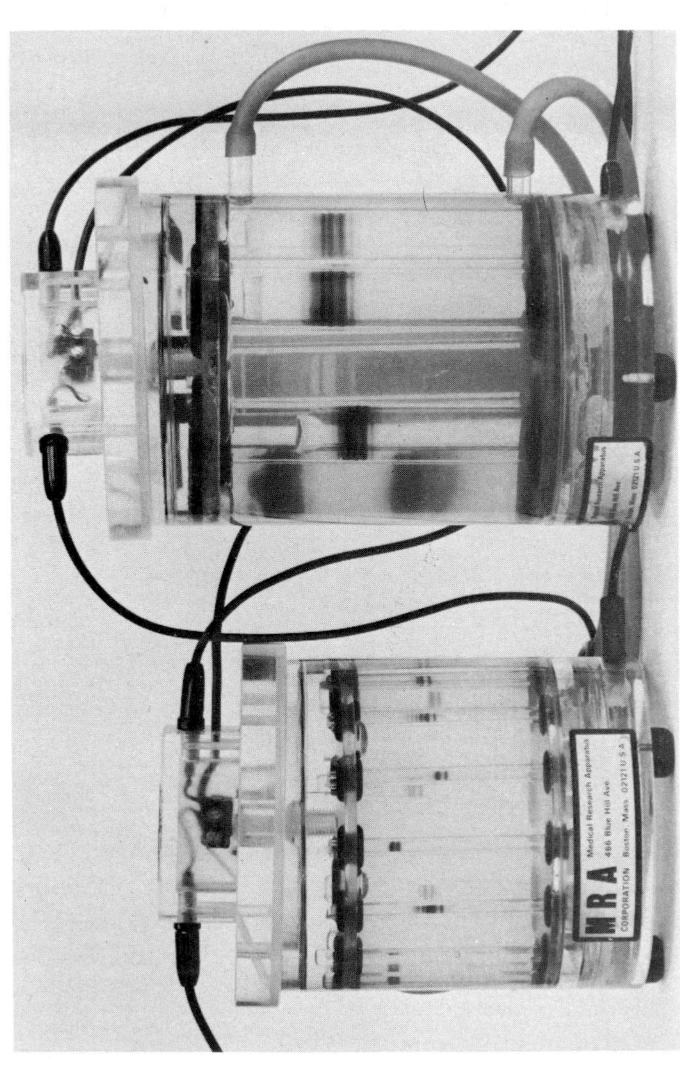

Fig. 2.17. Analytical (left) and preparative (right) apparatus for gel IEF. The analytical apparatus holds 12 gels of 10 × 0.3 cm (I.D.). The preparative apparatus holds six gels: two of 20 ml capacity, three of 50 ml capacity and one indicator gel of 2 ml capacity. The central core is interchangeable between the two units. The salient features of both apparatus include efficient gel cooling and small electrolyte compartments. (From Righetti and Drysdale 1973, by permission of the New York Academy of Sciences.)

$C = 4\%$) (nomenclature of Hjertén, 1962, see page 434) are used to minimise molecular sieving effects. After polymerisation gels are stored at 4°C before use. The maximum sample load in the larger tubes is approximately 200 mg protein. Samples may be applied directly to the top of the gel or may be incorporated into the gel solution before polymerization. Resolution is comparable to that given by the analytical system. As in other gel systems, so much protein can be focused in a zone that uncolored proteins are often clearly visible as opaque discs (see Fig. 2.17). The methodology is essentially that described in detail for the analytical system (§3.2.7). However, because of the poorer heat transfer in the thicker gels, power inputs should not be increased proportionally to gel volumes. Satisfactory results are usually given by running at an initial voltage of 100 V for 12 hr, then for a further 12–18 hr at 250 V. An adaptation of this apparatus, shown in Fig. 2.17, is commercially available from MRA, Boston, Mass.

One of the major drawbacks of preparative IEF in polyacrylamide gels is in sample detection and recovery after IEF. Sectioning the gel and eluting with buffer can give incomplete sample recoveries and high dilutions. As an alternative, Suzuki et al. (1973) have described an ingenious method for sample retrieval after IEF in gels. They seal one end of a gel tube with dialysis tubing and introduce a narrow layer of a dense buffer. A polymerising solution of polyacrylamide is floated on top of the buffer to form a short plug of polyacrylamide gel. The desired section of the focused gel is now minced and packed into the tubes on top of the stacking gel in dense buffer solution. The tube is then filled with buffer and the sample recovered by electrophoresis in the same apparatus used for IEF. This allows quantitative sample recovery, in a highly concentrated form, while also eliminating ampholytes which pass through the dialysis membrane. A similar apparatus has also been described by Chrambach et al. (1973).

2.2.5. IEF in polyacrylamide gel slabs

In addition to Sephadex beds, it is also possible to conduct preparative IEF in slabs of polyacrylamide gel. Because of the superior heat dissipation in slabs, this approach may be preferable to preparative IEF in large rods of polyacrylamide which are difficult to cool efficiently. Moreover, focusing in an acrylamide slab may give slightly superior resolution to Sephadex beds (Graesslin and Weise 1974). The apparatus required is similar to that for analytical IEF in thin slabs of polyacrylamide, the thickness of the gel being altered by use of appropriately larger spacer frames or gaskets. Samples may be applied directly to the gel surface as a streak or on an absorbent paper rectangle. Alternatively, samples may be applied in small slits or troughs, either in paper or in free solution. The resolution given by this method is comparable to that given in thinner analytical gels. Focused proteins on quartz plates may be detected by direct UV scanning with a Zeiss spectrometer. It is also possible to assess the position of proteins from rapidly stained longitudinal strips taken from both edges of the sample track (see Fig. 2.18). The proteins can be recovered from the gel by elution or by electrophoretic procedures (p. 520). Although many

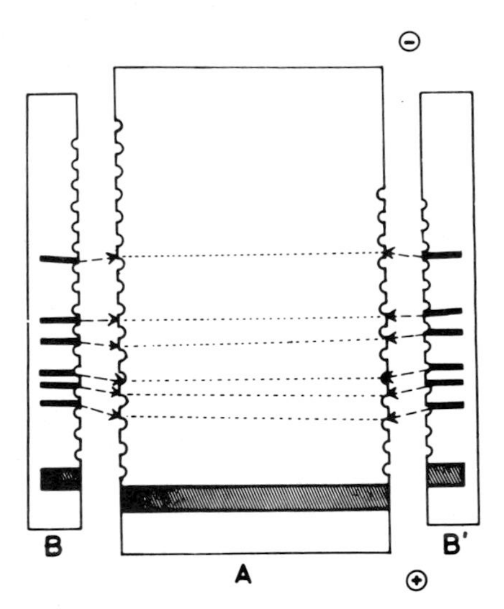

Fig. 2.18. Technique for localization of separated proteins in preparative thin-layer IEF. A, unstained middle part of the gel slab. B and B′, Coomassie blue stained gel strips. The places of corresponding protein bands are indicated by marker holes. (From Graesllin and Weise 1974, by permission of W. de Gruyter.)

investigators have reservations about the efficiency of eluting proteins from polyacrylamide, Graesslin and Weise (1974) obtained 74% recovery of total proteins by elution with gentle agitation.

2.2.6. Concluding remarks – load capacity

The limited survey presented here clearly attests to the ingenuity of investigaors in designing, building or modifying apparatus to capitalise on the excellent resolution offered by IEF in adapting the technique for both analytical and preparative purposes. A recent article by Fawcett (1975a) makes excellent reading for other suggestions for constructing simple but effective systems for preparative IEF.

We have seen that in some methods it is possible to fractionate large amounts of proteins, in the gram scale. In the field of IEF, it has generally been believed that sucrose density gradients could only carry a limited amount of protein. As a representative figure, in the 440 ml LKB column, a maximum load of only a few hundred mg protein is recommended. However, in view of some recent results and theoretical considerations by Rilbe and Pettersson (1975) (see also §1.3.8) we think that this point of view should be changed. According to these authors, the theoretical mass content of a protein zone (m) in a linear sucrose density gradient ranging from 0 to 0.5 g/cm^3, cannot exceed the following inequality:

$$m < 0.625\ Vr^2\ \mathrm{g/cm^3}$$

where V is the total column volume and r is the zone breadth. For a 100 cm^3 column, one has:

$$m < 0.625\ r^2\ \mathrm{g}$$

Thus, the load capacity rises with the square of the zone breadth. Rilbe and Pettersson (1975) have verified this experimentally. In a 110 ml column, for an r value of 0.125, according to the above inequality, one should be able to load a maximum of 1074 mg protein in a single zone. In fact, working with myoglobin, these authors have loaded 1050 mg in the column, and approximately 800 mg (that is, 76% of the applied sample and 74% of the theoretical maximum) were confined within the main peak (MbI) (see also p. 356).

Janson (1972) has reported the fractionation of 7.3 g of cytophaga Johnsonii cytoplasmic extracts in a 440 ml sucrose density gradient. Fawcett (1975a) using a narrow-range (pH 6.3 to 7.3) Ampholine, in the 110 ml LKB column, has applied up to 1 g of protein from red cell haemolysates (see also §1.3.8). These high loads in liquid systems would appear to compare well with gel systems. However, at these high sample inputs, protein precipitates may form and sediment along the density gradient. Therefore, high loads in liquid systems are of limited applicability, that is when the protein of interest focuses away from the zone of heavy precipitates. Thus, with regard to protein carrying capacity, gel systems appear to be superior to liquid systems. Although precipitates may form, they are confined to the pI and usually do not interfere with the behaviour of other proteins.

On the basis of these considerations, and of the equipment presently available, we think IEF has the prerequisite to become one of the leading techniques for large scale separations of macromolecules.

Analytical IEF

A good analytical system should conveniently provide high resolution separations at a low cost in time and materials. Ideally, the system should not be too demanding in experimental technique and should allow simultaneous processing of multiple samples and a convenient means for assessing and recording data. Although the original systems for IEF in vertical density gradients were used for both analytical and preparative purposes, they were not very well suited for routine screening and analyses of multiple samples. The smallest column then commercially available had a volume of 110 ml and required substantial amounts of material for analysis. Moreover, a typical experiment usually took several days to complete. Electrolysis periods were of 2–3 days' duration and the subsequent analyses of fractions eluted from the gradients were tedious and laborious. Several attempts were made to develop smaller columns for more rapid and convenient analyses. Many ingenious methods were devised but few were of general applicability. In addition to the inconvenience in analysing gradients after fractionation or with flow cells, most systems were not so suitable because of some problems associated with the use of sucrose density gradients. These problems included convective mixing and isoelectric precipitation during electrolysis and loss of resolution by diffusion or by mixing on elution. Resolution was also lost by collecting fractions with a greater volume than that of focused zones. Although many of these difficulties may be overcome by e.g. scanning the column in situ or by using flow cells to monitor eluates, few if any

Subject index p. 587

systems in sucrose density gradients met the day to day requirements for a convenient rapid, small-scale method for simultaneous fractionation of multiple samples.

This section will first deal with analytical IEF in liquid media and will then cover more thoroughly the field of IEF in gel media, since this last system, in the analytical scale, has at present attained wide popularity. It is somewhat difficult to draw a line between analytical and preparative techniques. In theory, any analytical technique which allows sample recovery could also be called preparative. The term 'preparative' depends also on the experimental needs, either in the laboratory scale or at the industrial level. Some of the systems for IEF in liquid media, described in next section, which we have grouped under 'analytical' techniques, could also be used for laboratory scale, preparative applications. We have therefore drawn an arbitrary line between analytical (in the μg to mg scale) and preparative (in the mg to g scale) techniques.

3.1. IEF in small density gradient columns

Parallel to the development of IEF in polyacrylamide gel, several research groups explored other designs and apparatus for IEF in small-scale density gradients. Weller et al. (1968) built two small U columns for small scale IEF, one with a diameter of 1.0 cm and a working volume of 11 ml, the other with a diameter of 1.9 cm and a working volume of 33 ml. In these U-tubes, one of the arms (the dummy arm) contains only the electrolyte (usually catholyte) and the other (separational arm) the sucrose density gradient to perform the separation and the anolyte. The U-tube is usually cooled in a cold room or immersed in a tank with cold water. When the system is at equilibrium, the anode is removed and a siphon inserted in the ground joint of the separation arm. When solution is added to the dummy arm, a siphon is formed and the column emptied dropwise in a series of test tubes.

A similar technique for rapid IEF of multiple samples on a microscale has been described by Godson (1970). He performs IEF in J-tubes, made of glass of 1.1 cm outer diameter and 0.9 cm inner diameter, having a total volume of 10 ml. The J-tubes fit standard acrylamide gel electrophoresis apparatus, so that up to eight electrofocusing columns can be run at once (see Fig. 3.1). The long arm (25 cm) contains the sucrose density gradient and acts as the separation chamber. The short arm (16 cm) contains a cushion of heavy sucrose solution and, usually, terminates with the cathode solution. The entire J-tube is almost completely immersed in the lower electrolyte chamber, which is stirred with a magnetic bar and cooled by an outer cooling jacket. The two extremities of the J-tube are connected to the anolyte and catholyte by short platinum wire loops. At the end of the experiment, the column is fractionated by fitting a conical glass funnel to the top of the long arm of the J-tube, and pumping heavy sucrose into the short arm. Similar U-tube systems have been described also by Koch and Backx (1969).

In the last few years, Rilbe's group has been developing increasingly short and strong density gradients in conjunction with stable natural pH gradients. Rilbe (1970) has described a parallelepipedic quartz column 14 cm long with a volume of 11.2 ml. Focusing in this column could be completed in 6–7 hr, but the price of the quartz column and its accessories was very high. In a search for commercially available apparatus, Rilbe's group found that a spectrophotometric cell designed for flowing solutions and equipped with mantles for thermostating medium satisfied practically all demands to be put on a column for IEF. Rilbe (1973b) has described for use in IEF the flow cuvette type code 167-QI from Hellma GmbH (Müllhein, Baden, West Germany) (see Fig. 3.2). The central compartment, useful as a column for IEF, has the dimensions $4 \times 10 \times 35\ \mathrm{mm}^2 = 1.4$ ml. Thus focusing in this cell requires a very small volume and can be completed in a very short time, usually less than 2.5 hr, inclusive of preparations of solutions for the density gradient and of absorption scanning. Since IEF is performed in a spectrophotometer, scans can be made during or after focusing simply by equipping the spectrophotometer with a cell elevator and a fixed horizontal slit. To this purpose, Rilbe has used a Vitatron Universal Photometer. In addition to scans, spectra can also be taken of individual protein bands: this allows the study of chemical

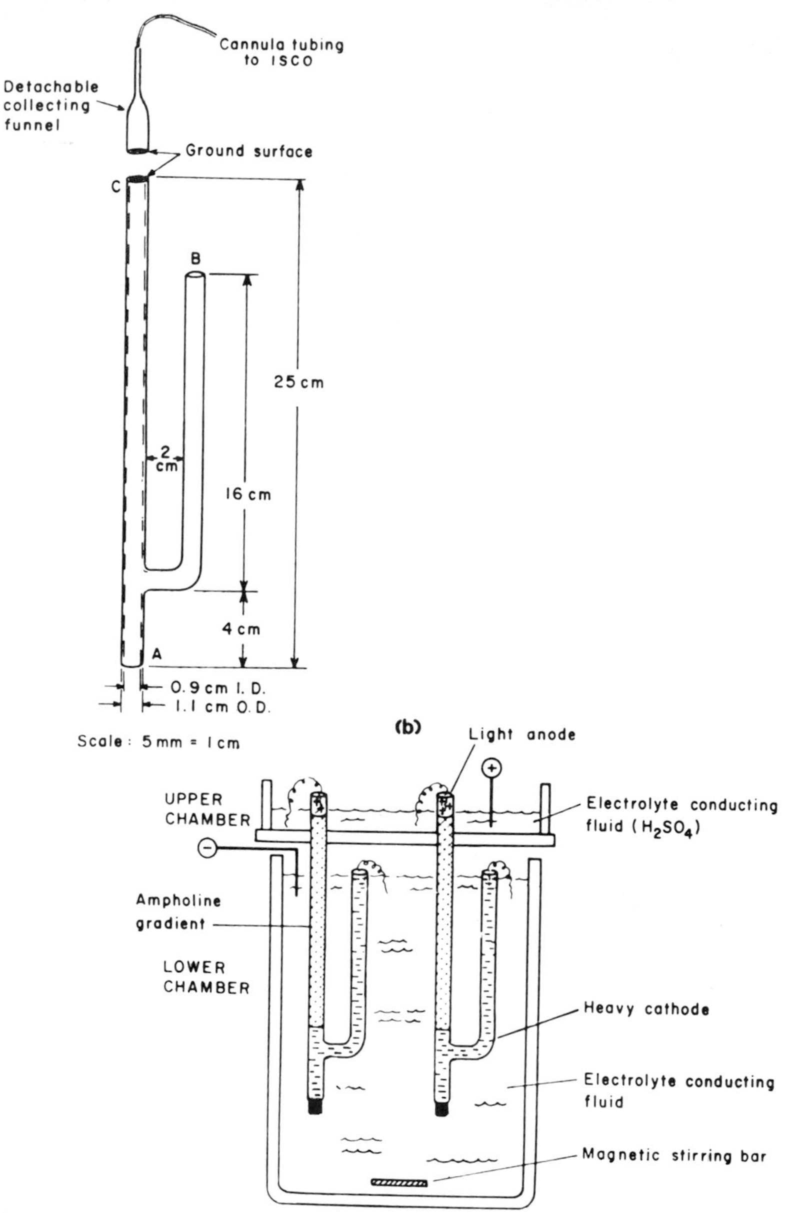

Fig. 3.1. J-tubes for isoelectric focusing in a acrylamide gel apparatus: (a) dimension of the J-tube and arrangement for collecting; (b) J-tubes in their running position. (From Godson 1970, by permission of Academic Press.)

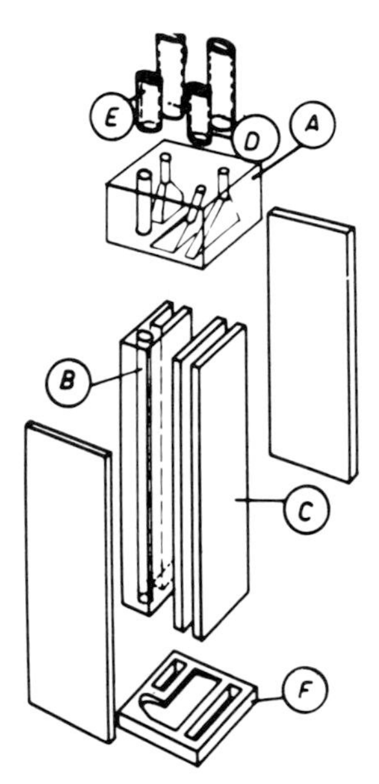

Fig. 3.2. Exploded view of the cell 167-QI. A, top piece; B, vertical channel to the bottom of the column; C, central piece with column and two cooling mantles; D, tube for insertion of top electrode; E, tube for insertion of bottom electrode; F, bottom piece. There is a tunnel connection, not visible, between the two outer grooves for cooling water. (From Rilbe 1973b, by permission of the New York Academy of Sciences.)

reaction of proteins when reactive ionic species are migrated through already focused proteins. Thus, by injecting a small amount of sodium dithionite at the cathode, Rilbe (1973b) was able to monitor the conversion of ferric sperm whale myoglobin into the ferrous form. At this point, the addition cf a small amount of potassium ferricyanide at the cathode reconverts the ferrous into the ferric form, as seen by the specific spectra of the two forms. The evaluation of the pH course in this small volume cells would appear to be a very difficult proposition. Yet Fredriksson (1972) has described a microfractionation method which allows the recovery of 60 μl fractions from the flow-cell used for IEF. For pH measurements, the Radiometer E 5021 microelectrode unit (Radiometer, Copenhagen, Denmark) was used. The pH sensitive glass membrane is shaped as a horizontal capillary tube that, via a vertical polyethylene tube, can be filled simply by suction. In this electrode, as little as 20 μl of sample volume are required for pH measurements.

3.2. *Analytical IEF in gel media*

3.2.1. *IEF in granulated gels and in agarose matrices*

Sephadex gels have proved very useful for preparative (see §2.2.1 and §2.2.2) and analytical procedures for IEF. This medium effectively reduces convective disturbances and zone instability. It also has the advantage of allowing essentially unrestricted migration of proteins in the external liquid phase. The major drawback was due to variability between different batches of Sephadex, some of them presenting a rather high electrosmotic flow. This problem has been solved by Winters et al. (1975) who have found that Sephadex contains charged, low-molecular weight contaminants which severely interfere with the pH gradient formation in IEF (see pp. 463–465). By simply pre-washing the gel with distilled water, they obtain a suitable Sephadex for IEF. Radola (1973a,b) has also explored the possibility of using granulated polyacrylamide gels, such as Bio-Gel P-60, minus 400 mesh. Practical aspects for focusing in layers of granulated gels will be given further on.

A very attractive anticonvective medium for IEF is agarose since, at the concentrations normally used (1 %), it allows practically unhindered migration of macromolecules in the multimillion molecular weight range. However, early attempts of IEF in agarose matrices (Riley and Coleman 1968; Catsimpoolas 1969a,b) were unsuccessful due to the presence, even in highly purified commercial product, of sulphate and carboxyl groups, which generated a severe electrosmotic flow. There was also a concomitant dissolution of the agarose matrix at the electrodes, and especially at the anode, in the presence of strong acids. Quast (1971) suggested to use agarose treated with anion-exchange resin, to increase the Ampholine concentration in the gel and to raise its viscosity by addition of sucrose. Johansson and Stenflo

(1971) have recommended the purification of commercial agarose by ion-exchange chromatography in DEAE-Sephadex and the incorporation in the gel of methylcellulose of high viscosity (7000 cps). These gels proved useful in immunoelectrophoresis, but were not tested in IEF. More recently, Lååas (1972) described the production of virtually charge-free agarose by alkaline desulphatization in the presence of sodium borohydride, followed by reduction with lithium aluminium hydride in dioxan. However, to our knowledge, this type of agarose has not been used in IEF. It appears that the procedure described does not allow complete removal of the reducing agents, which then interfere with the IEF separation.

A recent modification described by Johansson and Hjertén (1974) appears to be very promising for IEF in agarose gels. These authors purify commercial agarose by treatment with a strong anion-exchanger, QAE-Sephadex. This reduces the sulphur content in agarose to only traces. Electrosmotic flow is further diminished by incorporating in the gel 1 % (w/v) linear polyacrylamide polymer, or polyethylene oxide with a mean molecular weight of about 4×10^6 daltons. In these gels, focusing is achieved in less than 2 hr, even when using low voltage gradients (10 V/cm). However, residual endosmosis precluded runs of more than 5–6 hr. For best results, highly purified agarose should be used as a starting material. Johansson and Hjertén (1974) have found the best product to be agarose from l'Industrie Biologique Française. Similar results have recently been obtained by Weise et al. (1975) who purified agarose by the same procedure, but incorporate methylcellulose in the gels. If these techniques can be adopted for routine analysis, they will be extremely useful in IEF, especially when used in combination with an immuno-electrophoresis technique in the second dimension.

3.2.2. *Properties and structure of polyacrylamide gels*

Results from many laboratories have clearly indicated the considerable advantages of gel electrofocusing with its simplicity, remarkable resolution and versatility. In one simple step, IEF in gels overcame many of the problems of convective mixing, isoelectric precipitation

Subject index p. 587

and focused patterns. The technique requires only simple apparatus and experimental procedures and allows simultaneous fractionation and convenient analyses of multiple samples. Only a few micrograms of protein are required and equilibrium focusing conditions can be achieved in a few hours. Resolution is also usually superior to that given by conventional procedures for IEF in sucrose density gradients and analysis of banding patterns is greatly simplified. Also, many of the procedures previously developed for gel electrophoresis for quantitating banding patterns for fractionating gels to assay biological activity, radioactivity, etc., can be adapted for gel electrofocusing (GEF). Finally, many of the techniques for IEF in polyacrylamide gels can be used in conjunction with other procedures based on other physicochemical properties for more complete characterisation of protein mixtures.

It soon became apparent, however, that the same attractive anti-convective properties of polyacrylamide gels that allow such excellent resolution also introduce a new element – molecular sieving – which must not be overlooked. All polyacrylamide gels will retard the electrophoretic migration of most molecules – even buffer salts (Rodbard and Chrambach, 1970, 1971). Obviously, meaningful and reproducible results can only be obtained if all components reach their pI in the time allotted for the experiment. In theory, it should be possible to focus any molecule that will migrate electrophoretically in the gel. Although the small carrier ampholytes may migrate rapidly to equilibrium positions large molecules such as proteins may require considerably longer periods.

The structure of polyacrylamide gels has recently been elucidated by Rüchel and Brager (1975) with the aid of the scanning electron microscope (SEM). The SEM image of gel without cross-linker reveals longitudinal structures having long, thin fibrils, connected by slender filaments (Fig. 3.3). These fibrils, which in a polyacrylamide gel become organized as leaflets or membrane structures, represent most probably the sieving elements, since in gels of increasing acrylamide concentration they show a parallel increase in thickness. At concentrations of about 5% acrylamide, these structures are more like a sponge

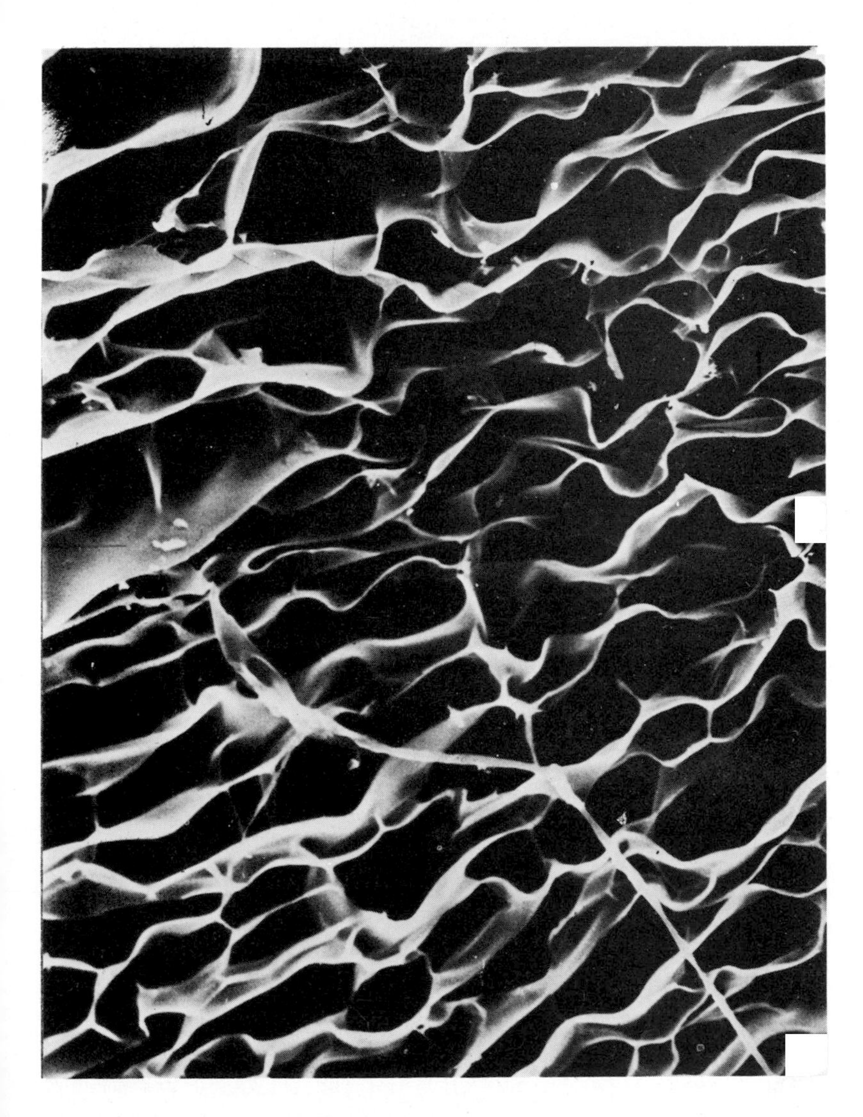

Fig. 3.3. Scanning electron microscope image of a polyacrylamide gel of low monomer concentration (2.5 % acrylamide). Notice the loose arrangement of sheets or fibers in the gel matrix. Magnification: 600 ×. (From Rüchel and Brager 1975, by permission of Academic Press.)

with numerous passageways. These findings are in good agreement with theoretical considerations on the length and structure of polyacrylamide fibers ('1-D' gels of Rodbard and Chrambach (1970) and 'chains' of Richards and Lecanidou (1974)). They also have some important implications on the structure of non-sieving gels.

3.2.3. Highly cross-linked polyacrylamide gels

Twenty percent cross-linked gels were first introduced by Ornstein (1964) and Davis (1964) in their discontinuous buffer systems, as 'upper' or 'stacking' gel. Even though not explicitly stated in the literature, the finding by these authors that such highly cross-linked gels exhibited a larger pore size than obtainable by gels of low percent cross-linking, was implicit in their application. Fawcett and Morris (1966), by using gels of high degree of cross linking in gel filtration, found a progressive increase in mean pore radii in gels of increasing % cross linker. This aspect was further investigated by Rodbard et al. (1972) who showed that, as the % cross linker increases from 10 to 50%, the retardation coefficients of macromolecules with widely differing molecular weights progressively approach each other and approach zero asymptotically.

Theoretical considerations concerning highly cross-linked gels lead to the assumption that the fiber length of the gel decreases with increasing rate of cross-linking (Fawcett and Morris 1966) and might be almost reduced to points ('O-D' gels of Rodbard and Chrambach (1970) or 'knots' of Richards and Lecanidou (1974)). This has been confirmed by the SEM work of Rüchel and Brager (1975). They have found that, by progressively increasing the % of cross linker in the gel, the fiber or membrane like structure progressively decreases, and bulb-like structures appear. A gel of pure bisacrylamide consists of multitudinous spherical units (knots) in random aggregation (Fig. 3.4).

These highly cross-linked gels allow the relatively unrestricted passage of multimillion molecular weight particles (Rodbard et al. 1972), therefore they appear particularly attractive for use in IEF, where equilibrium conditions have to be achieved and molecular sieving has to be kept to a minimum. Unfortunately, gels containing

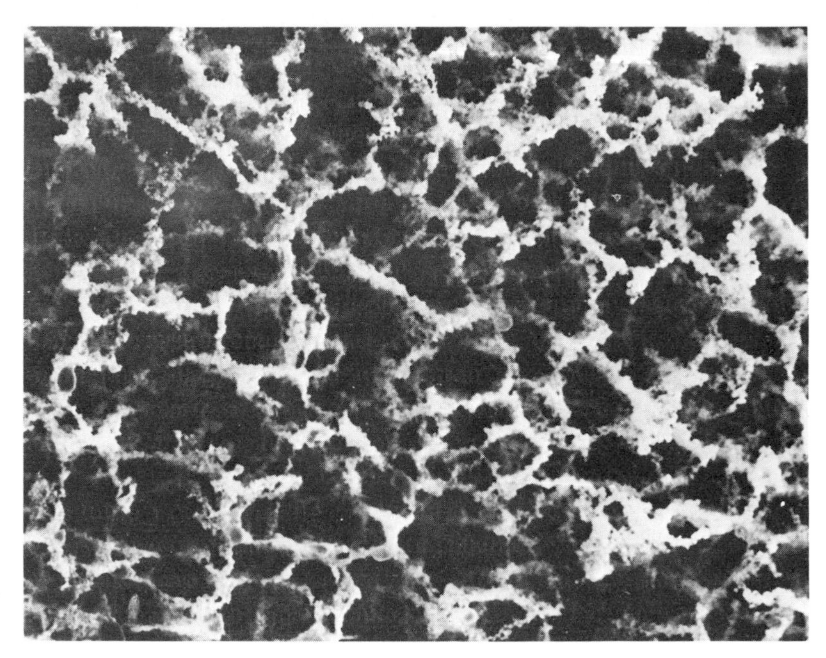

Fig. 3.4. Image of pure polymerized cross-linker (initial concentration 2% w/v of N,N′-methylenebisacrylamide). This gel becomes white and inelastic during polymerization. It can be seen that the membrane structure is lost (see Fig. 3.3) and that the gel appears as a random aggregation of spherical units. This is concomitant with a large increase in pore size. Magnification: 2250 ×. (From Rüchel and Brager, by permission of Academic Press.)

more than 15% N,N′-methylene bisacrylamide (Bis) are opaque and tend to be rather brittle. They are therefore not well suited for most analytical procedures. These problems may be overcome by using diallyltartaramide (DATD) rather than Bis as cross linker. High %*C* (DATD) gels are optically clear, are also more elastic than high %*C* (Bis) gels and can be sliced with adequate precision. DATD also has another advantage in that its gels may be dissolved in periodic acid (Chrambach and Rodbard 1972), a useful property which greatly facilitates many procedures for sample detection after focusing (see § 3.3.1). Baumann and Chrambach (1975) report that gels with 5% *T* and 15% C (see §3.2.6) (as DATD) are non-restrictive for molecules up to 0.5×10^6 daltons. However, gels of this composition have a smaller porosity than gels with comparable levels of Bis and they are actually more restrictive than gels with low %C e.g. $T = 4\%$ $C = 4\%$ (Bis) (Drysdale, unpublished). We have as yet no explanation for this apparently anomalous result. It is unlikely to be due to the presence of ampholytes during gel polymerisation since we have obtained the same result on electrophoresis without ampholytes. Hopefully, this problem will soon be resolved or perhaps another cross-linker combining the desirable properties of Bis and DATD will be found. The development of a transparent, mechanically stable gel permitting unrestricted migration of molecules several million in molecular weight would be a major advance for IEF in continuously polymerised gels.

3.2.4. IEF in polyacrylamide gels

The development of analytical techniques for IEF in polyacrylamide gel and the advantages of this method over IEF in liquid media were described in a flurry of papers published independently and almost simultaneously (Awdeh et al. 1968; Dale and Latner 1968; Fawcett 1968; Leaback and Rutter 1968; Riley and Coleman 1968; Wrigley 1968; Catsimpoolas 1969a). These papers clearly demonstrated the considerable potential of this new technique. However, during the transition from IEF in all liquid media to gels, few systematic studies were made to optimise factors such as gel composition and electrolysis conditions. Many of these early systems, particularly those using tubes, were plagued by a marked instability in the pH gradients which somewhat restricted their general applicability. Much of this particular problem may have been due to the use of inappropriate apparatus. However, as will be discussed later, (§4.9), this troublesome phenomenon appears to have a multifactorial basis and other factors such as gel composition and electrolysis conditions may also contribute to a decay in the pH gradient (Righetti and Drysdale 1974). These considerations require that the method used for IEF in gels allows the development of sufficiently stable pH gradients to allow all molecules of interest to reach their pI. Unfortunately, this was not always achieved in some systems where the decay in the pH gradient often necessitated premature termination of the experiment. Fortunately, this problem has now been sufficiently contained so that proteins can be focused at their pI and stable banding patterns obtained long before the decay in the pH gradient becomes significant. However, in certain situations, notably in the most basic pH ranges, it is still difficult to achieve stable pH gradients in polyacrylamide gels, although quasi-equilibria are usually obtainable.

With these minor reservations, we feel that IEF in polyacrylamide gels presently offers the best method for analytical IEF. Procedures have now been fairly well standardised and the technique has found wide acceptance in many laboratories.

3.2.5. Apparatus

Best results with GEF in cylinders or slabs of polyacrylamide gel are given with apparatus specifically designed for this purpose. Suitable apparatus for both is readily made or is available commercially. Although systems for GEF in cylinders or slabs have features in common with apparatus for gel electrophoresis there are major design differences that are crucial for reliable and reproducible results by GEF. Perhaps the most important differences are in the electrode chambers and in heat dissipation. Most electrophoresis apparatus have large electrolyte volumes to minimize changes in buffer pH during electrolysis. Such large chambers, however, create excessive convective mixing which, in GEF, may disrupt the connecting pH gradient between electrodes and the gel. This convective mixing may be reduced in IEF cells by minimizing electrolyte volumes and the distance between electrodes and the extremities of the gel. Fawcett (1970) carried this idea to its logical conclusion by inserting individual electrode wires directly into the top of the gel tube, while Awdeh et al. (1968) working in slabs, applied electrodes directly to the surface of the gel.

Heating effect must also be more carefully controlled in IEF than in electrophoresis. Considerable resistance (up to 4 kilo Ω) may develop across a 10 cm gel during IEF. Because this resistance is not uniform throughout the gel, local 'hot spots' may arise in areas of low conductivity. This local heating must be efficiently dissipated to minimize convective mixing and possible heat denaturation of proteins migrating through these zones. Unfortunately, most electrophoresis cells do not allow adequate gel cooling for the higher potential differences used in GEF and special cooling procedures are, therefore, necessary. This is best achieved in tube apparatus by thermostating the tubes in a central compartment with a circulating coolant. It is usually insufficient merely to immerse the tubes in a cold liquid or to run the apparatus in air in the cold room because of poor heat dissipation. These points are demonstrated in Figs. 3.5 and 3.6. Fig. 3.5

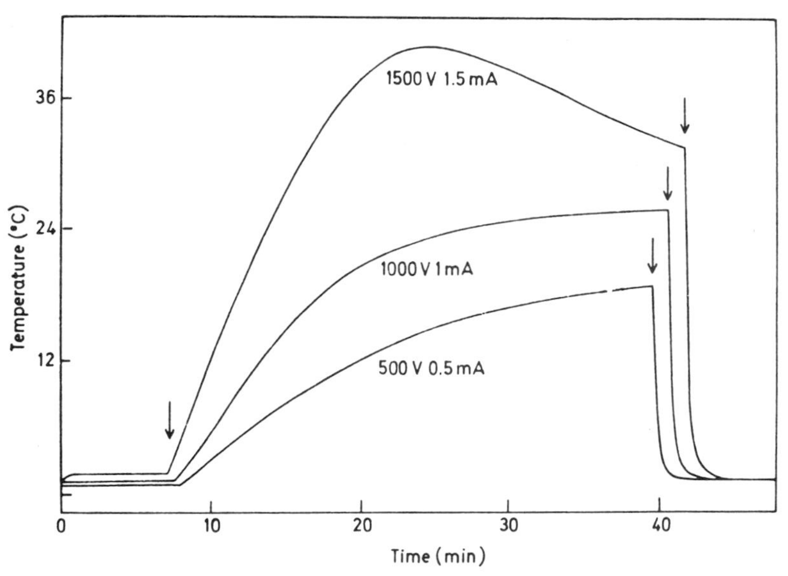

Fig. 3.5. Comparison of air cooling and water cooling. The same gel was run sequentially at 0.25 W (lower curve), 1 W (middle curve) and 2.25 W (upper curve). During the first 8 min water cooling was applied. Between 8 and 40 min the coolant was removed (as indicated by arrows) and the gel dissipated heat into the air. At approx. 40 min, the cooling water was allowed to circulate again. Experiments ware performed in 3 mm inner diameter plastic tubes, with a wall of 0.8 mm thickness. (From Righetti and Righetti 1975, by permission of Butterworths.)

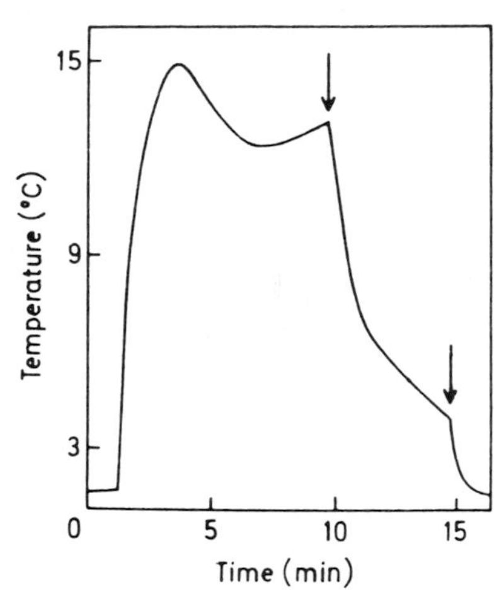

Fig. 3.6. Comparison of direct and indirect gel cooling. A gel cast in a 3 mm inner diameter plastic tube (0.8 mm wall thickness) was immersed in non circulating coolant at 1°C. When 500 V were applied to it, its temperature rose by 15°C. When water circulation was started, (first arrow) the gel was cooled to approx. 5°C. When the power supply was turned off (second arrow) the gel equilibrated with the coolant in approx. 30 s. (From Righetti and Righetti 1975, by permission of Butterworths.)

compares the effect of air cooling and water cooling on the internal temperature of gel cylinders subjected to the same electrical loads. The heat dissipation in air was very much less than that in water. Fig. 3.6 shows the internal temperature in gels cooled by either a stationary liquid or a circulating liquid. It is obvious that effective cooling is only achieved in systems with circulating coolant or some other means for efficient heat transfer.

3.2.6. Gel composition

Selection of an appropriate gel composition is another key consideration for IEF in polyacrylamide gel. Two factors are particularly important: gel porosity and the concentration of ampholyte in the gel. These are discussed below. Ideally, one should use the highest gel porosity consistent with satisfactory mechanical strength. Both properties are largely determined by the relative amounts of acrylamide and cross-linker which are usually expressed by the relationship (Hjertén 1962):

$$T = \frac{\text{g acrylamide + cross linker}}{\text{100 ml solution}}$$

$$C = \frac{\text{g cross-linker}}{\%T}$$

Rodbard and Chrambach (1970) have discussed in detail many of the factors affecting the porosity and sieving of polyacrylamide gels. In general, the pore size (p) in gels containing low levels of cross linker (e.g. $C = 1\text{–}5\%$) is related to T by:

$$p \propto 1/\sqrt{T}$$

while in gels containing high levels of cross-linker (e.g. $C = 15\text{–}25\%$), the relationship is closer to

$$p \propto 1/\sqrt[3]{T}$$

Bis is the most commonly used cross-linker, though other substances such as DATD may be used for special purposes. The lowest acrylamide concentration that will form a satisfactory gel for IEF is about $T = 3\%$, $C = 4\%$ (Bis), but these gels are very soft and fragile. Gels of composition $T = 5\%$, $C = 3\%$ (Bis) (Vesterberg 1971b) or $T = 4\%$, $C = 4\%$ (Righetti and Drysdale 1971) have been found suitable for many purposes. Proteins of molecular weight 500,000 and lower will usually focus in about 6 hr in 10 cm gels of these compositions. The upper size of proteins that can be focused in these gels is about 1.5×10^6 daltons, but such large proteins often require at least 12 hr to reach their equilibrium positions. On the other hand, smaller proteins of 100,000 daltons or less may focus in less than 4 hr. These gels are rather soft, and require careful handling. However, since most proteins are smaller than 200,000 daltons, more robust gels e.g. $C = 5\%$, $T = 4\%$ may be preferred and are, in fact, to be recommended. These gels are easily handled for staining and densitometry.

Another important requirement for GEF is the development of sufficiently stable pH gradients to ensure that proteins overcome restrictive effects of the gel to reach their equilibrium points. Also, since the position of a focused protein will depend on the actual pH gradient, it is obviously desirable to develop reproducible pH gradients. Failure to develop stable and reproducible pH gradients can lead to considerable difficulties in correlating banding patterns from different experiments or with different batches of carrier ampholyte of the same nominal pH. This is especially true when measuring the position of a protein in the gel without regard to the actual pH gradient developed in the gel. Such variability may be particularly vexing to investigators accustomed to characterising substances separated by electrophoresis by their relative R_f values. For example, Fig. 3.7 shows the banding patterns of the same sample of haemoglobins run under identical conditions except for the level of ampholyte in the gel. Both samples were resolved into several components. The pH gradient given in the gel containing 1% (w/v) ampholyte was uneven and contained a long plateau. In contrast, the pH gradient formed with 4% (w/v) was smooth and linear. When assessed by the relative positions of the components in both gels, it might appear that the samples were different. It is only when the banding patterns are correlated with the actual pH gradients in the gels, that the anomaly is resolved.

We have found that a certain minimum concentration of 2% (w/v) carrier ampholyte is required for the development of stable pH gradients in polyacrylamide gel. Below this level and particularly below 1% (w/v) pH gradients in gels are often unstable (Righetti and Drysdale 1971, Finlayson and Chrambach 1971). Considerable local heating may occur in these gels. This suggests that the dependency on carrier ampholyte concentration is probably not just a consequence of increased viscosity such as occurs with glycerol or sucrose in stabilizing the pH gradient. A more likely explanation is that a certain level of ampholyte is required to provide sufficient overlap in adjacent ampholyte zones for good conductivity throughout the gel. This point may be readily understood from the patterns of focused ampholytes in different pH ranges presented earlier (Figs. 1.7–1.9). A minimum concentration of ampholytes is also required to buffer large amounts of protein in a pH zone. Ideally, the carrier ampholytes should provide

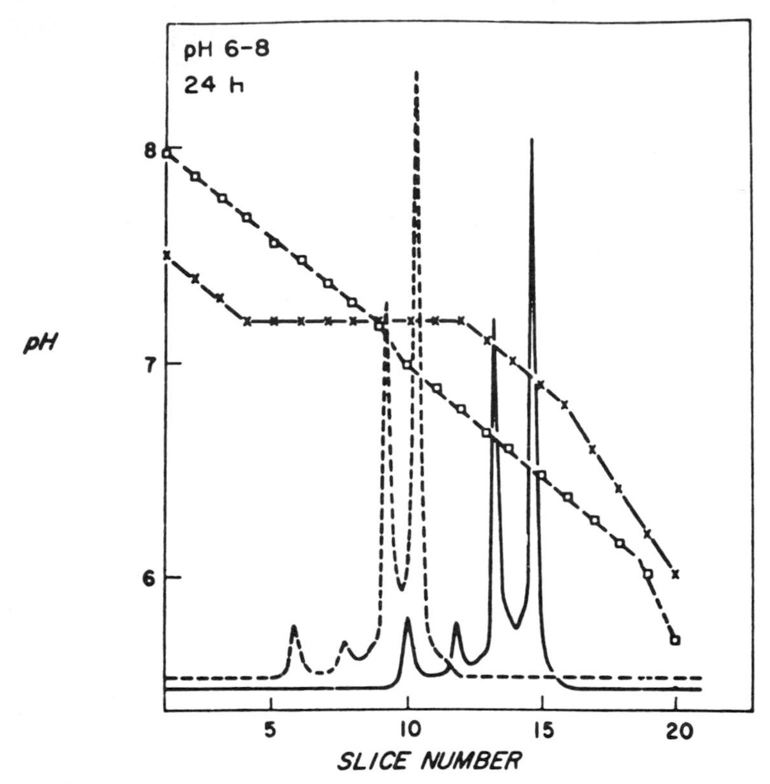

Fig. 3.7. Fractionation of rat haemoglobins in the pH range 6–8 at different ampholyte concentrations. Samples of haemolysates (100 μg protein) were fractionated in 4% acrylamide gels containing either 1 or 4% Ampholine pH range 6–8. After a total electrolysis period of 24 hr, the gels were scanned at 576 nm in a recording spectrophotometer (———, 1% Ampholine; ------, 4% Ampholine and subsequently sectioned for determination of the pH gradient (×———×, 1% Ampholine; □———□, 4% Ampholine). Notice that a level of 1% Ampholine is too low to ensure stability and reproducibility of pH gradients. (From Righetti and Drysdale 1971, by permission of Elsevier.)

a buffering capacity of at least 0.3 μ equiv/mg at pH values close to their pI (Vesterberg 1973a).

In addition to ensuring adequate levels of ampholytes in different pH ranges, additional precautions are also required in pH gradients that do not include ampholytes buffering near pH 7 but which, nevertheless, must include pH 7 between the electrodes. In such cases, much of the electrical potential drop will occur at this point, at the expense of other parts of the gradient which may be underfocused. This situation may be rectified by adding a small amount of pH range 3–10 or 6–8 to the gel ampholytes (Haglund 1971).

3.2.7. IEF in gel cylinders

Most methods for IEF in gel cylinders use gel dimensions similar to those used in analytical 'disc' gel electrophoresis; i.e. 5–15 cm in length and 0.2–0.5 cm in diameter. Although some authors advocate the use of conventional electrophoretic apparatus, we think that many of these systems would benefit from more efficient heat control and different electrode arrangements.

Apparatus Fawcett (1969) described a simple method for overcoming heating problems by thermostating gel tubes in a chilled, circulating electrolyte solution. With this arrangement, the cold finger and stirrer should be shielded with plastic to prevent electrical shocks. Fawcett also recognized the importance of minimizing the electrolyte volumes and devised the ingenious solution of inserting a platinum electrode directly into a layer of Ampholine on top of each gel tube.

Righetti and Drysdale (1971) developed an apparatus whose salient features included efficient gel cooling, small electrolyte volumes and close juxtaposition of electrodes and gels. This apparatus is depicted schematically in Fig. 3.8. The gel tubes are held between the electrodes in a separate compartment through which coolant is circulated. The outer wall of the central core extends about 1 inch above the top end plate to form an electrolyte chamber. The lower electrolyte chamber is a separate unit which forms the base of the apparatus. Platinum wire electrodes are used. The top electrode is on the inner face of the lid while the bottom electrode is set into the inner face of the lower electrolyte chamber. The extremities of the tubes pass through rubber grommets in the end plates to make contact with the electrolytes. The electrodes are positioned so that they are less than 1 cm from the tube extremities when the apparatus is assembled. An appropriately placed

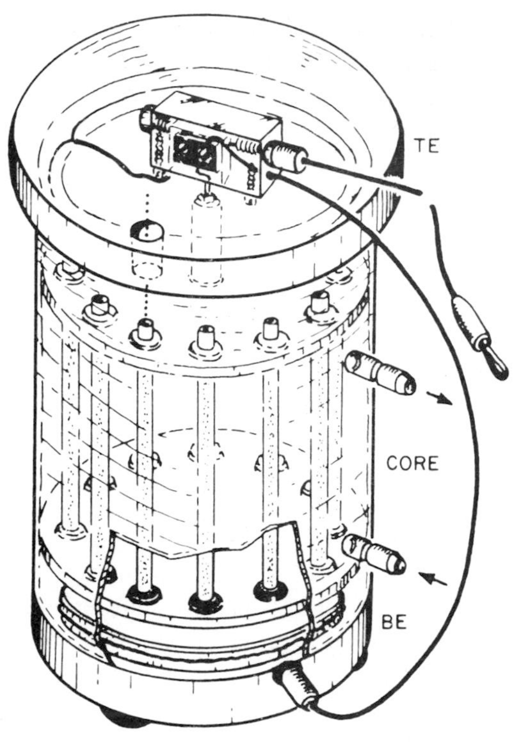

Fig. 3.8. Schematic drawing of the analytical gel-tube apparatus of Righetti and Drysdale (1971, 1973). TE, top electrode compartment fitted with a safety switch. BE, bottom electrode compartment. The central core is interchangeable with the preparative unit. (Courtesy of Medical Research Apparatus.)

aperture in the lid (which rotates freely around a central nylon post) allows samples to be applied directly to the gel surface without raising the lid. This aperture also allows the top compartment to be flushed with a gas if required. Small vents are also cut from the rim of the base for gas escape or flushing. Finally, the lid also contains a safety switch which is wired across both electrodes, and only passes current when the lid is properly seated on the central post. This safety switch disconnects current from both electrodes whenever the lid is lifted or the apparatus accidentally knocked over.

The apparatus usually holds 12 tubes (10 cm long and 0.3 cm inner diameter, volume approximately 1 ml) but other sizes may be used by changing the rubber grommets. We routinely use gels 3 mm in diameter, since these gels can be cooled efficiently and be scanned in most recording spectrophotometers. Tubes may be of plastic, glass or quartz. Glass tubes dissipate heat more efficiently, but this need not be critical with properly cooled thin plastic tubes. Plastic is less subject to electroendosmosis such as may occur in a charged glass surface. Gel adherence in plastic is less than in glass. However, unlike electrophoresis, complete wall adherence is not critical in IEF provided wall separation does not allow electrolyte passage. Many proteins will focus as rings rather than uniform disks in plastic tubes, and consequently are more readily identified by biological activity or staining procedures. Moreover, the reduced wall adherence in plastic allows focused gels to be extruded without reaming merely by applying air pressure from a rubber bulb. Poor wall adherence in plastic tubes also simplifies gel fractionation with moving plungers (see p. 484). Quartz and glass tubes are useful for scanning gels directly in the UV or visible regions. These tubes may be fashioned to fit snugly into the cuvette carrier in the linear transport system of a recording spectrophotometer. This type of apparatus may also be used for preparative purposes by substituting a larger core with larger tubes. Analytical and preparative models of this design are shown in Fig. 2.17.

3.2.8. Methodology

This section will deal with some of the more practical aspects of analytical IEF in cylinders of polyacrylamide gel. Many of the procedures such as gel preparation will, of course, also apply to thin slab technique. Because most systems using gel cylinders are basically similar, much of the information may be generally applicable. However, since we cannot cover all variations, we have restricted the detailed methodology to that for our own system (Fig. 3.8).

Reagents required: recrystallized acrylamide, recrystallized N,N′-methylene bisacrylamide (Bis) or diallyl tartaramide (DATD) (cross-linkers); N,N,N′,N′-tetramethyl ethylenediamine (TEMED); ammonium persulphate and riboflavin (catalysts); carrier ampholytes; optional gel additives, e.g. sucrose, sorbitol, glycerol, urea.

The following stock solutions should be prepared:

Gel solution	30 g acrylamide and 1.2 g Bis in 100 ml distilled water
Carrier ampholytes	40% (w/v) solutions (except for pH 2.5–4 and pH 9–11 Ampholine, which are 20% (w/v) solutions)
TEMED	1 in 100 dilution in distilled water
Ammonium persulphate	2% (w/v) solution prepared fresh weekly
Riboflavin	4 mg/100 ml distilled water
Sucrose or sorbitol	25% (w/v) solution
Glycerol	25% (v/v) solution

Care should be taken in handling acrylamide since it is a neurotoxin in its monomeric form. Contact with skin should be prevented. There is usually no need to maintain separate solutions of acrylamide and Bis or DATD except when using high cross-linked gels. Most of these solutions have shelf lives of several months. Solutions of riboflavin and acrylamide are sensitive to light and should be stored in a bottle wrapped in aluminium foil (a brown bottle does not give much protection from light).

3.2.9. Gel preparation

Thoroughly clean gel tubes, rinse in distilled water and dry. If glass tubes are used, they should be siliconized or treated with methyl cellulose to reduce electroendosmotic effects (Hjertén, 1970). Seal one end of the gel tube with a wet piece of dialysis tubing. The tubing should be stretched taut across the end and held in place with a tight fitting rubber sleeve. A thin section of rubber or plastic tubing with an inner diameter equal to the outer diameter of the tube will usually form a satisfactory watertight seal. The dialysis tubing should be applied immediately prior to the preparation of gels to prevent it from drying out. Alternatively, the end of the tube may be sealed with Parafilm or some similar product. The seal should be replaced with a suitable wick or dialysis tubing before the gel is used.

1) For the 12 gels, add to a 50 ml conical flask with a side arm:

water	3.6 ml
glycerol (or sucrose, or sorbitol)	6.0 ml
acrylamide solution	1.7 ml
TEMED (1/100)	0.1 ml
carrier ampholytes	0.6 ml

This solution will give gels of compositions $T = 4.4\%$; $C = 3.85\%$; ampholyte concentration of 2% (w/v) and glycerol (or sucrose, or sorbitol) concentration of 12.5%. For alternative compositions, alter the ratio of the various solutions appropriately.

2) Stopper the flask and degas thoroughly on a water or vacuum line for about 5 min. If possible, connect the vacuum line to a nitrogen tank, via a three-way tap, to equilibrate the Erlenmayer with nitrogen once the air has been removed.

3) Add 0.1 ml of 2% ammonium persulphate and mix rapidly.

4) Immediately apply the degassed gel solution to the gel tubes with a long-tipped Pasteur pipette controlled by a rubber bulb. Expel all air bubbles from the tip of the pipette and add the gel solution from the bottom of the gel tube, gradually withdrawing the pipette as the gel tube is filled. The tip of the pipette should always be kept beneath the surface of the gel solution to prevent formation of air bubbles in the gel tube. If air pockets develop, they are best removed by sharply tapping the tube while holding it in a vertical position. Place the gels upright in a suitable holder. If no special holder is available the gel tubes may be supported inside short tubes in a test tube rack. Fill to within 0.5 cm of the top of the gel tube. Ideally all tubes should be filled to the same height.

5) Carefully float a layer of water onto the gel solution. This requires a light touch. One simple method is to apply the water through a narrow gauge needle from a syringe. The tip of the needle should be held against the side of the tube at the meniscus of the gel solution and gradually raised as the water layer forms. Those with a steady hand

may prefer to apply the water layer more rapidly with a Pasteur pipette. Ideally one should try to avoid 'bombing' the gel surface when applying the water overlay. However, formation of a perfectly flat gel surface is not as critical in gel electrofocusing as it is in gel electrophoresis since samples entering the gel in a skew band will eventually form flat discs at their equilibrium positions in the vertical pH gradient.

6) Set the gels aside to polymerise at room temperature. This should take no more than 20 min to ensure maximal polymerization of the acrylamide.

7) Seal the open end of the gel tube with Parafilm and store in the refrigerator until required. Precast gels may be kept for several weeks before use if properly sealed.

Notes

1. All gel reagents should be of the highest quality. In particular, acrylamide should be recrystallized to remove free acrylic acid, a common contaminant in reagent grade acrylamide. Acrylic acid will generate fixed charges in the gel causing electroendosmosis and eventual instability of the pH gradient. In addition, unreacted monomeric acrylamide can react with α-amino, sulphydryl and phenolic hydroxyl groups at elevated pH. This reaction preceeds very slowly at pH 9, but is quite fast (30 min) at pH 11 (Dirksen and Chrambach 1972). The acrylamide may be recrystallized from chloroform and Bis from acetone, according to Loening (1967). Several satisfactory preparations of purified acrylamide are commercially available.

2. Gels may also be photopolymerised in glass or quartz tubes by using riboflavin at a final concentration of 0.656 mM. The tubes should be placed at 10 cm distance from fluorescent tubes e.g. Philips TL 20W/SS for one hour at room temperature. However, gels will not photopolymerise with riboflavin above pH 7.5. Consequently, chemical polymerisation should be used or gels should be cast without ampholytes and the ampholyte later diffused into the polymerised gel.

3. The porosity and mechanical strength of the gels may be altered

by varying the amount of acrylamide solution and/or compensating volume changes of water.

4. Other gel additives e.g. sucrose, urea, non-ionic detergents may also be added at step 1.

5. The addition of TEMED is not essential since the carrier ampholytes serve a similar function in gel polymerisation (Riley and Coleman 1968).

6. Several factors can affect the polymerisation and physical properties of acrylamide gels (see Rodbard and Chrambach 1971; Chrambach and Rodbard 1972). Experimental procedures should, therefore, be standardized as much as possible. Particular attention should be paid to the following variables: (1) purity of reagents, (2) degassing conditions, (3) amount and type of catalyst added, (4) temperature of polymerization, (5) duration of polymerisation. Perhaps the most troublesome factor is the amount of persulphate required as catalyst. Different preparations may vary in their potency and an appropriate level should be determined by trial and error. Only the minimum amount required for satisfactory gelling should be used since persulphate may oxidise carrier ampholytes and slightly alter the resulting pH gradient. Excess persulphate also generates potentially troublesome free-radicals (Brewer 1967; Fantes and Furminger 1967; Mitchell 1967) which, if not removed prior to sample application, may create artifacts, e.g. by oxidizing cysteine residues to cysteic acid. However, according to Dirksen and Chrambach (1972), even riboflavin as a catalyst has oxidizing properties on thiol groups. Free radicals seem to persist in polyacrylamide gels, even after prolonged electrolysis (Peterson 1971). Only pre-electrophoresis of thioglycolate into the gel, in amounts equivalent to the catalyst added (1 to 5 mM), results in reducing conditions in the gel (Dirksen and Chrambach 1972). Riboflavin alone, as a catalyst, produces gels which are softer and contain more unreacted monomer than equivalent gels polymerized with persulfate. In disc electrophoresis, Chrambach and Rodbard (1972) suggest, for higher polymerization efficiency, at neutral or acidic pH, the use of combinations of all three catalysts (riboflavin, persulfate and TEMED). It would also seem that storage of gels in the cold room

overnight, before use, improves the gel consistency. Probably, during this time, the polymerization reaction slowly proceeds to completion (T. Wadström, personal communication).

7. Highly cross-linked gels with DATD. These gels do not polymerise as readily in the presence of ampholytes as gels formed with Bis. Consequently, it is advisable to initiate gel polymerisation before adding the ampholytes. By arranging a polymerisation time of 10–20 min (without ampholytes), it is possible to obtain satisfactory gels by adding the ampholytes at about two-thirds of the time required for polymerisation. After adding the ampholytes, the solution should be degassed again before casting into gels.

8. pH range of Ampholine. It should be noted that the pH gradients given by LKB Ampholine may not always be linear or have a uniform field strength over the indicated pH range (Vesterberg 1975). Consequently, it may be advisable to fortify deficient regions with ampholytes of the appropriate pH range. For example, Vesterberg (1973c) recommends the addition of pH 9–11 Ampholine to the wide range pH 3.5–10 Ampholine to extend the linearity of the gradient in the alkaline region. He also recommends the addition of pH 4–6 and 5–7 Ampholine to improve the distribution of the field strength. It is also possible to produce specially tailored pH gradients by expanding one section of the gradient by the addition of the appropriate pH range Ampholine. This often allows a better separation of components of particular interest while displaying all components in the wider pH range.

Loading apparatus Wet the rubber grommets in the cooling chamber of the apparatus and insert the gel tubes so that the maximum length of gel is contained in the chamber and the extremities of the gel tubes project equally from the end plates. Seal unused positions with stoppers or solid gel rods. Place the central core with the tubes on top of the lower electrode vessel and circulate coolant at 2–4°C through the chamber. We routinely use circulating coolant from a water bath with a built-in refrigerator. Other sources such as refrigerated coolants from fraction collectors are often convenient alternatives, provided the level of antifreeze does not damage the plastic chamber. A simple immersible pump in conjunction with an ice bath in a styrofoam or other insulated container is also perfectly adequate.

Electrode Select electrode arrangement according to experimental plan. Since most proteins are kept in solution in their anionic form, it is usually preferable to select the cathode (–) as the upper electrode. This arrangement will facilitate the rapid entry of the protein into the gel. If, on the other hand, the sample is to be applied in its cationic form, electrodes and electrolytes should be reversed. (See also section on sample application).

Electrolytes Many different electrolytes have been used for IEF in cylinders of polyacrylamide gel and most are probably adequate. Dilute solutions of ethylenediamine, ethanolamine, triethylamine, or sodium hydroxide have been used for catholytes with solutions of phosphoric acid or sulphuric acid as anolyte. Even ampholytes have been used. Perhaps the most important requirements are that sufficient hydrogen and hydroxyl ions are generated to titrate the ampholytes and so repel them from the electrode vessels. On the other hand, the electrolytes should not be so strong as to preclude application of samples in appropriately buffered media onto the tops of the gels. We have found that solutions of 10 mM phosphoric acid and 20 mM sodium hydroxide are satisfactory electrolytes for most pH ranges in gel cylinders. For reasons given previously, (§3.2.5), only the minimum amount of electrolyte should be used to establish electrical contact between gel tubes and electrodes. In our system, about 50 ml are required for each electrode vessel.

3.2.10. Sample application

Samples may be applied directly to the top of the gel or incorporated into the gel solution before polymerisation. Top loaded samples may usually be added at any period during the electrolysis period. However, due consideration should be given to the probable pH range through which the sample must migrate after application. For example, at the very early stages of electrolysis, the pH range of the gel will be close to median pH of the ampholyte pH range. Shortly after electrolysis commences, a sigmoidal pH gradient develops whose extremities are close to those of the carrier ampholytes. Thus in the case of the pH range 3–10, the cathodic end of the gel will be close to the median pH, i.e. pH 7 before electrolysis, but close to pH 10 shortly after electrolysis commences. Samples sensitive to high pH should, therefore, be applied at the early stages. Proteins that are stable throughout the pH range in the gel but are unstable for other reasons should be applied after the pH gradient has been established. The proteins will then carry more of the current so that they will take less time to reach their pI.

Samples (10–50 μl) should be applied in a dense buffer to lie as a flat layer on top of the gel. Care must be taken to prevent sample contact with the electrolyte. This may be achieved by displacing the liquid inside the gel tube with a solution of 1% ampholyte immediately before applying the sample. The sample may then be introduced either in a stronger solution of ampholyte, e.g. 2–4%, or in a buffered solution containing 3–10% glycerol or sucrose. The wide range pH 3–10 of ampholytes is often a satisfactory buffer medium for many proteins. For small sample volumes, the added ampholytes will usually not appreciably alter the developing pH gradient in the gel. However, for uniform results, all gels in a series should receive the same additions. Should the sample precipitate, other pH ranges should be explored, or the sample should be applied in a favorable buffer. Ideally the buffer should have a low ionic strength, e.g. less than 100 mM to facilitate the rapid entry of sample into the gel. The volume of the sample is not critical. Multiple applications of dilute samples may be made at intervals, providing sufficient time elapses to ensure that components from the last application have penetrated the gel. The addition of a trace of bromophenol blue or another dye that does not bind to the sample proteins greatly facilitates visualization of application of colorless samples. Most samples can be accurately and conveniently dispensed from capillary pipettes with a microapplicator controlled by a thumb screw or similar device.

Internal loading For larger volumes of dilute samples, it may be advisable to incorporate the sample directly into the gel before polymerisation. Appropriate corrections should, of course, be made in gel formulations to take account of the added sample volume. Internal loading should only be considered if samples are not adversely affected by gel constituents or by the heat of the polymerisation reaction. Gel temperatures may reach as high as 40°C during polymerisation (Righetti and Righetti 1975). Consequently, it is advisable to cool polymerising gels with water to prevent adverse effects on heat-labile proteins.

Sample load Best resolution is usually given at low sample levels on the gel. With sensitive dyes for staining proteins, e.g. Coomassie Brilliant Blue, less than 1 μg protein can be readily detected in a 3 mm gel cylinder. Consequently, load levels of 1–10 μg per focused band are usually satisfactory. Substantially higher loads may be applied to detect minor components. With many proteins, band densities of up to 800 $\mu g/band/cm^2$ of gel are possible.

3.2.11. Electrophoresis conditions

The standard gel tube 10 × 0.3 cm (I.D.) wall 0.17 cm, cooled by a circulating liquid at 0–4°C, will tolerate an electrical load of up to 1 watt without appreciable temperature change (less than 3°C). However, because this heat is not always uniformly distributed throughout the gel, we usually limit the electrical load to 0.25 watts/gel. This may be arranged by running gels initially at a constant current of 0.5 mA/gel until the voltage required to drive this current increases to 300 or 400 volts. This usually takes between 0.5–1 hr depending on the number of gels. The voltage should then be stabilized at this level when the current will decline to a constant plateau as the pH gradient stabilizes. With pulse power supplies an average power of 0.2 watts/gel at a potential difference of 400 V gives satisfactory results. Higher potential differences and faster focusing periods may be obtained at higher electrical loads or by lower cooling temperatures. For example, Park (1973) obtained equilibrium focusing of hemoglobins in less than 1 hr by cooling gels at −10°C and applying a direct potential difference of 800 volts (see p. 449).

3.2.12. Micro-isoelectric focusing

The presently available techniques for IEF in gel tubes or in slabs allow detection of few μg proteins. However, when lower amounts of sample are available, it could be desirable to have systems which allow detection of protein in the nanogram range. One such a system has been described by Grossbach (1972). He performs IEF in quartz capillaries of 50 μ, 100 μ and 300 μ inner diameter and 65 mm long (from Heraeus Quarzschmelze, Hanau, Germany). The gels are made to contain 7% acrylamide, 2% Ampholine, 20% sucrose and an amount of sample varying from 20 to 70 nanogram of each protein component. The sample is mixed with the monomer solution before polymerization and the quartz tubes are filled by capillary action. After polymerization, the extremities of the capillary are filled with 2% Ampholine solution, by means of a micropipette operated by a micromanipulator. The capillaries are then connected to a conventional apparatus for disc electrophoresis by means of a bored silicone rubber stopper and a short length of glass tubing. IEF is performed at room temperature in a voltage gradient of 100 V for 10 to 120 min. Immediately after the run, the gels are pushed out of the capillaries, by means of a tightly fitting steel-wire, into a drop of 20% TCA on a depression slide.

The resolution obtained in these gels is comparable to that achieved in standard 3–5 mm gel tubes. However, in the capillary system, a shift of the pattern of proteins along the gel column is regularly observed and, upon prolonged focusing, the zones migrate to one end of the column and finally run out of the gel. This might be due to the high electroendosmotic flow along the walls of the quartz tube. This flow is probably enhanced by the high ratio of surface/gel volume.

Gainer (1973) has described a micro method capable of detecting proteins at the 10^{-10} to 10^{-9} g levels. He polymerizes 7.5% or 5.5% acrylamide gels, containing 2% Ampholine, in glass capillary tubes (Corning 7740, 0.58 mm inner diameter, 1.15 mm outer diameter 7 cm long. After polymerisation, the gels are connected to a micro-electrophoresis apparatus (Gainer 1971) so that the upper parts of the tubes are in the anodic chamber containing 0.3% H_2SO_4, while the lower parts are immersed in the cathodic chamber containing 0.5% ethanol-diamine. The gels are subjected to a pre-run for 30 min, to remove excess persulphate and then the sample (5–50 nanograms of each protein component) is layered on top of the gel. Equilibrium is reached in approximately 3 hr at 120 V at room temperature. The stained protein bands on the micro gel appeared very sharp (less than 0.2 mm in thickness) and it was possible to detect as little as 10^{-10} g protein.

Gainer (1973) has found good correlation between protein patterns and pI's obtained with 'micro' and 'macro' gels. However, he has reported consistently higher pI values for basic proteins in micro gels, as compared to macrogels. It might be that the manipulation of micro gels allows for less absorption of atmospheric CO_2, so that micro gels may be more suitable for IEF in basic pH ranges. One advantage of Gainer's procedure is that in his system the protein can be loaded on top of the gel, and does not have to be incorporated into the gel, and thus exposed to persulphate, as in Grossbach's procedure. Also, Gainer's system seems to be free from the high electroendosmotic flow reported by Grossbach. This could be due to the more favourable glass surface/gel volume ratio. If this is so, it might be advisable, when performing micro IEF, to choose capillaries with an inner diameter of no less than 500–600 μ. Additional information on micro-IEF can be found in the book 'Micromethods in Molecular Biology' by V. Neuhoff (1973) (pp. 49–56).

3.2.13. IEF at sub-zero temperature

This could be a useful addition to presently available systems to study reaction intermediates. Often these intermediates have only a short half life at the temperatures commonly used in conventional electrophoresis, but their lifetime could be considerably extended by IEF at sub zero temperatures, to a point at which their detection becomes feasible. Thus Park (1973) has studied the formation of mixed tetramers of HbA-HbS, HbA-HbC, HbA-HbC Harlem and HbS-HbC. To extend the lifetime of the mixed tetramers Park (1973) has developed a system which allows the gels to run at approximately $-10°C$. Six

Subject index p. 587

percent acrylamide gels are polymerised with 25% ethylene glycol and they are submerged in the bottom buffer chamber which contains a frozen slush of sulfurous acid and 25% ethylene glycol. The gels are prefocused for at least 3 hr at a final voltage of 800 V. These low-temperature systems considerably extend the lifetime of the mixed tetramers and reduce bands widths, allowing resolution into distinct bands in less than 5 min. To ensure rapid equilibration with surrounding coolant, Park has used rather thin gels (2 mm inner diameter) with a wall thickness of only 0.5 mm. The temperature at the gel core, as probed with a thermistor polymerised in place, was usually only 2–3°C higher than the coolant temperature. However, in a gel system, the temperature cannot be lowered below approximately −15°C, otherwise the gel structure will undergo irreversible changes.

In this system, Park has been able to focus CO-haemoglobins partially oxidised with $K_3(FeCN)_6$. Seven to nine distinct species were separated. Given the finding that $\alpha_2^+\beta_2$-CO and $\alpha_2CO\beta_2^+$ have different isoelectric points, one would expect to see nine tetrameric species in such solutions. It must be emphasized that these nine bands are actually only seen when working in this low temperature system. At 4°C, Bunn and Drysdale (1971) in partially oxidised Hb solutions, were able to separate only four species.

Other intermediate species of liganded haemoglobin (with CO or NO) could possibly be studied, provided even lower temperatures are reached. For instance, at −30°C the dissociation half-time for the fourth CO ligand in Hb would be several hours (Gibson 1959). However, for such low temperatures gels would be unsuitable, since the gel matrix would be damaged. However, a system at −30°C could be run in a density gradient containing 50% ethylene glycol as supporting medium. Since this would be preferentially an analytical system, small columns (10–20 ml total column) could be used, such as U-tubes or other systems described in §3.1.

3.2.14. Thin layer slab technique

Isoelectric focusing in thin slabs of polyacrylamide offers several advantages. Perhaps the most important is the excellent comparison and convenience offered by analysing multiple samples in parallel tracks of a single gel rather than in several individual tubes. Possible modification of sample by contact with electrolytes, an ever present hazard with gel tubes, is minimized in gel slabs since samples may be applied at any point onto the gel surface. In addition, thin layer technique simplifies many of the methods used to analyze focused patterns such as conventional staining, autoradiography, immunofixation, and zymogram techniques involving overlays (see later). Developed gels may also be dried onto plates for convenient assessment or storage. Finally the thin layer technique usually allows better control of heat dissipation and is less subject to pH gradient instability. Perhaps the only advantage of gel rods over gel slabs run horizontally is the possibility, in the former system, of running oxygen-sensitive enzymes under strictly anaerobic conditions.

Suitable apparatus for isoelectric focusing in thin slabs of polyacrylamide gel is now available from several sources. The gel may be held either vertically or horizontally between cooling plates. We prefer horizontal systems since the gel is subjected to less mechanical stress. Simpler electrode arrangements are possible and there is more flexibility in methods for sample application. Gels are usually only 1–2 mm thick and separations may be conducted either along the long or the short axis. Gels are usually cooled through their bottom surface and the electrodes are usually placed directly on the gel surface. Evaporation is minimized with a close fitting lid over the gel. A typical apparatus design for thin layer is shown in Fig. 3.9. Practical advice for the use of a typical thin layer system is described below. As with tube systems, minor variations are required for different apparatus, but specific procedures are usually adequately covered in instruction manuals.

Gel composition The same considerations for selecting optimal composition (§3.2.6) also hold for focusing in thin slabs. However, for any given gel composition, molecular sieving seems to be less troublesome in thin slabs than in tubes. This fortunate circumstance may be because the thinness of the slab prevents uniform polymerization on the gel surface and/or because proteins may run in a surface film on

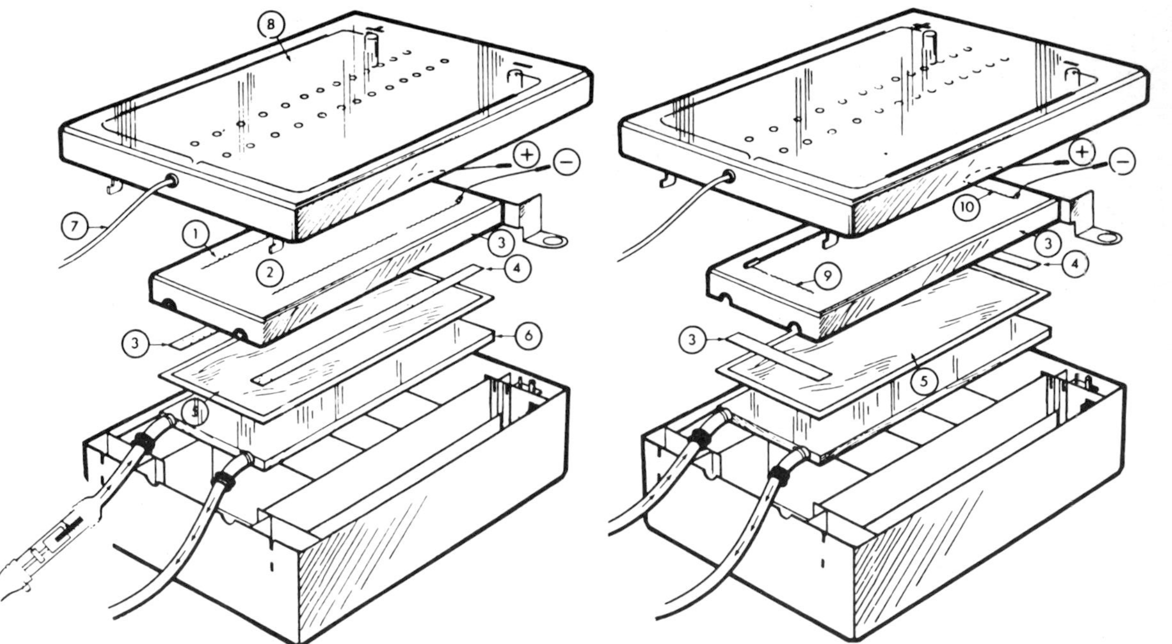

Fig. 3.9. Drawing of the LKB 2117 Multiphor. Left: arrangement for IEF in the short side (10 cm). Right: arrangement for IEF in the long side (24 cm). (1) and (2), anodic and cathodic platinum wires, respectively, for runs in the short side. (3), platinum electrode holder and gel cover lid. (3′) and (4), anodic and cathodic filter paper strips, respectively. (5), thin-layer polyacrylamide gel cast on a 1 mm thick glass slab. (6), cooling block. (7), cable connection to power supply. (8), safety cover lid. (9) and (10), anodic and cathodic platinum wires, respectively, for runs on the long side. (Courtesy of LKB Produkter AB.)

the gel surface (Jeansson et al. 1972). Whatever the reason, it is rarely necessary to use a gel weaker than $C = 5\%$, $T = 3\%$ or $C = 4\%$, $T = 4\%$ for focusing molecules up to 1×10^6. For more typical proteins, more robust gels of 5%T and 4%C may be preferred. Table 3.1 lists a typical gel composition for IEF in thin layer slabs. The various Ampholine ranges should be mixed as indicated in this table, in order to obtain stable and linear pH gradients as well as an

TABLE 3.1

Gel composition for different pH ranges*.

Stock solutions	Volume (ml) of stock solutions for gels having the following pH ranges			
	3.5–9.5	2.5–6	5–8.5	7.5–10.5**
Acrylamide/Bis	10	10	10	10
Riboflavin	0.4	0.4	0.4	
Persulphate***				0.4
Sucrose	30	30	30	
Sorbitol				30
Ampholine pH 3.5–10	2.8			
Ampholine pH 2.5–4		1.2		
Ampholine pH 4–6	0.2	1.2		
Ampholine pH 5–7	0.2	0.6	1.5	
Ampholine pH 7–9			1.5	0.2
Ampholine pH 9–11	0.4			3.6
H_2O	16	16.6	16.6	15.8

The solution should be mixed thoroughly followed by de-aeration for 1 or 2 min.

* From Davies (1975), with slight modifications.

** According to Vesterberg (personal communication) a better gel is obtained in this pH region using $T = 6\%$, $C = 4.5\%$ in which case the volumes given in this table should be changed accordingly.

*** Ammonium persulphate should be added after de-aeration.

For composition of stock solutions see p. 440.

even distribution of the field strength between the electrodes. As shown in Table 3.1, when focusing above pH 7, persulphate should be used as a catalyst and sorbitol as an additive instead of, respectively, riboflavin and sucrose. The total gel volume given in Table 3.1 (60 ml) is suitable for focusing in the LKB Multiphor 2117 apparatus. For other chambers and volumes, change the solutions in Table 3.1 accordingly. The gel mixture reported in this table gives gels of 5.2% *T*, 3.8% *C*, 2% Ampholine and 12.4% sucrose or sorbitol. The composition of the stock solutions has been reported on p. 440).

In most horizontal systems, the gel is usually cast between two glass plates. One of these plates may be plastic if a template for sample application is required. The plates are usually separated by 1 or 2 mm by a rubber gasket or by a spacer frame lightly coated with water repellent grease. The plates are clamped together to form a watertight chamber (Fig. 3.10). One of the two plates may also be used as a template to imprint indentations into the gel for convenient sample application. These indentations should not extend into more than 30% of the gel thickness. Deeper troughs create conductivity problems and may cause local skewing of the pH gradient. Should this happen, the trough should be filled with a slurry of Sephadex and gel ampholyte to improve the conductivity. When only a few analyses are required or when different analyses are planned for several samples, the gel chamber may be subdivided by suitable frames.

Gels are usually cast in a vertical position (see Fig. 3.11). Both plates must be thoroughly clean. However, one may be lightly siliconised before use to facilitate its separation from the gel surface. The gel solution should be applied to the chamber with a fine tipped pipette or with a syringe. It is advisable to apply the first portion by tilting the assembly on a lower corner to avoid air pockets. As the chamber fills, it should be gradually lowered to the horizontal. The height of the gel should correspond to the distance between electrodes. After filling the chamber to the appropriate level, the rubber gasket is closed, to prevent air from entering the chamber. Since the gel chamber is completely sealed by the rubber gasket, there is no need to add water to the top of the gel solution. In chambers with an open top, the gel solution

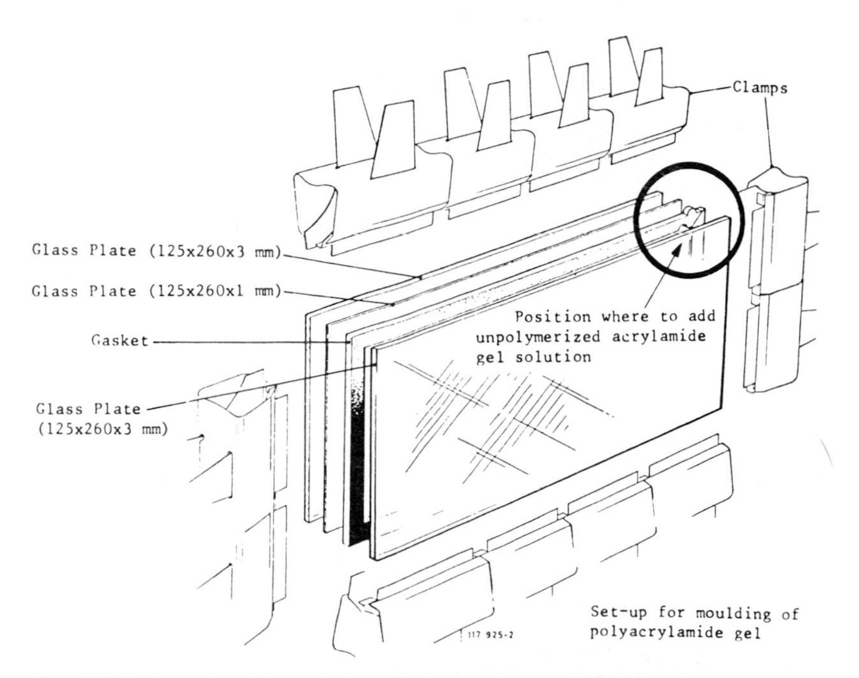

Fig. 3.10. Polyacrylamide moulding cell for the LKB 2117 Multiphor apparatus. (From Davies 1975, by permission of Butterworths.)

should be overlaid with water containing traces of catalyst, to give a flat surface and uniform polymerization.

When the gel has completely polymerized, the gel chamber should be placed in a horizontal position in the refrigerator or on a cooling block. The resulting contraction usually causes a partial separation of the gel from the cover plate which simplifies its subsequent removal. Shortly before use, the cover lid should be carefully separated. To do this, the gel chamber should be placed on a flat surface and the top plate pried apart from the gel surface by gentle leverage with a spatula against the bottom plate (see Fig. 3.11). It is extremely important not to disturb the adhesion of the gel to the bottom plate, since this may alter the electrical conductivity at that point and cause a local distortion in the pH gradient.

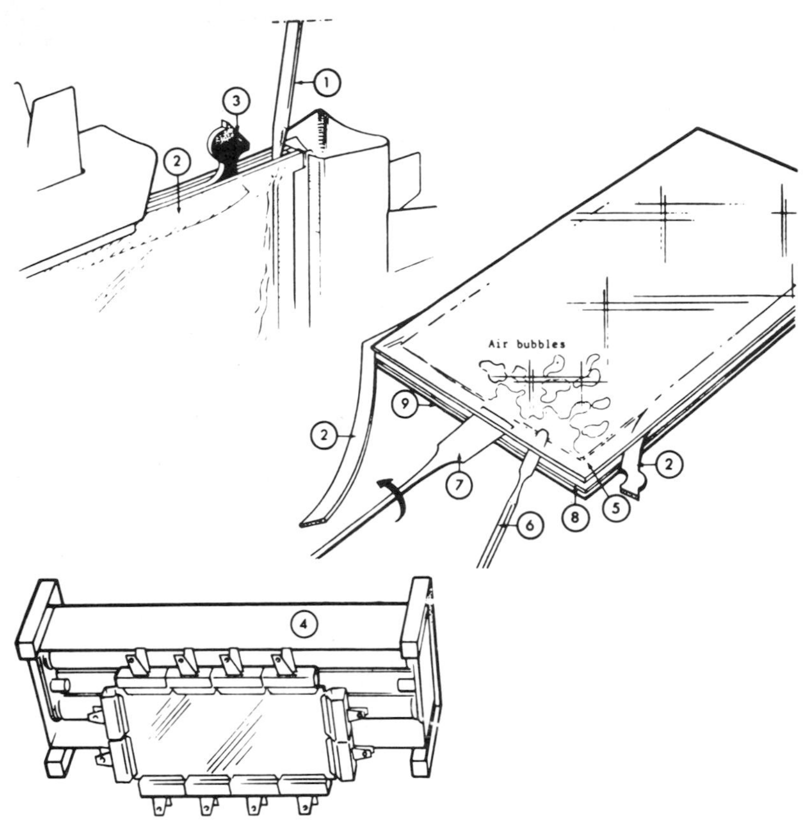

Fig. 3.11. Sequential procedure for filling a moulding cell (upper left), photopolymerizing the polyacrylamide gel (lower left) and removing the glass cover slab from the thin layer gel (right side). This last operation is best done with two spatulas: while slightly pressing down the upper part of the polyacrylamide gel with a thin spatula (6), a broader spatula (7) is slowly twisted as indicated by the arrow. As air bubbles begin to penetrate the space between the upper gel surface and glass plate (5), the broader spatula is gently twisted to a vertical position and the upper glass plate (5) removed. During this operation care should be taken not to loosen the gel from the lower glass plate (8) as this would cause severe band distortion during subsequent IEF. (Courtesy of LKB Produkter AB.)

Electrodes and electrolytes The disposition of electrodes for most systems of isoelectric focusing in thin slabs usually follows the model of Awdeh et al. (1968) who obtained satisfactory results by placing carbon electrodes directly on the gel surface. However, rather than carbon electrodes, most systems now use platinum electrodes which make contact with the gel through suitably impregnated paper wicks. These wicks should have a high liquid retention and be of the same length as the gel and less than 5% of the width of the gel. Non volatile acid and base are used as anolyte and catholyte. Most authors recommend stronger electrolyte solutions than are commonly used for gel tubes apparatus. Solutions of 1 M phosphoric acid and 1 M sodium hydroxide have been found satisfactory for most purposes. Alternatively, as suggested in the Bio Rad technical bulletin 1030 (1975), a solution containing a mixture of NaOH and $Ca(OH)_2$ could be used as catholyte. Calcium hydroxide will form insoluble calcium carbonate and prevent the build up of carbonate in the basic electrolyte, thus eliminating the migration of these ions into the gel.

Although strong acid and base are useful for most pH ranges, better results may often be obtained by using other electrolytes such as ampholytes, whose pH ranges encompass that of the gel ampholyte. Thus Vesterberg (1975) used 1% Ampholine pH range 4–6 and 6–8 as anolyte and catholyte, respectively, to maximize the spread of a pH 5–7 Ampholine in a gel. Suggested formulations for anode and cathode solutions are given in Table 3.2. As shown in this table, when focusing in the pH range 2.5–6, the recommended cathode solution is Ampholine in the pH range 5–7. When focusing in the pH range 7.5–10.5, the recommended anode solution is Ampholine pH 7–9

TABLE 3.2

Suitable electrode solutions for IEF in gel slabs (Davies 1975).

pH range	Cathode	Anode
3.5–9.5	1 M NaOH	1 M H_3PO_4
2.5–6	0.5% Ampholine pH 5–7	1 M H_3PO_4
5–8.5	0.1–1 M NaOH	0.1–1.0 M H_3PO_4
	or 1% Ampholine pH 8–10	*or* 1% Ampholine pH 5–7
7.5–10.5	1 M NaOH	0.1% Ampholine pH 7–9

Subject index p. 587

(Davies 1975). The wicks should be wetted with these solutions before application to the gel. Care should be taken to ensure that the wicks are uniformly wetted. A convenient and satisfactory method is to place the wicks on a glass plate and pipette a standard amount of electrolyte between the interface of the wick and the plate. This ensures uniform wetting and the removal of any air bubbles that might be trapped in the wick. Excess liquid should be removed by blotting.

The gel plate should now be placed on top of the cooling block. To ensure good contact with a fast heat transfer it is essential to have a water layer between the gel plate and the cooling face. This may be achieved by wetting both the cooling surface and the gel plate with a solution of non-ionic detergent. Alternatively, kerosene or light paraffin oil can be used. The gel plate should be lowered slowly over this solution so that a straight water front, without trapped air, forms between the two faces (see Fig. 3.12). The electrode is now placed in position in preparation for electrolysis.

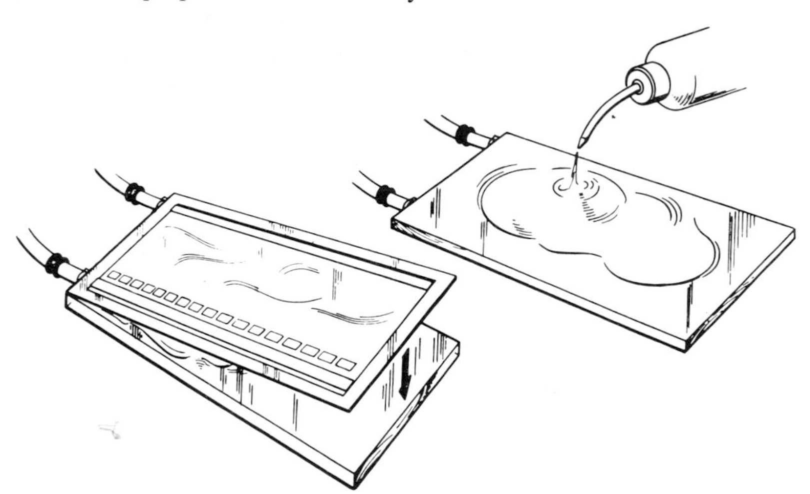

Fig. 3.12. Setting of the polyacrylamide gel plate on the cooling block. The cooling block is first wetted with a solution of non-ionic detergent, or with kerosene or light paraffin oil, and then the gel plate is gently lowered on the cooling block, so that air bubbles are completely excluded. This ensures uniform cooling. (Courtesy of LKB Produkter AB.)

Pre-cast plates Recently LKB has introduced pre-cast polyacrylamide gel plates, called 'Ampholine PAG plates'. These plates, which at present are available only in the wide pH range 3.5–10, present some interesting features. They are of the same overall dimensions as the gels cast in the LKB chamber, however they are only 1 mm thick. This improves thermal equilibration with the cooling block and allows faster staining and destaining of the focused bands. Instead of being cast against a glass plate (1 mm thick) as a support, they are strongly bound to a very thin (less than 100 μ) plastic film which allows easy handling of the thin gel slab and better heat exchange with the cooling chamber. Also, since these gels adhere firmly to the plastic backing, they can be stained and destained without removal of the plastic film, thus preventing possible tear of the gel matrix – a hazard often encountered when handling thin layer slabs. Another advantage is that, when analyzing only a few samples, the gel slab can be sliced with a sharp razor blade and the unused portion of the gel stored for subsequent use.

Selection of electrode positions Because samples may be introduced at any position along the gel track, considerably more latitude is permissible in designation of electrode positions with thin slab techniques than with gel cylinders. However, since sample slots are provided in some models and are usually positioned close to the edge of the gel, electrode polarity should be chosen in light of the earlier considerations (pp. 445–447). Focusing may be conducted across either the short of the long axes.

Sample application Samples may be applied in a variety of ways to thin layer slabs. Some form of template, either with preformed indentations or basins that restrict lateral spread is advisable. As many as 20 tracks, each 1 cm wide, may be handled in the LKB gel plate. Small indentations in the gel are convenient for sample application. However, samples may be applied either directly to the gel surface, or on some suitable absorbent. Fig. 3.13 indicates some of the ways for applying samples to thin layer plates. The choice of absorbent is important. Wadström and Smyth (1975a) have observed that many proteins are not readily eluted from the commonly used cellulose filter

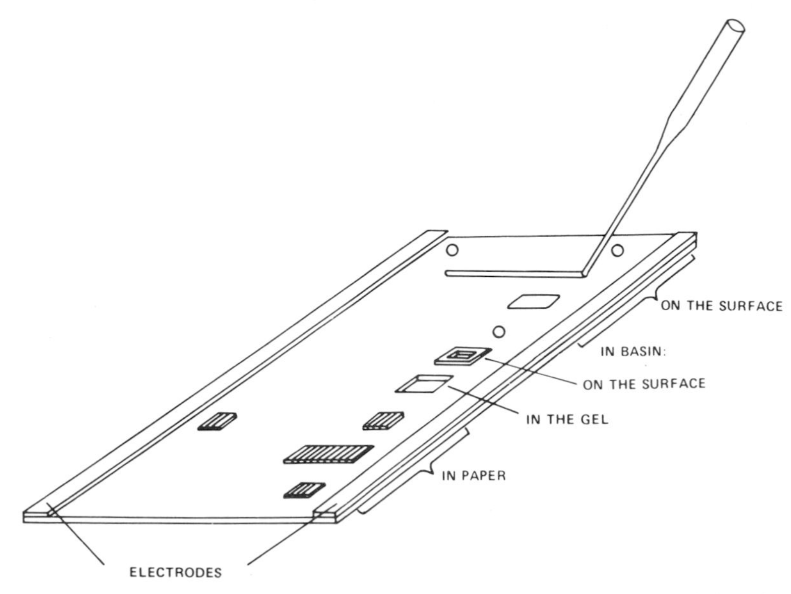

Fig. 3.13. Different ways of applying a protein solution to the gel for isoelectric focusing. From left to right: soaked into a square of chromatography paper; into a rectangle of the same material, and into two squares, one placed close to each electrode; in a basin in the gel; in a basin on the gel surface; as a droplet left on the gel; as a droplet spread over a rectangular area; as a streak; and finally as two droplets, one close to each electrode. (From Vesterberg 1973c, by permission of LKB Produkter AB.)

papers, in which case cellulose acetate or dessicated polyacrylamide may be more suitable (see Table 3.3). Samples of unknown concentration may be applied from a strip of paper cut in the form of an acute-angled triangle. Focused zones appear as long thin lines of linearly increasing concentration, thereby facilitating the estimate of appropriate sample loads for subsequent experiments. Dilute samples may be loaded from a rectangular paper or from a suitably sized trough in the gel. As much as 150 μl may be applied in a trough, provided the sample is equilibrated with the same ampholyte concentration as the gel to prevent local discontinuities in the pH gradient.

Electrolysis conditions Many widely different electrolysis conditions have been recommended for IEF in thin layers. As always, the

TABLE 3.3

Absorbent papers and cellulose acetate membranes employed as sample applicators (Wadström and Smyth 1975a).

Absorbent paper (type)	Grade	Weight* (g/m^2)	Thickness (mm)
Whatman	1	87	0.16
	3	185	0.16
	3 MM	180	0.32
Schleicher and Schüll	2043 B	120	0.23
Munktells	00	80	0.25
	0B	90	0.20
	3	90	0.25
	5	130	0.35
	2 OR	70	0.20
	1300	90	0.10
LKB electrofocusing strip		520	1.20
Beckman	Blotter	230	0.50
Gelman	Absorbent	310	0.80
Millipore	AP100420R	290	0.90
Cellulose acetate		*Type*	
Millipore		Filter membrane (0.45 μ)	
Beckman		Electrophoresis membrane	
Gelman		Sepraphore III	

* Data from Morris and Morris (1964) or determined by the authors.

power delivered to the gel should not exceed the cooling capacity. Generally, an average power of 10 watts is adequate for standard gels 25 × 10 cm and 2 mm thick. For constant voltage regulation, a voltage of 50 V/cm after an hour electrolysis will usually give focusing patterns in 4–6 hr. For power supplies capable of regulating power, we have obtained satisfactory results with a potential difference (DC) of 80 V/cm (at equilibrium) at a regulated power of 10 watts. However, several authors have obtained rapid patterns by using very high voltages with regulated power supplies. For example Söderholm and Wadström (1975) and Righetti and Righetti (1975) reported equilibrium focusing in only 1 hr by applying as much as 50 W to a 25 × 10 × 0.2 cm gel. With the wide range Ampholine, linear pH gradients were established between pH 3.5 and 9.5 after only 10–15 min. The gradient remained fairly stable for at least one hour. At 2.6 mW/mm^3, the gel temperature rose 20°C above the temperature of the cooling block. The temperature rise was however, not uniform. After 10 min, a hot zone was formed near the anode, but it later disappeared. When the pH gradient had been established, the temperature near the electrodes varied between 20 and 25°C. The authors wisely stress the importance of knowing and controlling the temperature at all points on the gel throughout the experiment. Thus, although these very high voltages and rapid banding patterns may be valuable for studying heat stable proteins, it would appear to be limited to studying substances that are unaffected by such large temperature rises.

For additional details for focusing in thin layer slabs the reader is referred to articles by Davies (1975), Karlsson et al. (1973) and Vesterberg (1975).

The following general aspects should also be considered:

1) The efficiency of gel cooling. The efficiency of heat transfer per cm^2 surface area will depend on the materials used to construct the cooling block and the thin slab plate.

2) More efficient cooling is achieved in slabs of high surface area/gel volume.

3) The thicker the gel, the more heat will be generated for any given applied potential difference. Ideally, one would like to focus in a thin film of gel, but such gels are difficult to make. The lowest convenient thickness we routinely use is about 1 mm.

4) Temperature of coolant. The lower the temperature of the coolant, the higher the electrical load that can be applied.

5) Concentration of ampholytes. The current carried in the gel is directly related to the level of ampholytes. As mentioned before, a minimum level of 2% ampholyte is, however, required for most purposes to allow the development of smooth and stable pH gradients. Gel additives such as sucrose may increase the resistance in a gel by increasing the viscosity.

6) The gel path. The important value in determining electrical load is the potential drop per cm. Thus higher potential differences may be tolerated when running the samples lengthwise rather than across the width of the gel.

7) Variation in conductivity. As mentioned previously, the electrical field need not be uniformly distributed throughout the gel. In the wide pH range Ampholine, pH 3.5–10, the areas close to the electrodes heat more than the remainder of the gel in the initial stages of IEF (Davies 1975). As the pH gradient develops these warm zones broaden and move towards the middle of the gel. At very high electrical loads, these 'hot spots' correspond to areas of condensation on the inner surface of the electrode lid.

Storage of thin layer gels Gel slabs may be conveniently dried and stored for a permanent record. Once the gel has been destained, it is rinsed in distilled water, to remove excess alcohol and acetic acid, and then is equilibrated in 25% (v/v) glycerol. After careful transfer to a glass or plastic surface, the gel is covered by dialysis tubing and allowed to dry in air or in a ventilated oven at 37°C. The dialysis tubing should be stretched taut across the gel surface and air bubbles should be eliminated. Originally we used regular dialysis tubing (Visking), cut open and stretched on the gel. However, these tubings are rather thick, and they tend to form corrugations upon drying and thus to crack the gel. A much better solution is to cover the gel with an ultra-thin dialysis membrane, such as the ones used in kidney units in hospitals (T. Wadström, personal communication).

3.2.15. IEF in thin layers of granulated gels

This is the analytical version of the Radola technique, already discussed on p. 405. Granulated gels offer some advantages over continuously polymerized polyacrylamide gels. With granular gels, there are no risks due to polymerization catalysts that are known to produce artefacts in polyacrylamide gels (see p. 443). Gel porosity is not a limiting factor. There is no steric hindrance on focusing very large molecules.

The methodology is essentially the same described previously for preparative applications (see p. 405). Glass plates of several sizes (20 × 10 cm; 20 × 20 cm or 40 × 20 cm) are used and coated with a layer of Sephadex G-75 (superfine) of 0.6 to 1 mm thickness for analytical purposes or 2 mm for small-scale preparative purposes. A schematic drawing of the apparatus is shown in Fig. 3.14. The thick gel

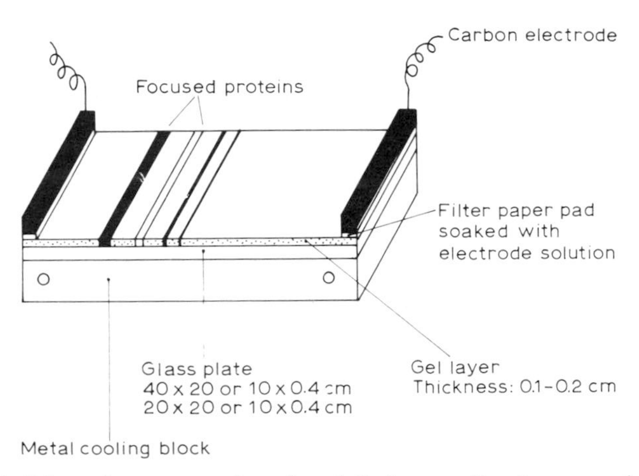

Fig. 3.14. Schematic representation of analytical or small-scale preparative IEF in layers of granulated gels. Notice that the filter paper pads and the carbon electrodes are laid flat on the gel surface. (From Radola 1975, by permission of Butterworths.)

slurry, containing 1% Ampholine, is dried in air until the gel coat does not move when the plate is inclined at an angle of 45°. This corresponds to a loss of approximately 25% water from the gel slurry. The sample is applied usually at the middle or at the anodic side, soaked in pieces of filter paper or as a drop or a streak on the gel layer. The plate is then placed on the cooling block of the Desaga 'Double Chamber'. The electrical field is applied through flat carbon electrodes sitting onto pads of a thick paper (such as type MN866 with a surface weight of 650 g/m^2, Machery, Nagel and Co., Duren, W. Germany) soaked with 1 M sulphuric acid at the anode and 2 M ethylene diamine at the cathode (Radola 1971, 1973a,b, 1975). Proteins are located after focusing by the print technique with a filter paper (type MN827, surface weight 270 g/m^2, Machery, Nagel and Co.) gently rolled on the gel layer. The prints are stained with one of the following stains: Light Green SF, Coomassie Brilliant Blue R-250, Coomassie Violet R-150 and Coomassie Brilliant Blue G-250.

3.3. Detection methods in gels

3.3.1. Staining procedures

Special procedures have to be used for staining proteins separated by IEF since the carrier ampholytes form insoluble complexes with many protein stains (see p. 368). This problem can be overcome by first precipitating focused proteins in acid, e.g. 5% trichloroacetic acid (TCA), and eluting the acid soluble ampholytes with exhaustive washing. After removal of ampholytes, proteins can be detected by conventional staining procedures. Several direct staining methods have now been developed to circumvent this laborious and time consuming method. Their success appears to be in discriminating between complexes of dyes and proteins on the basis of their differential solubility in alcoholic solution and/or their temperature and pH stability. Several convenient direct staining procedures are described below. These methods generally give satisfactory results with commercial sources of ampholytes. However, because of batch variability in some preparations, the validity of these methods should always be checked by staining gels without added sample. Occasionally, ampholyte species are found which cannot be distinguished from proteins by any of the usual procedures (Otavsky and Drysdale 1975; Wadström and Smyth 1975b). Presumably these represent high molecular weight aggregates of ampholytes. In such cases, it is especially important to run control blank gels to determine this background 'noise'. However, this should not be a problem any longer, since today LKB screens all the Ampholine batches for TCA precipitable and Coomassie Blue-stainable material, and discards the positive batches (H. Davies, personal communication).

As with gel electrophoresis, many of the staining procedures differ considerably in sensitivity. Such differences may be useful in analysing patterns with a wide range of band densities on the gel. For example, the major components may first be identified by using a dye with a low binding capacity and a low colour value such as bromophenol blue or Light Green SF. Minor components may subsequently be visualised by overstaining with a more sensitive stain such as Coomassie Brilliant Blue. This two stage procedure offers the best chance of visualising all components since large amounts of proteins in closely spaced bands may not be distinguishable with sensitive methods, due to 'filling-in' with the more sensitive stain. Several methods of different sensitivity are described below. Generally, most efficient staining and destaining is achieved in large volumes to allow rapid equilibration. A simple staining apparatus may be constructed by suspending perforated glass or plastic tubes in the appropriate solution in a covered tank. Efficient mixing may be achieved with a bar magnet driven by a supporting magnetic stirrer. Charcoal or ion exchange resins may be used to absorb dye or ampholytes for faster destaining.

When staining thin slabs, it is advisable to leave the gel on the glass plate for support. A simple method is to run rubber bands around the paper wicks to hold these in place (Fig. 3.9). This arrangement allows dye or destain solutions to bathe both sides of the gel and also facilitates subsequent handling of gels for photography, autoradiography, etc.

Fast Green FCF (*Riley and Coleman 1968*) For detecting bands containing more than 10 μg of protein in 3 mm gels. Immerse gels for 4–8 hr in an aqueous solution of 0.2% Fast Green FCF (Fisher Scientific, N.J., U.S.A.) in an aqueous solution containing 45% ethanol-10% acetic acid. Destain in an aqueous solution of 10% acetic acid-25% ethanol. Fast Green FCF stained bands tend to fade away with time.

Light Green SF This method has been used by Radola (1973a) to stain gel patterns by the paper print technique. After taking the gel print, the paper is washed 15 min in TCA. TCA is subsequently removed by rinsing for a few minutes with a mixture of methanol:-water:acetic acid (33:66:10, v/v/v) and then the print is stained for 15 min by immersion in the same mixture containing 0.2% Light

Green SF. For destaining the above mixture is used in the absence of dye. This stain is 3 to 5 times less sensitive than either Coomassie Brilliant Blue R-250 or Coomassie Brilliant Blue G-250. However, Radola (1973a) has used it successfully to resolve protein patterns in regions of very high protein loads, where staining with Coomassie Brilliant Blue would have produced blurred and unresolved zones.

Bromophenol Blue After Awdeh (1969). Immerse gels for 3 hr in an aqueous solution of 0.2% Bromophenol Blue in 50% ethanol–5% acetic acid. Destain in 30% ethanol, 5% acetic acid in water. Gels dehydrate considerably in this staining solution and should be handled carefully. Their affinity for fingers, tissue paper or other dry surfaces is high, so beware!

Coomassie Brilliant Blue R-250 Several direct methods have been described for staining with this popular and sensitive dye. Four variations are offered here.

1) After Hayes and Wellner (1969).
After IEF the gels are fixed in 5% TCA–5% sulphosalicylic acid solution for at least 1 hr. They are then washed in 3 liters of water (at least 1 hr) to remove most of the excess acid. Staining is accomplished by transferring the gels to a solution of 0.066% Coomassie Brilliant Blue R-250 in 0.2 M Tris–HCl buffer, pH 7.7. Staining is allowed to proceed for 3 to 4 hr. Background stain is removed by washing the gels in dilute buffer (Tris–HCl, 0.001 M, pH 7.7) for 24 to 48 hr.

2) After Spencer and King (1971).
This is a valuable method for detecting focused protein bands rapidly. Proteins absorb dye from a weak solution (0.01%) in 5% trichloroacetic acid, 5% sulphosalicylic acid and 20% methanol in water. Although less sensitive than other procedures with Coomassie Brilliant Blue, background staining is very low and only one staining solution is required. Banding patterns may usually be detected after about 1 hr. The intensity of the bands may be enhanced by increasing the level of dye to 0.05% but at the expense of higher backgrounds.

3) After Vesterberg (1972).
This method stains proteins with Coomassie Brilliant Blue without

interference from ampholytes by heating the gels at 60°C for 15 min in a solution of 0.1% dye in 28% methanol, 11% TCA and 3.5% sulphosalicylic acid, in water. Destaining is also effected at 60°C in 25% ethanol −8% acetic acid in water. This method is more convenient with gel slabs rather than cylinders. Although rapid, stained patterns often tend to fade. In this method, the dye tends to precipitate on the gel surface and to be trapped in gel cracks. A modification of this procedure has been described by Söderholm et al. (1972). After the experiment, the gel is immediately transferred to a bath containing 2% (w/v) sulphosalicylic acid, 11% (w/v) TCA and 27% (v/v) methanol in distilled water, at 65°C. After 20 min, the gels are washed twice in destaining solution (8.5% acetic acid–27% ethanol in distilled water) at 20°C and then stained with 0.1% dye dissolved in destaining solution.

4) After Righetti and Drysdale (1974).
We have developed a method that combines high sensitivity and low background and is equally suited for slabs and cylinders. Gels are immersed for 4–6 hr in a solution of 0.05% Coomassie Brilliant Blue and 0.1% cupric sulphate in acetic acid–ethanol–water (19 : 25 : 65). Gels are destained for 4 hr in the same solution but containing only 0.01% Coomassie Brilliant Blue and finally in acetic acid–ethanol–water (10 : 10 : 80). Gels should be handled with gloves since this method stains fingerprints!

For alternative procedure with Coomassie Brilliant Blue see Malik and Berrie (1972). All of the above procedures may also be used after fixing gels in trichloracetic acid provided the acid is first removed by washing the gel in water.

Reisner et al. (1975) have recently described a direct staining procedure with Coomassie Brilliant Blue G-250 that does not require alcohol. They found that the dye remains in its leuco form in concentrations of perchloric acid that effectively percipitate proteins but that it changes to its blue form when coupled to protein. They stain gels directly in a 0.04% solution of Coomassie Brilliant Blue G-250 in 3.5% perchloric acid. Proteins stain blue against a faint orange background.

Table 3.4 summarizes different protein staining techniques used after thin layer isoelectric focusing in granulated gels, by the method of the paper print (Radola 1973a).

3.3.2. Densitometry of focused bands

The dimensions of most gel cylinders used in IEF are compatible with scanning devices in most spectrophotometers. Equipment normally used for densitometric evaluation of patterns is obtained by disc-gel electrophoresis. The gels used in the apparatus of Righetti and Drysdale (1971), may be scanned in a 10 cm cuvette in a Gilford Model 240 recording spectrophotometer fitted with a linear transport device. Quartz or glass gel tubes fit directly into the cuvette holder, so that the focused gels can be scanned in the tube without being extruded. Fawcett (1969) adapted a Amicon SP 800 spectrophotometer either for scanning gels at a constant wavelength or for obtaining spectra of individual components.

Catsimpoolas (1973c) has devised the technique of in situ analytical scanning IEF. He performs IEF in a quartz cell held vertically in the chamber of a modified Gilford linear transport device. The focusing column can be monitored continuously and the developing patterns observed. This system gives more reliable analysis of banding patterns in sucrose density gradients, as it permits the detection of closely spaced peaks that might diffuse together during elution in the absence of an electric field. In situ scanning also indicates the end-point in IEF, which is important when the pH gradient becomes unstable on prolonged electrolysis (plateau phenomenon, Chrambach et al. 1973).

Densitometric evaluation in thin layers has been carried out by Radola (1973b) with a Schoeffel SD 3000 spectrodensitometer (Schoeffel Instruments, Westwood, N.J., USA). The Zeiss MPQ II chromatographic spectrophotometer allows densitometry by transmittance, reflection and fluorescence emission and is compatible with gel cylinders, polyacrylamide thin layers, granulated flat beds and paper or cellulose acetate membranes up to dimensions of 20 × 20 cm. This scanning spectrophotometer can also be used to obtain spectra of individual bands in the electropherogram.

Subject index p. 587

TABLE 3.4

Protein staining in thin-layer isoelectric focusing by the paper print technique (*Radola*, 1973*a*). Prior to staining the carrier ampholytes were removed by washing with trichloroacetic acid (each washing 15–30 min). The trichloroacetic acid was washed out of the paper for a few minutes with a mixture of methanol–water–glacial acetic acid (33 : 66 : 10, v/v/v). The paper print was stained for 15–30 min with occasional stirring and then kept for 5–15 min in several changes of the destaining solution. The first two washings were rejected, the other were decolourized with activated charcoal and reused.

Dye	Washing with 10% (w/v) trichloroacetic acid No. times	Dye concn % (w/v)	Methanol : water : glacial acetic acid (v/v/v)		Time factor for destaining**	Sensitivity*** μg protein
			Staining solution	Destaining solution		
Amido Black 10 B	3	0.2	50 : 50 : 10	66 : 33 : 10	10–20	5–10
Bromophenol Blue	0 or 1*	0.1	66 : 33 : 10	40 : 60 : 10	2–5	3–5
Coomassie Brilliant Blue G-250	3	0.2	50 : 50 : 10	33 : 66 : 10	3–5	0.5–1
Coomassie Brilliant Blue R-250	3	0.2	50 : 50 : 10	33 : 66 : 10	3–5	0.5–1
Coomassie Violet R-150	0 or 1*	0.2	50 : 50 : 10	33 : 66 : 10	2–3	2
Light Green SF	0 or 1*	0.2	33 : 66 : 10	33 : 66 : 10	1	3

* Removal of the carrier ampholytes was not necessary but destaining was more rapid and uniform when a single washing was applied.

** For Light Green SF the time necessary to get a completely destained background was 20–30 min.

*** μg protein detectable when the sample was applied as a 18 mm zone.

One of the limiting factors in densitometric evaluation of a stained protein pattern is the resolution afforded by the scanning instrument. The absolute resolution of a scanner can be defined as the minimal distance between two discrete points, whose tracing is a valley touching the baseline (Zeineh et al. 1975). By this criterion, with the best scanners presently available, the resolution limit is 230 μm. That is, peaks must be spaced 230 μm apart or more, in order for the recording pen to reach the baseline between the two peaks. In practice, with most scanners, the resolution limit is much higher than that (1 mm or more). Although narrower slits may be used to improve the resolution, Zeineh et al. (1975) have reported that, with slits narrower than 100 μm, the resolution is not improved, but lost, since the beam diverges after passing through the slit. Very narrow slits are also difficult to use, especially in the UV region, because the amount of energy reaching the sample is too low to allow good densitometry. Therefore, presently available scanners do not do justice to the high resolving power of IEF where, especially in thin layer gels, gel scanners reaching a resolution of 50 μm are often needed.

To overcome that, Zeineh et al. (1975) have built a soft laser scanning densitometer. This instrument has as a light source a polarised, monochromatic, coherent soft laser light, of 633 nm wave length, with waves vibrating in the same plane as the slit. This red wavelength is useful for quantitation of Coomassie Blue stained protein patterns or for immuno-precipitin lines or turbidity assays (Ponceau S, or red dyes stained proteins will be transparent to the radiation). An additional UV soft laser beam at 320 nm for fluorescence or UV absorption of protein bands is also available. This laser scanner does not use the adjustable slit system of conventional scanners and the beam width is internally controlled. The laser beam can be reduced to a micro-spot having a diameter as narrow as 3–10 μm. This eliminates the interference problems produced by the slit and minimizes spherical aberration produced by the tube containing the stained gel. Also, the recorder pen goes back to the baseline when the spacing between adjacent bands is more than the beam width. The microbeam is also able to resolve a protein pattern in parallel and uniform gaussians, even if the

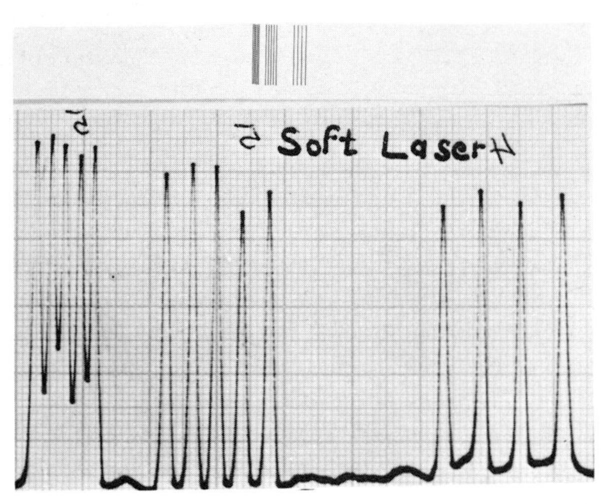

Fig. 3.15. Actual size picture of the illustration plate (top) and its scanning (bottom) by the soft laser densitometer. From left to right each group has 5, 5 and 4 lines and the interspacings of each group are 65, 160 and 320 μm. The left group was not resolved by any conventional scanners. (From Zeineh et al. 1975.)

actual band profile is skewed, zigzagged or arc-shaped. Two adjacent bands 80 μm apart are fully resolved on this system, as seen in Fig. 3.15. Zeineh et al. (1975) have anticipated that, if needed, the resolution of the laser can be improved to a few hundred angstroms, by modifying the presently available equipment.

3.3.3. Specific stains and zymograms

Most of the specific stains developed for detecting proteins after gel electrophoresis can usually be adapted for use in IEF. Some of these methods depend on a specific stain for a cofactor such as a metal or prosthetic groups e.g. haem, or other component, e.g. carbohydrate, lipid etc. Others depend on identifying the position of a protein by its biological activity. Some examples of the various possibilities are given below.

Histochemical staining: In many cases, specific stains may be used to detect substances bound to the protein. In the case of metalloproteins, such as transferrin (Latner 1973), ceruloplasmin (Latner

1973) and ferritin (Drysdale 1970), the protein zone may be located by a specific color reaction for the metal.

Alternatively, proteins which bind specific metals can often be detected by instrumental neutron activation analysis. The sample, in sealed ampoules of high purity quartz, is irradiated in a reactor with thermal neutrons. Measurement of gamma ray activity of the long lived radionuclides is performed by using a GeLi detector (Ortec) coupled to a multichannel analyser (Intertechnique). Schmelzer and Behne (1975) have used this method to detect trace elements bound to serum proteins, such as selenium, chromium, zinc, silver, scandium, iron, cobalt and antimony.

Proteins with cofactors or prosthetic groups such as haem or flavin may also be detected by specific stains. Conjugated proteins offer other possibilities. For example, lipoproteins may be revealed directly by staining the lipid with Sudan black B (Kostner et al. 1969) and glycoproteins with the periodic acid Schiff stain (PAS) for carbohydrate without removal of ampholytes (Catsimpoolas and Meyer 1969).

Proteins which bind ligands can be revealed after IEF by incubating the gel with an appropriate radioactive ligand. Thus concanavalin A can be detected with ^{14}C-α-methyl-D-glucoside, the intrinsic factor from gastric juice with ^{14}C-cyanocobalamine and thyroxine-binding-globulin with ^{131}I-thyroxine. Keck et al. (1973a) described an interesting approach for the detection of immunoglobulins. After IEF, the focused immunoglobulins were copolymerised with glutaraldehyde in situ, and were thus trapped and immobilized within the gel matrix. The fixed antibodies were then detected by their binding radiolabelled or fluorescent-labelled antigens. Table 3.5 summarizes some common specific stains used in IEF.

Many substances may be detected by their biological activity after IEF. By most criteria, IEF is a very mild procedure for enzyme fractionation. Although separations occur in salt-free media, the polyvalent carrier ampholytes have a stabilizing or protective effect on many proteins and help to maintain them in solution. In some cases, carrier ampholytes may afford even greater stabilization than inorganic salts as has been found for α-haemolysin, protease, hexosami-

TABLE 3.5

Some specific stains used in situ in gel IEF.

Sample	Stain	Authors
Transferrin	2,4-dinitroso-1,3-napthalene-diol (Fe^{+++} reduced with hydroquinone)	Latner (1973)
Apotransferrin	Fe^{+++}-nitrilotriacetate	Hovanessian and Awdeh (1975)
Ceruloplasmin	p-phenylenediamine (oxidase activity)	Latner (1973)
Haptoglobins	haemoglobin/benzidine/ H_2O_2 (peroxidase activity)	Latner (1973)
Haemoglobin	benzidine/H_2O_2 (peroxidase activity)	Latner (1973)
Lipoproteins	Oil Red in 61 % EtOH	Latner (1973)
	Sudan Black B	Kostner et al. (1969)
		Godolphin and Stinson (1974)
Ferritin	Prussian Blue reaction	Drysdale (1970)
Glycoproteins	PAS stain	Hebert and Strobbel (1974)
Immunoglobulins	Glutaraldehyde fixation and then detection with: a) 125-I-labelled antigen; b) fluorescent-labelled antigen; c) radiolabelled polysaccharide antigen	Keck et al. (1973a)
Chromatin nonhistone proteins	Labelling with ^{32}P	McGillivray and Rickwood (1974)
Concanavalin A	α-methyl-D-glucoside	Akedo et al. (1972)
RNA	toluidine blue, methylene blue, or acridine orange	Drysdale and Righetti (1972); Drysdale and Shafritz (1975)
Peptides	pre-labelling with dinitrofluorobenzene	Kopwillem et al. (1973)
Haemopexin	binding of haem	Latner and Emes (1975)
Heparin	Toluidine blue	Nader et al. (1974)
Thyroxine-binding-globulin	^{131}I-thyroxine	Latner and Emes (1975)
Intrinsic Factor (gastric juice)	^{14}C-cyanocobalamine	Latner (1973)

nidase (Vesterberg et al. 1967) and for DNA-dependent RNA polymerases (Drysdale, unpublished observations). In some cases, enzyme stabilization may be due to the chelation of inhibitory heavy metals such as Cu^{2+}, Hg^{2+}, and Pb^{2+} by the ampholytes.

Many enzymes may be detected by a suitable histochemical stain. Fig. 3.16 shows patterns of the isozymes of lactate dehydrogenase

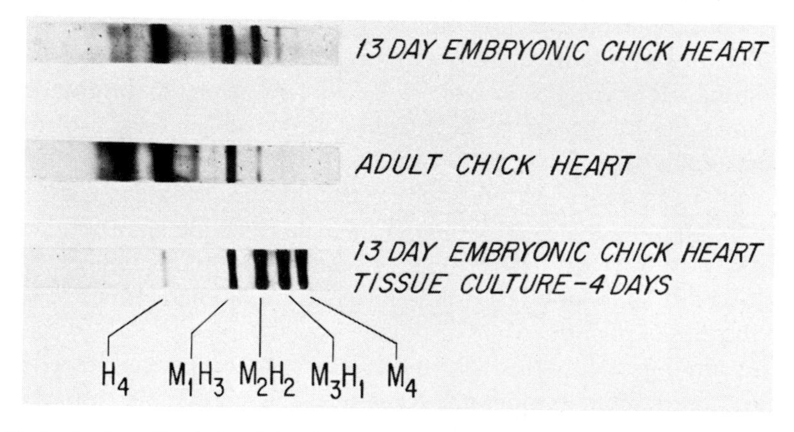

Fig. 3.16. Histochemical staining for lactate dehydrogenase (LDH) isozymes separated from crude supernatant fractions from chicken liver by IEF in gel cylinders. IEF in pH range 3.5–10. (From Righetti and Drysdale 1973, by permission of the New York Academy of Sciences.)

separated from crude supernatant fractions from chicken liver by IEF in gel cylinders. Although this enzyme represented only a small fraction of the proteins in the tissue extracts, it was readily detected by its biological activity. In this case, the enzyme activity was revealed by the reduction of pyrimidine nucleotides formed during the oxidation of lactic acid. The formation of the reduced nucleotides was demonstrated by the reduction of tetrazolium salts to form an insoluble colored precipitate. Appropriate steps must, of course, be taken to counteract the buffering effects of the ampholytes in cases where the enzyme focuses at a pH remote from its optimal pH range. Such corrections can usually be effected by carrying out the incubation in strongly buffered solutions or by a brief pretreatment to equilibrate the gel pH. Since the average molarity in gels containing 2% (w/v) ampholyte is probably in the order of 20–50 mM, it is usually fairly easy to alter the pH with solutions of 100–200 mM. Wadström and Smyth (1975a) have discussed various ways to overcome adverse pH conditions in the gel after IEF.

Occasionally, enzymes may lose activity after IEF due to chelation of necessary metal cofactors. For example, Latner et al. (1970) found a 90% loss of activity of alkaline phosphatase on prolonged electrofocusing. However, this enzyme activity could be largely restored by incubation with the zinc cofactor. Additional problems may also arise from anodic oxidation or perhaps electrolytic reduction but these may usually be prevented by the addition of appropriate additives to the electrolyte solution (§4.2). Although some antioxidants may be added directly to sucrose density gradient solutions, many inhibit the polymerisation of polyacrylamide gels and should therefore be added to the tops of preformed gels or to the electrolytes.

In addition to direct histochemical stains with low molecular weight reactants, several 'zymogram' techniques have been developed in which larger reactants are brought into contact with focused enzymes by overlaying the focused gel with another gel impregnated with the substrate or other reactants. Zymogram techniques are best suited for detecting substances separated in thin slabs. Several proteolytic enzymes have been identified by this method. Vesterberg and Eriksson (1972) demonstrated staphylokinase activity by layering fibrinoclot containing plasminogen over a thin gel slab with the focused enzyme. The staphylokinase converted the plasminogen into plasmin so that the fribrinolytic activity corresponding to the staphylokinase activity appeared as a clear spot on the opaque fibrin plate. Arvidson and Wadström (1973) incorporated casein into an overlaying agar gel to detect staphylococcal proteases after gel electrofocusing while Vesterberg (1973b) detected pepsin activity by the hydrolysis of albumin in an agar overlay. In such cases, the enzyme activity is seen as a clear zone against a blue background when the agar plate is subsequently stained for protein (Fig. 3.17) (Wadström and Smyth 1975b). Fig. 3.18

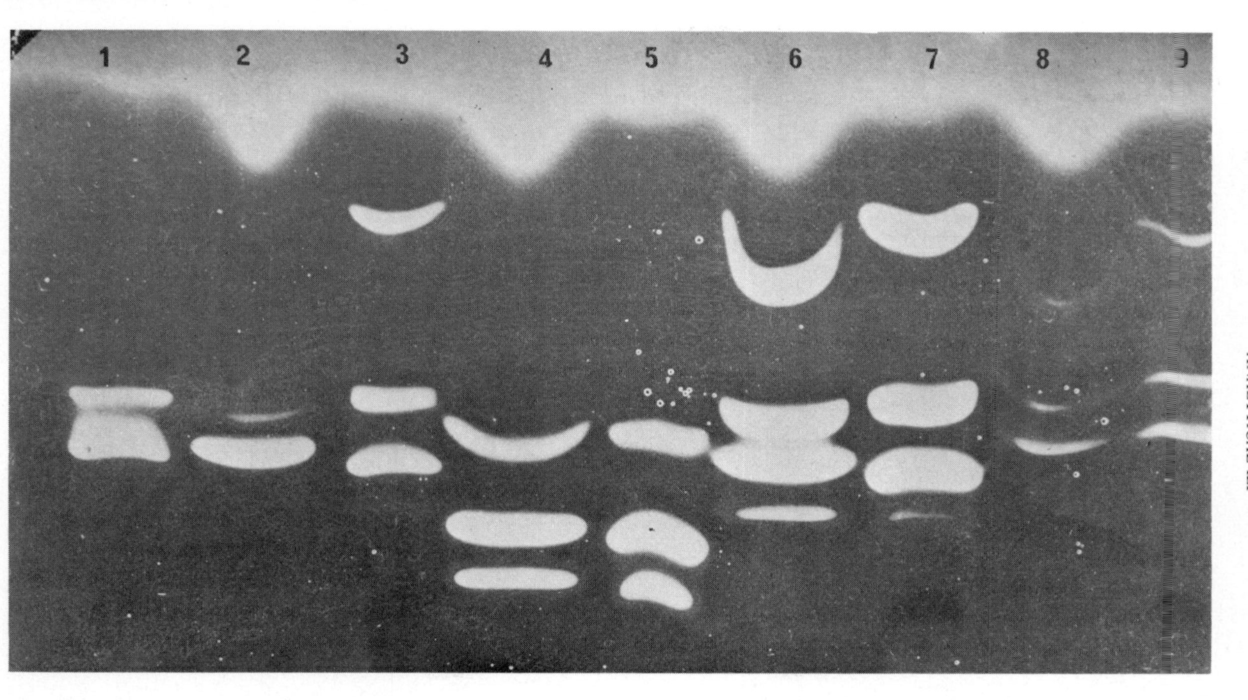

Fig. 3.17. Protease zymogram on casein agar. After IEF in a polyacrylamide gel slab the gel is overlayered with a layer of casein agar. Extracellular proteases of *Serratia liquefaciens* (S.l.) and *Serratia marcescens* (S.m.). From left to right: *S.l.* strain 2; *S.l.* strain 3; *S.l.* strain 26; *S.m.* strain 6; *S.m.* strain 7; *S.m.* strain 8; *S.m.* strain 10 and *S.m.* strain 11. (From Wadström and Smyth 1973b, by permission of ASP Biol. Med. Press B.V.)

Fig. 3.18. Separation of extracellular staphylococcal haemolysins in a gel containing 4 M urea and 2% Ampholine pH 3.5–10. The agar overlayer contains rabbit erythrocytes. From left to right: *S. aureus* (S.a.) strain Newman I; *S.a.* strain G128; *S.a.* strain M18; *S.a.* strain Wood 46 UWO; *S.a.* strain Plommet; *S.a.* strain V8; *S.a.* strain Wood 46 purified α-haemolysin (50 × conc.). As shown in this figure and in Fig. 3.17, IEF can be very useful in bacterial taxonomy. (From T. Wadström, unpublished.)

shows a zymogram obtained after focusing extracellular staphylococcal haemolysins. The agar overlayer contains rabbit erythrocytes and the enzyme zones are seen as clear zones against a red background. As before, appropriate steps may be required to arrange for a suitable gel pH for the reaction. Tables 3.6 and 3.7 list some zymogram

TABLE 3.6

Zymogram methods used with isoelectric focusing and employing low molecular weight substrates.

Enzyme	Substrate	Authors
Lactate dehydrogenase	Lactate/PMS/tetrazolium MTT	Dale and Latner (1968) Chamoles and Karcher (1970a, b) Righetti and Drysdale (1973)
β-Glucuronidase	8-Oxy-chinolin-D-glucuronide/Fast black K salt	Coutelle (1971)
L- and D-amino acid oxidase	L-Leu or D-Phe/PMS/triphenyltetrazolium	Hayes and Wellner (1969)
Esterase	α-Naphthyl acetate/Fast red TR salt	Bianchi and Stefanelli (1970)
Procarboxypeptidase A	n-Carbo-β-napthoxy-L-Phe/Diazo blue B/trypsin	Kim and White (1971)
Alkaline phosphatase	α-Naphthyl phosphate/4-aminodiphenylamine diazonium sulphate	Righetti and Drysdale (1971)
Alkaline phosphatase	β-Naphthyl phosphate/Fast blue BB salt *or* 4-methyl umbelliferone phosphate	Smith et al. (1971)
Acid phosphatase	4-Methyl-umbelliferyl phosphate	Leaback and Rutter (1968)
NAD/NADP dehydrogenase	Substrate/NAD-NADP/PMS/tetrazolium salt MTT	Humphryes (1970)

TABLE 3.6 (continued)

Zymogram methods used with isoelectric focusing and employing low molecular weight substrates.

Enzyme	Substrate	Authors
Peroxidase**	Urea peroxide/guaiacol *or* o-toluidine and mesidine	Delincée and Radola (1970, 1971) Rücker and Radola (1971)
Creatine phosphokinase	creatine phosphate/PMS/ nitroblue tetrazolium	Thorstensson et al. (1975)
Catechol oxidase	4-methyl catechol/ p-phenylenediamine	Dubernet and Ribérau Gayon (1974)
Esterases	naphtyl acetate/fast garnet GBC salt	Young and Bittar (1973)
Lipoxidase	linoleic acid/KI	Catsimpoolas (1969b)
Neuraminidase	2-(3-methoxyphenyl)- N-acetyl neuraminic acid	Groome and Belyavin (1975)
Trypsin and α-chymotrypsin inhibitor	N-acetyl-D, L-phenyl-alanine-β-naphthyl ester/ trypsin/α-chymotrypsin	Kaiser et al. (1974)
β-N-acetyl glucosaminidase	methyl umbelliferyl glycosides	Leaback and Robinson (1974)
N-acetyl-β-DHexosaminidase	naphthol-AS-BI-N-acetyl-β-D-glucosamide/fast garnet GBC salt	Hayase et al. (1973) Hayase and Kritchevsky (1973)

* Abbreviations: PSM, phenazine methosulphate: Leu, leucine; Phe, phenylalanine; MTT, 3-(4,5)-dimethyl-thiazolyn-2,5 diphenyl tetrazolium bromide.

** Print technique with Whatman No. 1 chromatography paper impregnated with buffered substrate.

techniques employing low molecular weight and high molecular weight substrates to detect focused enzymes after IEF. The use of buffered agarose overlayers containing appropriate substrates is particularly attractive for thin-layer gel IEF. Table 3.8 summarizes some overlay techniques used in Wadström's laboratory (Wadström and Smyth 1975a). Alternatively, instead of an agarose overlayer, one could use the paper print technique described by Radola

TABLE 3.7

Zymogram methods used with isoelectric focusing and employing high molecular weight substrates (Wadström and Smyth 1975a).

Enzyme	Substrate	Zymogram technique	Authors
Thin-layer gel technique			
Staphylokinase*	Fibrin clot containing plasminogen	'Substrate' gel	Vesterberg and Eriksson (1972)
Cellulase	Carboxymethyl cellulose	Paper print	Eriksson and Pettersson (1973)
α-Amylase	Starch-iodine	Starch-film print	Beeley et al. (1972)
Pepsin	Albumin/Coomassie Blue	Immersion	Vesterberg (1973b)
Subtilopeptidase, trypsin, δ-chymotrypsin, crude bacterial proteases	Vitamin-free casein	Agarose overlayer	Arvidson and Wadström (1973)

* Plasminogen activator produced by *Staphylococcus aureus*.

TABLE 3.8

Enzymes studied by the use of buffered agarose overlayers containing appropriate substrates (Wadström and Smyth 1975a).

Enzyme	Substrate*	Buffer/ions*
α-Amylase	Phadebas[R] amylase test (Pharmacia) 1 tablet/10 ml	0.05 M Tris–HCl, pH 7.4
Deoxyribonuclease	Salmon sperm (Koch-Light) *or* Calf thymus DNA (Koch-Light) 1 % (w/v)	12.5 mM $CaCl_2$ and 12.5 mM $MgCl_2$
Penicillinase	0.2 % Soluble starch (Merck) Benzylpenicillin (AB Kabi) 20 mM Lugol's solution: dist. H_2O = 1 : 1 (v : v)	0.1 M KH_2PO_4, pH 5.9
Phospholipase C	Soya bean lecithin (Sigma) 1.5 % (w/v) emulsified	0.15 M NaCl in 0.02 M Tris–HCl, pH 7.4, 1 mM $CaCl_2$ and 1 mM $ZnCl_2$
Protease	Vitamin-free casein (NBC) 1 % (w/v), pH 7.4, containing 2 mM (KH_2PO_4)	1 mM $CaCl_2$
Staphylococcal haemolysin	Rabbit, human or sheep erythrocytes (3 %, v/v)	0.15 M NaCl in 0.02 M Tris–HCl, pH 7.4
Elastase	Elastin (Worthington) 0.5 % (w/v)	0.01 M Tris–HCl, pH 8.8

* Final concentrations of substrate and buffer ions in 1.0 or 1.5 % (w/v) agarose overlayers.

(1973a) or a contact print with Celloclear (Chemetron, Milano, Italy). Celloclear is a transparent cellulose acetate membrane (supplied in various sizes and thicknesses) with very large pore size, which allows quick equilibration of high and low molecular weight substrates. The contact print and, if needed, staining and destaining, are performed in a very short time. The membrane can be easily handled and is very resistant to tear (Righetti, P.G. and Molinari, G., unpublished).

When no suitable histochemical stain is available, biological activities may be detected in gel eluates. A particularly convenient method for fractionating cylindrical gels is shown in Fig. 3.19. Gels are cast in a plastic syringe tube. After focusing, a No. 20 snub nosed needle is placed over the tapered end and a rubber disc over the top of the gel. The gel may now be fractionated into 10 μl aliquots by a plunger driven by a Hamilton Repeating Dispenser. By using plastic tubes where gels do not adhere to the tube walls, band distortion is minimized. Alternatively, cylindrical polyacrylamide gels can be fractionated with an egg slicer device (Heideman 1965; Chrambach 1966) or with an array of steel blades (Matsumura and Noda 1973) or with a gel slicer equipped with an iris diaphragm (Peterson et al. 1974). The last system allows sectioning in 1 mm thick slices with a precision of 13 to 34% without freezing or other pretreatment of the gel. Possibly, some of these devices (such as the egg slicer or the set of razor blades) could be used for sectioning of slab gels, with only minor modifications.

3.3.4. Autoradiography

Autoradiography is a highly sensitive detection method, which allows visualization of trace amount protein, not usually detectable with common protein stains. By this technique, for instance, O'Farrell (1975) has detected as many as 1100 protein spots in an *E. coli* cell sap (see p. 494). A protein species representing as little as 10^{-4} to 10^{-5} of one percent of the total protein can thus be detected and quantitated. Autoradiography is usually best performed in flat gel slabs, possibly already dried, since the technique essentially consists of overlayering the gel plate, containing the radioactive spots, with an X-ray film, usually Kodak No-Screen Medical Film, in an X-ray exposure holder.

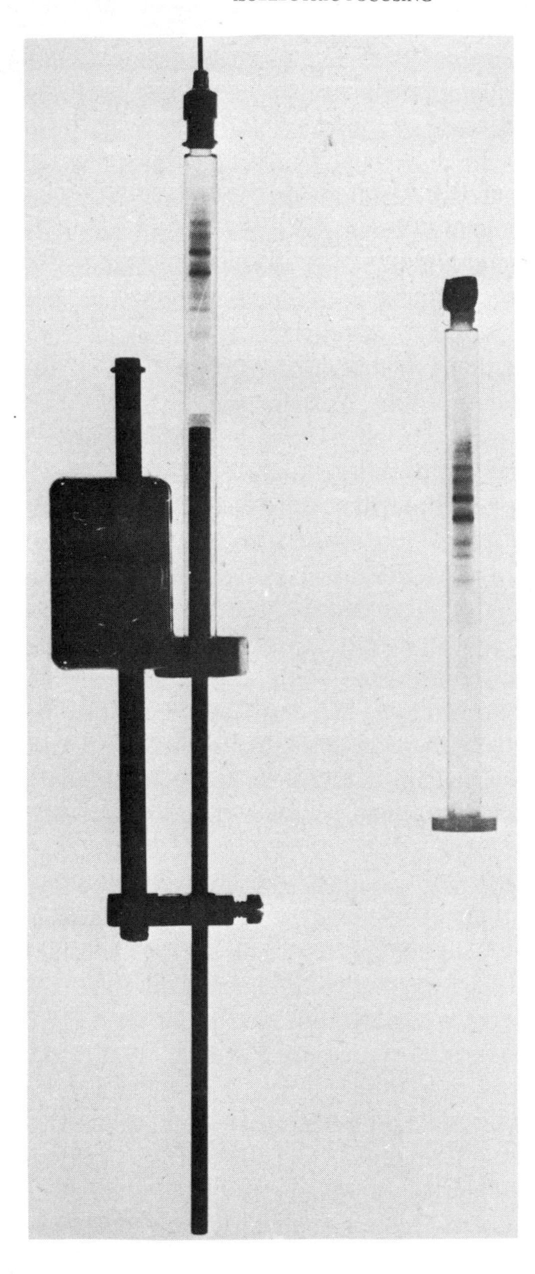

Fig. 3.19. Fractionating device for gel rods. The gel tubes have a round collar at one end and a standard syringe taper tip at the other. After focusing, a syringe needle is attached to the tapered tip and the gel is extruded in uniform fractions through the needle with the aid of a Hamilton repeating dispenser. Note that no band distortion or compression occurs. (From Bagshaw et al. 1973, by permission of the New York Academy of Sciences.)

After the proper contact time, the film is developed with Kodak X-ray Developer.

Usually, proteins are radio-labelled in vivo. However, in the case of antibodies, they can be located by reaction with a radio-labelled hapten, as shown by Williamson (1971) who examined the heterogeneity of anti-DNP antibodies (DNP : 2,4-dinitrophenyl) by separating them by IEF and then incubating the gel for a short time with ^{131}I-α-N-(4-hydroxyphenacetyl)-ε-N-2,4-dinitrophenyl-lysine. These methods require that the molecule used to react with the protein be much smaller than the protein itself, so it can penetrate faster into the gel than the protein can diffuse. Thus, these methods are restricted to those cases in which the specificity is for a small ligand. In cases in which the ligand is another macromolecule, Keck et al. (1973a,b) have devised the ingenious method of first immobilizing the focused antibodies in the gel with glutaraldehyde, and then locating the focused bands by using radioiodinated antigens (see also p. 473). Other proteins can be located by reacting them with radiolabelled specific antibodies. After washing out the excess radioactive macromolecule from the gel, the gel itself is dried and then subjected to autoradiography. An example of this technique is given in Fig. 3.20, which shows an autoradiograph of mouse-anti-DNP-antibodies localized with ^{125}I-DNP-ovalbumin.

One of the major drawbacks of autoradiography is the length of time required to fully develop an autoradiograph. While gamma emitters, such as ^{135}I or ^{131}I, may only need a contact time of few hours, ^{14}C or ^{3}H-labelled proteins might require days or weeks of exposure. Thus, the bidimensional map of *E. coli* proteins shown in Fig. 3.23 had to be exposed for 825 hr (O'Farrel 1975).

To overcome that, LKB has built the 2105 Radiochromatogram Camera. This camera is basically a spark chamber, for determining the distribution of radioactivity on chromatography or electrophoresis media. It functions as a triode, with an anode, a grid made of tungsten meshes and a cathode. The cathode is, in fact, the radioactive source itself, which is inserted into the spark chamber on a tray. The cathode emits electrons, which produce sparks between the anode and the grid,

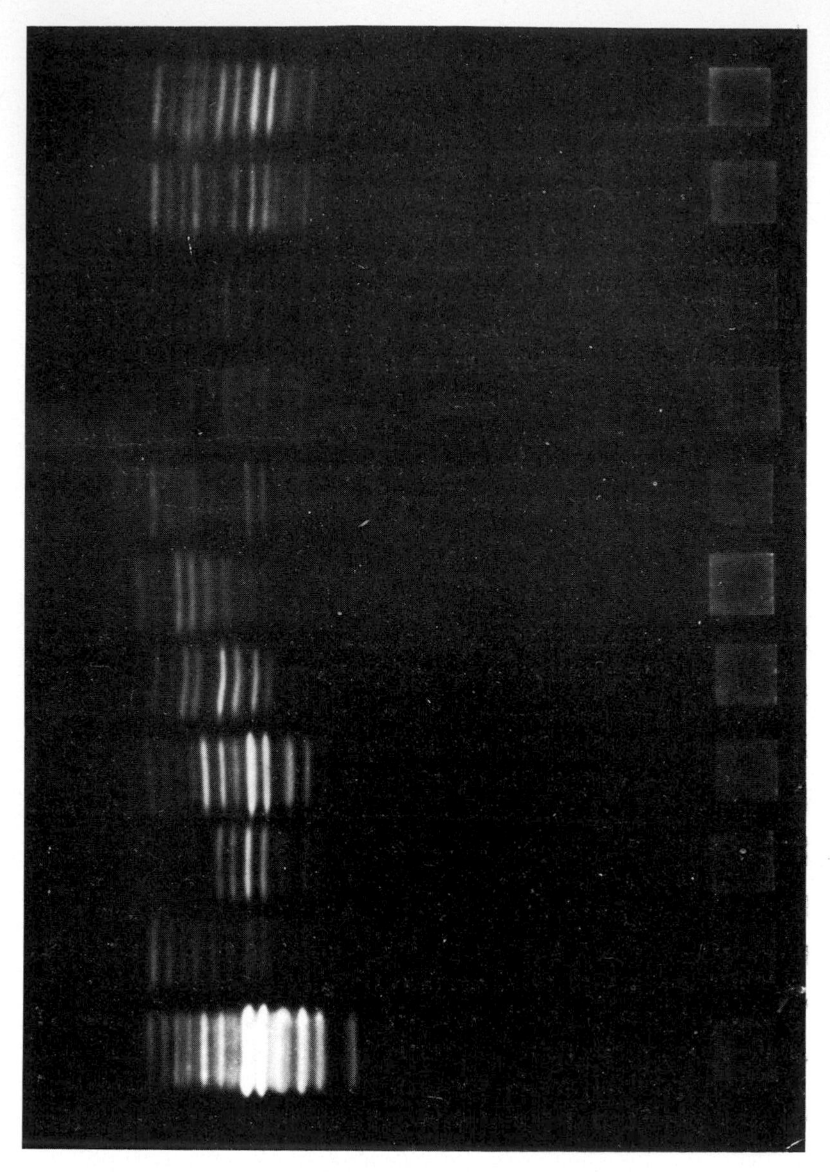

Fig. 3.20. IEF of mouse anti-DNP-antibodies in a thin-layer polyacrylamide gel. After IEF, the bands are immobilized in the gel first by precipitation in 18% sodium sulfate solution and then by cross-linking with 50% solution of glutaraldehyde (see Keck et al. 1973a, b). The bands are then located by autoradiography after treatment with ^{125}I-DNP-ovalbumin. (From K. Keck, unpublished.)

when a sufficiently high voltage is applied between them. At the anode, a Polaroid camera is used to photograph the sparking activity, giving a two-dimensional record of the distribution of radioactivity on the surface of the cathode (Asard 1974). The resolution of the Radiochromatogram Camera is given as 1 mm for ^{3}H and ^{14}C and 2 mm for gamma isotopes. The camera has also a high sensitivity, the lower detection limit being 100 pCi/mm^2 for ^{3}H and 10 pCi/mm^2 for ^{14}C. Perhaps one of the greatest advantages of this instrument is that it is a much faster method than conventional autoradiography (approximately 1000 times faster), thus allowing detection of radioactive spots within reasonable working times. A similar instrument is also available from Birchover Instruments Ltd. (Radiochromatogram Spark Chamber).

3.3.5. Immunotechniques

IEF followed by any type of immunoelectrophoresis has been considered a bidimensional technique and as such will be dealt with further on (see §3.4.1). IEF followed simply by immunodiffusion, possibly within the same matrix used for IEF, will be considered a detection method and will be briefly treated here. Detection by immunoprecipitation of a specific antigen, after IEF, is best done in small gel rods, possibly 1 mm or so in diameter, to reduce to a minimum the amount of antiserum needed for the reaction. After IEF, the gel is extruded in the proper amount of antiserum in a test tube and immunodiffusion allowed to proceed for few hours (Carrel et al. 1969). Precipitin bands are seen as opalescent rings on the gel surface. The polyacrylamide gel matrix does not seem to hamper the antigen–antibody cross-reaction. This might be due to the fact that, in gel cylinders, proteins tend to focus as hollow rings, with most of the protein packed on the gel surface and little in the gel core. This can easily be seen by focusing a colored protein, such as haemoglobin, and then cutting the gel in its proximity. Thus, the protein is readily available for cross-reaction with the antibody.

An alternative to this method consists in cutting the gel rod into discs which are then crushed and transferred to sample wells punched into agar gel. A central trench in the agar slab is then filled with antiserum and diffusion is allowed for 48 hr (sectional immunoelectrofocusing; Catsimpoolas 1969c). Alternatively, the intact gel rod can be embedded in agar. Trenches are then cut parallel to the polyacrylamide gel and filled with appropriate antiserum (Spragg et al. 1973). If focusing has been performed in thin-layer equipment, the gel slab can be overlayered with strips of paper soaked with appropriate antibody (Cotton and Milstein 1973) or a gel strip can be cut and embedded in agar between parallel trenches filled with antiserum (Dewar and Latner 1970).

3.4. Two-dimensional procedures

By combining IEF with other procedures, it is usually possible to obtain more information than is available from either procedure above. Ideally, each dimension should separate proteins according to different parameters. Some of the more popular techniques used in conjunction with IEF are given in Fig. 3.21. Judicious use of these methods will often define several physico-chemical parameters in a distinct manner for ready reference or comparison.

3.4.1. Immunoelectrofocusing

With the higher resolving power of IEF, particularly in gels, one can usually obtain additional information from immunoelectrophoretic analyses by initially separating proteins by IEF rather than by electrophoresis. The feasibility of this approach was demonstrated from the early work of Riley and Coleman (1968) and Catsimpoolas (1969c) who performed IEF in agarose gels on glass slides. After IEF, they detected the separated antigens by the usual method of diffusion of antibodies from a central trough. Although these early experiments suffered from pH gradient instability because of electroendosmosis, it should now be fairly simple to standardise this procedure with the more suitable preparations of 'charge-free' agarose now available (see p. 425). Alternatively, very high porosity polyacrylamide gels may prove adequate. Of course, as with so many other biological assay, the pH of the gel

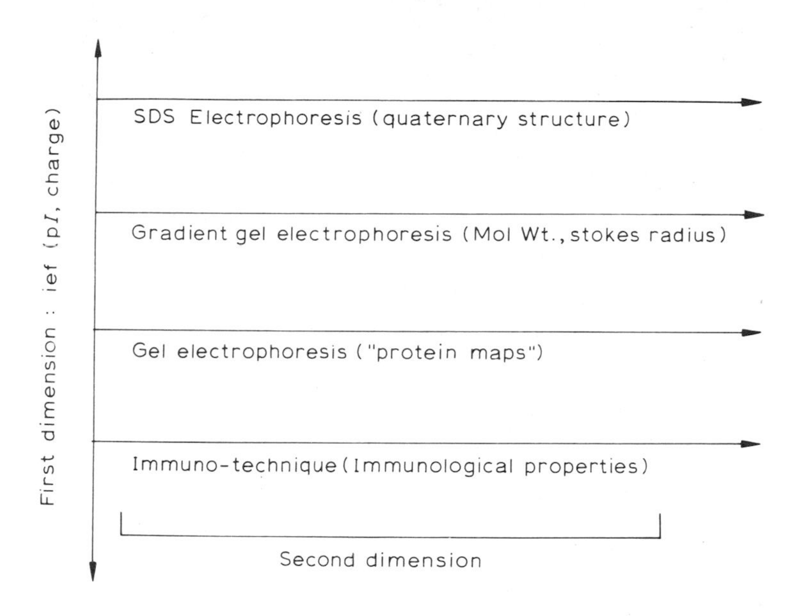

Fig. 3.21. Summary of various two-dimensional procedures. (From Righetti and Drysdale 1974, by permission of Elsevier.)

must be appropriately adjusted to compensate for possibly unsuitable pH prevailing in the gel. A brief washing (2 min) in a 50–100 mM phosphate buffer, pH 7.4 is usually adequate. Alternatively the focused gels may be incubated directly in an appropriately buffered solution containing a specific antiserum (Carrel et al. 1969; Powell et al. 1975).

Other variations of immunoelectrofocusing have also been described. Some of the more common are illustrated in Fig. 3.22. For example, after IEF in rods or strips of polyacrylamide gel, the focused gel can be embedded in an antibody-inpregnated agarose gel, through which the focused proteins are subsequently caused to migrate by electrophoresis (crossed immunoelectrofocusing). Antigen–antibody complexes form conical precipitating zones whose dimensions give a good measure of antigen concentration (Laurell's rocket technique 1966). In addition, it is often possible to investigate possible immuno-

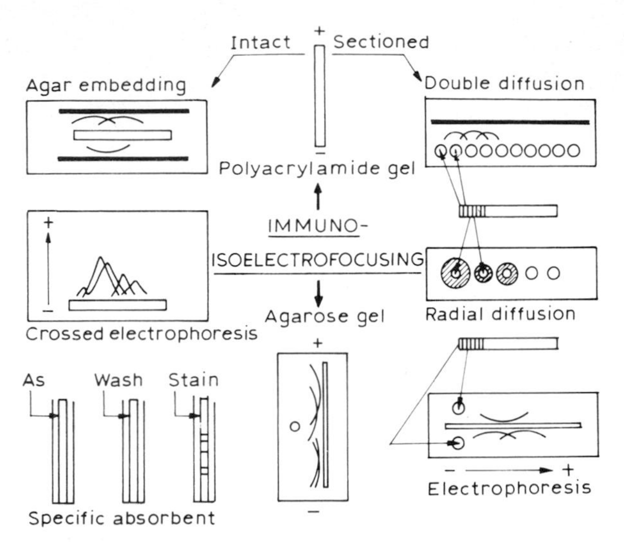

Fig. 3.22. Various forms of immunoisoelectric focusing analysis in gels. (From Catsimpoolas 1973c, by permission of the New York Academy of Sciences.)

logical relationships of closely spaced antigens by this technique (Alpert et al. 1972).

However crossed immunoelectrofocusing, since its description has been beset by a number of practical problems, which have discouraged its application. Only recently Söderholm and Smyth (1975) have made a thorough methodological study of the various factors effecting the technique. For reproducible results, when working with polyacrylamide gel slabs, these authors stress the following points:

a) *Preparation and cutting of the acrylamide strip.* For best results, the gel slices should have even breadth with straight and even edges. This is obtained by covering the gel first with a polyethylene film, and then by cutting the strip with the aid of a plexiglass slab with a slit of appropriate breadth placed on top of the film, and with the aid of a sharp scalpel driven into the gel along the edges of the slit.

b) *Transfer of acrylamide strip to agarose gels.* The best method is to place a long razor blade vertically against one edge of the gel slice, whilst a second blade is slid horizontally under the gel so that about

2/3rds of the slice is resting on the blade. The razor blade bearing the slice is then inverted so that contact is made between the protein-bearing surface of the polyacrylamide gel and the agarose gel.

c) Breadth of the polyacrylamide strips. When the breadth of the slices is reduced so that they are as thin as they are thick, the problems of gel transfer and positioning are increased to such an extent that any benefit gained in resolution is far outweighed by practical considerations. Therefore, Söderholm and Smyth (1975) suggest to cut 4 to 5 mm broad slices, which are readily manipulated.

d) Electrode solutions used for electrofocusing. As suggested by Skude and Jeppsson (1972) the areas of the gel that had been in contact with the electrode strips should be removed, otherwise the strong acid and base in these gel regions would cause uneven conductivity and band distortion in the agarose gel. Alternatively, when applicable, the electrode wicks should be soaked in Ampholine, as suggested by Davies (1975) (see also Table 3.2).

e) Acrylamide agarose contact. Of the various contact methods (e.g. moulding-in of a polyacrylamide strip into antibody-containing agarose, or moulding-in of the polyacrylamide gel strip into agarose intermediate, antibody-free agarose spacer gel, etc.) the laying-on technique reported above gave the most satisfactory results. In fact, in all moulding-in experiments, a water front builds up at the junction between the acrylamide and agarose, resulting in leakage of protein sideways and blurring of patterns.

Focused antigens can also be detected in squashed gel segments by double diffusion, radial immunodiffusion or immunoelectrophoresis (Fig. 3.22, Catsimpoolas 1973c). When working with gel rods, the fractionating device shown in Fig. 3.19 is very convenient since it also homogenizes the gel.

3.4.2. Isoelectric-focusing gel electrophoresis

Several methods have been described for mapping proteins in a two dimensional procedure using GEF in the first dimension and gel electrophoresis in the second. Dale and Latner (1969) were the first to demonstrate the potential of this procedure in detecting quantitative and/or qualitative changes in protein maps from normal and pathological sera. After the separation of proteins by IEF in rods of polyacrylamide, the focused gels are embedded into the top of a slab of polyacrylamide and subjected to electrophoresis through the width of the gel. The focused proteins and ampholytes migrate into the gel slab where their electrophoretic mobility is determined by both their charge and their molecular size. The ampholytes are thus effectively separated from the proteins and a characteristic map obtained. The gel slabs can be dried down onto supporting plates, after washing in 7% gelatin and 3% glycerol, to provide a permanent record. Dale and Latner (1969) were able to identify many of the known components of serum in these plates by running standards or by specific histochemical or immunological procedures.

Wrigley (1968b) and Wrigley and Shepherd (1973) used a somewhat similar procedure to analyse wheat grain proteins. On gel electrophoresis, about 20 components may be detected in wheat gliadin preparations, while a similar number are revealed by GEF. However, by combining these procedures, nearly twice as many components were revealed. A similar bi-dimensional procedure has been used by Macko and Stegeman (1969) and by Stegeman et al. (1973) to map potato proteins. These authors too have found that the combined technique gives a more complex pattern and is more specific for a given potato variety than either electrophoresis or electrofocusing alone.

3.4.3. Isoelectric focusing-electrophoresis in gel gradients

The combined procedures of GEF and gel electrophoresis clearly demonstrated the value of this two dimensional procedure for genetic and taxonomic studies. However, electrophoresis in gels usually separates proteins on the basis of their size and charge and the two may occasionally work against one another to obscure or nullify a separation (Rodbard et al. 1974). Kenrick and Margolis (1970) adopted an earlier two dimensional electrophoretic procedure to eliminate this complication and to separate focused proteins in the second dimension almost solely on the basis of their molecular size. They used a concave 4.5–26% gradient polyacrylamide gel slab in which the focused proteins migrated electrophoretically to a quasi-equilibrium position according to their molecular size or Stokes radius. Thus, in addition to increased resolution, this method gives estimates of the pI and the approximate molecular weight of the native protein.

An interesting variation of this technique has been described by Felgenhauer and Pak (1975). They perform IEF in the first dimension in a granulated gel layer (Sephadex G-75 superfine) and electrophoresis in the second dimension in a linear gradient of 3–30% acrylamide. The use of granulated gels in the first dimension allows rapid equilibration of large proteins, such as ferritin, α_2-macroglobulin and β-lipoprotein, and quick detection of the focused bands by the paper print technique (Radola 1973a; see also p. 464). During IEF, the Sephadex gel is spread on a thin plastic sheet (0.15 mm). This allows easy cutting of the Sephadex strip and transfer to the polyacrylamide gel slab for the second dimension run. During electrophoresis in the gel gradient, the gel chamber is run horizontally and paper wicks used to provide electrical contact with the electrode reservoirs. As in the previous case, provided that each component is allowed to reach its exclusion limit (Felgenhauer 1974), a quasi-equilibrium is achieved which allows a good estimation of molecular size (Andersson et al. 1972; Rüchel et al. 1973).

3.4.4. Isoelectric focusing-SDS gel electrophoresis

The next logical step to studying the pI and molecular size of native proteins was to obtain additional information about quaternary structure and polypeptide molecular weights. SDS gel electrophoresis (Shapiro et al. 1967; Weber and Osborn 1969) is now generally accepted as a most effective way of assessing the molecular weight of proteins or their constituent polypeptides. Two variations are possible in the 2 dimensional method. 1) to focus native proteins and then to denature them for separation in the second dimension. 2) to map the total peptide population by denaturing the proteins before the first dimension separation by gel electrofocusing. In the former case, IEF is performed in the absence of denaturant, but the focused proteins are subsequently denatured by appropriate preincubation of the gel before

SDS gel electrophoresis in the second dimension. In the latter case, proteins are first denatured in urea or other suitable means and IEF is performed in the presence of high levels of urea (e.g. 6 M) to maintain the protein in a denatured state. These gels usually require shorter pretreatment periods with SDS before the electrophoretic separation in the second dimension. Both methods have their advantages and can give much valuable information and provide an excellent comparison of complex mixtures of proteins. Barrett and Gould (1973) and MacGillivray and Rickwood (1974) applied this technique to characterise non-histone proteins in chromatin. The accuracy of the method for these proteins was given as ± 3000 in molecular weight and ± 0.2 pH units for pI (MacGillivray and Rickwood 1974).

An example of the resolution possible with this system is given in Fig. 3.23 which shows an analysis of proteins from *E. coli* (O'Farell 1975). The cells were labelled with ^{14}C-amino acids. They were then lysed and treated with DNase and RNase to solubilise nuclear proteins. The soluble native proteins were then subjected to IEF in rods of polyacrylamide gel ($T = 4\%$, $C = 5\%$). After focusing, the gels were bathed for 30 min in 0.0625 M Tris–HCl buffer, pH 6.8, containing 2.3% SDS. They were then incorporated into the top of a preformed polyacrylamide slab containing 0.1% SDS and an exponential gel gradient formed from 9.25% to 14.4% acrylamide. Fig. 3.24 depicts the construction of a suitable apparatus for this procedure. After SDS gel electrophoresis, the gel slab was fixed, dried onto a plate and subjected to autoradiography. On the original autoradiograph it was possible to detect over 1000 spots. With this method it was possible to detect and quantitate a protein species representing as little as 10^{-6} to 10^{-7} of the total cell proteins. This approach has obvious potential for characterising cell proteins and for comparative genetic analysis.

Although it is conventional to use SDS to dissolve proteins for separation in the second dimension by SDS gel electrophoresis, it may be possible also to analyse proteins dissolved in SDS by IEF in the first dimension. This approach might be useful for analysing proteins that can only be extracted with SDS but which can subsequently remain soluble in other non-ionic denaturants such as urea.

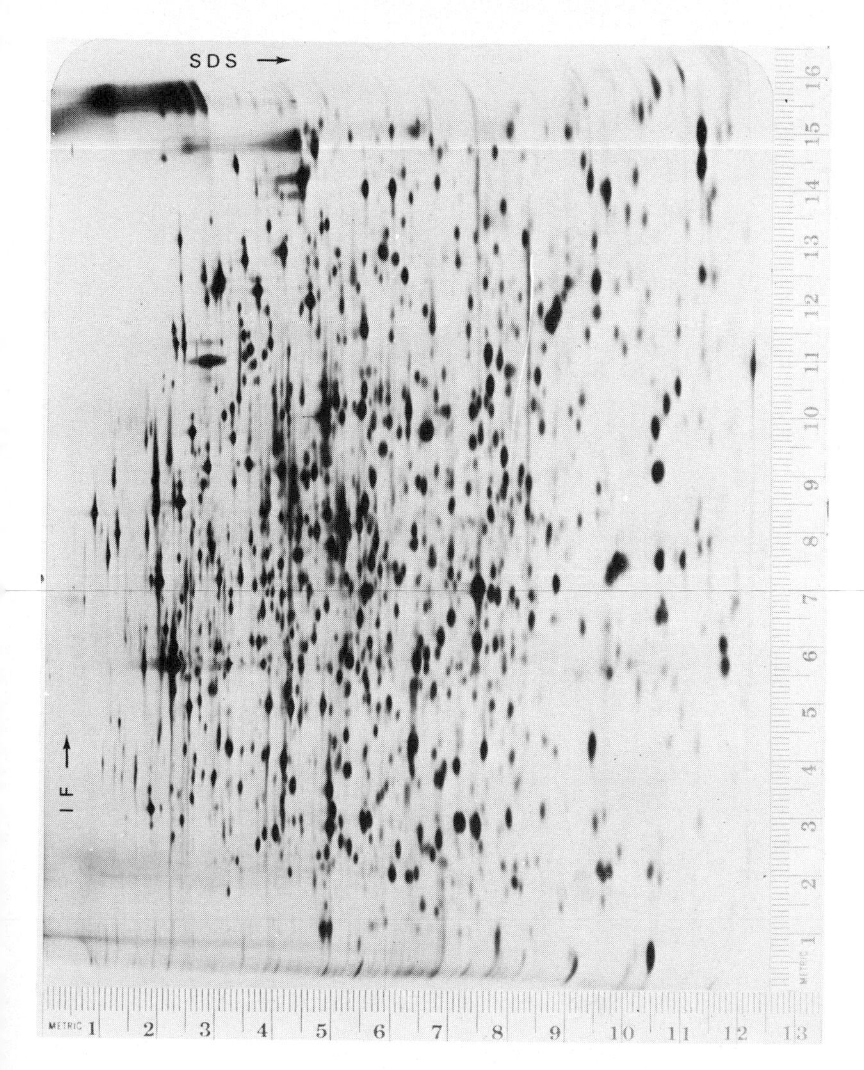

Fig. 3.23. Separation of *E. coli* proteins. *E. coli* was labelled with ^{14}C-amino acids *in vivo*. The cells were lysed by sonication, treated with DNase and RNase and dissolved in 9.5 M urea, 2% Nonidet P-40, 2% Ampholine pH 5–7 and 5% 2-mercatoethanol. 25 μl of sample containing 180,000 cpm and approximately 10 μg protein were loaded in the gel in the first dimension (IEF). The gel in the second dimension (SDS) was a 9.25 to 14.4% exponential acrylamide gradient. At an exposure of 825 hr, it is possible to count 1000 spots on the original autoradiogram. (From O'Farrel 1975, by permission of Williams and Wilkins Co.)

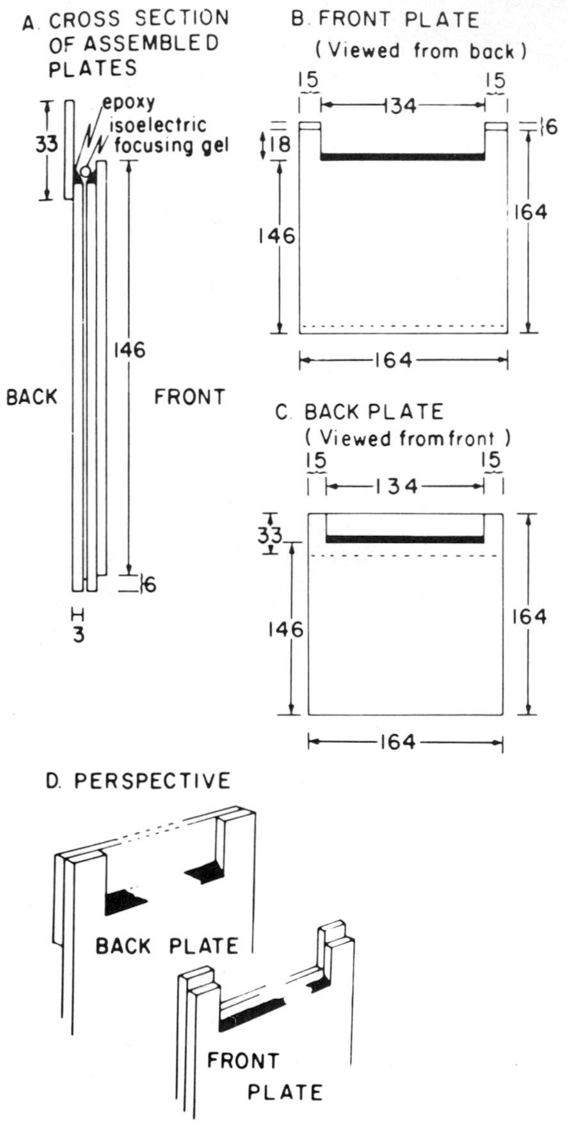

Fig. 3.24. Construction scheme of a slab gel plate. After IEF in the first dimension, the gel rod is embedded on top of a polyacrylamide gel slab for SDS-electrophoresis in the second dimension. All measurements are given in mm. (From O'Farrel 1975, by permission of Williams and Wilkins Co.)

Miller and Elgin (1974) have shown that proteins pretreated with SDS band at the same pI as the untreated proteins if focused in a gel containing 6 M urea. Apparently, the SDS protein complex dissociates in urea under the electric field to give an SDS-free protein which behaves indistinguishably from the protein treated only with urea.

Stegeman et al. (1973) have made a detailed study of genetic and physiological changes of potato proteins by IEF followed by SDS gel electrophoresis. While the tuber proteins show very high charge heterogeneity in IEF, their size distribution in SDS-electrophoresis is very restricted, which suggests the existence of a parent protein common to this population of charge-heterogeneous proteins.

3.4.5. IEF in urea-gradient gels

An interesting extension to the analysis of quaternary protein structure, is the use of urea concentration gradients in gels (Hobart 1975). Urea gradient thin layer gels are poured between glass plates arranged so that the urea gradient is oriented at 90° to the final direction of current flow. The sample is applied uniformly across the gel, in a long paper strip parallel to the electrodes. In this way the sample will focus in regions of progressively increasing urea concentration. The information gained by this type of experiments is summarized in Fig. 3.25. In the case of covalently linked structures (which cannot be dissociated by urea), if denaturation is not accompanied by a change in pI, the protein band will be straight (Fig. 3.25a). If there is a pI change, it may occur sharply at a critical urea concentration (Fig. 3.25b) or over a wide urea range (Fig. 3.25c). Trace acetylation of covalent structures, will give rise to a set of bands which will remain parallel (Fig. 3.25d). In the case of non-covalently associated subunits, the native band may give rise to a number of free subunit bands in high urea regions (Fig. 3.25e). In the case of isozymes made up by random combination of few dissimilar subunits (e.g. LDH), the native isozyme pattern will first give rise to a complex zone with interdigitating bands then at higher urea levels emerge into a zone where only the dissociated subunits are seen (Fig. 3.25f). If the native protein is an oligomer of identical subunits, the denatured pattern will be indistinguishable from that of a

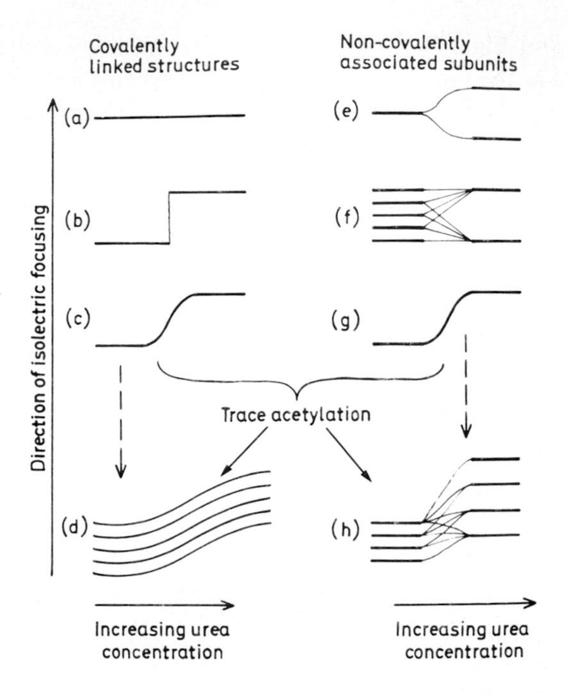

Fig. 3.25. Diagrammatic representation of possible patterns of polypeptide bands obtained by thin-layer gel IEF of native or trace-acetylated proteins in the presence of discontinuous urea concentration gradients. (From Hobart 1975, by permission of Butterworths.)

covalent structure (Fig. 3.25g). However, identical subunits can be made dissimilar by random charge changes (such as in trace-acetylation). In this case the bands of such artificial isozymes will give rise to a complex pattern (Fig. 3.25h) similar to that described in Fig. 3.25f.

3.5. *Transient state IEF (TRANSIF)*

Catsimpoolas (1973c,d,e,f) has developed some elegant techniques for continuous analytical scanning IEF which allow estimates of the first and second moments of the concentration profile throughout the course of the IEF experiment. This method gives quantitative information on parameters of the experimental aspects of IEF and gives valuable insight into other physico-chemical properties of amphoteric molecules. He coupled a scanning device to an on line digital data acquisition and processing system. This allows continuous monitoring and recording of peak positions, peak area, segmental pH gradient and pI. Unlike IEF, which is a steady state system, TRANSIF is essentially a kinetic method which is assumed to consist of three stages: initial focusing (IE), defocusing (DF) and refocusing (RF). The first stage involves the electrophoretic migration of amphoteric molecules to their pI with the attainment of steady-state focusing. The second stage (DF) is concerned with analyses of diffusion in the absence of current and the third stage (RF) studies the reapproach to the steady state when the electrical field is reapplied.

TRANSIF gives valuable information about the minimal focusing time for a protein zone. It also provides quantitative data on the shifts or instability of the pH gradient, such as the so called 'plateau phenomenon' or cathodic drift (see p. 525). In agreement with our previous findings (Righetti and Drysdale 1971), Catsimpoolas (1973c) found that the plateau phenomenon usually involved a progressive migration of the more basic ampholytes toward the cathode while the anodic zones remained essentially unaltered.

Catsimpoolas has used the DF stage to estimate diffusion coefficients of proteins. In following the rate of zone spreading in the absence of an electric field, the measurement of variance (σ) of the diffusing zone as a function of time and gel concentration yields a linear relationship in which the slope corresponds to the apparent diffusion coefficient (D) of the protein. Measurements of diffusion coefficients given by this method are, however, substantially higher than those given from ultracentrifugation data, perhaps because of interaction with the gel matrix. It is also possible to obtain information by TRANSIF about the effective molecular radius and molecular weights of proteins by using the retardation coefficient measured at different gel strengths. The computer program for TRANSIF data evaluation has been published by Catsimpoolas and Griffith (1973). Fig. 3.26 summarizes the various parameters measurable by TRANSIF (Catsimpoolas 1975b).

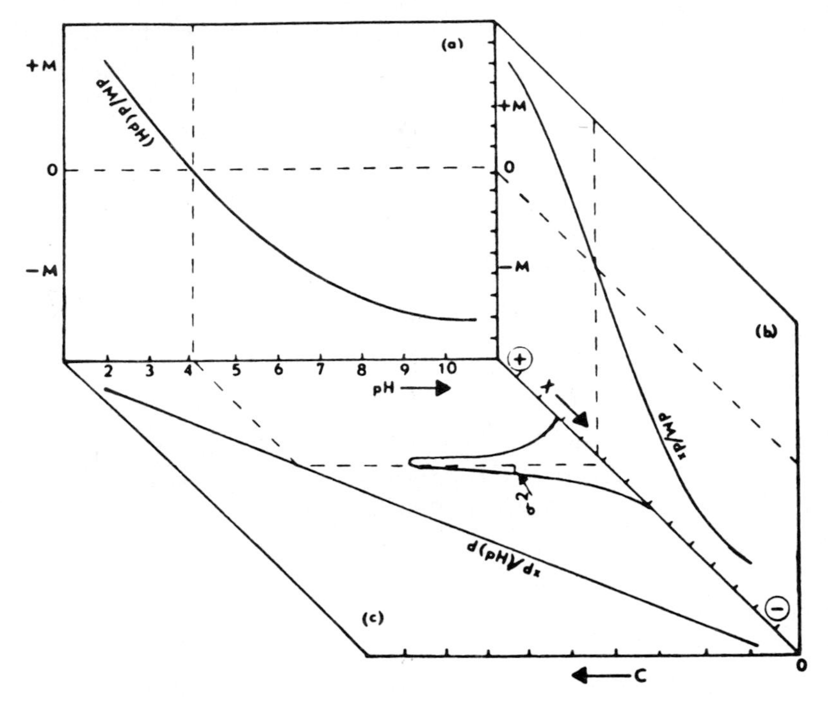

Fig. 3.26. Schematic diagram of the relationships among pH, mobility, distance and protein concentration in isoelectric focusing. The migration of a protein towards its isoelectric point is a function of its pH-mobility curve, dM/d(pH) (panel a) and the pH gradient curve d(pH)/dx (panel c). The observed mobility dM/dx (panel b) is the product of the above two parameters. X, distance; M, mobility; pI, isoelectric point; C, solute concentration; σ^2, variance; $\sigma^2 = D/(dM/dx)E$; $dM/dx = [dM/d(pH)]\ [d(pH/dx]$; D, diffusion coefficient; (dM/dx)E = velocity; E, field strength. (From Catsimpoolas 1975b, by permission of ASP Biol. Med. Press B.V.)

General experimental aspects

4.1. Isoelectric precipitation

One great drawback of IEF in liquid media is the flocculation of less soluble proteins at their pI. Precipitates, once formed, aggregate into larger particles which may sediment slowly through the column. Thus, separations are often spoiled by precipitates sticking to the walls and disturbing the density gradient or by contaminating adjacent zones. Sometimes a precipitate will redissolve as it moves away from its pI, only to reprecipitate when forced back to its pI. Consequently, a quasi-equilibrium may occur with a protein existing in different physical states over a wide pH range.

Such problems may be prevented by decreasing the protein load or by increasing the level of carrier ampholyte. Neither remedy is entirely satisfactory since both lead to an economically unfavourable protein/Ampholine ratio. A better approach is to use additives (see §4.2). Should this fail, there are some methods to circumvent problems of isoelectric precipitation in vertical, density gradient stabilized columns. One such a method is to select the electrode polarity in such a way as to ensure that the precipitate will focus and collect near the bottom of the column, while the protein of interest will focus away from it, and possibly close to the top of the column. In this situation, sample collection by column drainage will be a difficult proposition, since the precipitate may clog the drain tubing and the sample of interest may considerably diffuse. A better solution is to withdraw the sample from

Subject index p. 587

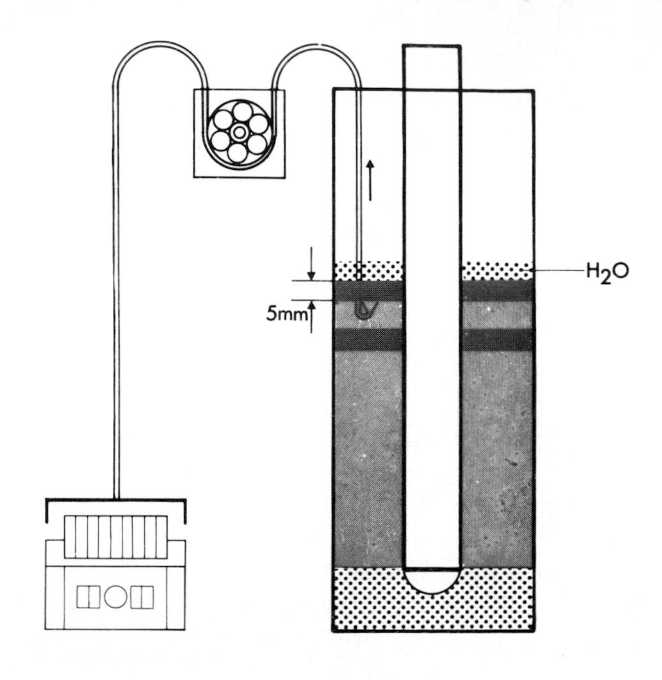

5mm
H_2O

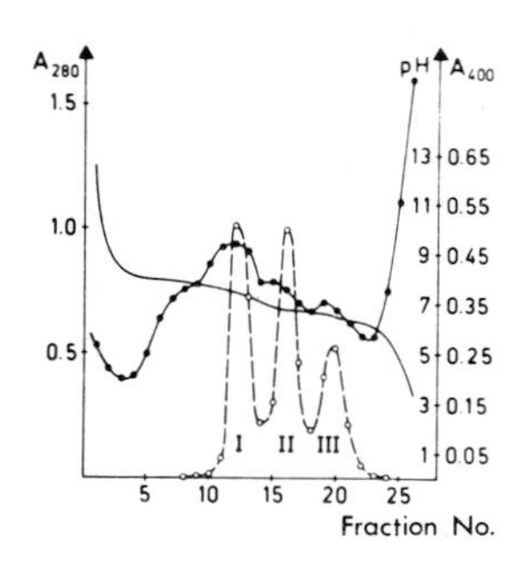

A280
pH
A400
1.5
1.0
0.5
13 0.65
11 0.55
9 0.45
7 0.35
5 0.25
3 0.15
1 0.05
I II III
5 10 15 20 25
Fraction No.
pH :
A280 :
A400 :

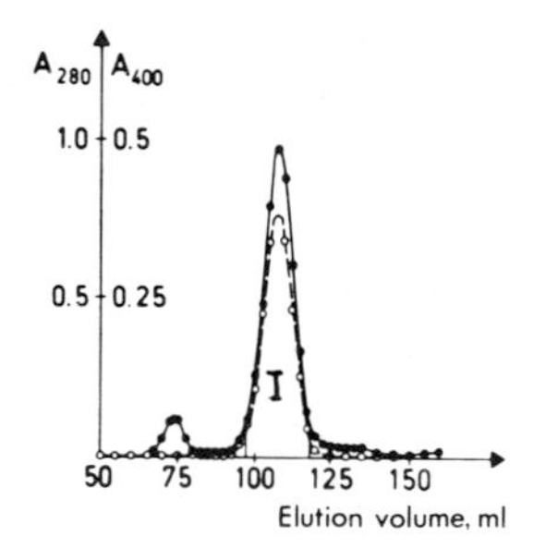

A280 A400
1.0 0.5
0.5 0.25
I
50 75 100 125 150
Elution volume, ml

Fig. 4.1(a)

(b)

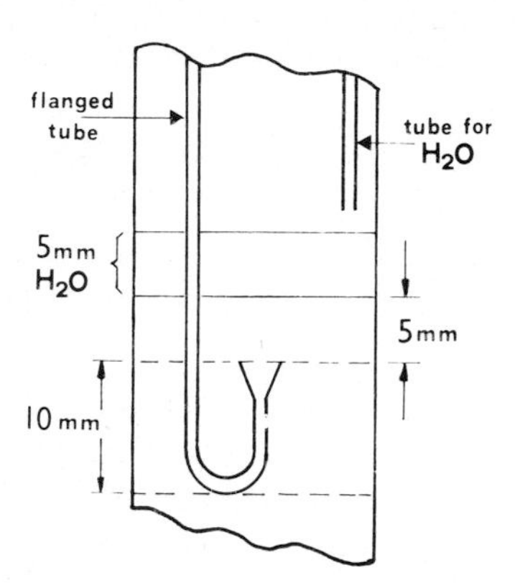

Fig. 4.1. Gradient elution from the Ampholine Electrofocusing Column in the presence of precipitates. It is assumed that the protein of interest focuses away from the precipitate. The column is then emptied from the top, through nipple (10) (see Fig. 2.2) by a stepwise procedure. The end of a capillary tubing is flanged (60°) and U-bent. This capillary is introduced 5 mm below the surface of the gradient and a 5 mm water layer is layered on top of it with the aid of a second tubing (b. As shown in (a), this layer is eluted with the aid of a peristaltic pump operated at a flow rate of 100–120 ml/hr. This stepwise procedure is then sequentially repeated. Examples of the separation achieved are given in the two graphs. (Courtesy of LKB Produkter AB.)

the column top. This operation, which is easily done in the two LKB columns, is schematically represented in Fig. 4.1. When electrofocusing is completed, the power supply is switched off and the bottom valve connecting the dense electrode solution to the central platinum wire is closed (see Fig. 2.2). From the upper nipple, used to fill the density gradient into the column, a capillary tubing is inserted 1 cm below the upper electrode and the light electrode solution is pumped out. A second tubing is now used for sample removal. Before insertion, the tubing is flanged, with the aid of a 60 degree flanging tool, by gently heating the end of the capillary with a match or lighter. By the same heating procedure, a U-bend is formed in the tubing 1 cm from the flange. This tubing is now inserted in the column, 5 mm below the surface of the liquid to be eluted. To ensure that all sample is washed off the column walls, a 5 mm layer of distilled water is introduced, with the aid of the first capillary, before each fraction is removed. Then the U-bent capillary is used to withdraw the 5 mm thick sample layer and the 5 mm water layer overlying it. This is done with the aid of a peristaltic pump at a flow rate of 120 ml/hr. Then the U-shaped tubing is lowered another 5 mm, a water layer of 5 mm is layered with the first capillary and the operation repeated sequentially until all the fractions of interest are eluted. For further details, see the LKB instruction manual I-8100-EO4.

By this method, Janson (1972) has been able to focus 7.3 g protein in the 440 ml LKB column and to recover from the top, well separated, three isozymes of α-glucosidase from *Cytophaga johnsonii*. Interestingly, in this case, IEF was used as the first preparative step in the purification procedure and yet, in this single step, a 300-fold purification (based on removal of inactive proteins) was achieved with a yield of 20% in α-glucosidase activity.

Sometimes, however, one is faced with the opposite problem, i.e. the removal not of a protein soluble at its pI, in the presence of precipitates, but of the protein itself precipitated at its pI. Rathnam and Saxena (1970) have devised an ingenious solution to this problem. The protein bands precipitated during IEF are removed by the use of an indwelling perforated glass coil (see Fig. 4.2a). The perforations in the coil directly opposite the limb have greater diameter and their diameter decreases in size as they approach the limb. This arrangement permits equalization of suction at each perforation around the circumference of the coil. The limb is attached to the coil at a slightly obtuse angle (see Fig. 4.2b) which facilitates vertical upward and downward motion of the coil in a horizontal plane. As shown in Fig. 4.2b, the coil is placed near the upper electrode or, better, in the predetermined vicinity of the precipitate before the assembly of the column. Following the appearance of the precipitate, this is removed by suction via a tubing connected to the coil. This operation can be performed, with due precaution, while the column is under voltage, with no disturbances for other parts of the gradient. Should the upper electrode solution be lowered below the upper platinum electrode, more can be pumped in with the aid of a second tubing. Actually, the two operations can be performed simultaneously.

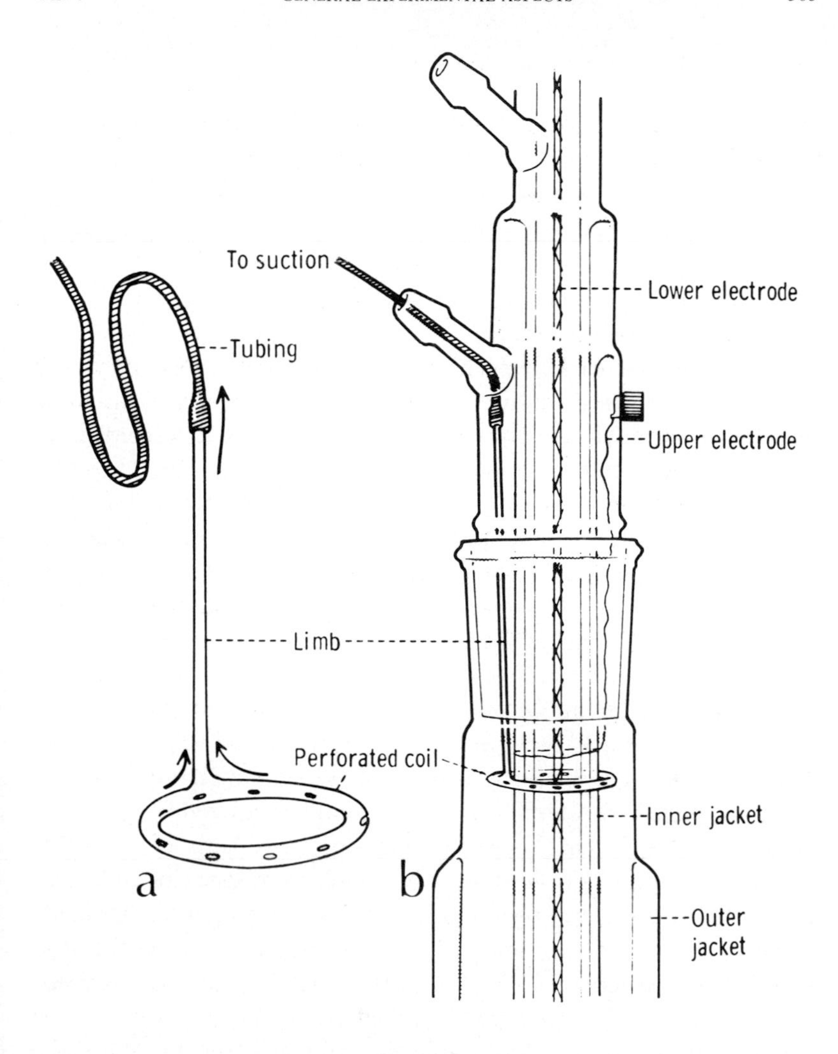

Fig. 4.2. Removal of an isoelectrically precipitated sample from the Ampholine Electrofocusing Column. In this case it is assumed that the precipitate is the sample itself. When mounting the column, a perforated glass coil (a) is placed either near the upper electrode or in the predetermined vicinity of the precipitate (b). Once a precipitate is formed, it is removed by suction via a tubing attached to the limb of the perforated coil. (From Rathnam and Saxena 1970: by permission of Williams and Wilkins Co.)

4.2. Additives

Isoelectric precipitation can often be prevented or lessened by the proper use of additives (Vesterberg 1970). In addition to stabilising or solubilizing focused bands, some additives have the added advantage of reducing differences in osmolarity along the gradient caused by focused ampholytes. Also, by increasing the osmolarity, additives reduce the risk of an unacceptably large surface film of water.

Urea and formamide up to concentrations of 4 M have been used to solubilize proteins. However, these substances may markedly alter the physico-chemical nature of proteins, particularly multimeric proteins and should be used with caution. Non-ionic detergents have also been successfully used, in concentrations ranging from 0.1 to 5%. Detergents have proven especially useful for maintaining the solubility of focused membrane bound proteins. Many such detergents are now available under various trade names, e.g.: Tween 80, Emasol, Brij 39, Triton X-100 (Sigma Chemical Co.). Serum lipoproteins have been maintained in solution by using 33% ethylene glycol (Kostner et al. 1969), or tetramethyl urea (Gidez et al. 1975). When focusing slightly soluble peptides or antibiotics, 50% dimethyl sulphoxide (DMSO) can be added to gels or liquid media (Righetti and Righetti 1974). Alternatively, as suggested by Rosengren (1975), dimethyl formamide might be preferred, since DMSO easily undergoes redox reactions. Wadström (1975) has suggested the combined use of additives in some instances. For instance, he found a mixture of urea and Triton X-100 to be very useful in the analysis of protein components of the red blood cell membrane. The same results have been reported also by Bhakdi et al. (1975). Wadström (1975) has also stressed the importance of using highly purified non-ionic detergents, since many preparations contain impurities which adversely affect the separation. In the case of Triton X-100, he suggests the scintillation grade product from Serva.

As reported by Jacobs (1971), additional problems may arise from the oxidation of cysteine and methionine to cysteic acid and methionine sulphoxide. This can be prevented by performing IEF in the presence of antioxidants, such as thiodiglycol or ascorbic acid. The latter seems to be more effective, and also prevents possible modifications of tyrosine and arginine residues (Jacobs 1973). In the case of sulphydryl-dependent enzymes, their activity can be preserved by working in the presence of a low concentration (about 10^{-4} M) of thiol compounds (Wadström and Hisatsune 1970a,b). For this purpose, 2-mercaptoethanol, 2,3-dimercaptopropanol and dithiothreitol (or its isomer, dithioerythritol) (Cleland 1964) have been effectively used.

Working with milk-fat globule membranes and with human erythrocyte membranes, Allen and Humphries (1975) found it useful to incoporate in the sucrose density gradient zwitterionic surfactants, such as:

$$R{-}\overset{\displaystyle CH_3}{\underset{\displaystyle CH_3}{\overset{|}{\underset{|}{N^+}}}}{-}CH_2{-}COO^- \qquad (R{=}C_{10} \text{ to } C_{16})$$

of the alkylbetaine type (Empigen BB, supplied by the Marchon Division of Albright and Wilson Ltd.) or:

$$R{-}\overset{\displaystyle CH_3}{\underset{\displaystyle CH_3}{\overset{|}{\underset{|}{N^+}}}}{-}CH_2{-}CH_2{-}SO_3^- \qquad (R{=}C_{12})$$

of the sulphobetaine type (sulphobetaine DLH, supplied by Textilana Corp.). These compounds are very effective in solubilizing membranes at low concentrations (between 5 to 30 mM; 0.1 to 0.5% w/v) and under very mild conditions. Enzymatic activities are not, in general, destroyed by the action of these surfactants (an exception is erythrocyte acetylcholinesterase which is inactivated by Empigen but not by sulphobetaine DLH). This retention of biological activity with zwitterionic surfactants is in contrast to the denaturant effect of anionic surfactants such as sodium dodecyl sulphate (SDS). These zwitterionic surfactants are compatible with IEF: they are usually required in only low concentrations, they are electrically neutral, and the pK_a values of the charged groups lie on either side of the pH range of 3 to 10 which is usually employed.

Alper et al. (1975) have used a similar approach in the analysis by gel IEF of the polymorphism of the sixth component of complement. They incorporated 0.2 M taurine in the gel. Taurine (NH_3^-—CH_2—CH_2—SO_3^-) is zwitterionic and has no net charge within the pH range of the run; it therefore serves to raise the osmotic pressure of the gel without contributing to its electrolyte concentration or to the viscosity of the liquid phase.

4.3. Sample application

We have already discussed at length sample application in geld rods (see §3.2.10), in gel slabs (see §3.2.14) and, briefly, also in vertical columns stabilized by sucrose density gradients (see §2.1.1). In this last case, more methodological details will be given here.

The sample can be added to the column either before or after the pH gradient has formed. In the former case, the sample can be uniformly distributed over the entire column length or added in a narrow zone where it is expected to focus. Often this last method is preferred, since it allows shorter focusing times and avoids harmful contact of the sample with anolyte and catholyte. In this last case, the following procedure should be used. Fill the density gradient in the column until it has reached the level where the sample is expected to focus. Now, fill three test tubes with the following volumes of density gradient: 0.5 ml (1st tube); 3.0 ml (2nd tube); 0.5 ml (3rd tube) for the 110 ml column, or 1.0 ml (1st tube); 10.0 ml (2nd tube); 1.0 ml (3rd tube) for the 440 ml column.

Add the sample to the second tube, mix gently and then dissolve in it few sucrose crystals until the density of the solution in the second tube is between the higher density of the first tube and the lower density of the third tube. The density of the second tube is satisfactory when a droplet of solution from the first tube will sink in it, while a droplet of solution from the third tube will float on its surface. At this point, the sample containing-solution can be pumped in the column and layering of the density gradient terminated (see Fig. 4.3).

It is often desirable, however, to add the sample after the pH gradient is formed, to minimize contact time with Ampholine, for instance in the case of metallo-enzymes or in the case of components which are inactivated or unstable at their pI, for instance, when focusing intact cells. To facilitate this, Sherbet and Lakshmi (1973) have modified the LKB 8100-1 column by fusing a side arm with a special septum to the middle of the column. After prefocusing, the sample is injected into the middle of the column and the separation takes place in only 2 to 3 hr (see also ch. 5).

Alternatively, and preferably, a simpler method can be used. After the pH gradient has been formed, insert a capillary tubing, via the upper nipple of the column, to the level in the column where it is desired to introduce the sample. Withdraw 4 ml of gradient solution from the 110 ml LKB column and 12 ml from the 440 ml LKB column. Divide the solution into three tubes and add the sample to the second tube, just as described above. After the proper density adjustments in the sample tube, this solution is pumped back into the column and the capillary tubing removed (Fig. 4.3). Focusing is then continued until termination of the experiment.

4.4. Choice of pH gradient – production of narrow pH gradients

When the pI's of the components of interest are not known, it is better to carry out preliminary experiments in the wide pH range (pH 3.5–10 or 2.5–11) to find out a proper narrow pH range encompassing all the pI's of the components to be separated. To save time and materials, this operation should be done on an analytical scale, either on gel rods or on thin-layer equipment. As previously mentioned (see §2.2.6), narrow pH ranges are to be preferred, especially in the preparative scale, because they optimize separation of closely related species, and allow a much higher sample input than broad pH ranges.

The standard narrow pH ranges commercially available usually

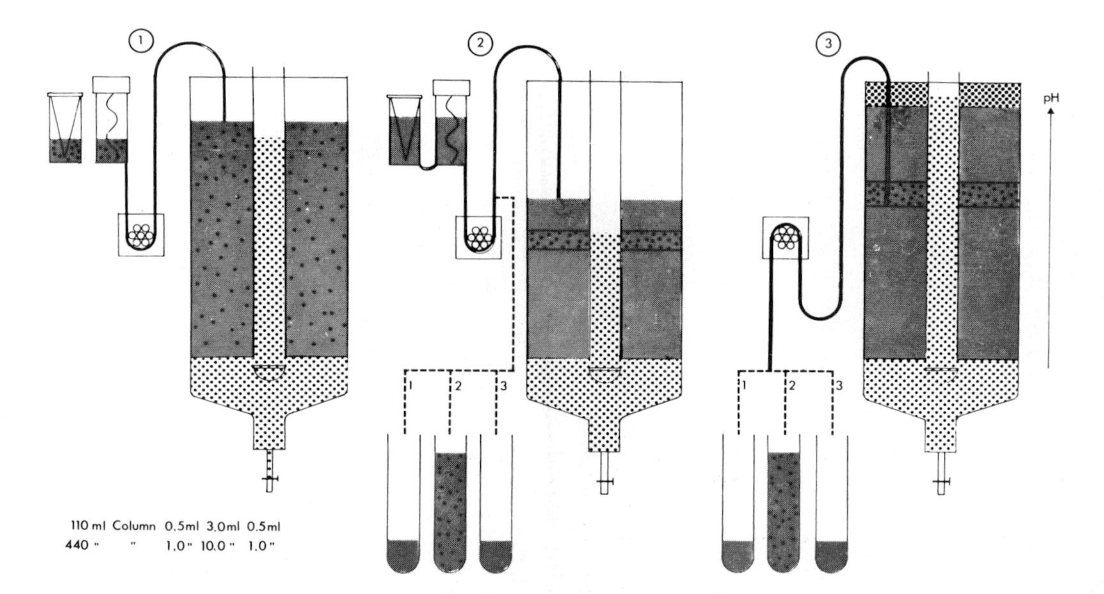

Fig. 4.3 Sample load in preparative IEF. (1) the sample is uniformely distributed throughout the density gradient. (2) the sample is loaded in a narrow zone, close to its expected pI, during the formation of the pH gradient. In this case, at the chosen level in the column, three test tubes are filled with solution from the gradient mixer (0.5 ml, 3.0 ml and 0.5 ml respectively for the 110 ml column or 1.0 ml, 10 ml and 1.0 ml respectively for the 440 ml column). The sample is dissolved in the middle tube and its density adjusted as explained in the text. The sample solution is then layered in the density gradient and the column filled with the remaining gradient solution. (3) the sample is loaded after the formation of the pH gradient. Upon focusing the Ampholine in the density gradient, harvest the desired amount of gradient solution as explained in (2). Add the sample to the middle tube, adjust its density, as described in the text, and pump it back into the column. (Courtesy of LKB Produkter AB.)

encompass a nominal range of two pH units (except for pH ranges 2.5–4 and 3.5–5). However, it is often desirable, in the case of very difficult separations, to obtain even narrower pH ranges, encompassing only one or half a pH unit. In the past, LKB had made available very narrow pH ranges, covering only half a pH unit, but, due to their high price, they have been discontinued. We describe here two methods for the production of these very narrow pH ranges. In the first procedure, it is assumed that the exact focusing position of the sample in the column is not known, therefore the sample is focused using an appropriate nominal pH range of two pH units. When the column is eluted, the fractions covering the pI range of the components of interest are collected and used for a second, narrow range, electrofocusing experiment. In this way, not only the separation between the sample components is maximized in the second run, but also all unwanted sample fractions are eliminated in the first run.

In the second procedure, it is assumed that the pI range of the sample proteins is already known. A preliminary electrofocusing experiment is carried out, without sample, using a suitable standard Ampholine solution with a nominal range of two pH units. When the column is eluted, the fractions in the required narrow pH range are collected. The sample is then added to these fractions and a second electrofocusing carried out. In this procedure the sample proteins are exposed to isoelectric focusing only once and the risk of denaturing the sample is reduced.

In both procedures, it is important to remember that the Ampholine concentration in the first run should be adjusted to the fraction volume of the column which will be utilized in the second experiment. Thus, if the sample proteins are expected to lie within a range of 1 pH unit (and, therefore, one expects to utilize 50% of the column content for the second run) and Ampholine concentration of 2% should be used in the first run. Similarly, if the sample proteins are expected to focus in a range of 0.5 pH units (and, therefore, one expects to utilize 25% of the column content for the second run) an Ampholine concentration of 4% should be used in the first run. With present day Ampholine, we believe that the narrowest range obtainable is 0.5 pH units, since narrower ranges would probably contain too few amphoteric species and give poor pH gradients with conductivity gaps.

Examples of the production of these very narrow pH ranges have been reported by Righetti and Drysdale (1971) and by Alpert et al. (1973). Production of these pH gradients is often important for clinical applications. Thus, a range covering the pH 7–7.5 would be most helpful in the screening of normal and pathological haemoglobins, since most haemoglobins are isoelectric in this pH range. Also, a range covering the pH 4.2–4.7 would be useful in the analysis of the genetic polymorphism of the plasma α_1-antitrypsin (Allen et al. 1974; Arnaud et al. 1975). It would in fact optimize the separation between the bands of α_1-antitrypsin and the albumin bands, which lie very close to each other. Alternatively, to improve separations, Allen et al. (1974), working with the LKB 2117 Multiphor, focused their samples not over the short length (10 cm) but over the longer dimension (24 cm) (see Fig. 3.9). Interestingly, in this last case, they have resolved proteins with pI's differing by as little as 0.0025 pH units, i.e. well beyond the resolution limit of 0.01 pH units found by Vesterberg and Svensson (1966).

4.5. Measurement of pH gradients

Accurate measurements of the pH gradient developed after IEF are of utmost importance in analyzing and characterizing fractionation and banding patterns. Although satisfactorily reproducible pH gradients may be obtained with any given carrier ampholyte preparation, variability in batches of ampholytes and in experimental procedures may give slightly different pH gradients. Such variability can, however, be readily corrected by correlating banding patterns with the actual pH gradient developed in each experiment.

In addition to its value in correlating banding patterns, measurement of the pH gradient also provides a direct estimate of the isoelectric point of individual components. Usually, the pI value will correspond to the isoionic point which may be defined as that pH which does not change on addition of a small amount of pure protein to the solution (Cannan 1942). This definition also holds for a focused protein zone, where the buffering capacity is determined by the carrier ampholyte.

Several factors can affect estimates of the isoelectric point and it is essential to standardize procedures to ensure good reproducibility. Since the dissociation of amphoteric molecules is temperature dependent, it is usually desirable to focus at constant temperature and to estimate the pH gradient or individual isoelectric points at the same temperature. It should be noted that p*K*'s, and therefore pI's, are temperature dependent. The pI difference for the same protein, measured at 25°C and at 4°C, may be as high as 0.5 pH units. This difference is usually more pronounced in alkaline regions and when a protein has a pI close to the p*K* of some of its functional groups. For example, Schantz et al. (1972) reported a pI of 7.26 for staphylococcal Enterotoxin A, at 4°C, while the pI measured for the same protein by Chu et al. (1966) at 25°C was 6.8. Since this protein contains seven histidine residues, from the Van 't Hoff equation, assuming a ΔH for the imidazole ionization of 7 Kcal (Cohn and Edsall 1943), an increase of 0.39 pH units, in going from 25°C to 4°C, was calculated, which accounts for the discrepancy observed. Similar temperature dependent pI jumps have been reported by Vesterberg and Svensson (1966) for myoglobins, which focus in neutral pH ranges.

Several other factors in addition to temperature can affect estimates of pI from IEF experiments. For example, the presence of solutes such as sucrose, glycerol or urea may alter the dielectric constant of the solution and thus the degree of dissociation of ionisable functions. In addition, solutes such as sucrose or urea may substantially alter the conformation and so the apparent pI of proteins or other amphoteric macromolecules. Such shifts in pI may be useful for conformational studies (Ui 1973).

Methods for estimating pH gradients

After IEF in sucrose density gradients The pH of fractions eluted from sucrose density gradients or similar anticonvective media may be measured directly from the column eluate in a flow cell (Jonsson et al. 1969; Secchi 1973; Strongin et al. 1973) (see Fig. 4.4) or after

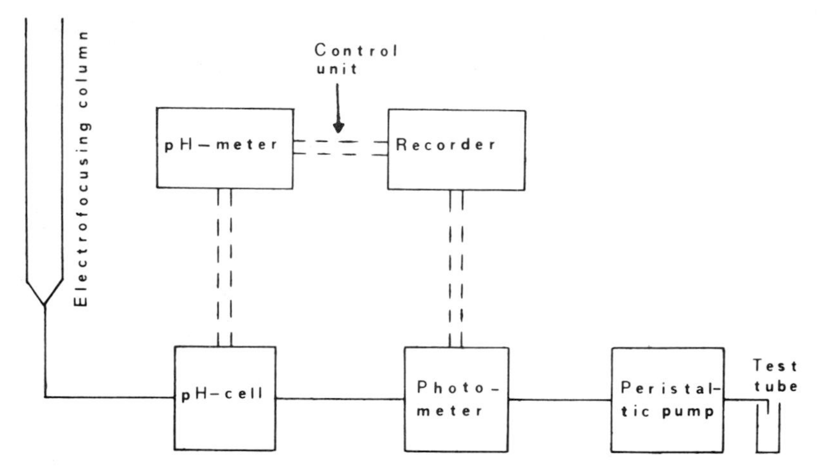

Fig. 4.4. Block diagram of series-coupled UV and pH flow cells, with simultaneous readings on the same recorder. For further details, see Secchi (1973). (From Righetti 1975b, by permission of Marcel Dekker, Inc.)

collecting convenient fractions, e.g. 1 ml. The latter method is simpler and more commonly used. Because of the small sample volume, a combination microelectrode is advisable.

After IEF in gel The pH gradient developed in slab or cylindrical gels may be measured directly from the surface with a special micro-electrode or from gel eluates. Several suitable microelectrodes are now available for use with polyacrylamide or Sephadex gels. For flat gels, a surface electrode with coplanar glass and reference electrode with either flat face or spear tip may be used. In either case, it is advisable to have the probes of both electrodes as close as possible to minimize errors due to differences in conductivity throughout the gel.

One of the first electrodes developed for pH measurements on gel surfaces is the antimony electrode of Beeley et al. (1972). This electrode (which was originally developed by Kleinberg (1958) to probe the pH of dental plaques) has a spear tip of approximately 1 mm in diameter. pH determinations are obtained by pressing the reference electrode at any position on the gel, and by leaving the gel caught at regular intervals with the antimony electrode. Readings of electromotive force are obtained in the mV scale, and converted into pH units by means of a calibration graph. Beeley et al. (1975) report frequent use of this electrode over several years with no changes in the calibration curve. The antimony electrode might be better standardized against Ampholine, previously calibrated against standard buffers.

Recently Drysdale (in collaboration with Ingold Inc., Lexington, Mass.) has developed a flat membrane microelectrode, which overcomes the problem of the high surface area of previously available models (Ingold type LOT 403-30-M8, with a membrane diameter of 8 mm). This y-shaped electrode contains the reference unit in one arm and the measuring membrane in a parallel arm. The two electrodes are coplanar, and are only 3 mm apart to offset variations in conductivity along the gel. The actual pH membrane has a diameter of 2.5 mm. With this small membrane area, surface pH readings can be performed not only in gel slabs, but also in gel cylinders. In the latter case, the electrode has a spear tip, 1 mm in length. This allows measurement of the internal pH rather than surface pH which may be slightly altered by smearing when the gel is extruded from the tube. The electrode can be assembled with a bridge and a sliding arm over a calibrated scale, so that accurate positioning of the pH readings are obtained. A sheet of graph paper beneath the gel plate can also serve as a useful template and guide. With this electrode, pH readings at 5 mm intervals can be taken, allowing for approximately 20 points on the pH curve. For reference, one could punch a 1 or 2 mm hole, with a gel puncher for immunodiffusion, on the side of each pH reading in the gel. Thus the gel will always bear the actual position of each single pH measurement and therefore the pH curve can be accurately related to the gel length irrespective of subjective gel shrinkage or swelling. Fig. 4.5 shows the Ingold flat membrane microelectrode mounted on the bridge.

Often, satisfactory estimates of pH gradients may be obtained from gel eluates. Gels may be sectioned linearly and the ampholytes eluted in water for pH measurements. As a general rule, the volume of eluate should not be more than 5 times that of the gel section, to avoid excessive dilution of the ampholytes. Although distilled water may be used, it is usually advisable to add low levels of salt, e.g. 10 mM KCl

Fig. 4.5. Ingold flat-membrane microelectrode mounted on a sliding arm over a calibrated scale. This y-shaped electrode contains the reference unit in one arm and the measuring membrane in a parallel arm (left arm). The two electrodes are coplanar, and are only 3 mm apart. The pH membrane has a diameter of 2.5 mm. (From J.W. Drysdale, unpublished.)

to ensure adequate conductivity. Care must be taken to minimize absorption of atmospheric CO_2 which can substantially reduce the pH of the basic pH ranges. Absorption of CO_2 may be minimized by boiling the water immediately before use or by flushing with nitrogen. As always, conditions for measuring pH should be carefully standardized and quoted in estimates of pI.

An alternative to pH measurements along the gel length, especially when working with thin-layer equipment, is the use of a calibrated mixture of pH markers, covering the pH range of interest. Ideally, these pH markers should be colored proteins or dyes. In the case of protein markers, they should not be heterogeneous, or at least have a major, easily identifiable band, to avoid confusion in pH assessments. Studies on the use of pH markers have been reported by Conway-Jacobs and Lewin (1971), Nakhleh et al. (1972), Bours (1973) and Radola (1973a). Table 4.1 lists some of the most commonly used pH markers.

4.6. *Removal of carrier ampholytes after IEF*

After an IEF experiment it is often desirable to remove the carrier ampholytes recovered with the protein fraction. Since most of the ampholytes in commercial sources have a mean molecular weight (M.W.) between 400 and 600 daltons, and since only a very small percent reaches a M.W. around 5000 daltons (Gasparic and Rosengren 1975), it would appear to be a simple matter to remove them by dialysis. However, working with ^{14}C-carrier ampholytes, Vesterberg (1970) has found that a complete removal of Ampholine by dialysis requires at least 32 hr, even when dialysing against 0.1 M phosphate buffer pH 7.0, containing 0.5 M NaCl.

Nilsson et al. (1970) obtained a very efficient removal of carrier ampholytes by precipitation and subsequent washings of the protein with ammonium sulphate solutions. For protein precipitation, ammonium sulphate of analytical grade is used in concentrations ranging from 60 to 100% saturation, at an appropriate pH for the stability of the separated protein. The precipitate is then washed three times with the same ammonium sulphate solution. Where applicable, this procedure will remove essentially all of the carrier ampholytes since they remain soluble even at 100% salt saturation.

TABLE 4.1a

pH markers for isoelectric focusing.

Protein	pI at 25°C	Reference	pI at 25°C	Reference	pI at 4°C	Reference
Cytochrome c	9.28 ± 0.02	Radola (1973a)				
Ribonuclease	8.88 ± 0.03	Radola (1973a)				
Myoglobin (sperm whale)						
major component	8.18 ± 0.02	Radola (1973a)	8.18 ± 0.04	Nakhleh et al. (1972)		
minor component	7.68 ± 0.02	Radola (1973a)				
Myoglobin (horse)						
major component	7.33 ± 0.01	Radola (1973a)	7.45 ± 0.04	Nakhleh et al. (1972)	7.58 ± 0.02	Bours (1973)
minor component	6.88 ± 0.02	Radola (1973a)	7.15	Nakhleh et al. (1972)	7.22 ± 0.05	Bours (1973)
Bovine hemoglobin A	6.80	Conway-Jacobs and Lewin (1971)				
Carbonic anhydrase (bovine)					6.18 ± 0.02	Bours (1973)
Conalbumin	5.88 ± 0.02	Radola (1973a)				
β-lactoglobulin B	5.31	Radola (1973a)			5.45 ± 0.02	Bours (1973)
β-lactoglobulin A	5.14 ± 0.01	Radola (1973a)			5.35	Bours (1973)
Bovine insulin	5.32 ± 0.02	Conway-Jacobs and Lewin (1971)				
Albumin (bovine)						
Cohn fraction V	4.90	Conway-Jacobs and Lewin (1971)			4.95 ± 0.02	Bours (1973)
Ovalbumin	4.70	Conway-Jacobs and Lewin (1971)				
Horse spleen ferritin I	4.50 ± 0.02	Radola (1973a)				
Horse spleen ferritin II	4.38 ± 0.02	Radola (1973a)				
Horse spleen ferritin III	4.23 ± 0.03	Radola (1973a)				

TABLE 4.1b
pH markers for isoelectric focusing.

Dye	pI at 25°C	Reference
Tris(5-OH)iron(II)[a]	7.15	Nakhleh et al. (1972)
Bis(5-OH)-4-OH iron(II)[b]	6.82	Nakhleh et al. (1972)
Bis(4-OH)-5-OH iron(II)[c]	6.24	Nakhleh et al. (1972)
Congo red	5.80	Conway Jacobs and Lewin (1971)
Tris(4-OH)iron(II)[d]	5.45	Nakhleh et al. (1972)
Evans blue	5.35	Conway-Jacobs and Lewin (1971)
Methyl blue	3.60	Conway-Jacobs and Lewin (1971)
Fast green FCF (major component)	3.05	Conway-Jacobs and Lewin (1971)
Patent blue V	3.00	Conway Jacobs and Lewin (1971)

[a] Tri(5-hydroxy-1,10-phenantroline)iron(II).
[b] Mixed complex between tris(5-hydroxy-1,10-phenantroline)iron(II) and tris(4-hydroxy-1,10-phenantroline)iron(II).
[c] mixed complex between tris(4-hydroxy-1,10-phenantroline)iron(II) and tris(5-hydroxy-1,10-phenantroline)iron(II).
[d] tris(4-hydroxy-1,10-phenantroline)iron(II).

Vesterberg (1969) obtained a good separation of proteins from carrier ampholytes by gel filtration in Sephadex G-50 fine. He performs the separation in columns equilibrated with 0.1 M phosphate buffer, containing 0.5 M NaCl, or 0.1 M ammonium bicarbonate, or 0.1 M imidazole acetate. These buffers are usually kept at a pH of about 7. The last two buffers have the advantage of being volatile, which allows them to be removed by lyophilization. Whichever method is used, it is important to have a high ionic strength in the buffer (0.1 M or more, or 0.5 M salt concentration) to dirupt weak complexes which might form between Ampholine and proteins by electrostatic interaction.

Alternatively, Brown and Green (1970) have proposed the use of a mixed-bed ion-exchange resin (Bio Rad AG 501-X8) for the separation of proteins from carrier ampholytes. While they found partial overlapping of protein and Ampholine peaks by gel filtration on Sephadex

G-25 beds, they have reported complete Ampholine removal by using ion-exchange chromatography.

Ampholine can also, of course, be removed by electrophoresis. This is the method used by Suzuki et al. (1973) and by Stathakos (1975) to recover proteins from polyacrylamide gels, after IEF in preparative scale. The protein, with carrier ampholytes, is electrophoresed out of the gel into a chamber built with a dialysis membrane on the floor, through which Ampholine escapes (see also §2.2.4 and §2.2.5). A very high protein recovery (in a concentrated form and Ampholine-free) is thus achieved (see Fig. 4.6).

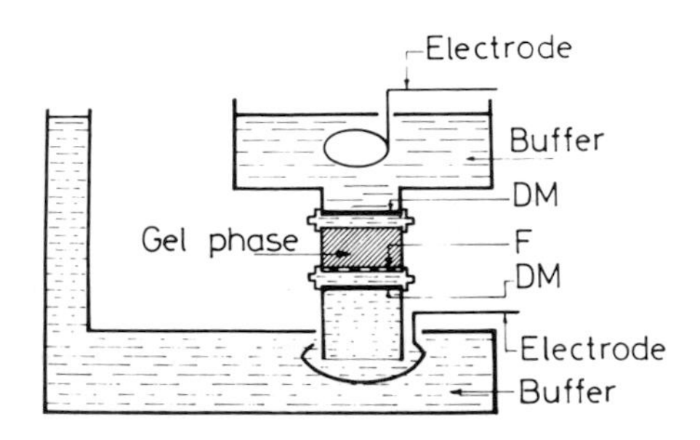

Fig. 4.6. Schematic drawing illustrating the electrophoretic recovery of focused bands. DM, dialysis membrane; F, filter membrane. The protein is recovered into the small chamber between F and DM, while the carrier ampholytes escape through the dialysis membrane. (From Stathakos 1975, by permission of ASP Biol. Med. Press B.V.)

Separation of carrier ampholytes from proteins could also be achieved by ultrafiltration, or by the use of the hollow-fiber technique (Bio Rad).

4.7. Possible modification of proteins after IEF

This subject has been dealt with briefly under the heading 'additives' (§4.2) and will be dealt with more extensively at the end of the book (§5.13). As previously seen, the protein moiety can be modified by oxidation of cysteine and methionine to cysteic acid and methionine sulphoxide. Furthermore, partial loss of tyrosine and arginine residues of bovine ribonuclease was observed during IEF (Jacobs 1971, 1973). Many of these modifications can be prevented by deaerating the solutions used in IEF under nitrogen and by the use of antioxidants such as ascorbic acid. Funatsu et al. (1973) have also reported the partial loss of glutamic acid from reduced ricin D during IEF.

Modifications often occur not in the protein moeity per se, but in its prosthetic groups. This is often the case, for instance, with glycoproteins (§5.4) especially if they focus in pH regions unfavorable to the stability of the polysaccharide chain. In the case of metallo-proteins (§ 5.7) additional heterogeneity can be elicited not only by partial removal of the metal (especially in the case of copper-containing proteins) but also by anodic oxidation or cathodic reduction of the metal still bound to the protein.

Perhaps one of the most controversial arguments in the field of IEF is whether or not artifactual heterogeneity can be elicited in a protein by binding of carrier ampholytes. If this were the case, indeed, it would shed doubts on most cases of microheterogeneity found by IEF and not readily detectable by other techniques. An example of such controversy comes from experiments with bovine serum albumin (BSA). It must be emphasized, however, that BSA represents an extreme case, since BSA is known to present a high affinity for a multitude of ligands (Steinhardt et al. 1972). By IEF in the pH range 4–6, Kaplan and Foster (1971) have presented evidence that a low-pI peak in BSA represents albumin molecules strongly bound to a minor constituent or an impurity in the ampholyte mixture. Similar evidence for complex formation between BSA and ^{14}C-labelled carrier ampholytes has been presented by Wallevik (1973). On the other hand, these results have not been substantiated by Salaman and Williamson (1971) and by Spencer and King (1971) when focusing BSA both in the presence and in the absence of 6 M urea. Actually, in these last two publications, the possibility of artifactual heterogeneity is rejected. Another example of protein–ampholyte complex has been described by Frater (1970) when working with an acidic, low sulphur protein component from wool.

In the light of the present information, it appears that evidence for protein–Ampholine interaction is rather scanty and some, strong evidence to the contrary is now available. Thus, Baumann and Chrambach (1975a), working with human growth hormone and ovine prolactin, in the presence of ^{14}C-labelled Ampholine, have been able to exclude any interaction between protein and Ampholine. Protein binding to excess Ampholine was negligible. Ampholine binding to excess protein, if any, was below the detection limit of 0.2 moles Ampholine per mole protein. Similar results have been obtained by Dean and Messer (1975) working with purified proteins (albumin, ferritin and β-glucuronidase) and with serum proteins. Thus, in the light of these recent findings, it appears that artifacts in IEF due to ampholyte–protein interactions are at worst a rare occurrence.

One elegant way to check for artifacts is to apply, to a pre-focused gel slab, the same amount of sample both, at the anode and at the cathode. If the anodic- and cathodic-travelling samples showed the same protein banding patterns and pI distribution, then artifacts were excluded. If the two patterns were different, then ampholyte–protein interaction maybe sought as an explanation. While the former assumption is generally valid, the latter is not. Often, the explanation for pattern diversity for anodic- and cathodic-migrating species has to be found in the nature of the sample itself. For instance, as shown by Wadström and Smyth (1975b) in the case of bacterial cytoplasmic protein extracts, they are strongly adsorbed at the applicator when applied to the cathode, while they are unaffected if applied at the anode. Usually, addition of 4 M urea to the gel and to the applicator minimizes this effect. Often, in crude extracts, the presence of large amounts of nucleic acids can give rise to blurred patterns and to pattern diversity for anodic- and cathodic-species. This can often be prevented by pre-treatment with DNase (Wadström and Smyth 1975b).

Sometimes, the nature of the sample is such that meaningful results can only be obtained by focusing it either from the anode or from the cathode. Thus, working with zeins isolated from maize, Soave et al. (1975) found that, while the anodic pattern was highly heterogeneous, the cathodic pattern hardly showed any protein band at all. The two patterns could not be reconciliated even when performing IEF in 6 M urea or when applying the sample in a free solution, in a pocket on the gel surface. It was then found that at the cathode, due to unfavorable pH conditions, the zeins aggregated and precipitated at the application point, even in the presence of 6 M urea in the sample and in the gel. Thus, in this case, meaningful results could only be obtained by applying the sample at the anode.

4.8. Power requirements

Since the introduction of the technique of electrophoresis, supporting media, buffer systems, electrophoretic cells and cooling methods have all undergone extensive changes. Little attention, however, has been paid to the source of electrical potential. Among typical power sources used in electrophoresis, the only differences in output are in the range of power provided and in the type of regulation: constant voltage or constant current.

Recently, a second generation of power supplies has appeared, capable of operating either in the constant voltage, or in the constant current mode, and with automatic switch-over from one mode of operation to the other. With the advent of IEF, there has been a need for a third generation of power supplies, operating at constant wattage. Since the migration rate of the sample and the sharpness of focused bands at their pI's is proportional to the applied voltage, it is desirable to have the applied field strength as high as possible, while maintaining the joule heating (which is proportional to the square of the applied voltage) at an acceptable level. This maximum tolerable power, defined as optimum power by Schaffer and Johnson (1973), will depend on the conductivity of the system and on its efficiency to dissipate joule heating, and can be determined by inserting a thermic probe in the separation column. Once this optimum power, for a given system, has been determined, that system will run on an isotherm for the duration of the experiment.

Schaffer and Johnson (1973) have built a regulator which transforms an unregulated DC power supply into a constant wattage power unit. Söderholm and Lidström (1975) have built a constant wattage power pack, capable of delivering a maximum of 3000 V, 300 mA and 300 W, by applying a wattage regulator, which feeds the signal to a reversible synchronous motor (5 rpm), used as a servomotor, to a constant voltage power supply.

There are now several constant wattage power supplies available on the market. One is the ISCO model 492, which is capable of delivering a maximum of 1000 V, 150 mA and 150 W. Another is the LKB model 2103, which delivers a maximum of 2000 V, 200 mA and 110 W. A third constant wattage power supply is available from Savant (model CWS 300) and can be operated at a maximum of 300 V, 200 mA and 60 W. All these power supplies are built with dual outputs, thus allowing simultaneous separations with two cells. It is also possible to transform any unregulated DC power supply into a constant wattage power supply, suitable for IEF, by using the Constant Power Controller CPC-300 from Techtum Instruments AB. This control unit can be adapted to any unregulated power supply capable of delivering up to 3000 V and 300 mA. The maximum power output is 180 W. One useful feature, common to all these units, is that they can be operated not only in the constant wattage mode, but also in the constant voltage and constant current modes. Thus, it is possible to use these power supplies not only for IEF and conventional electrophoresis, but also for isotachophoresis (where constant current is a must).

We should like to emphasize, however, that the so called 'pulsed' power supplies do not really provide constant wattage when applied to IEF (Chrambach et al. 1973; Righetti and Righetti 1975) except for the Regulated Pulsed Power supply from MRA, which maintains a constant average power by automatically regulating the pulse frequency at a predetermined potential difference (Drysdale 1974).

With the availability of constant wattage power supplies it is now possible to perform high speed IEF by the use of higher wattages, under thermally controlled conditions. For instance, we have already seen that, in the case of the 110 ml and 440 ml LKB columns, separations often required 48 to 72 hr by applying wattages of 5 W and 10 W, respectively (see p. 395). Now, by using a constant wattage power supply, Lundin et al. (1975) have shown that it is possible to obtain equilibrium conditions in 15–20 hr by applying a maximum of 15 W and 1600 V to the LKB 8100-1 column and a maximum of 30 V and 2000 V to the LKB 8100-2 column. Under these conditions, the temperature was quite constant and only a few degrees above the coolant temperature, as measured with a pin point thermistor at different levels in the two columns. When using these higher wattages, Lundin et al. (1975) recommend that the experiment should terminate within 20 hr, to avoid deformation and possible breakdown of the pH gradient upon prolonged focusing. In the case of the LKB Multiphor 2117 thin layer equipment, wattages as high as 20–25 W can be applied with no distortion of the pH gradient and no heat damage to the protein. In this last case, the experiment should be terminated within two hours, to avoid excessive cathodic drift.

4.9. Instability of pH gradients

The 'cathodic drift' (Righetti and Drysdale 1971) or 'plateau phenomenon' (Chrambach et al. 1973) seems incompatible with the concept of isoelectric focusing, yet it is an ever-present hazard that the users of the technique must learn to cope with. We have already seen that, by using appropriate conditions, especially in gel media, it is possible to achieve equilibrium focusing before the pH gradient begins to decay (see p. 430).

The pH drift during IEF, which has been studied by Baumann and Chrambach (1975b), Fawcett (1975b) and by Haglund (1975) seems to have a multifactorial origin. Seven hypothesis have been advanced to explain pH gradient decay: 1) electrophoretic migration of isoelectric Ampholine; 2) electroendosmosis; 3) formation of a zone of pure water in the neutral region of the pH gradient; 4) 'isoelectric focusing' of water (considered to be an ampholyte) with resulting accumulation of water in the neutral gel region and backflow towards the periphery of the gel; 5) selective deficiency or progressive destruction of neutral ampholytes; 6) progressive gain or loss of charged ligands by Ampholine.

From experiments with ^{14}C-labelled Ampholine and $^{3}H_2O$ Baumann and Chrambach (1975b) have concluded that the major force involved in pH gradient decay appears to be electrophoresis of Ampholine, resulting in Ampholine depletion from the gel. It is possible that the movement of Ampholine out of the gel takes place 'en bloc', especially in the cathodic direction, possibly due to chemical interaction among Ampholine species. Also, chemical modification of Ampholine by reaction with acid or base cannot be ruled out.

The experiments using tritiated water indicate that electroendosmosis is a possible, albeit minor, force in pH gradient decay. However, electroendosmosis could play a major role in the predominantly unidirectional (cathodic) drift of Ampholine. Diffusion of Ampholine out of the gel, and of water or electrolytes into the gel, seems to be only a minor phenomenon. Moreover, the hypothesis postulating entry of 'amphoteric water' from the electrolyte reservoirs into the gel and its condensation at the neutral region is not supported by the recent findings of Baumann and Chrambach (1975b).

Whichever the explanation, it is important to remember that, when using high voltage techniques for quick focusing (Righetti and Righetti 1975; Söderholm and Wadström 1975), conditions which enhance the cathodic drift, the experiments should be terminated in 15–20 hr for the sucrose density gradient stabilised columns, and in 2 hr for the gel slabs, to avoid excessive decay of the pH gradient (Lundin et al. 1975; see also pp. 461–462).

Applications of IEF

This section is a survey of some interesting applications of IEF in areas of interest in fields, such as biochemistry, molecular biology, immunology and clinical chemistry. It cannot possibly cover all the various fields of applications nor list the numerous publications in each selected topic. Some of the topics have been selected as examples of pitfalls and limitations encountered when using IEF to purify and characterize some classes of proteins, such as glycoproteins, lipoproteins and metalloproteins. In these cases, one should be aware of possible artefactual heterogeneity due to partial loss of polysaccharides, lipids or metals, respectively. Other topics have been selected as examples of the very high resolution afforded by IEF as compared to other techniques.

5.1. IEF of cells, subcellular particles, bacteria and viruses

The cell membrane plays a unique role in embryonic development and organization, in the interaction of cells in vivo and in vitro, and in many transport processes. It has recently been found possible to utilize differences in the surface charge of cell membranes to separate different cell types and to characterise the nature of the surface charge by IEF.

Sherbet et al. (1972) were among the first to study the cell surface by isoelectric equilibrium analysis. They separated cells by IEF in linear gradients of sucrose (55–10 %), glycerol (70–10 %) or Ficoll (30–5 %) in the LKB 110 ml column, loaded with 20 to 200 × 10^6 total cells and

Subject index p. 587

run for approximately 30 hr. They found that the cells reached a true isoelectric equilibrium. Complex formation with Ampholine did not contribute to the banding pattern since the assessed pI values were independent of the level and range of Ampholine used. The cell pI was not significantly affected by the supporting medium, since focusing in either sucrose, or glycerol or Ficoll density gradients gave essentially the same results. The measured pI's were also independent of the buoyant density of the cells, since reversal of polarity made no difference to the cell pI. This result excludes the possibility that the cells were collecting at a point where the rate of sedimentation matched the rate of electrophoretic mobility.

All the eukaryotic cell types analyzed showed pI value ranging from 5.6 to 6.85, these values being consistent with the view that cells bear a net negative charge at their surface. Interestingly, all cell types analyzed, including those derived from pure lines, gave two peaks. The main peak (80 to 90% of the total) represented viable cells and exhibited higher pI values; the minor peak (having a pI of about 5, irrespective of the original cell pI) being composed of non viable cells. Thus it would appear that the surface charge increases when cells become non-viable. This rearrangement or conformational change on the cell surface may represent shielding of thiol groups (p*K*'s ranging from 9.1 to 10.8), with consequent lowering of the cell surface pI.

Sherbet and Lakshmi (1973) have also been able to probe charged groups on cell surfaces. For instance they have found that *E. coli* cells have a surface pI of 5.6. However, when treated with ethyleneimine to esterify the carboxyl groups, these cells showed a pI of 8.55. This pI would appear to exclude the presence of ε—NH_3^+ of lysine side chains ($pK = 10.4$); –OH groups of aromatic side chains ($pK = 10$) and guanidyl groups of arginine side chains ($pK = 12.5$). It is, however, compatible with the presence of a few weakly acidic groups, such as thiols. The presence of thiols is in fact confirmed by treating the cells with 6.6′-dithiodinicotinic acid carboxypyridine disulphide. On the other hand, when these cells are treated with formaldehyde as a means to block the amino groups, they show a pure carboxyl surface, with a pI of 3.85. This pI would appear to exclude the presence of phosphate groups in the layer probed by IEF. It would thus appear that IEF is able to characterise the cell surface to a depth of about 60 Å, while cell electrophoresis detects groups present within the 10 Å zone of shear. Since the bacterial cell wall is approximately 200 Å thick, IEF is able to probe approximately one third of its depth. And since in this region no phosphate groups are detected, it would appear that in the cell wall the phospholipids are located at a depth of more than 60 Å from the bacterial surface.

It should be noted that, in all cells examined by IEF, the pI's are considerably higher (2 to 3 pH units) than the respective pI's calculated by cell electrophoresis (Gittens and James 1963). It must be born in mind, however, that the pI as determined by electrophoresis is ionic strength dependent, and usually decreases with increasing ionic concentrations. In addition, in the low ionic strengths prevailing in IEF, the particle behaves as though it had a smooth surface. By contrast, at the higher ionic strengths in electrophoresis, the surface becomes pitted so that some charged groups may not be able to exert their electrophoretic effect (Bull 1964).

One drawback of this type of analysis is that the cells are kept in the IEF column for a long time (approx. 30 hr) and this could cause denaturation and loss of cell viability. More recently, Sherbet and Lakshmi (1973) have modified the LKB 8101 column, by fusing a side arm with a special septum to the middle of the column. After prefocusing, the sample is injected into the middle of the column and the separation takes place in only 2 to 3 hr.

Another interesting approach to the separation of cells and subcellular organelles is the technique of continuous flow, carrier free IEF, developed by Werner's group in Frankfurt (Seiler et al. 1970a,b). The apparatus is described in the section 'Preparative IEF' (§2.1.7). Using this technique, Just et al. (1975) have reported excellent separations of mixed red blood cell (RBC) populations and of subcellular particles. They were able to separate mixtures of human, mouse and rabbit RBS's in pH gradients 3–10 and 5–7, at an RBC injection rate of 3 to 7 $\times$ 10^7 cells/min and with a flow-through time of only 7 min. To prevent cell clumping, they have used 1 mM EDTA.

Just et al. (1975) have also reported preliminary data on the separation of the light mitochondrial fraction of the rat liver into lysosomes and mitochondria. In this case, they used polyanionic substances (such as heparin, chondroitin sulfate, dextran sulfate and polyanethol sulfonic acid) to prevent aggregation. The method is mild and allows recovery of fully active and non-aggregated fractions, it is quick and it has a resolving power superior to conventional differential or buoyant density centrifugation.

Talbot (1975) has attempted fractionation of viruses by IEF in his trough. Working with foot-and-mouth disease viruses (acid labile picornaviruses) he obtained two fractions, one with pI 6.9 and the other with pI 4.3. The two forms are interconvertible, and represent pH dependent viral degradation products. Unfortunately, these viruses focus at an unfavourable pH, which causes the viral RNA to be released, with the formation of an aggregate containing three of the four protein species found in the capsid.

Tables 5.1 and 5.2 summarize the results obtained in the field.

5.2. IEF of hormones

Several reports have dealt with the purification and characterization of hormones. The pituitary gonadotropins have been widely studied (Bettendorf et al. 1968; Reichert 1971). The follicle stimulating hormone (FSH) from horse (Braselton and McShan 1970), sheep (Sherwood et al. 1970) and human (Saxena and Rathnam 1968) pituitary glands has been analyzed by IEF. The fractionation of luteinizing hormone (LH) from porcine (Anderson and O'Grady 1972) and human (Maffezzoli et al. 1972; Rathnam and Saxena 1970) pituitary glands by IEF has also been described. Rathnam and Saxena (1971) have succeeded in purifying LH subunits. Reports on the analysis of human chorionic gonadotrophin (HCG) (Maffezzoli et al. 1972; Graesslin et al. 1971) and of human growth hormone (HGH) (Moritz 1969) have been published.

Data are available in the literature on the isolation of thyroid stimulating hormone (thyrotropin, TSH) from the anterior pituitary

TABLE 5.1

Isoelectric equilibrium analysis of *E. coli* cells after different treatments and of viruses[1].

Cell type[5]	Observed pI	Probable ionogenic groups in isoelectric zone[4]		Calculated pI[2]	No. of net negative charges per cell × 10^{-6} [3]
		Present	Excluded		
E. coli cells (untreated controls)	5.60 ± 0.10	COOH; NH_2 pK = 7.5	Acidic groups of pK < 3.2; ε-NH_3^+ of lysine	5.38	0.37
AW-EI-*E. coli*	8.55 ± 0.04	NH_2 pK = 7.5	NH_2 groups of phospholipids; phenolic -HO; guanidyl groups	8.7	+0.41
HCHO-*E. coli*	3.85 ± 0.06	COOH groups[7]	Acidic groups with pK < 3.2	3.5	0.84
E. coli-CPDS	4.28 ± 0.05	As in *E. coli* control	Thiols[6]		
E. coli AW-EI-CPDS	7.47 ± 0.18	As in AW-EI cells	Thiols[6]		
Foot- and mouth-disease virus	6.9 and 4.3				

[1] All values from Sherbet and Lakshmi (1975) except for foot- and mouth-disease virus, taken from Talbot (1975). [2] See Table 5.2 (footnote 3). [3] See Table 5.2 (footnote 4). [4] The isoelectric zone is estimated to extend to a depth of 6.0 nm below the cell surface. [5] AW-EI-*E. coli*, acid-washed ethyleneimine-treated *E. coli* cells; HCHO-*E. coli*, *E. coli* cells treated with formaldehyde; AW-EI-CPDS, AW-EI cells treated with 6,6′-dithiodinicotinic acid (CPDS); *E. coli*-CPDS, *E. coli* cells treated with CPDS. [6] The total number of thiol groups on the *E. coli* surface equals the number of net negative charges introduced by reacting them with 6,6′-dithiodinicotinic acid, which is approx. 0.35×10^6 groups. Only a proportion of these (less than 3 %) are dissociated at physiological pH. [7] COOH groups of glucoronic and neuraminic acids β-COOH of aspartic acid and γ-COOH of glutamic acid.

TABLE 5.2

Isoelectric equilibrium analysis of some mammalian cell types and subcellular organelles[1].

Cell type	Observed pI	Probable ionogenic groups in isoelectric zone[2]		Calculated pI[3]	No. of net negative charges per cell $\times 10^{-6}$ [4]
		Present	Excluded		
Polyoma-transformed BHK–21	6.40 ± 0.19	Sialic acid COOH, α-COOH; thiol; α-NH_2	ε-NH_3^+ of lysine; guanidyl groups	5.6–6.4	6.0
Yoshida ascites sarcoma	6.35 ± 0.19				36.4
Ehrlich's ascites	5.60 ± 0.01				100.7
HeLa	6.85 ± 0.11	Sialic acid COOH; α-COOH; thiol; α-NH_2	ε-NH_3^+ of lysine; guanidyl groups; phosphate	5.8–6.7	3.9
γ-Globulin-treated HeLa	6.36 ± 0.14				16.7
Normal rat liver	6.50 ± 0.10				27.9
Human RBC[5]	5.60 ± 0.05				
Mouse RBC	5.80 ± 0.10				
Rabbit RBC	6.00 ± 0.15				
Rat liver mitochondria	7.40				
Rat liver lysosomes	7.20				

[1] The values for RBC'S, mitochondria and lysosomes have been taken from Just et al. (1975). All the other values are from Sherbet and Lakshmi (1975).

[2] The isoelectric zone is estimated to extend to a depth of 6.0 nm below the cell surface.

[3] pI values were calculated from surface pK of ionogenic groups corrected using the Hartley – Roe equation, $pI = -\log(\Sigma k'a \times k'w/\Sigma k'b)^{\frac{1}{2}}$ where $k'a$ and $k'b$ are corrected dissociation constants of the strongest anionogenic and cationogenic groups, respectively. $k'w = 6.81 \times 10^{-15}$ at 20°C.

[4] The number of net negative charges on the cell surface was calculated as described by Sherbet et al. (1972) using the equation $Q = PDr^2K \times 3.3\ 10^{-3}/e$, where Q is the number of net negative charges per cell. The potential, P, on the surface of the cells is estimated as the potential difference between a solution at neutral pH and a solution isoelectric with the cell surface and is given by $(7 - I) \times 2.303\ RT/F$, where R is the gas constant (8.315 Joules/°C, T the absolute temperature, F the Faraday (96 500°C). The value of $2.303\ RT/F = 0.0592$ at 25°C. D is the dielectric constant of water (78.54 at 25°C), r the radius of the cell in cm, K the Debye–Hückel function $(0.327 \times 10^8 I^{\frac{1}{2}})$ at 25°C where I is the ionic concentration due to Ampholine assumed to be 0.01 M, and e the electronic charge $4{,}8 \times 10^{-10}$ e.s.u.

[5] RBC = red blood cell.

gland of humans, monkeys (Hummel et al. 1972) and oxen (Fawcett et al. 1969).

One interesting, common feature is that, as with so many other proteins, even highly purified hormones appear heterogeneous by IEF. Thus Fawcett et al. (1969) have isolated four forms of bovine TSH, having pI's in the pH range 8.2–8.8, the γ form being the most active and stable form. It appears that the other three forms, α, β and δ, are conversion or degradation products of the γ form, though proteolytic activity during the extraction procedure does not seem to be responsible. However, most pituitary hormones are glycoproteins, therefore a partial loss in their polysaccharide moiety might be responsible for some of the observed heterogeneity.

Several reports have appeared on the characterization and purification of insulin (Percival and Dunckley 1969; Lewin 1969; Percival et al. 1970; Shapcott and O'Brien 1970; Salaman and Williamson 1971) as well as derivatives, such as acetylated (Lindsay and Shall 1971), carbamylated, methylthiocarbamylated (Lindsay et al. 1972), iodinated (Welinder 1971; Lambert et al. 1972) and semisynthetic (Borras and Offord 1970). Preparations of crystalline insulin (human, porcine, bovine), all appear heterogeneous on IEF. Kohnert et al. (1972, 1973a) have been able to assign the various bands to proinsulin, desamidoinsulin, insulin, insulin-like material and intermediate forms (Fig. 5.1). These last components, which represent the most acidic bands, probably correspond to the intermediate forms lacking the Arg_{31}-Arg_{32} residues and the Lys_{59}-Arg_{60} sequence (Kohnert et al. 1973a). Differences in metal content can also influence the insulin-banding spectrum as demonstrated by Lewin (1969) for insulin zinc protamine and insulin zinc semilente. Finally, high molecular weight aggregates of about 28,5000 daltons may also contribute to the heterogeneity in insulin (Kohnert et al. 1973b).

5.3. *IEF of peptides*

There are only a few reports on the isolation of peptides by isoelectric focusing, probably because overlap in the size distribution of peptides

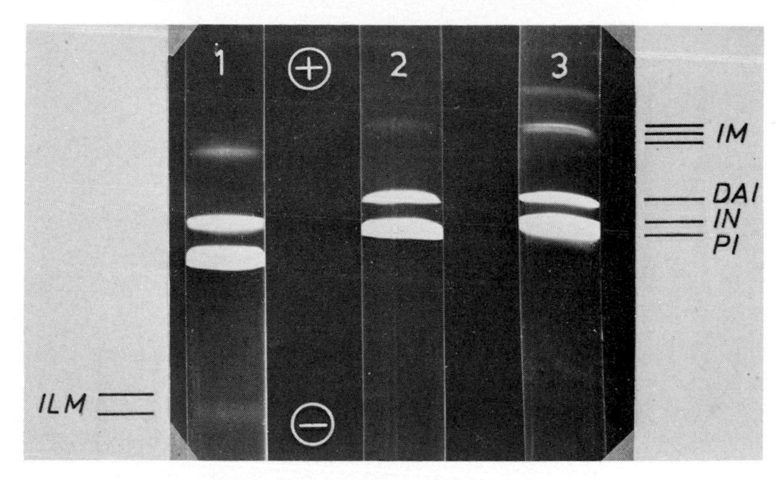

Fig. 5.1. IEF pattern of (1) human insulin (crystallized once); (2) porcine insulin (crystallized twice); (3) bovine insulin (crystallized twice) in 7.5% acrylamide gels containing 2% Ampholine, pH 3.5–10. Total electrolysis period: 9 hr. PI, proinsulin; IM, intermediate forms; DAI, desamido insulin; IN, insulin; ILM, insulin-like material. (From Kohnert et al. 1973a, by permission of Elsevier.)

and ampholytes complicated their separation by gel filtration or by differential solubility in acids. Also, only peptides containing aromatic residues can be easily distinguished from ampholytes by UV absorbancy. Catsimpoolas (1973c), by analytical scanning IEF, has been able to show separations of chromophoric (UV-absorbing) amino acid derivatives and dipeptides, such as L-lysil-L-tyrosine and L-hystidil-L-tyrosine.

Kopwillem et al. (1973) used gel IEF and analytical isotachophoresis to analyze peptides in the amino acid sequence 125–156 from the human growth hormone, synthesized by the Merrifield technique. In order to detect the various products, the peptide was synthesized with a DNP-His residue so that the focused bands were visible as bright yellow zones under UV light (Fig. 5.2).

Analyses and isolation of peptides may be facilitated with the low M.W. Ampholine synthesized by Gasparic and Rosengren (1975). These ampholytes have a mean molecular weight around 300 and may

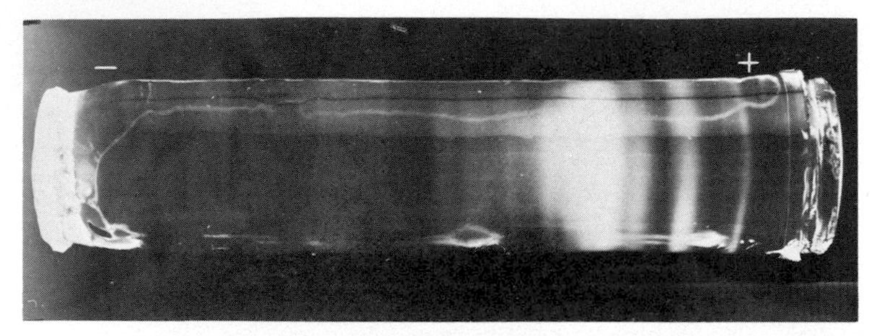

Fig. 5.2. Preparative gel IEF of peptide sequence 125–156 of human growth hormone, synthesized with the Merrifield solid-phase method. The 5th residue in the fragment is a DNP-histidine. After IEF, the peptide bands are revealed by excitation with UV-light (340 nm). The gel (6% acrylamide, 2% Ampholine pH range 3.5–10) was cast in an LKB Uniphor column. (From Kopwillem et al. 1973, by permission of Pergamon Press.)

be useful for preparative separations of peptides having an apparent molecular weight above 2000.

In the case of amphoteric antibiotics of low solubility in water, Righetti and Righetti (1974) perform IEF in gels containing 50% dimethylsulphoxide (DMSO). After focusing, the gel is extruded into distilled water: the DMSO is rapidly leached out and the focused antibiotic zones precipitate at their pI as opalescent bands which can be quantitated by a scan at 600 nm and recovered fully active.

5.4. IEF of glycoproteins

IEF has proved useful for the isolation and characterization of glycoproteins from different sources. The purification of glycoproteins from soybean cotyledons (Catsimpoolas and Leuthner 1969) and of glycoprotein I from *Phaseolus vulgaris* (Pusztai and Duncan 1971) have been reported. Meachum et al. (1971) have determined the M.W., amino acid and carbohydrate composition, quaternary structure and pI of invertase from Neurospora Crassa. Grasbeck (1968) has demonstrated the microheterogeneity of the complex between human intrinsic factor and cyanocobalamin.

$8s\alpha_3$ glycoprotein from human serum (Haupt et al. 1971) and plasma α_1-acid glycoprotein from chimpanzee (Li and Li 1970) and from humans have been purified and characterized by IEF. In the latter case, the binding affinity of progesterone to the glycoprotein has also been measured (Van Baelen and De Moor 1970; Ryan and Westphal 1972). Mucoids from the red cell membrane (Wintzer and Uhlenbruck 1970) and sialylglycopeptides from human platelet membranes (Pepper and Jamieson 1969) have also been isolated by IEF. Anderson and Jackson (1972) have reported the isolation of glycoproteins from bovine achilles tendon and their interaction with collagen.

In additiorr to the examples reported, there is the class of pituitary gonadotropins, most of which are glycoproteins (see §5.2).

The basis for heterogeneity in many glycoprotein may prove difficult to determine. The microheterogeneity demonstrated by Hayes and Wellner (1969) by IEF analyses of electrophoretically homogeneous preparations of L-amino acid oxidase which provided one of the earliest and most dramatic examples of the remarkable resolution given by GEF is an interesting case. This glycoprotein, which contains galactose, mannose, N-acetylglucosamine, fucose and sialic acid, is usually obtained from pooled venoms from *Crotalus adamanteus*. On disc electrophoresis this preparation was found to contain three components, which all seemed homogeneous when isolated and rerun. However, when analyzed by gel IEF, this same preparation was resolved into at least 18 enzymatically active components, isoelectric between pH 5.2 and pH 8.4 (Fig. 5.3). Each of the three electrophoretic components separated into different multiple components on IEF. This heterogeneity does not seem to be due to partial loss of FAD, nor to different redox forms of FAD. Part of the heterogeneity represents genetically different polypeptide chains, but may also be attributed to differences in the amount of carbohydrate present in the glycoprotein (Wellner and Hayes 1973).

Sometimes heterogeneity in carbohydrate may be present in the sample as isolated. However, artefactual heterogeneity can also occur during IEF. Goldstone and Koening (1974) found that two acidic glycoproteins from rat kidney lysosomes, arylsulphatase and β-glucu-

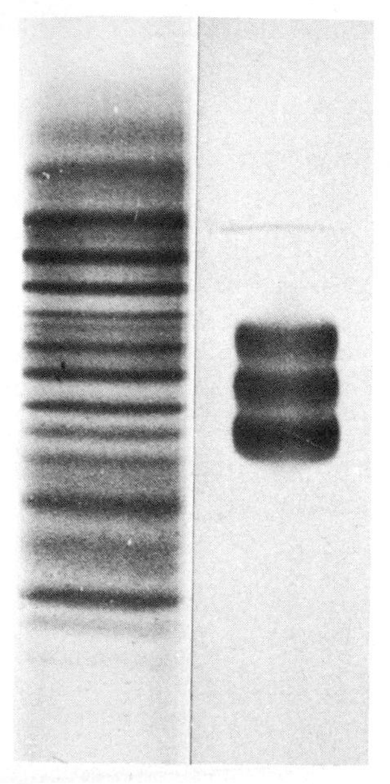

Fig. 5.3. Disc-electrophoresis (right) and gel IEF (left) of L-amino acid oxidase from snake venom. The three electrophoretic bands are resolved by IEF into 18 enzymatically active bands. IEF was performed in 10-cm gels containing 2% Ampholine pH range 3.5–10. (From Wellner and Hayes 1973, by permission of the New York Academy of Sciences.)

ronidase, undergo autolytic cleavage of labelled N-acetyl neuraminic acid and peptide. N-acetyl neuraminic acid was released more rapidly than peptide during incubation at 37 or 4°C at pH 5. Such autolysis could be minimized by adding 0.1% p-nitrophenyloxamic acid to the media used for extraction and electrofocusing, and by maintaining an alkaline pH (pH 8.8–9) during extraction and dialysis. With these precautions a main peak with pI 4.4 and a second one around pH 6.1–6.5 were obtained by IEF. However, with lysosomal extract exposed to pH 5, the pI's of both peaks rose approximately one pH unit, and a greater degree of heterogeneity was observed. During incubation at mildly acidic pH, these glycoproteins loose not only N-acetyl neuraminic acid, but also neutral and amino sugars, and peptides. Thus, although much of this degradation seems to be due to sample history before IEF, it is clear that it could also be partially elicited by IEF if the glycoproteins band in pH regions which adversely affect the stability of their sugar moiety.

Similar results have been reported by Gräsbeck et al. (1972) in a study of the vitamin B_{12}-binding protein in human saliva. The native protein was found to have a M.W. of 73,000 daltons. Depending on the method of purification and treatment, this M.W. was reduced to 66,000 and, sometimes, to 51,000 daltons. Concomitant with this M.W. loss, there was a decrease of stokes radius, a decrease of frictional ratio and an increase in diffusion coefficient. These losses in carbohydrate, which in this case seem to be generated only during the purification procedure, are then reflected in the isoelectric pattern and in the heterogeneity observed during IEF.

5.5. *IEF of lipoproteins*

GEF has proven useful for analyzing and characterizing both native lipoproteins and their derived apoproteins. Analysis of lipoproteins requires the presence of non-ionic detergents such as Tween 80, Emasol, Brij 39, Triton X-100 (Kostner et al. 1969), or tetramethyl urea (Gidez and Murnane 1975) to maintain their solubility. Kostner et al. (1969) prestained human serum lipoproteins with Sudan Black and obtained about 8 distinct bands after focusing in gels containing 33% ethylene glycol. The relationships of these various forms to those obtained by other methodologies was not determined. As with many other systems, considerable caution must be exercised in interpreting focusing patterns of lipoproteins. Multiple forms in lipoproteins presumably could arise through differences in bound lipids, in the apoproteins, or in associated carbohydrate. In addition, an artificial heterogeneity could conceivably arise from variations in dye binding when analysing prestained samples.

Scanu and colleagues have also used IEF for analyzing both the lipoproteins and their component polypeptides in human serum (Albers and Scanu 1971; Scanu et al. 1973). A typical separation of apo

Subject index p. 587

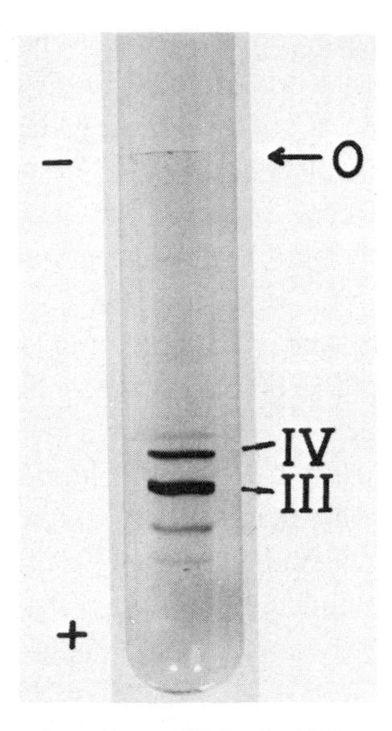

Fig. 5.4. Gel isoelectric focusing of apo HDL_2 in 6 M urea. 'O' indicates the sample application point (cathode). (From Scanu et al. 1973, by permission of the New York Academy of Scijnces.)

HDL_2 by GEF is shown in Fig. 5.4. The pattern given by GEF is considerably more complicated than that given by SDS gel electrophoresis, indicating heterogeneity in polypeptide chains of similar size. Gidez et al. (1975) have found rat serum apolipoproteins also to be more heterogeneous on GEF than is indicated by native or SDS gel electrophoresis. Fig. 5.5 compares electrophoretic and electrofocusing analyses of unfractionated rat apo HDL and subfractions separated by gel filtration. The unfractionated material resolved into about 7 fractions on polyacrylamide gel electrophoresis but about 20 on GEF. In order to define the multiple components more clearly, these investigators used a 2 dimensional procedure with gel electrofocusing followed by SDS gel electrophoresis. The basis for the additional complexity seen in apolipoproteins by IEF has not yet been established. However, it is significant that the complexity revealed in a single electrofocusing analysis of human apolipoproteins is of the same order

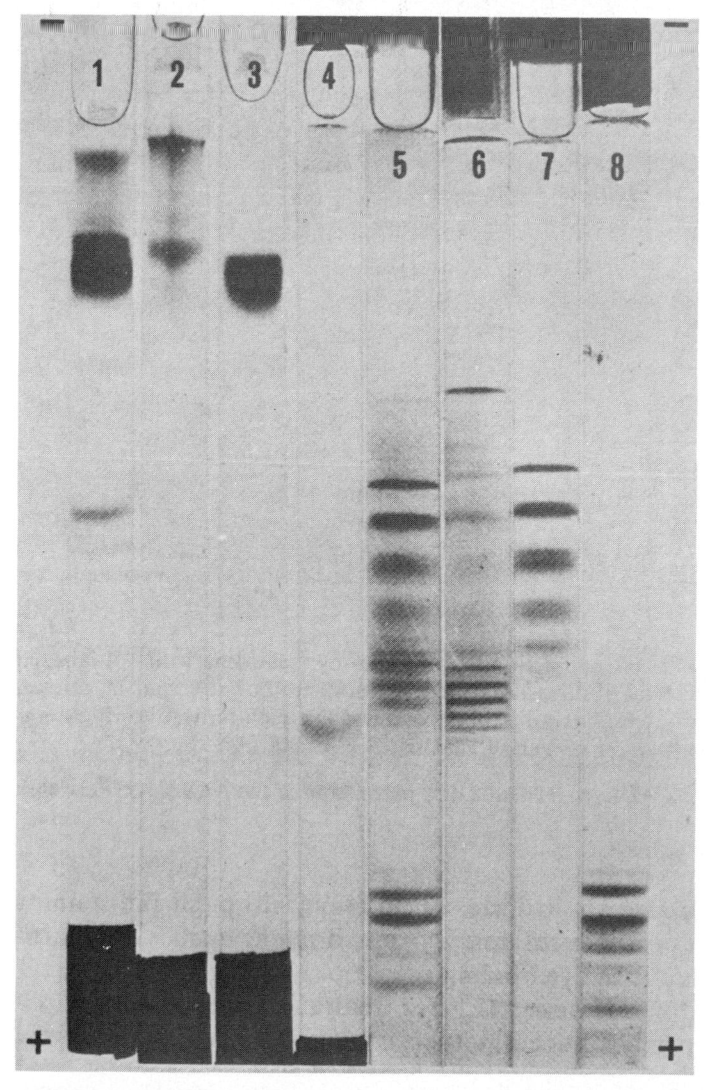

Fig. 5.5. Comparison of fractionation of rat apolipoproteins by SDS gel electrophoresis and by gel electrofocusing. Left to right: 1, apo-HDL unfractionated; 2-4, sub-fractions of apo-HDL separated by gel filtration (all analyzed by SDS-gel electrophoresis); 5–8, gel IEF patterns of samples 1–4. (From Gidez et al. 1975.)

as might be expected from combined techniques of gel filtration and ion-exchange chromatography (Scanu et al. 1973).

The present situation may be summarized as follows: a) for lipoproteins, IEF is more useful as an analytical procedure because of the tendency of lipoproteins to precipitate at their pI's; b) IEF may have wider application for both analytical and preparative purposes with apolipoproteins; c) best separations of the apolipoproteins are obtained by focusing in gels containing chaotropic agents such as 6–8 M urea to prevent aggregation, but at the risk of generating an artifactual heterogeneity by carbamylation of the polypeptide chain.

5.6. *IEF of membranes*

At present, gel filtration and gel electrophoresis, both in the presence of sodium dodecyl sulphate (SDS), have been the methods of choice for fractionating and characterizing membrane components. Recent experiments, however, indicate that IEF may provide a useful alternative. Jamieson and Groh (1971) have separated human erythrocyte and lymphocyte populations by IEF in sucrose density gradients. Similar procedures were used to isolate and characterize plasma membranes from human blood platelets (Barber and Jamieson 1970). Bonsall and Hunt (1971a,b) employed IEF to study interactions of human red blood cell (RBC) membranes with sodium trinitrobenzenesulphonate and surfactants. Several attempts have also been made to analyze disaggregated membrane components by IEF. Merz et al. (1972) solubilized RBC ghosts in 8 M urea, 20 mM EDTA and 0.2% 2-mercaptoethanol, and fractionated the extract in 2.5% acrylamide gels, containing 12.5% sucrose, 1% Ampholine pH 3–10 and 8 M urea. Approximately 40 components, isoelectric between pH 5.90 and pH 8.25 were obtained by this method. A similar number of components is given by SDS-electrophoresis in polyacrylamide gels (Righetti et al. 1973). Bhakdi et al. (1974) have also characterised EDTA-extractable proteins from erythrocyte membranes by IEF in urea gels and combined this in a two dimensional procedure to assign a molecular weight as well as a pI to each polypeptide. In a later paper, Bhakdi et al.

(1975) have focused hydrophobic proteins in erythrocyte membranes by solubilizing the proteins in Triton X-100 and focusing in gels containing 1 % Triton X-100 and 8 M urea. This latter method seems better suited for displaying both types of membrane proteins in a single IEF experiment (see Fig. 5.6).

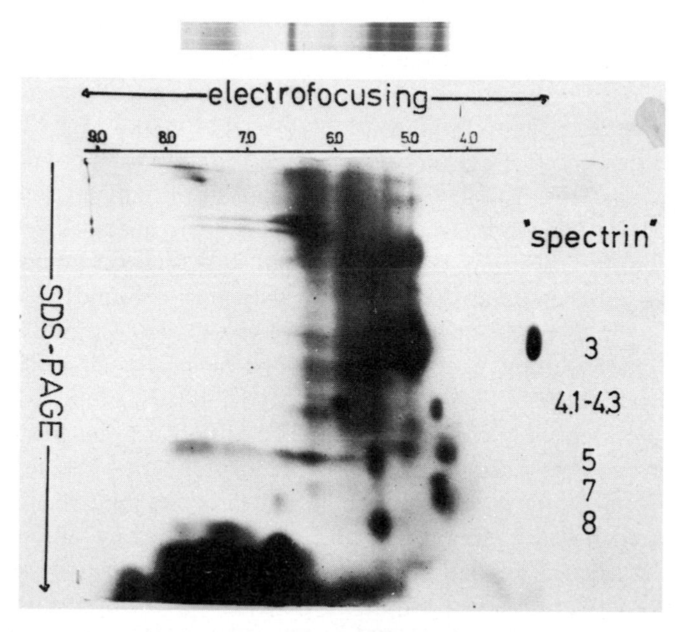

Fig. 5.6. Two-dimensional separation of a Triton X-100 extract of human erythrocyte membrane proteins. IEF: left to right; SDS-gel electrophoresis: top to bottom. The pH gradient is indicated on the upper scale. The numbers to the right refer to known protein bands on the red blood cell ghost. The lake-like material on the lower left of the gel represents residual Ampholine used in the first dimension. (From Bhakdi et al. 1975, by permission of ASP Biol. Med. Press B.V.)

5.7. IEF of metallo-proteins

IEF has proved a double edged sword for analysis of some metallo-proteins. It is well recognized that the presence of metal ions may change the surface charge of a protein to produce multiple forms. Such differences in metal content or in the redox state of a metal may be readily detected by IEF. However, the chelating properties of Ampholine for metal ions may introduce an artifactual heterogeneity in many metalloproteins. This heterogeneity may be present in the sample before IEF, but may also be elicited during IEF by metal chelation. Both cases may occur together. For example, rabbit transferrin can be separated into three molecular species, corresponding to the iron-free protein (pI 6.0), the one iron mole-protein complex (pI 5.5) and the two iron mole-protein complex (pI 5.1) (Van Eyk et al. 1969). All three forms occur in vivo. Essentially the same results have been obtained by Hovanessian and Awdeh (1975) in the analysis of human transferrin by thin-layer IEF. In this case, actually, little if any metal-Ampholine binding occurred during the experimental time, since in human sera the fully saturated transferrin species was always found to be the major component. It must also be emphasized that Ampholine has very little chelating properties for iron, especially at acidic pH (Galante et al. 1975; §1.7.4). However, Medellin (1972) found that all forms are eventually converted into apotransferrin upon prolonged electrolysis (64 hr). This iron sequestering phenomenon is concentration dependent, the equilibrium being shifted by the relative amounts of transferrin and Ampholine. For instance, using trace amounts of doubly labelled, ^{59}Fe- and ^{125}I-transferrin, and 64 hr electrolysis time, Medellin (1972) observed only a single peak of ^{125}I-apotransferrin. All of the iron had been removed and was found in a second peak of ^{59}Fe-Ampholine, the pI of this complex being 3.70. At high levels of transferrin, under conditions which allow the detection of the three forms, Medellin (1972) actually separated five forms, three stable and two unstable. Since one of the two latter forms is a half-saturated species, it is tempting to speculate that this species exists in two conformational states, depending on which site is occupied by iron.

Another interesting case is the separation of rabbit liver metallothionein (Nordberg et al. 1972). Upon electrofocusing, this protein splits into two bands, one containing only Cd (pI 3.9) and the other containing both, Cd and Zn (pI 4.5).

In addition to the possibility of metal removal, there is also the possibility of multiple band formation due to different redox states of the metal bound to the protein. This case is illustrated by met-myoglobin (met-Mb). Vesterberg (1967) and Satterlee and Snyder (1969) reported the fractionation of met-Mb into nine components, of which six were in the met form (Fe^{+++}) and three were in the reduced form (Fe^{++}). No pI differences were found between ferro-Mb and MbO_2, implying that the binding of oxygen to ferro-Mb has no influence on its pI. An interesting point is that the transition from the ferric to the ferrous form is due to the IEF procedure. At present, it is not clear whether this represents an electrolytic reduction or chemical modification by a gel constituent. The latter seems possible since Quinn (1973) has reported that reduction occurs when met-Mb is incubated with Ampholine and either persulphate or riboflavin. On the other hand, also the first mechanism could occur, since Flatmark and Vesterberg (1966) observed autoreduction of cytochrome C during IEF in sucrose density gradients.

A more complex pattern is obtained with hemoglobin (Hb). In preparations containing partially oxidized Hb, one should be able to detect three species, corresponding to oxy-Hb ($\alpha_2\beta_2$), met-Hb ($\alpha_2^+\beta_2^+$) and an intermediate, 50% oxidized component. In fact four bands were isolated (see Fig. 5.7) corresponding to $\alpha_2\beta_2$ (pI 6.95), $\alpha_2^+\beta_2^+$ (pI 7.15) and two intermediate bands, which, on the basis of their spectra, were identified as $\alpha_2\beta_2^+$ and $\alpha_2^+\beta_2$ (Bunn and Drysdale 1971).

Of course, not all metalloproteins are afflicted by these complications of metal chelation or alteration of the redox state. Much depends on the stability of the binding of the metal to the apoprotein and on the accessibility of the ampholytes to the metal. The iron storage protein, ferritin, is an interesting case of a metalloprotein whose isoelectric spectrum is not altered by metal chelation or by change in redox state. This protein normally contains variable amounts of iron. Some molecules contain little, if any, iron whereas others may contain anywhere up to 2500 atoms of iron. Unlike transferrin, this extensive heterogeneity in iron content is not, fortunately, reflected in the isoelectric spectrum of ferritin, otherwise an overwhelming complexity

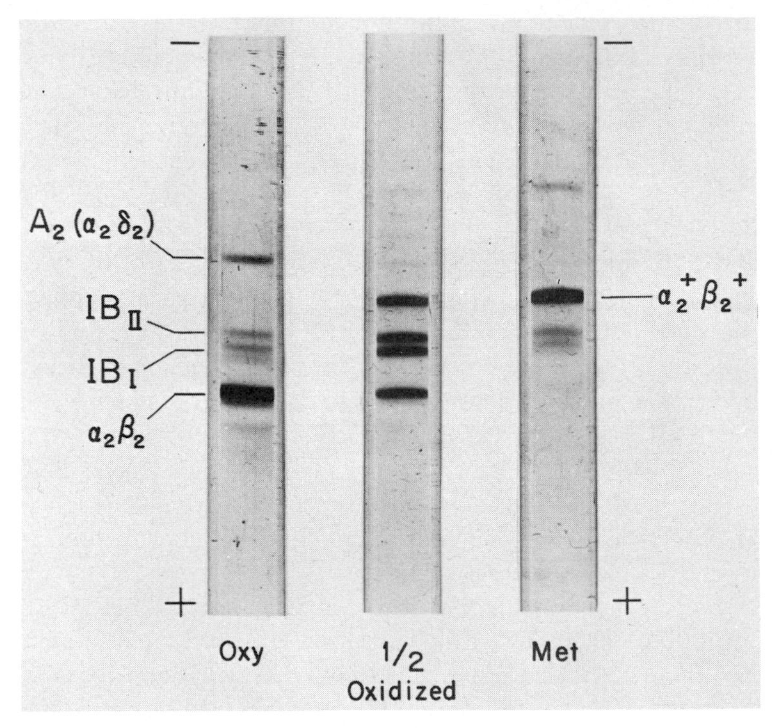

Fig. 5.7. Gel IEF patterns demonstrating partially oxidized haemoglobins IB_I and IB_{II}, separated from oxyhaemoglobin ($\alpha_2\beta_2$) and methaemoglobin ($\alpha_2^+\beta_2^+$). Ampholine pH range 6–8. (From Drysdale et al. 1971, by permission of ASP Biol. Med. Press B.V.)

might be expected. It has been clearly shown that chemically prepared iron-free apoferritin gives essentially the same pattern as the ferritin preparation from which it is derived. Further, the differences in iron content in ferritins are retained after IEF. The reason for this apparently surprising result is that the iron in ferritin is present as a ferricoxyhydroxide which does not alter the net charge. Moreover, the iron is contained in micelles within the multimeric spherical protein shell and the channels between the subunits through which iron passes are presumably too small to allow ampholytes to penetrate and chelate the iron.

GEF helped to clarify the relationships of tissue specific ferritins. Many tissue ferritins have been shown to differ in primary structure as shown by analyses of amino acid composition and peptide mapping. As shown in Fig. 5.8, many tissue ferritins differ slightly in their mobility on gel electrophoresis. Most of the electrophoretic bands are, however, rather broad and have considerable overlap. When examined by gel electrofocusing (Fig. 5.9), all of these ferritins were resolved into multiple components, with several common to most tissues. Those ferritins which were most difficult to resolve electrophoretically, e.g.

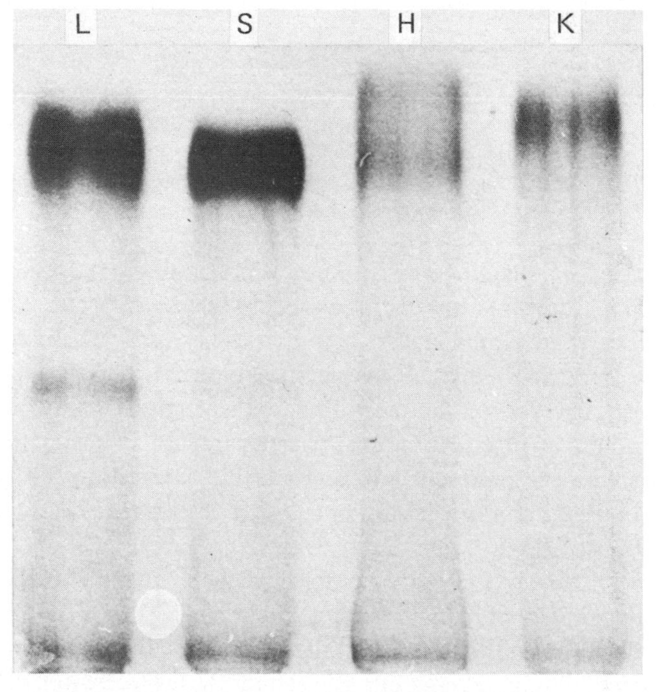

Fig. 5.8. Electrophoresis in 4.5% acrylamide gel slab of ferritin isolated from human liver (L), spleen (S), heart (H) and kidney (K). The gel was stained for protein with Ponceau S and counterstained for iron with $K_3Fe\ (CN)_6$. The major bands represent the ferritin monomers, i.e. single shells. (From Drysdale et al. 1975a, by permission of ASP Biol. Med. Press B.V.)

Subject index p. 587

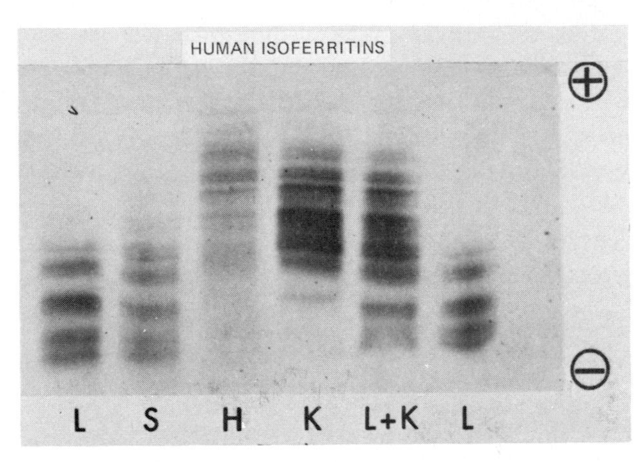

Fig. 5.9. Isoelectric focusing in 4% acrylamide gel, pH range 4–6, of ferritins from human liver (L), spleen (S), heart (H), kidney (K) and a mixture of liver and kidney (L + K). After equilibrium focusing, the gel was stained for protein using Coomassie Brilliant Blue R-250. (From Drysdale et al. 1975a, by permission of ASP Biol. Med. Press B.V.)

liver and spleen, shared many common isoferritins. On the other hand, those which were readily distinguished by electrophoresis, e.g. liver and heart, shared only a few common isoferritins. These findings indicate that the different tissue ferritins are not distinct species as previously thought but families of related isoferritins, many of which are common to several tissues. Present evidence indicates that the tissue isoferritins represent hybrid molecules consisting of different proportions of 2 or more dissimilar subunits (Drysdale 1974; Drysdale et al. 1975a, b).

Ferritin is one of the largest proteins to be analysed by GEF. The molecule consists of a multimeric protein shell of MW 444,000 and with a diameter of 120 Å. To complicate matters, the shell can aggregate to form dimers, trimers, etc. Consequently, the formation of stable pH gradients is of paramount importance to overcome molecular sieving effects and to achieve equilibrium focusing. Although this protein may be successfully focused in $C = 4\%$, $T = 4\%$ cylindrical gels, it migrates to equilibrium much more rapidly in thin slabs of the same nominal composition, possibly because of a higher porosity in the surface film of the gel slab (p. 451) (Drysdale 1974).

5.8. *IEF of immunoglobulins*

The high resolving power of IEF has been put to good use in the analysis of the enormous heterogeneity of immunoglobulins. Two general types of heterogeneity occur in this family of proteins: a) multiple protein species representing the products of distinct structural genes; these may have closely related sequences (alleles at one or more loci), b) microheterogeneity generated by post-synthetic modifications of a biosynthetically homogeneous protein. The extensive heterogeneity of γ-globulins is a good example of the first case.

For instance, in the case of inbred CBA/H mice, injected with 3-nitro-4-hydroxy-5-iodo-phenyl acetyl (NIP) coupled to bovine γ-globulin, it has been estimated that the minimum number of anti-NIP molecules which can be synthesized is 8000. These antibodies will focus mostly in the pH region 5.5 to 7.5. Since in thin layer IEF the resolving power is of the order of 0.005 pH units, the maximum theoretical number of resolved bands would be 400 (Williamson 1973). Despite this impressive resolution, it is still inadequate to ensure homogeneity of a single band. For example, Hoffman et al. (1972) fractionated anti-X antibody 2883 by preparative IEF. As shown in Fig. 5.10, these fractions ran as discrete single bands, and appeared to be homogeneous when rerun on analytical IEF in gels. However, each of these apparently pure fractions, when mildly reduced and alkylated, generated a heterogeneous population of light and heavy chains (Hoffman et al. 1972) (Fig. 5.11). Thus the purification of a single antibody represents a formidable task. The conclusions are that for a successful purification of antibodies by IEF the initial antibody population should not be too complex. This can be achieved by starting with a restricted immuno-response; in this case IEF can be used to isolate antibody species which are homogeneous by several criteria. The heterogeneity of myeloma proteins is a good example of the latter case. A myeloma protein is the immunoglobulin product of

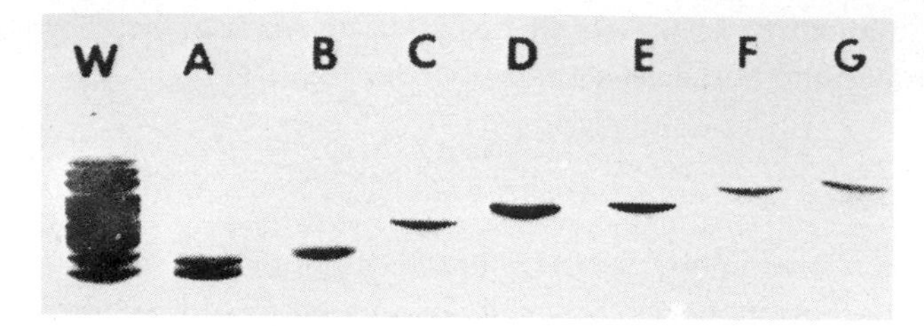

Fig. 5.10. Isoelectric spectra of anti-Xp antibody 2883 and fractions A to G taken from a sucrose density gradient IEF separation. The cathode is at the bottom. Analytical IEF was performed in disc gels. Staining of protein was carried out with Coomassie Brilliant Blue R-250. W, unfractionated antibody. (From Hoffman et al. 1972, by permission of Williams and Wilkins Co.)

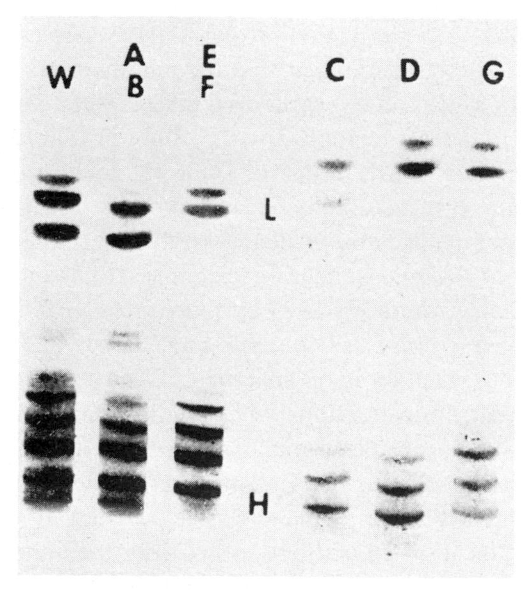

Fig. 5.11. Isoelectric spectra of mildly reduced and alkylated antibody 2883 and fractions A to G. From left to right: W, unfractionated antibody; A and B, pooled fractions; E and F, pooled fractions; fraction C, fraction D and fraction G. Light chains (L) focus at lower pI values than heavy chains (H). (From Hoffman et al. 1972, by permission of Williams and Wilkins Co.)

a single neoplastic plasma cell clone. The individual plasmacytoma synthesizes a single molecular species of immunoglobulin, which is subsequently altered in charge, giving rise to the characteristic micro-heterogeneous isoelectric spectrum of the serum protein. Much of this heterogeneity has been found to be due to loss of amide groups from glutamine and asparagine residues (Awdeh et al. 1970; Williamson et al. 1973). Microheterogeneity in some immunoglobulins may also arise from differences in carbohydrate content.

A new, interesting method comes to us from the immunologists. After IEF in polyacrylamide gels, Keck et al. (1973a) precipitate the immunoglobulin bands in 18% sodium sulphate, and then cross-link the protein chains with glutaraldehyde. This is an old trick used by X-ray crystallographers to prevent breaking of their crystals (Quiocho and Richards 1964). The aldehyde forms a continuous co-polymer, thus trapping and immobilizing the protein within the gel matrix, but without hampering the binding activity of antibodies. The antibody bands are then located by treating the plates with ^{125}I-labelled specific antigens. The excess radiolabelled macromolecule can then be washed from the plates without disrupting the focused pattern. The method can be extended to the use of fluorescent-labelled proteins and radio-labelled polysaccharide antigens as locators. Williamson (1973) has reviewed the applications of IEF in immuno-chemistry.

5.9. *IEF of tissue extracts*

Due to the high resolving power of IEF, it is possible to obtain protein patterns from the same tissue at different growth stages or from different tissues within the same organism, without any need for a previous purification step. These types of studies are particularly useful in developmental biology and they are preferably performed in thin-layer equipment, for accurate comparisons. Thus Bours and Brahma (1973) have compared, by flat-bed IEF, the soluble lens crystallins from several embryonic stages and also from adult cow lens cortex. They have found a gradual change in crystallin composition between the lens proteins of 1.5 to 7 month old embryos. In particular, during the various embryonic stages, they have detected a decrease of γ-crystallins along with an increase in β-crystallins and a slight increase in α-crystallins. When compared with the adult lens nucleus, the cortex showed a sharp decrease in concentration of γ-crystallin components. In addition, the composition of γ-crystallins of embryonic lens was found similar to that of the adult lens nucleus, but different from that of the adult lens cortex. Also in all embryonic stages two pre-α-crystallins could be detected.

All the various crystallin fractions analyzed (from chick lens, cow lens cortex, human lens cortex and nucleus) displayed a microheterogeneity, the α-crystallins being isoelectric between pH 4.8 and 5.1, the β-crystallins between pH 5.7 and 6.9 and the γ-crystallins between pH 7.0 and pH 8.1 (Bours 1971). On the other hand, the patterns of chick lens crystallins and chick cornea protein were found to be closely similar in appearance and pI's (Bours and Van Doorenmaalen 1970). An example of crystallin separation is shown in Fig. 5.12. Brahma and Van der Saag (1972) have used the same technique to study the biosynthesis of soluble proteins in early amphibian development. Axolotl gastrulae were grown in a radioactive protein hydrolysate for 16 hr. The soluble proteins were fractionated by IEF in thin layer and then revealed by protein stain and autoradiography. As seen in Fig. 5.13, many proteins incorporate the label, however there is a group of higher pI proteins which had been actively synthesized, since they stain faintly with Coomassie Blue but give marked bands by autoradiography.

5.10. IEF as a probe of interacting protein systems

IEF promises to be a valuable tool for the study of interacting protein systems. It can provide information on the interactions between protein molecules or subunits and between proteins and small molecules, including data on any resulting conformational changes that accompany these interactions. Ultimately, these findings may offer a better insight into the molecular basis of cooperative interactions in allosteric proteins and in larger biological systems.

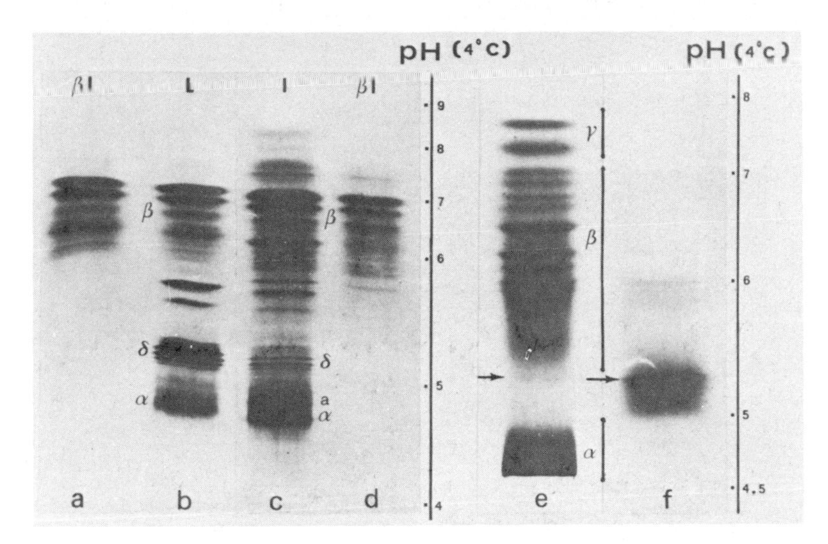

Fig. 5.12. Isoelectric focusing patterns of various crystallins from chick lens, chick iris and bovine cortex. References applied are β-crystallin fractions from chick lens (βL) and from chick iris (βI) and a pre-α-crystallin fraction obtained after free isoelectric focusing. (a) chick lens β-crystallins, also containing a low amount of chick lens γ-crystallin, (βL) 100 μg; (b) total chick lens crystallins (L) 200 μg; (c) chick iris proteins (I) 200 μg; (d) chick iris β-crystallins (βI) 100 μg, in a 5% acrylamide gel of 8 × 16 cm and 2% Ampholine carrier ampholytes of pH range 3–10 composed of pH (3–5):(5–7):(7–10) = 1.2:1:1. A fraction containing pre-α-crystallin (indicated by arrow) and β-crystallin, isolated by free isoelectric focusing, is isofocused in a 5% acrylamide gel with 2% Ampholine carrier ampholytes of pH range 3–10 composed of pH (3–5):(5–7):(7–9):(3–10) = 2:2:1:1; (e) lens cortex extract of a 14-week old calf, 200 μg (reference); (f) pre-α-crystallin free isofocusing fraction, selected between pH 5.18–5.35, 300 μg. The pH is measured at 4°C with a combined flat membrane micro-electrode. The scales along these plates show the pH values along the gels; pre-α, pre-α-crystallin; β, β-crystallins; δ, δ-crystallin; a, albumin. (From Bours 1975, by permission of ASP Biol. Med. Press B.V.)

Hemoglobin has proved a particularly favourable model for this type of study. For example, Park (1973) and Bunn (1973) were able to demonstrate subunit exchange between several types of hemoglobins and the formation of mixed tetramers between human and canine hemoglobins. Many of these intermediate forms were best detected by

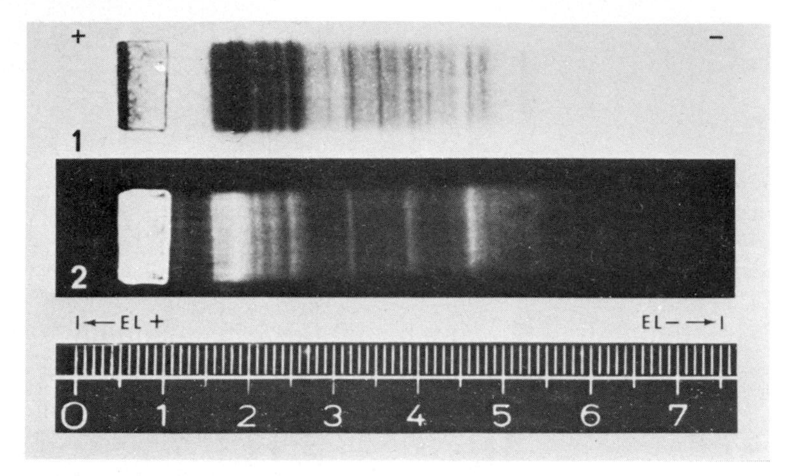

Fig. 5.13. Top: thin layer IEF of the radioactive soluble proteins from axolotl gastrulae in 5% acrylamide gel containing 2% Ampholine pH range 3.5–10. The sample was made radioactive by labelling with ^{14}C-protein hydrolysate during 16 hr. The protein bands were stained with Coomassie Brilliant Blue R-250 after fixation in 14% TCA. Bottom: autoradiogram of the upper gel. The print was made from Kodak No-Screen X-ray transparency. Distance between the electrodes was 7.5 cm, and the sample was 0.5 cm away from the anode. (From Brahma and van der Saag 1972, by permission of Academic Press.)

using deoxyhemoglobins under strict anaerobic conditions and at low temperatures. Although hybrids formed with oxyhaemoglobins, they dissociated rapidly during separation. Fig. 5.14 gives a nice example of this mixed tetramer formation.

Park (1973) has also used IEF to study the kinetics of dimer formation by performing the incubation of the various hemoglobins in solution under known conditions and using the gels to analyze the extent of reaction. In this way she was able to calculate the rate constant for the formation of mixed tetramers. Park also studied the dimerization plane in deoxy-Hb and in oxy-Hb and its pH dependance by IEF. These studies have shown that both liganded and unliganded haemoglobin cleave along the same plane at pH 7 as well as at pH 10.6. These observations on subunit exchange are useful as a tool for the elucidation of subunit organization in an oligomeric protein.

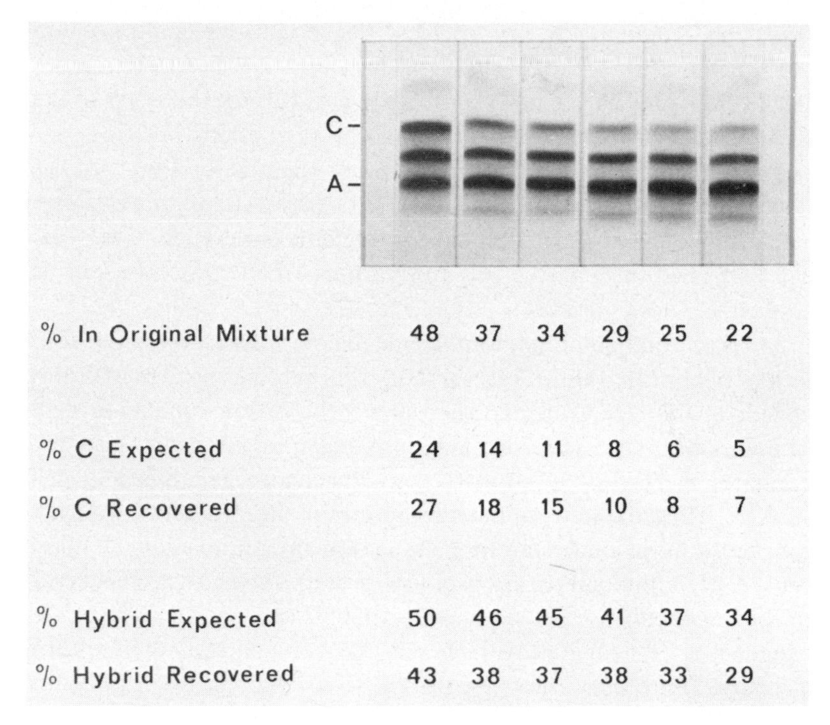

Fig. 5.14. Demonstration of the hybrid haemoglobin $\alpha_2\beta^A\beta^C$. Mixtures containing varying amounts of oxyhaemoglobin A and oxyhaemoglobin C were deoxygenated and then applied to the gels anaerobically. Sixty minutes later they were photographed and scanned at 555 nm. The amount of C and hybrid haemoglobins recovered (determined from the scanning data) is compared with that expected from the binomial distribution $a^2 + 2ab + b^2 = 1$. (From Bunn and McDonough 1974, by permission of the American Chemical Society.)

Haemoglobin has also been used to study interactions between macromolecules and small molecules. Thus Park (1973) and Bunn (1973) were able to demonstrate the conformational change associated with the Bohr effect when haemoglobin is oxygenated. By focusing partially oxygenated haemoglobin, the purple-violet deoxyhaemoglobin, pI 7.15, can be resolved from the red oxyhaemoglobin pI 6.95. If, after focusing Hb in a gel, ATP is run through it, it is possible to study the formation of a new species, ATP-bound-Hb, and to obtain the stoichiometry of the complex (in this case one ATP per tetramer). Park (1973) predicts also a favourable use of IEF for the study of the molecular basis of cooperative interactions in allesteric proteins.

The unique advantage of the focusing system is for those studies that combine equilibrium and kinetic processes. As an example, after a protein is focused, ligands, chemical reactants or interacting proteins can be passed through the gel, and various kinetic processes can be studied.

Another example of macromolecule–ligand interaction is the shift in pI's of concanavaline A observed upon binding of carbohydrates (Akedo et al. 1972). Concanavaline A, a carbohydrate binding protein from jack beans, possesses hemagglutinating and mitogenic activities, regulates the growth of transformed fibroblasts and binds to cell surfaces. Thus, the study of the interactions of this protein with sugars can be useful in understanding the action of concanavalin A upon cells. A crystalline preparation of this protein shows a spectrum of 7 to 8 components, isoelectric between pH 5.9 and pH 8.0. When concanavalin a is incubated with increasing amounts of D-mannose or α-methyl-D-glucoside, there is a progressive shift of the lower pI bands toward the pI 8.0 band, until, at 200 mM D-mannose, all the bands practically merge in the high pI component (see Fig. 5.15). As a control, D-galactose, which is known to bind only poorly to concanavalin A, causes considerably less pI shift. Since the ligand is an uncharged molecule, these pronounced pI shifts can only be explained in terms of conformational changes of concanavalin A. This in fact has been demonstrated by measuring circular dichroic spectra (Akedo et al. 1972).

In a similar manner, IEF has been found to be very useful in studying enzyme–substrate and enzyme–coenzyme complexes. This type of studies is illustrated in Fig. 5.16. When a pure preparation of folate reductase is analyzed by IEF, it bands as a single component having a pI of 6.5. Upon addition of half-saturating amounts of substrate, the pI 6.5 band splits into two bands, having very close pI's one of which corresponds to the enzyme–substrate complex. Upon addi-

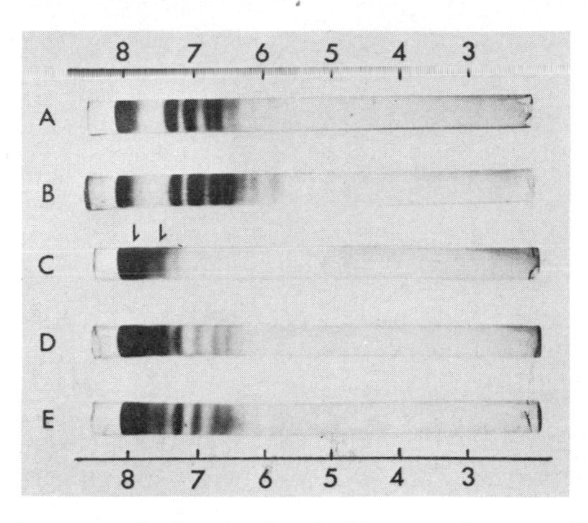

Fig. 5.15. IEF of concanavalin A preincubated with sugars. Concanavalin A (1.2 mg/ml) was preincubated with sugars (0.06 M) for 30 min at 37°C. A, with D-galactose; B, with no sugar (control); C, with D-mannose; D, with α-methyl-D-glucoside; E, with D-glucose. Proteins were stained with Coomassie Blue. Numbers indicate pH gradient along the gel. Arrows indicate protein bands produced by preincubating concanavalin A with specific sugars. (From Akedo et al. 1972, by permission of Academic Press.)

tion of coenzyme, two additional bands appear with pI's close to pH 4, one corresponding to E + TPN^+ and the other to E + TPNH. With time, these complexes dissociate and the free enzyme returns to its original focusing position.

IEF has also been found particularly useful in monitoring chemical modifications of proteins that affect their net charge and in determining the homogeneity of the product. Thus Bobb and Hofstee (1971) and Bobb (1973) have studied the progressive carbamylation (elimination of positive charges) and maleylation (substitution of positive with negative charges) of chymotrypsinogen A. By following the successive decrease in pI's upon introduction of increasing numbers of carbamyl or maleyl groups in the protein, it is possible to construct an 'isoionic titration' curve, relating pI's to the actual number of substituting groups introduced in the protein. Fig. 5.17 shows the pattern obtained by progressive maleylation of chymotrypsinogen A: it can be seen that the relative pI's decrease from the three native forms (pI's 8.8 to 9.6)

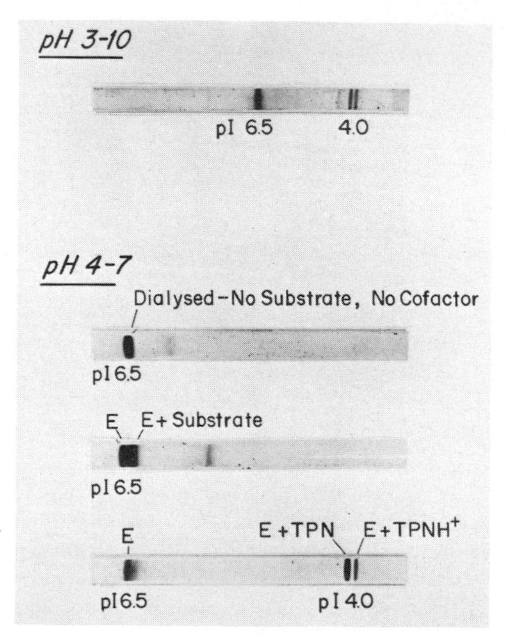

Fig. 5.16. IEF of folate reductase. The pure enzyme bands as a single band, having pI of 6.5. Upon addition of half-saturating amounts of substrate, the pI 6.5 band splits into two bands, having very close pI's one of which corresponds to the enzyme-substrate complex. Upon addition of coenzyme, two additional bands appear close to pH 4, one corresponding to E + TPN^+ and the other to E + TPNH. (From Righetti and Drysdale 1973, by permission of the New York Academy of Sciences.)

all the way down to pI 2.4 for the maximum (14) maleylation number. However, the product at each step of maleylation is highly heterogeneous.

5.11. Clinical applications

IEF has not yet found extensive use in routine clinical applications. There are many reasons for this. Perhaps the most immediate is the lag between research development and routine application of any new technique. In addition to other considerations of cost in apparatus,

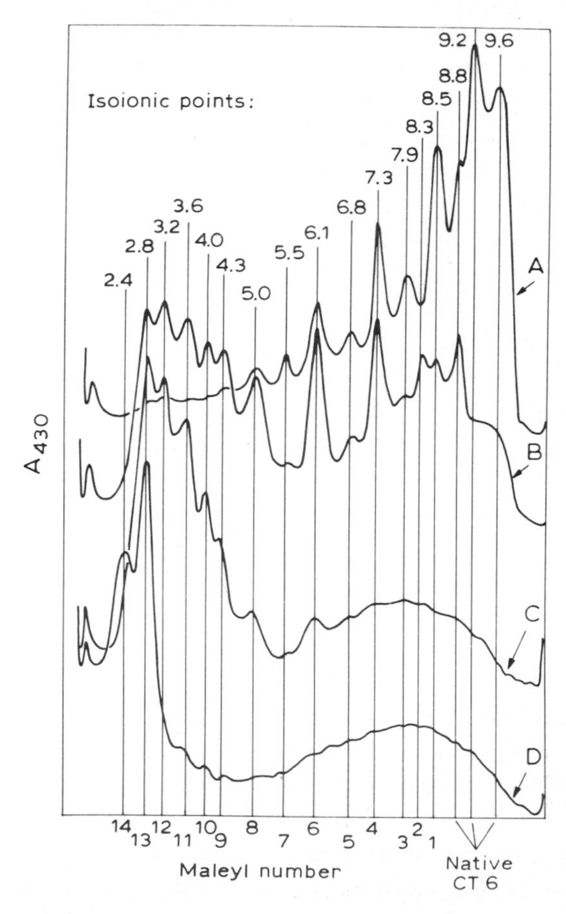

Fig. 5.17. Conversion of chymotrypsinogen A into maleylated derivatives as shown by gel IEF in the pH 3.5–10 region. The gels were stained with bromophenol blue and scanned at 430 nm. The isoionic points shown at the top of each curve were estimated from the position of the peak (determined in duplicate gels csanned at 280 nm) with respect to the pH gradient. In C and D, the bulge in the curve in the alkaline region is a staining artifact. These preparations had an average of 3.1 (A), 6.8 (B), 9.4 (C) and 13.7 (D) amino groups per molecule blocked by maleylation. (From Bobb 1973, by permission of the New York Academy of Sciences.)

reagents and time, there are likely to be additional complications with IEF in terms of interpretation and assessment of patterns. For example, analyses of serum by electrophoresis on paper or agar gels, usually reveals about 10 components. With higher resolution of gel electrophoresis with discontinuous pH gradients, or gradient pores, the number of detectable components may double or triple. Many, but not all, of these components have now been identified. However, with IEF in gels, the complexity may increase by an order of magnitude. The identification and quantitation of so many components now becomes a tremendous task and assessment of patterns requires considerable expertise.

From a cost analysis, it is certainly possible to consider routine clinical application of IEF with present day thin slab techniques. Apparatus is available in the market or could be produced cheaply. Several slabs may be run in series from common power supplies and coolant (Williamson 1973). Gel preparation is fairly simple. Moreover, precast gels containing ampholytes are now available to lighten laboratory loads (see p. 459). With the present cost of commercial ampholytes, the cost in ampholyte is a few pennies per sample. Should this be too much, large batches of ampholytes may be prepared easily and at a low cost in the laboratory (see §1.5). The time required for analysis should not pose a problem either. The entire operation can be completed in a day with procedures using high voltage and rapid staining. Thus, the major obstacle would seem to be in analyses of data. Some day soon, no doubt, someone will characterise the banding positions of the clinically important proteins separable and detectable by GEF or with two-dimensional techniques such as those of Dale and Latner (1969). Until then, clinical applications may be confined to simpler biological fluids or to isozymes, haemoglobins or other specific proteins that are more readily characterised.

Several such areas are presently finding increasing clinical use. For example, IEF has proved a valuable method for screening haemoglobinopathies (Drysdale et al. 1971; Bunn and McDonough 1974). Fig. 5.18 shows typical patterns given by some haemoglobin variants. The good separation of haemoglobins A, S and F is of particular

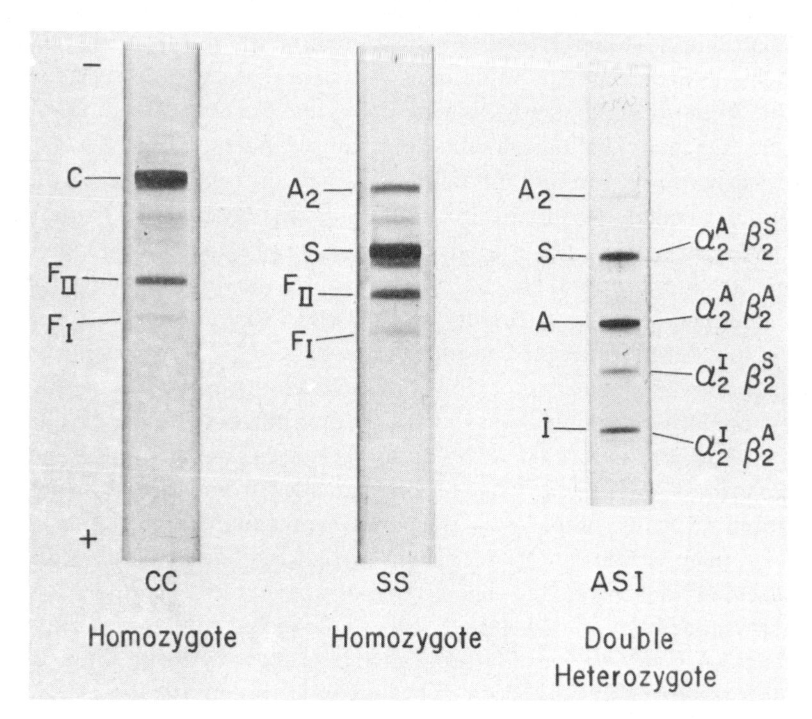

Fig. 5.18. Gel IEF patterns of human haemoglobin variants in the pH range 6–8. (From Drysdale et al. 1971, by permission of ASP Biol. Med. Press B.V.)

interest for the diagnosis of sickle cell disease. Focusing may be performed in either glass tubes or in thin slabs and results evaluated directly by densitometry. With high voltage techniques, as many as 100 samples may be comfortably analysed in one day by one person. IEF has been instrumental in detecting several new haemoglobin variants (Jensen et al. 1975). It is interesting to note that while haemoglobins with rather small differences, e.g. single amino acid substitutions such as haemoglobin Syracuse (β143 His → Pro) and Malmö (β97 His → Gln) (Jeppsson and Berglund 1972) can be distinguished by IEF from normal haemoglobin, others with more extensive differences in primary structure, e.g. A_2, C or E, are not readily resolved by IEF (Drysdale et al. 1971; Jensen et al. 1975). However, the differences in A_2, C and E tend to balance out and it is not surprising that their pI's are so similar. Sometimes, a variant may not be distinguishable from normal haemoglobin by IEF while the corresponding subunits are easily distinguishable. For example, haemoglobin Bethesda (β145 Tyr → His) is very difficult to distinguish from normal haemoglobin (pI 6.95), but the variant β subunit is easily separated from the normal β subunit. The explanation here may be that the p*K* of the imidazolium group of His is close to the pI of the multimeric protein. Consequently, it is only slightly protonated so that its substitution for Tyr makes little differences to the net charge at pH 7. However, the free β subunit has a considerably lower pI at which the imidazolium group is more fully protonated and so has a greater influence on the pI of the β subunit. Presumably haemoglobins Syracuse (β143 His → Pro) and Malmö (β97 His → Gln) can be distinguished from normal haemoglobin A because of the loss of the partially protonated His residue.

In their studies of haemoglobin Syracuse, Jensen et al. (1975) observed that while this haemoglobin was barely separated from normal haemoglobin when both were in the oxygenated form, the two haemoglobins were readily separated in the deoxy form. Since the substituted His residue does not contribute directly to the alkaline Bohr effect, the decreased Bohr effect of haemoglobin Syracuse indicates a conformational change beyond the site of the amino acid substitution. This conformational change may result from an interruption of the helical structure of the protein in that region.

5.12. Curiosities

Some puzzling but slightly encouraging results have been given by subjecting nucleic acids to GEF. From a consideration of the primary structure of nucleic acids, one would expect them to have a high negative charge above pH 3 and only reach an isoelectric state when the acidic phosphates were protonated. It is, therefore, very surprising to find that many nucleic acids form discrete banding patterns in the pH range 3–6. Experiments with well defined di- and trinucleotides clearly indicated that these substances focus at pH values which seem

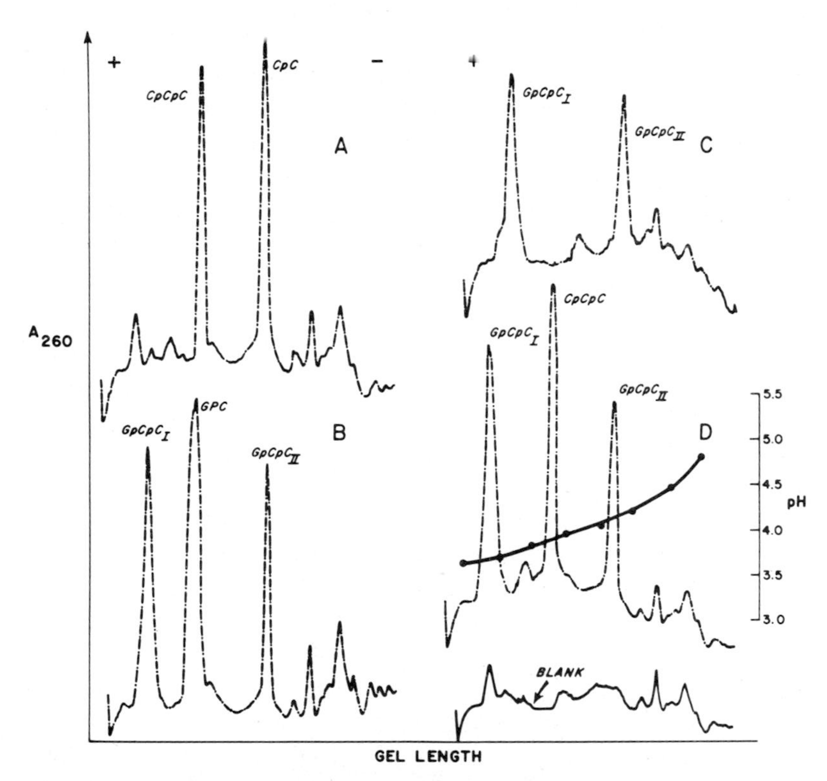

Fig. 5.19. Isoelectric focusing of mixture of di- and tri-nucleotides. A, IEF of CpC and CpCpC; B, IEF of GpC and GpCpC; C, IEF of GpCpC alone; D, IEF of CpCpC and GpCpC. Notice that GpCpC splits into two bands. (From Drysdale and Righetti 1972, by permission of the American Chemical Society.)

to represent true isoelectric points (Fig. 5.19). For example, CpC focused at pH 4.3 and was clearly separated from GpC and CpCpC, both of which focused near pH 3.9. These apparent pI values are entirely consistent with an isoelectric fractionation arising from protonation of ring nitrogens in the bases (C, pK_a 4.3, G, pK_a 2.4). Experiments with polynucleotides gave considerably more complicated results. For example, each of several highly purified isoaccepting species of tRNA was resolved into multiple components in the pH range 3–6. The patterns were highly pH dependent since the RNA banded in the same pH region with other pH ranges. Fig. 5.20 shows banding patterns given by isoaccepting species of fMet, Arg and Phe tRNA from *E. coli.* The multiple components were clearly visible as opaque regions, even at low sample inputs, and could be readily recovered from the gel. Subsequent analyses of isolated components indicated that the multiple forms were not degradation products, but

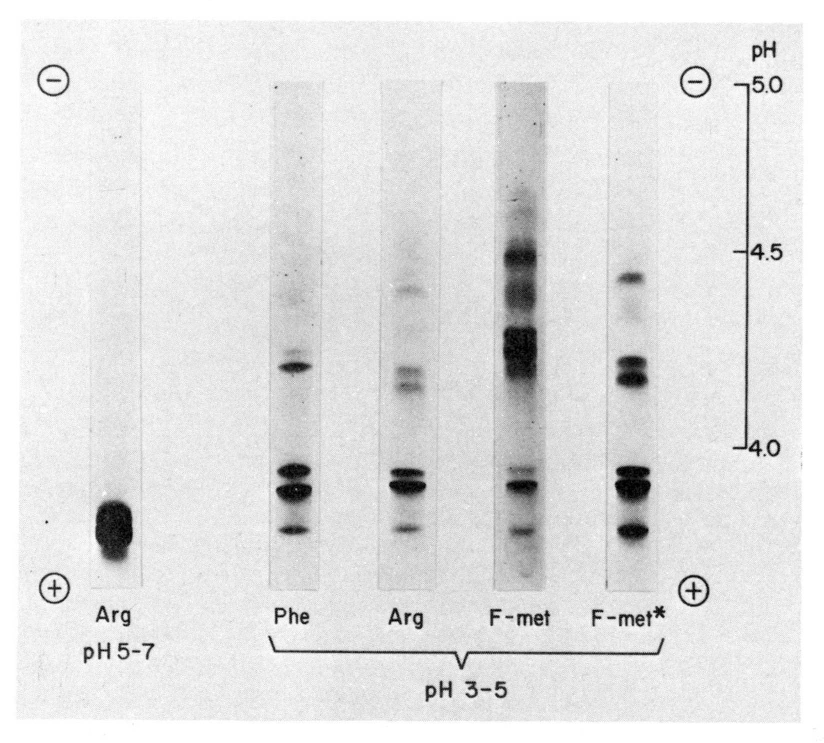

Fig. 5.20. Fractionation of isoaccepting species of Phe, Arg and fMet-tRNA by gel IEF. Samples 25 μl were subjected to gel electrofocusing for 6 hr in 5% acrylamide gels containing 2% Ampholine pH 3.5–5 or pH 5–7. In the case of fMet*, the sample was incorporated directly into the gel before polymerization. In all others, the samples were applied to the top of the gel in a buffering layer of 2% Ampholine. (From Drysdale and Righetti 1972, by permission of the American Chemical Society.)

molecules in different degree of denaturation, the extent of which increased progressively with decreasing pI. All of these forms could be restored essentially to their original activity by annealing at 65°C (Drysdale and Righetti 1972).

Studies with other cellular RNA species indicated that it was possible to separate tRNA, mRNA and ribosomal RNA species from one another by GEF. These experiments were conducted in high porosity gels and banding patterns were stable on prolonged focusing. Moreover, these RNA species did not band in order of their molecular weight. These results indicate that the patterns did not arise from molecular sieving. However, since the patterns also seem unlikely to represent a true isoelectric fractionation, it must be assumed that they are some part of methodological artifact – possibly from interaction with specific ampholytes in the acidic pH ranges. This notion is strengthened by the similarities in structure of the amines used to form the ampholytes and other amines, such as spermine, which are known to form complexes with nucleic acids. At high levels of ampholyte or in the presence of 6 M urea, the pattern was considerably simplified and all bands focused as the two most acidic components (Fig. 5.21). This result also suggests binding with ampholytes although other explanations are possible (Drysdale and Righetti 1972). Whatever the explanation, this type of fractionation may be useful in view of its apparent specificity. It appears to be capable of separating the major messenger RNA species for globin in rabbit reticulocytes and has resolved each messenger into multiple components. Some of these components differ in their content of poly (A) and in their translatability in messenger-dependent-cell-free systems (Drysdale and Shafritz 1975; Shafritz and Drysdale 1975).

5.13. On heterogeneity – facts and artifacts

The term 'homogeneous' is an operational definition indicating that only one molecular species can be detected with the applied analytical methods. In many cases, proteins which appear homogeneous by previous criteria have been found heterogeneous when examined by

Subject index p. 587

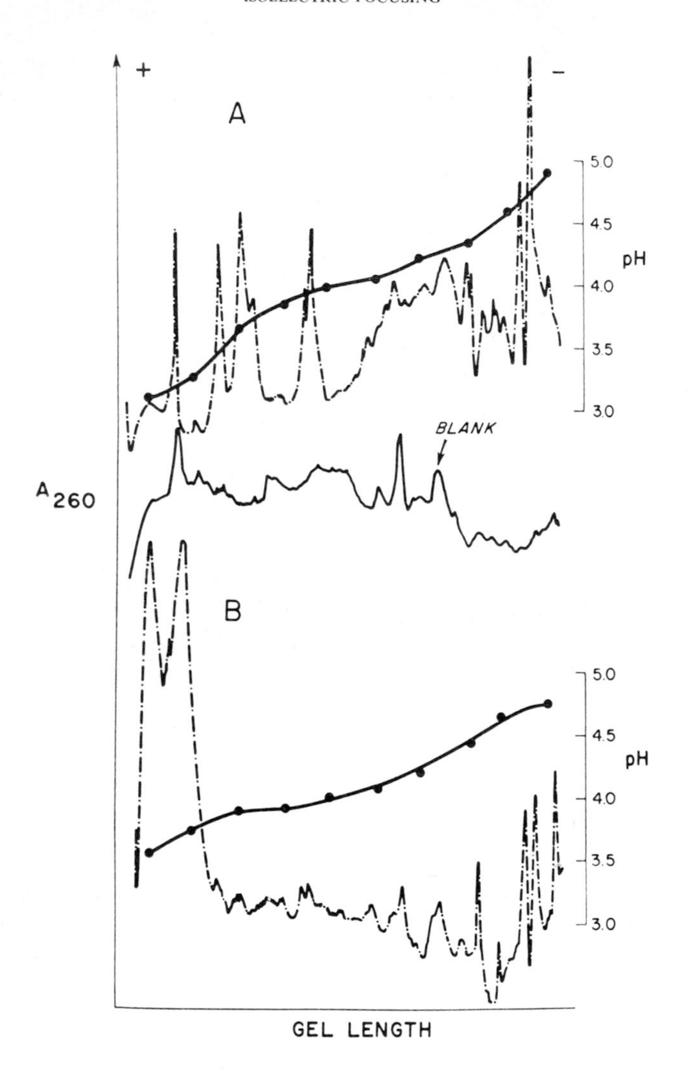

Fig. 5.21. Effect of urea on banding pattern of $tRNA_2^{Phe} \cdot tRNA_2^{Phe}$ (25 μl) was fractionated in gels containing 2% Ampholine, with or without 6 M urea. After 6 hr electrolysis, the gels were scanned at 260 nm. (A), control, no urea; (B), urea gel. (From Drysdale and Righetti 1972, by permission of The American Chemical Society.)

IEF. Such findings are often greeted with some concern, particularly when no explanation of the new-found heterogeneity is immediately forthcoming. However, as has been shown in part in previous sections, there are many different ways in which the surface charge of a molecule can be altered to change the pI of a molecule. Many of the possibilities for charge heterogeneity have been discussed by Epstein and Schechter (1968), Kaplan (1968) and Williamson et al. (1973). In addition to genetically determined differences in primary structure, one has to consider post-synthetic modifications to the polypeptide chain such as deamidation and deacetylation, interaction with ligands such as carbohydrate, lipid or nucleic acids in complex proteins or with substrates, cofactors and prosthetic groups. Often these ligands may have their own heterogeneity, e.g. differences in redox state in metals, which will also contribute to the heterogeneity of the complex. In addition to such factors which alter the net charge, one may also have to contend with position isomers or conformational differences in an otherwise homogeneous molecule. For example, molecules with the same number of groups, e.g. amide groups or disulphide bridges, but in different positions, might differ sufficiently to be separable by IEF.

It would, however, be naive to assume that all forms of heterogeneity seen in IEF are always present in the sample before IEF. Methodological artifacts are certainly possible and several have been well documented. Some of the major sources of heterogeneity are listed in Table 5.3 and some are discussed in more detail below. Some are avoidable with proper technique, others appear inherent in present-day

TABLE 5.3

Sources of heterogeneity.

1. Primary structure.
2. Post-synthetic modification in vivo or in sample preparation.
3. Prosthetic groups, carbohydrate, lipid, nucleic acid.
4. Ligand binding, substrates, cofactors.
5. Heterogeneity in ligands, differences in metal content or redox potential.
6. Conformers.

systems. Among the avoidable artifacts are heat denaturation of sample, incomplete focusing, interaction with gel constituents or electrolytes. Unavoidable artifacts fall in two categories: those arising from interaction with carrier ampholytes and those arising from subjecting an amphoteric molecule to unfavourable conditions of pH and ionic strength at its pI. Some of these possibilities are discussed below.

5.14. Avoidable artifacts in polyacrylamide gels

Sample modification by free radicals or other gel constituents Although interaction with polymerising catalysts is a potential problem, there have been few clear-cut examples of such artifacts in gel IEF. The greatest likelyhood would be with samples loaded internally or applied at the anode. Excess persulphate carries a high negative charge in the pH range 3–10 and will migrate rapidly to the anode. Thus, the interaction of persulphate or other charged species with the sample can usually be avoided by prerunning gels. Should this fail, residual free radicals may be removed by washing precast gels before equilibrating them with the appropriately diluted ampholyte solution by dialysis. Since the dimensions of the gel may alter on washing and dialysis, it is advisable to use gel slabs rather than gel tubes.

Sample modification by electrolytes This problem is most commonly encountered with gel tubes when applying the sample to the top of precast gels where it risks contact with the overlying electrolyte before it enters the gel. The best remedy is to ensure that the sample is adequately buffered and is applied beneath another less dense buffer which will separate it from the electrolytes (§3.2.10).

Incomplete focusing Generally caused by using gels with too restrictive pores and/or insufficient focusing periods. Gel porosity should be increased and/or electrophoresis period extended until patterns no longer change with time. Even if a slow pH decay occurs, the measured pI of focused bands should not alter. Equilibrium focusing in gel cylinders may be checked by demonstrating similar patterns from top loaded samples and internally loaded samples.

Molecular sieving is considerably reduced in thin slabs where sample run more freely in the surface film. Also, equilibrium focusing is more easily demonstrated in slabs with samples applied as a streak or at different ends of the gel.

5.15. General artifacts

Instability of molecule at its pI Most proteins focus in pH regions at which they are chemically unaltered. However, proteins with very low or very high pI's may be adversely affected by the extremes of pH and may be chemically modified. For example, some glycoproteins may focus in pH ranges where carbohydrate chains may be hydrolysed (see §5.4). Little can be done for this problem except to minimise focusing periods.

Conformational transitions Presumably, proteins, or other amphoteric molecules, may undergo conformational transitions as they migrate through different pH ranges. Conceivably, a molecule could exist in more than one conformation with only slow interconversion. Such conformers could have different pI's and so give the impression of heterogeneity. This situation should be suspected when the relative amounts of focused bands alter. With time, only the most stable form will remain but the interconversion may not go to completion in the experimental period. When appropriate, this problem may be prevented by running the sample under conditions which ensure a single conformation, e.g. 8 M urea.

Interaction with carrier ampholytes Although there have been few unequivocal demonstrations of artifactual heterogeneity due to complex formation with ampholytes, this possibility should always be considered. The complex banding patterns and the high apparent pI values given by nucleic acids strongly suggest complex formation with ampholytes (see §5.12). Such complex formation is often characteristically specific so that the resulting pattern depends largely on the likelihood of interaction of the appropriate species. Thus, one would expect the pattern to depend on the relative amounts of ampholyte and sample, the mode of sample application and the pH dependence of the interaction of sample and specific ampholytes. Complex formation should, therefore, be suspected when the focusing pattern changes markedly with sample load or ampholyte concentration. It may usually be detected simply by mixing different proportions of sample and ampholyte before sample application.

Metal chelation As noted earlier (§5.7), loosely bound divalent metals may be removed from metalloproteins with concomitant loss of activity and generation of additional protein species. Often, the extent of metal chelation is a function of the relative amounts of protein and ampholyte, the pH gradient traversed by the sample and the pH stability of the metalloprotein complex. Metal chelation may be reduced by increasing the protein/ampholyte ratio, applying the sample close to its pI and minimising the electrophoresis period (Galante et al. 1975).

Electrode modification Anodic oxidation or cathodic reduction may occur in samples that focus too close to the electrodes (§4.2). The easiest solution here is to select a pH range in which samples focus closer to the middle. Alternatively, or in addition, reducing agents may be added to the electrolytes or to the focusing gradients (§4.7).

Denaturation or precipitation There are two main causes of precipitation, isoelectric precipitation or heat denaturation. The former is more troublesome when focusing in sucrose density gradients and is much less of a problem in other media such as polyacrylamide or Sephadex beds, particularly in flat beds. Isoelectric precipitation may be reduced by reducing sample input, increasing ionic strength by using higher levels of ampholyte, or by additives such as taurine, low levels of urea or non-ionic detergents (§4.2). Heat denaturation is readily avoided by using more appropriate electrophoresis conditions, usually by working at lower temperatures or electrical loads (see p. 96).

5.16. Conclusion

It is hoped that this book will provide useful background information on IEF as well as practical advice to remove some of the black magic thought necessary for successful experiments. The superior resolution of IEF is now widely recognized and its potential has been clearly demonstrated in its many varied applications, both at the analytical and preparative level. One of the endearing features of IEF is that it is an equilibrium method, and therefore, is considerably less demanding in technique than many others. Given a chance, the system will correct itself. Experiments may be interrupted for long periods, then restored – often with little if any difference in the final result. Another attraction is that there is really only one variable, the pH gradient. In only one or two experiments it should be possible to resolve all molecules whose pI's differ by more than 0.01 pH units.

Many of the techniques for IEF have now been fairly well standardised, but there are still areas that need improvement. Perhaps the greatest need is for ampholytes that provide a truly uniform conductivity over all pH ranges. It would also be most desirable to characterise ampholyte preparations more fully than has been done in the past and to standardise all preparations. Although the exact composition, number and distribution of ampholytes in any given preparation may prove difficult, if not impossible, to control, it would be most helpful if the basic ingredients used in their synthesis were detailed as is the case for most other commercially produced reagents. In additions, it would be most helpful to have the actual pH gradient obtained experimentally, the conductivity profile and UV scans of focused gradients. One would also like to know the molecular weight distribution for each batch of ampholytes of the same nominal pH range.

The present cost of suitable ampholytes is often an inhibiting factor in the development of large scale preparative applications of IEF. It has now been clearly shown that several systems are capable of resolving mg and even g amounts of proteins with comparable resolution to that given in analytical systems. The major drawback to these applications is the cost of commercial ampholytes. An alternative could be, for the companies which sell carrier ampholytes (LKB, Serva and Bio Rad) to market less purified ampholytes, available in bulk quantities at cheaper prices. Another alternative, when using large amounts of carrier ampholytes, for preparative purposes, is the labo-

Subject index p. 587

ratory synthesis. The starting reagents are quite inexpensive and the experimental procedure rather simple. Moreover, one can subfractionate the ampholytes with most of the equipment presently used for IEF in columns, troughs or slabs for a custom tailored pH range. In addition to the considerations of cost, it would also be desirable to develop alternative types of carrier ampholytes that interfered less with protein detection and did not chelate metals.

Equipment and instrumentation for IEF is now perfectly adequate. In the analytical field, it would be most desirable to develop a thin but durable substrate such as a plastic backed cellulose product with large porosity and no electroendosmosis for faster, more convenient analyses than thin gels presently afford. In the preparative field, alternative anticonvective media that were easily standardised and allowed use of fractionation and recovery systems would be helpful.

References

Akedo, H., Mori, Y., Kobayashi, M. and Okada, M. (1972) Biochem. Biophys. Res. Commun. *49*, 107 113.

Albers, J.J. and Scanu, A.M. (1971) Biochim. Biophys. Acta *236*, 29–37.

Allen, J.C. and Humphries, C. (1975) In: Isoelectric Focusing; Arbuthnott, J.P. and Beeley, J.A., eds. (Butterworths, London) pp. 347–354.

Allen, R.C., Harley, R.A. and Talamo, R.C. (1974) Am. J. Clin. Pathol. *62*, 732–739.

Alper, C.A., Hobart, M.J. and Lachmann, P.J. (1975) *In:* Isoelectric Focusing, Arbuthnott, J.P. and Beeley, J.A., eds. (Butterworths, London) pp. 306–312.

Alpert, E., Drysdale, J.W., Isselbacher, K.J. and Schur, P.H. (1972) J. Biol. Chem. *247*, 3792–3798.

Alpert, E., Drysdale, J.W. and Isselbacher, K.J. (1973) Ann. N. Y. Acad. Sci. *209*, 387–396.

Anderson, F.B. and Jackson, D.S. (1972) Biochem. J. *127*, 179–186.

Anderson, F.B. and O'Grady, J.E. (1972) J. Endocrinol. *52*, 507–515.

Andersson, L.O., Borg, H. and Mikaelsson, M. (1972) FEBS Letters *20*, 199–202.

Arbuthnott, J.P. and Beeley, J.A. (1975) Isoelectric Focusing (Butterworths, London) pp. 367.

Arnaud, Ph., Chapuis-Cellier, C. and Creyssel, R. (1975) Prot. Biol. Fluids *22*, 515–520.

Arvidson, S. and Wadström, T. (1973) Biochim. Biophys. Acta *310*, 418–420.

Asard, P.E. (1974) LKB Application Note No. 192.

Awdeh, Z.L., Williamson, A.R. and Askonas, B.A. (1968) Nature *219*, 66–67.

Awdeh, Z.L. (1969) Sci. Tools 16, 42–43.

Awdeh, Z.L., Williamson, A.R. and Askonas, B.A. (1970) Biochem. J. *116*, 241–248.

Barber, A.J. and Jamieson, G.A. (1970) J. Biol. Chem. *245*, 6357–6365.

Barrett, T. and Gould, H.J. (1973) Biochim. Biophys. Acta *294*, 165–170.

Barrollier, V.J., Watzke, E. and Gibian, H. (1958) Z. Naturforsch. *13*b, 754–755.

Baumann, G. and Chrambach, A. (1975a) Anal. Biochem. *64*, 530–536.

BAUMANN, G. and CHRAMBACH, A. (1975b) *In:* Progress in Isoelectric Focusing and Isotachophoresis; Righetti, P.G., ed. (ASP Biol. Med. Press B.V., Amsterdam) pp. 13–23.

BEELEY, J.A., STEVENSON, S.M. and BEELEY, J.G. (1972) Biochim. Biophys. Acta *285*, 293–300.

BEELEY, J.A., STEVENSON, S.M. and BEELEY, J.G. (1975) *In:* Isoelectric Focusing; Arbuthnott, J.P. and Beeley, J.A., eds. (Butterworths, London) pp. 147–151.

BETTENDORF, G., BRECKWOLDT, M., CAYGAN, P.J., FOCK, A. and KUMASAKA, T. (1968) Gonadotropins, Proc. Work. Cinf. *3*, 13–23.

BHAKDI, S., KNÜFERMANN, H. and WALLACH, D.F.H. (1974) Biochim. Biophys. Acta *345*, 448–457.

BHAKDI, S., KNÜFERMANN, H. and WALLACH, D.F.H. (1975) *In:* Progress in Isoelectric Focusing and Isotachophoresis; Righetti, P.G., ed. (ASP Biol. Med. Press B.V., Amsterdam) pp. 281–291.

BIANCHI, U. and STEFANELLI, A. (1970) Atti Accad. Nat. Lincei Rend., Serie VIII, *93*, 539–542.

BLANICKY, P. and PIHAR, O. (1972) Coll. Czech. Chem. Commun. *37*, 319–325.

BLATON, V. and PEETERS, H. (1969) Prot. Biol. Fluids *16*, 56–60.

BOBB, D. and HOFSTEE, H.J. (1971) Anal. Biochem. *40*, 209–217.

BOBB, D. (1973) Ann. N.Y. Acad. Sci. *209*, 225–236.

BODDIN, M., HILDERSON, J., LAGROU, A. and DIERICK, W. (1975) Anal. Biochem. *64*, 293–296.

BODWELL, C.E. and CREED, G.J. (1973) Abstr. 9th Int. Congress Biochem., Stockholm, p. 25.

BONSALL, R.W. and HUNT, S. (1971a) Biochim. Biophys. Acta *249*, 266–280.

BONSALL, R.W. and HUNT, S. (1971b) Biochim. Biophys. Acta *249*, 281–284.

BORRAS, F. and OFFORD, R.E. (1970) Nature *227*, 716–718.

BOURS, J. and VAN DOORENMAALEN, W.J. (1970) Sci. Tools *17*, 36–38.

BOURS, J. (1971) J. Chromatogr. *60*, 225–233.

BOURS, J. (1973) Sci. Tools *20*, 2–14.

BOURS, J. and BRAHMA, S.K. (1973) Exptl. Eye Res. *16*, 131–142.

BOURS, J. (1975) *In:* Progress in Isoelectric Focusing and Isotachophoresis; Righetti, P.G., ed. (ASP Biol. Med. Press BV, Amsterdam) pp. 235–256.

BRAHMA, S.K. and VAN DER SAAG, P.T. (1972) Exptl. Cell Res. *75*, 527–530.

BRAKKE, M.D., ALLINGTON, R.W. and LANGILLE, F.A. (1968) Anal. Biochem. *25*, 30–39.

BRASELTON, W.E. and MCSHAN, W.H. (1970) Arch. Biochem. Biophys. *139*, 45–58.

BREWER, J.M. (1967) Science *156*, 256–257.

BROWN, W.D. and GREEN, S. (1970) Anal. Biochem. *34*, 593–595.

BULL, H.B. (1964) An Introduction to Physical Biochemistry (F.A. Davis Co., Philadelphia) p. 319.

BUNN, H.F. and DRYSDALE, J.W. (1971) Biochim. Biophys. Acta *229*, 51–57.

BUNN, H.F. (1973) Ann. N.Y. Acad. Sci. *209*, 345–353.

BUNN, H.F. and MCDONOUGH, M. (1974) Biochemistry *13*, 988–993.

CANNAN, R.K. (1942) Chem. Rev. *30*, 395–412.

CARREL, S., THEILKAES, L., SKVARIL, S. and BARANDUN, S. (1969) J. Chromatogr. *45*, 483–486.

CATSIMPOOLAS, N. (1969a) Anal. Biochem. *26*, 480–482.

CATSIMPOOLAS, N. (1969b) Arch. Biochem. Biophys. *131*, 185–190.

CATSIMPOOLAS, N. (1969c) Clin. Chim. Acta *23*, 237–238.

CATSIMPOOLAS, N. and LEUTHNER, E. (1969) Biochim. Biophys. Acta *181*, 404–409.

CATSIMPOOLAS, N. and MEYER, E.W. (1969) Arch. Biochem. Biophys. *132*, 279–285.

CATSIMPOOLAS, N. (1970a) Separ. Sci. *5*, 523–544.

CATSIMPOOLAS, N. (1970b) Clin. Chim. Acta *27*, 365–366.

CATSIMPOOLAS, N. (1973a) Isoelectric Focusing and Isotachophoresis, Ann. N.Y. Acad. Sci. *209*, 1–529.

CATSIMPOOLAS, N. (1973b) Separ. Sci. *8*, 71–121.

CATSIMPOOLAS, N. (1973c) Ann. N.Y. Acad. Sci. *209*, 65–79.

CATSIMPOOLAS, N. (1973d) Anal. Biochem. *54*, 66–78.

CATSIMPOOLAS, N. (1973e) Anal. Biochem. *54*, 79–87.

CATSIMPOOLAS, N. (1973f) Anal. Biochem. *54*, 88–94.

CATSIMPOOLAS, N. and GRIFFITH, A. (1973) Anal. Biochem. *56*, 100–120.

CATSIMPOOLAS, N. (1975a) Separ. Sci. *10*, 55–76.

CATSIMPOOLAS, N. (1975b) *In:* Progress in Isoelectric Focusing and Isotachophoresis; Righetti, P.G., ed. (ASP Biol. Med. Press B.V., Amsterdam) pp. 77–92.

CHAMOLES, N. and KARCHER, D. (1970a) Clin. Chim. Acta *30*, 337–341.

CHAMOLES, N. and KARCHER, D. (1970b) Clin. Chim. Acta *30*, 359–364.

CHRAMBACH, A. (1966) Anal. Biochem. *15*, 544–552.

CHRAMBACH, A. and RODBARD, D. (1972) Separ. Sci. *7*, 663–703.

CHRAMBACH, A., DOERR, P., FINLAYSON, G.R., MILES, L.E.M., SHERINS, R. and RODBARD, D. (1973) Ann. N.Y. Acad. Sci. *209*, 44–64.

CHU, F.S., THADHANI, K., SCHANTZ, E.J. and BERGDOLL, M.S. (1966) Biochemistry *5*, 3281–3289.

CLELAND, W.W. (1964) Biochemistry *3*, 480–482.

COHN, E.U. and EDSALL, J.T. (1943) Proteins, Amino Acids and Peptides as Dipolar Ions (Reinhold, New York).

CONWAY-JACOBS, A. and LEWIN, M.L. (1971) Anal. Biochem. *43*, 394–400.

COTTON, R.G.H. and MILSTEIN, C. (1973) J. Chromatogr. *86*, 219–221.

COUTELLE, R. (1971) Acta Biol. Med. Germ. *27*, 681–691.

DALE, G. and LATNER, A.L. (1968) Lancet, April 20, 847–848.

DALE, G. and LATNER, A.L. (1969) Clin. Chim. Acta *24*, 61–68.

DAVIES, H. (1970) Prot. Biol. Fluids *17*, 389–396.

DAVIES, H. (1975) *In:* Isoelectric Focusing; Arbuthnott, J.P. and Beeley, J.A., eds. (Butterworths, London) pp. 97–113.

DAVIS, B.J. (1964) Ann. N.Y. Acad. Sci. *121*, 404–427.

DEAN, R.T. and MESSER, M. (1975) J. Chromatogr. *105*, 353–358.

DELINCÉE, H. and RADOLA, B.J. (1970) Biochim. Biophys. Acta *200*, 404–407.

DELINCÉE, H. and RADOLA, B.J. (1971) Prot. Biol. Fluids *18*, 493–497.

DEWAR, J.H. and LATNER, A.L. (1970) Clin. Chim. Acta *28*, 149–152.

DIRKSEN, M.L. and CHRAMBACH, A. (1972) Separ. Sci. *7*, 747–772.

DRYSDALE, J.W. (1970) Biochim. Biophys. Acta *207*, 256–258.

DRYSDALE, J.W., RIGHETTI, P.G. and BUNN, H.F. (1971) Biochim. Biophys. Acta *229*, 42–50.

DRYSDALE, J.W. and RIGHETTI, P.G. (1972) Biochemistry *11*, 4044–4052.

DRYSDALE, J.W. (1974) Biochem. J. *141*, 627–632.

DRYSDALE, J.W. and SHAFRITZ, D. (1975) *In:* Progress in Isoelectric Focusing and Isotachophoresis; Righetti, P.G., ed. (ASP Biol. Med. Press BV, Amsterdam) pp. 293–306.

DRYSDALE, J.W., HAZARD, J.T. and RIGHETTI, P.G. (1975a) *In:* Progress in Isoelectric Focusing and Isotachophoresis, Righetti, P.G., ed. (ASP Biol. Med. Press BV, Amsterdam) pp. 193–204.

DRYSDALE, J.W., AROSIO, P., ADELMAN, T., HAZARD, J.T. and BROOKS, D. (1975b) *In:* Proteins of Iron Storage and Transport in Biochemistry and Medicine; Crighton, R.R., ed. (North Holland Publ. Co., Amsterdam) pp. 359–366.

DUBERNET, M. and RIBERAU-GAYON, P. (1974) Phytochemistry *13*, 1085–1087.

DU VIGNEAUD, V., IRWING, G.W., DYER, H.M. and SEALOCK, R.R. (1938) J. Biol. Chem. *123*, 45–55.

EARLAND, C. and RAMSDEN, D.B. (1969) J. Chromatogr. *41*, 259–261.

EPSTEIN, C.J. and SCHECHTER, A.N. (1968) Ann. N.Y. Acad. Sci. *151*, 85–101.

ERIKSSON, K.E. and PETTERSSON, B. (1973) Anal. Biochem. *56*, 618–620.

FANTES, K.H. and FURMINGER, I.G.S. (1967) Nature *215*, 750–751.

FAWCETT, J.S. and MORRIS, C.J.O.R. (1966) Separ. Sci. *1*, 9–26.

FAWCETT, J.S. (1968) FEBS Letters *1*, 81–82.

FAWCETT, J.S., DEDMAN, M.L. and MORRIS, C.J.O.R. (1969) FEBS Letters *3*, 250–252.

FAWCETT, J.S. (1970) Prot. Biol. Fluids *17*, 409–412.

FAWCETT, J.S. (1973) Ann. N.Y. Acad. Sci. *209*, 112–126.

FAWCETT, J.S. (1975a) *In:* Isoelectric Focusing; Arbuthnott, J.P. and Beeley, J.A., eds. (Butterworths, London) pp. 23–43.

FAWCETT, J.S. (1975b) *In:* Progress in Isoelectric Focusing and Isotachophoresis; Righetti, P.G., ed. (ASP Biol. Med. Press BV, Amsterdam) pp. 25–37.

FELGENHAUER, K. and PAK, S.J. (1973) Ann. N.Y. Acad. Sci. *209*, 147–153.

FELGENHAUER, K. (1974) Hoppe-Seylers Z. Physiol. Chem. *355*, 1281–1290.

FELGENHAUER, K. and PAK, S.J. (1975) *In:* Progress in Isoelectric Focusing and Isotachophoresis; Righetti, P.G., ed. (ASP Biol. Med. Press B.V., Amsterdam) pp. 115–120.

FINLAYSON, G.R. and CHRAMBACH, A. (1971) Anal. Biochem. *40*, 292–311.

FLATMARK, T. and VESTERBERG, O. (1966) Acta Chem. Scand. *20*, 1497–1503.

FRATER, R. (1970) J. Chromatogr. *50*, 469–474.

FREDRIKSSON, S. (1972) Anal. Biochem. *50*, 575–585.

FREDRIKSSON, S. and PETTERSSON, S. (1974) Acta Chem. Scand. *28*, 370–370.

FREDRIKSSON, S. (1975) J. Chromatogr. *108*, 153–167.

FUNATSU, M., HARA, K., ISHIGURO, M., FUNATSU, G. and KUBO, K. (1973) Proc. Japan Acad. *49*, 771–776.

GAINER, H. (1971) Anal. Biochem. *44*, 589–605.

GAINER, H. (1973) Anal. Biochem. *51*, 646–650.

GALANTE, E., CARAVAGGIO, T. and RIGHETTI, P.G. (1975) *In:* Progress in Isoelectric Focusing and Isotachophoresis; Righetti, P.G., ed. (ASP Biol. Med. Press B.V., Amsterdam) pp. 3–12.

GASPARIC, V. and ROSENGREN, A. (1975) *In:* Isoelectric Focusing; Arbuthnott, J.P. and Beeley, J.A., eds. (Butterworths, London) pp. 178–181.

GIANAZZA, E., PAGANI, M., LUZZANA, M. and RIGHETTI, P.G. (1975) J. Chromatogr. *109*, 357–364.

GIBSON, Q.H. (1959) Progr. Biophys. Biol. Chem. *9*, 1–53.

GIDDINGS, J.C. and DAHLGREN (1971) Separ. Sci. *6*, 345–356.

GIDEZ, L.I., SWANEY, J. and MURNANE, S. (1975) (personal communication).

GITTENS, G.J. and JAMES, A.M. (1963) Biochim. Biophys. Acta *66*, 237–249.

GODOLPHIN, W.J. and STINSON, R.A. (1974) Clin. Chim. Acta *56*, 97–103.

GODSON, G.N. (1970) Anal. Biochem. *35*, 66–76.

GOLDSTONE, A. and KOENING, H. (1974) Biochem. J. *141*, 527–535.

GRAESSLIN, D., TRAUTWEIN, A. and BETTENDORF, G. (1971) J. Chromatogr. *63*, 475–477.

GRÄSBECK, R. (1968) Acta Chem. Scand. *22*, 1091–1093.

GRÄSBECK, R., VISURI, K. and STENMAN, U.H. (1972) Biochim. Biophys. Acta *263*, 721–733.

GRASSMANN, W. (1950) Z. Angew. Chem. *62*, 1170–1176.

GROOME, N.P. and BELYAVIN, G. (1975) Anal. Biochem. *63*, 249–254.

GROSSBACH, U. (1972) Biochem. Biophys. Res. Commun. *49*, 667–672.

HAGLUND, H. (1971) *In:* Methods of Biochemical Analysis; Glick, D., ed. (Wiley, Interscience) vol. 19, pp. 1–104.

HAGLUND, H. (1975) *In:* Isoelectric Focusing; Arbuthnott, J.P. and Beeley, J.A., eds. (Butterworths, London) pp. 3–22.

HANNING, K. (1961) Z. Anal. Chem. *181*, 244–254.

HAUPT, H., BAUDNER, S., KRANZ, T. and HEINBURGER, N. (1971) Eur. J. Biochem. *23*, 242–247.

HAYASE, K. and KRITCHEVSKY, D. (1973) Clin. Chim. Acta *46*, 455–464.

HAYASE, K., REISHER, R. and MILLER, B.F. (1973) Prep. Biochem. *3*, 221–241.

HAYES, M.B. and WELLNER, D. (1969) J. Biol. Chem. *244*, 6636–6644.

HEBERT, J.P. and STROBBEL, B. (1974) LKB Application Note No. 151.

HEIDEMAN, M.L. (1965) Ann. N.Y. Acad. Sci. *121*, 501–524.

HJERTÉN, S. (1962) Arch. Biochem. Biophys. Suppl. 1, 147–151.

HJERTÉN, S. (1970) *In:* Methods of Biochemical Analysis; Glick, D., ed. (Wiley, Interscience) vol. 1*8*, pp. 55–79.

HOBART, M.J. (1975) *In:* Isoelectric Focusing; Arbuthnott, J.P. and Beeley, J.A., eds. (Butterworths, London) pp. 275–280.

HOCH, H. and BARR, C.H. (1955) Science *122*, 243–244.

HOFFMAN, D.R., GROSSBERG, A.L. and PRESSMAN, D. (1972) J. Immunol. *108*, 18–25.

HOVANESSIAN, A. and AWDEH, Z.L. (1975) *In:* Progress in Isoelectric Focusing and Isotachophoresis; Righetti, P.G., ed. (ASP Biol. Med. Press B.V., Amsterdam) pp. 205–211.

HUMMEL, B.C.W., BROWN, G.M., PAICE, J.C. and WEBSTER, B.R. (1972) Fed. Proc. 31, 276 (abstr.).

HUMPHRYES, K.C. (1970) J. Chromatogr. *49*, 503–510.

IKEDA, K. and SUZUKI, S. (1912) U.S. Patent 1,015–891.

JACOBS, S. (1971) Prot. Biol. Fluids *19*, 499, 502.

JACOBS, S. (1973) Analyst *98*, 25–33.

JAMIESON, G.A. and Groh, N. (1971) Anal. Biochem. *43*, 259–268.

JANSON, J.C. (1972) Ph.D. Thesis, University of Uppsala, Sweden.

JEANSSON, S., VESTERBERG, O. and WADSTRÖM, T. (1972) Life Sci. *11*, 929–937.

JENSEN, M., OSKI, F.A., NATHAN, D.G., and BUNN, H.F. (1975) J. Clin. Investig. *55*, 469–477.

JEPPSSON, J.O. and BERGLUND, S. (1972) Clin. Chim. Acta *40*, 153–158.

JOHANSSON, B.G. and STENFLO, J. (1971) Anal. Biochem. *40*, 232–236.

JOHANSSON, B.G. and HJERTÉN, S. (1974) Anal. Biochem. *59*, 200–213.

JONSSON, M. PETTERSSON, E. and RILBE, H. (1969) Acta Chem. Scand. *23*, 1553–1559.

JONSSON, M., PETTERSSON, S. and RILBE, H. (1973) Anal. Biochem. *51*, 557–576.

JUST, W.W., LEON-V., J.O. and WERNER, G. (1975) *In:* Progress in Isoelectric Focusing and Isotachophoresis; Righetti, P.G., ed. (ASP Biol. Med. Press B.V., Amsterdam) pp. 265–280.

KAISER, K.P., BRUHN, L.C. and BELITZ, H.D. (1974) Z. Lebensm. Unters. Forsch. *154*, 339–347.

KALOUS, V. and VACIK, J. (1959) Chem. Listy *53*, 35–37.

KAPLAN, N.O. (1968) Ann. N.Y. Acad. Sci. *151*, 382–399.

KAPLAN, L.J. and FOSTER, J.F. (1971) Biochemistry *10*, 630–636.

KARLSSON, C., DAVIES, H., OHMAN, J. and ANDERSSON, U.B. (1973) LKB Application Note No. 75.

KECK, K., GROSSBERG, A.L. and PRESSMAN, D., (1973a) Eur. J. Immunol. *3*, 99–102.

KEK, K., GROSSBERG, A.L. and PRESSMAN, D. (1973b) Immunochemistry *10*, 331–335.

KENRICK, K.G. and MARGOLIS, J. (1970) Anal. Biochem. *33*, 204–207.

KIM, W.J. and WHITE, T.T. (1971) Biochim. Biophys. Acta *242*, 441–445.

KLEINBERG, I. (1958) Brit. Dent. J. *104*, 197–204.

KOCH, H.J.A. and BACKX, J. (1969) Sci. Tools *16*, 44–47.

KOHNERT, K.D., ZIEGLER, M., ZUHLKE, H. and FIEDLER, H. (1972) FEBS Letters *28*, 177–178.

KOHNERT, K.D., SCHMID, E., ZUHLKE, H. and FIEDLER, H. (1973a) J. Chromatogr. *76*, 263–267.

KOHNERT, K.D., ZIEGLER, M., ZUHLKE, H. and WILKE, B. (1973b) Anal. Biochem. *53*, 650–653.

KOLIN, A. (1954) J. Chem. Phys. *22*, 1628–1629.

KOLIN, A. (1955) Proc. Natl. Acad. Sci. *41*, 101–110.

KOLIN, A. (1958) Methods Biochem. Anal.; Glick, D. ed. (Wiley, Interscience) *6*, pp. 259–288.

KOPWILLEM, A., CHILLEMI, F., RIGHETTI, A.B.B. and RIGHETTI, P.G. (1973) Prot. Biol. Fluids *21*, 657–665.

KOSTNER, G., ALBERT, W. and HOLASEK, A. (1969) Hoppe Seylers Z. Physiol. Chem. *350*, 1347–1352.

LÅÅS, T. (1972) J. Chromatogr. *66*, 347–355.

LAMBERT, B., SUTTER, B.Ch.J. and JACQUEMIN, Cl. (1972) Horm. Metab. Res. *4*, 149–151.

LATNER, A.L., PARSONS, M.E. and SKILLEN, A.W. (1970) Biochem. J. *118*, 299–302.

LATNER, A.L. (1973) Ann. N.Y. Acad. Sci. *209*, 281–298.

LATNER, A.L. and EMES, V. (1975) *In:* Progress in Isoelectric Focusing and Isotachophoresis; Righetti, P.G., ed. (ASP Biol. Med. Press B.V., Amsterdam) pp. 223–233.

LAURELL, C.B. (1966) Anal. Biochem. *15*, 45–52.

LEABACK, D.H. and RUTTER, A.C. (1968) Biochem. Biophys. Res. Commun. *32*, 447–453.

LEABACK, D.H. and ROBINSON, H.K. (1974) FEBS Letters *40*, 192–195.

LEWIN, S. (1969) Postgrad. Med. J. *45*, 729–730.

LI, Y.T. and LI, S.C. (1970) J. Biol. Chem. *245*, 825–832.

LINDSAY, D.G. and SHALL, S. (1971) Biochem. J. *121*, 737–745.

LINDSAY, D.G., LOGE, O., LOSERT, W. and SHALL, S. (1972) Biochim. Biophys. Acta *263*, 658–665.

LOENING, U.E. (1967) Biochem. J. *102*, 251–257.

LUDWIG, C.S. (1856) Sitzber. Akad. Wiss. Wien *20*, 539– .

LUNDAHL, P. and HJERTÉN, S. (1973) Ann. N.Y. Acad. Sci. *209*, 94–111.

LUNDIN, H., HJALMARSSON, S.G. and DAVIES, H. (1975) LKB Application Note No. 194.

LUNER, S.J. and KOLIN, A. (1970) Proc. Natl. Acad. Sci. *6*, 898–903.

MACGILLIVRAY, A.J. and RICKWOOD, D. (1974) Eur. J. Biochem. *41*, 181–190.

MACKO, V. and STEGEMAN, H. (1969) Hoppe Seylers Z. Physiol. Chem. *350*, 917–919.

MACKO, V. and STEGEMAN, H. (1970) Anal. Biochem. *37*, 186–190.

MAFFEZZOLI, R.D., KAPLAN, G.H. and CHRAMBACH, A. (1972) J. Clin. Endocr. *34*, 361–369.

MAHER, J.R., TRENDLE, W.O. and SCHULTZ, R.L. (1956) Naturwiss. *43*, 423–427.
MALIK, N. and BERRIE, A. (1972) Anal. Biochem. *49*, 173–176.
MATSUMURA, T. and NODA, H. (1973) Anal. Biochem. *56*, 571–575.
MEACHUM, E.D., Jr., COLVIN, H.J., Jr. and BRAYMER, H.D. (1971) Biochemistry *10*, 326–332.
MEDELLIN, J.M. (1972) Ph.D. Thesis, University of Cqlifornia at San Diego.
MERZ, D.C., GOOD, R.A. and LITMAN, G.W. (1972) Biochim. Biophys. Res. Commun. *49*, 84–91.
MILLER, D.W. and ELGIN, S.C.R. (1974) Anal. Biochem. *60*, 142–148.
MITCHELL, W.M. (1967) Biochim. Biophys. Acta *147*, 171–174.
MOLNAROVA, B. and SOVA, O. (1974) *A*bstr. Commun. 9th Meet. Fed. Europ. Biochem. Soc., Budapest, p. 432.
MORITZ, P.M. (1969) Prot. Biol. Fluids *16*, 499–521.
MORRIS, C.J.O.R. and MORRIS, P. (1964) Separation Methods in Biochemistry (Pitman and Sons Ltd., London) p. 55 and p. 687.
NADER, H.B., MCDUFFLE, N.M. and DIETRICH, C.P. (1974) Biochem. Biophys. Res. Commun. *57*, 488–493.
NAKHLEH, E.T., SAMRA, S.A. and AWDEH, Z.L. (1972) Anal. Biochem. *49*, 218–224.
NEUHOFF, V. (1973) Micromethods in Molecular Biology (Springer Verlag, Berlin) pp. 49–56.
NILSSON, P. WADSTRÖM, T. and VESTERBERG, O. (1970) Biochim. Biophys. Acta *221*, 146–147.
NORDBERG, G.F., NORDBERG, M., PISCATOR, M. and VESTERBERG, O. (1972) Biochem. J. *126*, 491–498.
O'FARREL, P. (1975) J. Biol. Chem. *250*, 4007–4021.
ORNSTEIN, L. (1964) Ann. N.Y. Acad. Sci. *121*, 321–349.
OTAVSKY, W. and DRYSDALE, J.W. (1975) Anal. Biochem. *65*, 533–536.
PARK, C.M. (1973) Ann. N.Y. Acad. Sci. *209*, 237–257.
PEPPER, D.S. and JAMIESON, J.A. (1969) Biochemistry, *8*, 3362–3369.
PERCIVAL, L.H. and DUNCKLEY, G.G. (1969) Proc. Univ. Otago Med. School *47*, 85–86.
PERCIVAL, L.H., DUNCKLEY, G.G. and PURVES, H.D. (1970) Austr. J. Exp. Biol. Med. Sci, *48*, 171–178.
PETERSON, R.F. (1971) J. Agr. Food Chem. *19*, 595–599.
PETERSON, J.I., TIPTON, H.W. and CHRAMBACH, A. (1974) Anal. Biochem. *62*, 274–280.
PETTERSSON, E. (1969) Acta Chem. Scand. *23*, 2631–2635.
PHILPOT, J.St.L. (1940) Trans. Faraday Soc. *36*, 38–46.
POGACAR, P. and JARECKI, R. (1974) *In:* Electrophoresis and Isoelectric Focusing in Polyacrylamide Gels; Allen, R.C. and Maurer, H., eds. (W. de Gruyter, Berlin) pp. 153–158.
POWELL, L., ALPERT, E., ISSELBACHER, K.J. and DRYSDALE, J.W. (1975) Brit. J. Haematol. *30*, 47–61.
PUSZTAI, A. and DUNCAN, I. (1971) Biochim. Biophys. Acta *229*, 785–794.

Quast, R. (1971) J. Chromatogr. *54*, 405–412.

Quinn, J.R. (1973) J. Chromatogr. *76*, 520–522.

Quiocho, F.A. and Richards, F.M. (1964) Proc. Natl. Acad. Sci. *52*, 833–839.

Radola, B.J. (1969) Biochim. Biophys. Acta *194*, 335–338.

Radola, B.J. (1973a) Biochim. Biophys. Acta *295*, 412–428.

Radola, B.J. (1973b) Ann. N.Y. Acad. Sci. *209*, 127–143.

Radola, B.J. (1975) *In:* Isoelectric Focusing; Arbuthnott, J.P. and Beeley, J.A., eds. (Butterworths, London) pp. 182–197.

Rathnam, P. and Saxena, B.B. (1970) J. Biol. Chem. *245*, 3725–3731.

Rathnam, P. and Saxena, B.B. (1971) J. Biol. Chem. *246*, 7087–7094.

Reichert, L.E., Jr. (1971) Endocrinology *88*, 1029–1044.

Reisner, A.H., Nemes, P. and Bucholtz, C. (1975) Anal. Biochem. *64*, 509–516.

Richards, E.G. and Lecanidou, R. (1974) *In:* Electrophoresis and Isoelectric Focusing in Polyacrylamide Gel; Allen, R.C. and Maurer, H.R., eds. (W. de Gruyter, Berlin) pp. 16–21.

Righetti, P.G. and Drysdale, J.W. (1971) Biochim. Biophys. Acta *236*, 17–28.

Righetti, P.G. and Drysdale, J.W. (1973) Ann. N.Y. Acad. Sci. *209*, 163–186.

Righetti, P.G., Perrella, M., Zanella, A. and Sirchia, G. (1973) Nature New Biol. *245*, 273–276.

Righetti, P.G. and Righetti, A.B.B. (1974) Abstr. Commun. 9th Meet. Fed. Europ. Biochem. Soc., Budapest, p. 432.

Righetti, P.G. and Drysdale, J.W. (1974) J. Chromatogr. *98*, 271–321.

Righetti, P.G. (1975a) Progress in Isoelectric Focusing and Isotachophoresis ASP Biol. Med. Press B.V., Amsterdam) pp. 425.

Righetti, P.G. (1975b) Separ. Purif. Methods *4*, 23–72.

Righetti, P.G. and Righetti, A.B.B. (1975) *In:* Isoelectric Focusing; Arbuthnott, J.P. and Beeley, J.A., eds. (Butterworths, London) pp. 114–131.

Righetti, P.G., Pagani, M. and Gianazza, E. (1975a) J. Chromatogr. *109*, 341–356.

Righetti, P.G., Righetti, A.B.B. and Galante, E. (1975b) Anal. Biochem. *63*, 423–432.

Rilbe, H. and Pettersson, S. (1968) Separ. Sci. *3*, 209–234.

Rilbe, H. (1970) Prot. Biol. Fluids *17*, 369–382.

Rilbe, H. (1971) Acta Chem. Scand. *25*, 2768–2769.

Rilbe, H. (1973a) Ann. N.Y. Acad. Sci. *209*, 11–22.

Rilbe, H. (1973b) Ann. N.Y. Acad. Sci *209*, 80–93.

Rilbe, H. and Pettersson, S. (1975) *In:* Isoelectric Focusing; Arbuthnott, J.P. and Beeley, J.A., eds. (Butterworths, London) pp. 44–57.

Rilbe, H., Forchheimer, A., Pettersson, S. and Jonsson, M. (1975) *In:* Progress in Isoelectric Focusing and Isotachophoresis; Righetti, P.G., ed. (ASP Biol. Med. Press B.V., Amsterdam) pp. 51–63.

Riley, R.F. and Coleman, M.K. (1968) J. Lab. Clin. Med. *72*, 714–720.

Rodbard, D. and Chrambach, A. (1970) Proc. Natl. Acad. Sci. *65*, 970–977.

RODBARD, D. and CHRAMBACH, A. (1971) Anal. Biochem. *40*, 95–134.

RODBARD, D., LEVITOV, C. and CHRAMBACH, A. (1972) Separ. Sci. *7*, 705–723.

RODBARD, D., CHRAMBACH, A. and WEISS, G.H. (1974) *In:* Electrophoresis and Isoelectric Focusing in Polyacrylamide Gel; Allen, R.C. and Maurer, H.R., eds. (de Gruyter, Berlin) pp. 62–105.

ROSE, C. and HARBOE, N.M.G. (1970) Prot. Biol. Fluids *17*, 397–400.

ROSENGREN, A. (1975) *In:* Progress in Isoelectric Focusing and Isotachophoresis; Righetti, P.G., ed. (ASP Biol. Med. Press B.V., Amsterdam) p. 390.

RÜCHEL, R., MESECKE, S., WOLFRUM, D.I. and NEUHOFF, V. (1973) Hoppe Seylers Z. Physiol. Chem. *354*, 1351–1368.

RÜCHEL, R. and BRAGER, M.D. (1975) Anal. Biochem., 68, 415–428.

RÜKER, W. and RADOLA, B.J. (1971) Planta *99*, 192–198.

RYAN, M.F. and WESTPHAL, U. (1972) J. Biol. Chem. *247*, 4050–4056.

SALAMAN, M.R. and WILLIAMSON, A.R. (1971) Biochem. J. *122*, 93–99.

SATTERLEE, L.D. and SNYDER, H.E. (1969) J. Chromatogr. *41*, 417–422.

SAXENA, B.B. and RATHNAM, P. (1968) Gonadotropins 3, 3–12.

SCANU, A.M., EDELSTEIN, C. and AGGERBECK, L. (1973) Ann. N.Y. Acad. Sci. *209*, 311–327.

SCHAFFER, H.E. and JOHNSON, F.M. (1973) Anal. Biochem. *51*, 577–583.

SCHANTZ, E.J., ROESSLER, W.G., WOODBURN, M.J., LYNCH, J.M., JACOBY, H.M., SILVERMAN, S.J., GORMAN, J.C. and SPERO, L. (1972) Biochemistry 11, 360–366.

SCHMELZER, W. and BEHNE, D. (1975) *In:* Progress in Isoelectric Focusing and Isotachophoresis; Righetti, P.G., ed. (ASP Biol. Med. Press B.V., Amsterdam) pp. 257–264.

SECCHI, C. (1973) Anal. Biochem. *51*, 448–455.

SEILER, N., THOBE, J.,and WERNER, G. (1970a) Hoppe Seylers Z. Physiol. Chem. *351*, 865–868.

SEILER, N./ THOBE, J. and WERNER, G. (1970b) Z. Anal. Chem. *252*, 179–182.

SHAFRITZ, D.A. and DRYSDALE, J.W. (1975) Biochemistry *14*, 61–68.

SHAPCOTT, D. and O'BRIEN, D. (1970) Diabetes *19*, 831–836.

SHAPIRO, A.L., VINUELA, E. and MAIZEL, J.V., Jr. (1967) Biochem. Biophys. Res. Commun. *28*, 815–820.

SHERBET, G.V., LAKSHMI, M.S. and RAO, K.V. (1972) Exptl. Cell Res. *10*, 113–123.

SHERBET, G.V. and LAKSHMI, M.S. (1973) Biochim. Biophys. Acta *298*, 50–58.

SHERBET, G.V. and LAKSHMI, M.S. (1975) *In:* Isoelectric Focusing; Arbuthnott, J.P. and Beeley, J.A., eds. (Butterworths, London) pp. 338–346.

SHERWOOD, O.D., GRIMEK, H.J. and MCSHAN, W.H. (1970) J. Biol. Chem. *245*, 2328–2336.

SKUDE, G. and JEPPSSON, J.O. (1972) Scand. J. Clin. Lab. Invest. *29*, suppl. 124, 55–58.

SMITH, I., LIGHTSTONE, P.J. and PERRY, J.D. (1971) Clin. Chim. Acta *35*, 59–66.

SOAVE, C., PIOLI, F., VIOTTI, A., SALAMINI, R. and RIGHETTI, P.G. (1975) Maydica *20*, 83–94.

SÖDERHOLM, J., ALLESTAM, P. and WADSTRÖM, T. (1972) FEBS Letters *24*, 89–92.

SÖDERHOLM, J. and LIDSTRÖM, P.A. (1975) *In:* Isoelectric Focusing; Arbuthnott, J.P. and Beeley, J.A., eds. (Butterworths, London) pp. 143–146.

SÖDERHOLM, J. and SMYTH, C.J. (1975) *In:* Progress in Isoelectric Focusing and Isotachophoresis; Righetti, P.G., ed. (ASP Biol. Med. Press B.V., Amsterdam) pp. 99–114.

SÖDERHOLM, J. and WADSTRÖM, T. (1975) *In:* Isoelectric Focusing; Arbuthnott, J.P. and Beeley, J.A., eds. (Butterworths, London) pp. 132–142.

SORET, C. (1879) Arch. Sci. Phys. Natl. (Genève) *2*, 48–56.

SPENCER, E.M. and KING, T.P. (1971) J. Biol. Chem. *246*, 201–208.

SPRAGG, J., KAPLAN, A.P. and AUSTEN, K.F. (1973) Ann. N.Y. Acad. Sci. *209*, 372–386.

STATHAKOS, D. (1975) *In:* Progress in Isoelectric Focusing and Isotachophoresis; Righetti, P.G., ed. (ASP Biol. Med. Press B.V., Amsterdam) pp. 65–75.

STEGEMAN, H., FRANKSEN, H. and MACKO, V. (1973) Z. Naturforsch. *28c*, 722–732.

STEINHARDT, J., LEIDY, J.G. and MOONEY, J.P. (1972) Biochestry *11*, 1809–1817.

STENMANN, U.H. and GRÄSBECK, R. (1972) Biochim. Biophys. Acta *286*, 243–251.

STRONGIN, A.J.A., BALDUEV, A.P. and LEVIN, E.D. (1973) Sci. Tools *20*, 34–35.

SUZUKI, T., BENESCH, R.E., YUNG, S. and BENESCH, R. (1973) Anal. Biochem. *55*, 249–254.

SVENSSON, H. (1948) *In:* Advances in Protein Chemistry; Anson, M.L. and Edsall, J.T., eds. (Academic Press, New York) vol. IV, pp. 251–295.

SVENSSON, H. and BRATTSTEN, I. (1949) Arkiv Kemi 1, 401–409.

SVENSSON, H. (1961) Acta Chem. Scand. *15*, 325–341.

SVENSSON, H. (1962a) Acta Chem. Scand. *16*, 456–466.

SVENSSON, H. (1962b) Arch. Biochem. Biophys. Suppl. 1, 132–140.

SVENSSON, H. (1967) Prot. Biol. Fluids *15*, 515–522.

TALBOT, P. (1975) *In:* Isoelectric Focusing; Arbuthnott, J.P. and Beeley, J.A., eds. (Butterworths, London) pp. 270–274.

TALBOT, P. and CAIE, I.S. (1975) *In:* Isoelectric Focusing; Arbuthnott, J.P. and Beeley, J.A., eds. (Butterworths, London) pp. 74–77.

THORSTENSSON, A., SJODIN, B. and KARLSSON, J. (1975) *In:* Progress in Isoelectric Focusing and Isotachophoresis; Righetti, P.G., ed. (ASP Biol. Med. Press B.V., Amsterdam) pp. 213–222.

TROITZKI, G.V., SAVIALOV, V.P. and ABRAMOV, V.M. (1974) Dokl. Akad. Nauk. USSR *214*, 955–958.

TUTTLE, A.H. (1956) J. Lab. Clin. Med. *47*, 811–816.

UI, N. (1973) Ann. N.Y. Acad. Sci. *209*, 198–209.

VALMET, E. (1969) Sci. Tools *16*, 8–13.

VALMET, E. (1970) Prot. Biol. Fluids *17*, 401–407.

VAN BAELEN, H. and DE MOOR, P. (1970) Ann. Endocrinol. *31*, 829–835.

VAN EYK, H.G., VERMAAT, R.J. and LEIJNSE, B. (1969) FEBS Letters *3*, 193–194.

VESTERBERG, O. and SVENSSON, H. (1966) Acta Chem. Scand. *20*, 820–834.

VESTERBERG, O. (1967) Acta Chem. Scand. *21*, 206–216.

VESTERBERG, O., WADSTRÖM, T., VESTERBERG, K., SVENSSON, H. and MALMGREN, B. (1967) Biochim. Biophys. Acta *133*, 435–445.

VESTERBERG, O. (1968) Svensk. Kem. Tidskr. *80*, 213–215.

VESTERBERG, O. (1969) Acta Chem. Scand. *23*, 2653–2666.

VESTERBERG, O. (1970) Prot. Biol. Fluids *17*, 383–387.

VESTERBERG, O. (1971a) Methods Enzymol. *22*, 389–413.

VESTERBERG, O. (1971b) Biochim. Biophys. Acta *243*, 345–348.

VESTERBERG, O. (1972) Biochim. Biophys. Acta *257*, 11–19.

VESTERBERG, O. and ERIKSSON, R. (1972) Biochim. Biophys. Acta *285*, 393–397.

VESTERBERG, O. (1973a) Ann. N.Y. Acad. Sci. *209*, 23–33.

VESTERBERG, O. (1973b) Acta Chem. Scand. *27*, 2415–2420.

VESTERBERG, O. (1973c) Sci. Tools *20*, 22–29.

VESTERBERG, O. (1975) *In:* Isoelectric Focusing; Arbuthnott, J.P. and Beeley, J.A., eds. (Butterworths, London) pp. 78–96.

VINOGRADOV, S.N., LOWENKRON, S., ANDONIAN, H.R., BAGSHAW, J., FELGENHAUER, K. and PAK, S.J. (1973) Biochem. Biophys. Res. Commun. *54*, 501–506.

WADSTRÖM, T. and HISATSUNE, K. (1970a) Biochem. J. *120*, 725–734.

WADSTRÖM, T. and HISATSUNE, K. (1970b) Biochem. J. *120*, 735–744.

WADSTRÖM, T. and SMYTH, C. (1973) Sci. Tools *20*, 17–21.

WADSTRÖM, T., MOLLBY, R., JEANSSON, S. and WRETLIND, B. (1974) Sci. Tools *21*, 2–4.

WADSTRÖM, T. (1975) *In:* Progress in Isoelectric Focusing and Isotachophoresis; Righetti, P.G., ed. (ASP Biol. Med. Press B.V., Amsterdam) p. 389.

WADSTRÖM, T. and SMYTH, C.J. (1975a) *In:* Isoelectric Focusing; Arbuthnott, J.P. and Beeley, J.A., (Butterworths, London) pp. 152–177.

WADSTRÖM, T. and SMYTH, C.J. (1975b) *In:* Progress in Isoelectric Focusing and Isotachophoresis; Righetti ,P.G., ed. (ASP Biol. Med. Press B.V., Amsterdam) pp. 149–163.

WALLEVIK, K. (1973) Biochim. Biophys. Acta *322*, 75–87.

WEBER, K. and OSBORN, M. (1969) J. Biol. Chem. *244*, 4406–4412.

WEISE, H.C., GRAESSLIN, D. and GLISMANN, D. (1975) *In:* Progress in Isoelectric Focusing and Isotachophoresis; Righetti, P.G., ed. (ASP Biol. Med. Press B.V., Amsterdam) pp. 93–98.

WELINDER, B.S. (1971) Acta Chem. Scand. *25*, 3737–3742.

WELLER, D.L., HEANEY, A. and SJOGREN, R.E. (1968) Biochim. Biophys. Acta *168*, 386–388.

WELLNER, D. and HAYES, M.B. (1973) Ann. N.Y. Acad. Sci. *209*, 34–43.

WILLIAMS, R.R. and WATERMAN, R.E. (1929) Proc. Soc. Exp. Biol. Med. *27*, 56–59.

WILLIAMSON, A.R. (1971) Eur. J. Immunol. *1*, 390–394.

WILLIAMSON, A.R. (1973) *In:* Handbook of Experimental Immunology; Weir, D.M., ed. (Blackwell, Oxford) Ch. 8, pp. 1–23.

WILLIAMSON, A.R., SALAMAN, M.R. and KRETH, W.H. (1973) Ann. N.Y. Acad. Sci. *209*, 210–224.

WINTERS, A., PERLMULTER, H. and DAVIES, H. (1975) LKB Application Note, No. 198.
WINTZER, G. and UHLENBRUCK, G. (1970) Hoppe Seylers Z. Physiol. Chem. 351, 834–838.
WRIGLEY, C. (1968a) Sci. Tools *15*, 17–23.
WRIGLEY, C. (1968b) J. Chromatogr. *36*, 362–365.
WRIGLEY, C.W. (1971) Methods Enzymol. *22*, 559–564.
YOUNG, C.W. and BITTAR, E.S. (1973) Cancer Res. *33*, 2692–2700.
ZEINEH, R.A., NIJM, W.P. and AL-AZZAWI, F.H. (1975) Amer. Lab. *7*, 51–58.

Subject index

Acetic acid 375, 376
acetylcholinesterase 507
N-acetyl glucosamine 537
N-acetyl neuraminic acid 538
α_1-acid glycoprotein 537
acrylamide 345, 440–444
acrylic acid 357–360, 362, 363, 442
agar 487, 488
agarose 424, 480, 488, 490
albumin 374, 476, 512, 521
alkaline phosphatase 476
L-amino acid oxidase 537
β-amino acids 357
anodic oxidation 367, 393, 570
anti-Markovnikov 357
antimony electrode 514
α_1-antitrypsin 512
arginine 371, 507, 520, 528
artificial pH gradients 347, 375
aryl sulphatase 537
ascorbic acid 507
asparagine 551
aspartic acid 351, 371, 375, 376
autoradiography 483–487

Bacto peptone 373
Bio-Gel P 405, 424
Biuret 368
Bohr effect 562
borohydride (sodium) 425
Brij 39 539
bromophenol blue 446, 466, 467
buffering capacity 349, 368, 372, 373, 436

Casein 374, 476
cathodic drift 499, 525
cathodic reduction 570
ceruloplasmin 472
chondroitin sulphate 401, 530
chloromethylphosphonate 361
chymotrypsinogen A 557
citric acid 375
collagen 537
concanavalin A 473, 556
conductivity 348, 368, 372, 373, 463
continuous-flow IEF 366, 404, 530
Coomassie Brilliant Blue G-250 465, 468
Coomassie Brilliant Blue R-250 447, 465, 466, 467, 552
Coomassie Violet R-150 366, 407, 465
copper 372, 521
crotonic acid 368
crystallins 552
cupric sulphate 468
cyanocobalamine 473, 537
cysteic acid 443, 506, 520
cysteine 443, 506, 520
cytidilyl-cytidine (CpC) 563

cytidilyl-cytidilyl-cytidine (CpCpC) 563
cytochrome c 545
Cytophaga Johnsonii 418, 504

Degree of ionization 348
delta pump 404
deuterium oxide 380
dextran 380
dextran sulphate 401, 530
diallyl tartaramide (DATD) 429, 434, 439, 440
dichloroacetic acid 375
diffusion coefficient 354, 499, 539
2,3-dimercapto propanol 507
dimethyl formamide 506
dimethyl sulphoxide (DMSO) 506, 536
DNase 494, 522
dinitrophenyl histidine 535
α-N-(4-hydroxyphenacetyl)-ε-N-2,4-dinitrophenyl lysine 485
dioxane 374
6,6′-dithiodinicotinic acid carboxypyridine disulphide 528
dithionite 423
dithiothreitol (dithioerythritol) 507

E. coli 483, 494, 528
EDTA 401, 529, 542
electroendosmosis 448, 525, 572
emasol 539
empigen BB 507
enterotoxin A 513
ethanolamine 382, 401, 445
ethylendiamine 382, 406, 445, 448, 464
ethylene glycol 380, 382, 450, 506, 539
ethyleneimine 528

Faraday 354
Fast Green FCF 466, 467
ferricyanide 423, 450
ferritin 387, 473, 493, 522, 545
Ficoll 380, 527
field strength 352, 353, 354, 380, 401, 444, 453
folate reductase 556
Folin's reagent 367
follicle stimulating hormone (FSH) 530
foot and mouth disease virus 530
formaldehyde 528
formamide 506
formic acid 375, 376
free-flow IEF 400
fucose 537

Galactose 537, 556
gas constant 354
gelatin 492
gliadin 492
γ-globulin 549
glucose caramelization 366
α-glucosidase 504
β-glucoronidase 522, 537
glutamic acid 349, 371, 375, 376, 521
glutamine 551
glutaraldehyde 473, 485, 551
glutaric acid 376
glycerol 374, 376, 380–382, 435, 440, 446, 463, 492, 513, 527
glycine 349, 385
$8s\alpha_3$-glycoprotein 537
guanylyl cytidine (GpC) 563

Haemoglobin 374, 447, 450, 512, 545, 553, 555
α-haemolysin 473
heparin 401, 530
hexamethylenetetramine (HMTA) 358, 360
hexosaminidase 473
highly cross-linked gels 428, 444
histidine 349, 373, 375, 513, 561
histidyl histidine 357

human chorionic gonadotropin (HCG) 530
human growth hormone (HGH) 522, 530
L-hystidyl-L-tyrosine 535

Imidazole 513, 519, 561
insulin 534
invertase 536
isoelectric precipitation 395, 419, 425, 501, 570
isopycnic centrifugation 347
itaconic acid 368

J-columns 420
Joule heat 352, 431, 447, 461, 523

Kerosene 458

Lactate dehydrogenase (LDH) 475, 497
lactic acid 376, 475
law of pH monotony 345, 351
Light Green SF 407, 465, 466
light paraffin oil 458
β-lipoprotein 493
lithium aluminum hydride 425
load capacity 355, 417
Ludwig-Soret effect 394
luteinizing hormone (LH) 530
lymphocytes 542
L-lysyl-L-tyrosine 535
lysine 349, 371, 528
lysosomes 401, 530, 537

α_2-Macroglobulin 493
maleic acid 368, 376
malic acid 376
malonic acid 376
mannitol 374, 380
mannose 537, 556
mass load 355, 417
2-mercaptoethanol 507, 542
metal chelation 372
metallothionein 544
methacrylic acid 368
methionine 506, 520
methionine sulphoxide 506, 520
methyl cellulose 397, 425
α-methyl-D-glucoside 473, 556
N,N′-methylene bisacrylamide (Bis) 429, 434, 439
Michael's reaction 357
micro-isoelectric focusing 448
mitochondria 401, 530
molecular sieving 425, 426
mRNA 565
multi-compartment electrolyzers 400, 411
multiphasic columns 411
myeloma protein 549
myoglobin 393, 396, 417, 423, 513, 545

Narrow pH gradients 509, 512
natural pH gradients 345, 347
Neurospora Crassa 536
ninhydrin 368
p-nitrophenyloxamic acid 538

Ovalbumin 396, 485

Paper print technique 407, 464
peak capacity 354
pentaethylene hexamine (PEHA) 358–362, 375, 368, 369
pepsin 375, 376, 476
perchloric acid 468
periodic acid 429, 473
persulphate 440, 443, 449, 454, 545
pevicon 411
phosphoric acid 375, 398
picric acid 366
plasminogen 476
plateau phenomenon 469, 499, 525
platelets 537, 542
polyanetholsulfonic acid 401, 530
polyethylene glycols 369

polyethylene oxide 425
polyethylene polyamines 357, 367
polyvinyl sulfate 401
ponceau S 471
potato proteins 492, 497
progesterone 537
prolactin 522
proline 560
pronase E 356, 408
propanoic acid 357
propansultone 361
propionic acid 376
propylene oxide 409

QAE-Sephadex 425

Red blood cells 401, 479, 529, 537, 542
refractive index 362
resolving power 353
retardation coefficient 428, 499
riboflavin 400, 453, 545
ribosomal RNA (rRNA) 565
ricin D 521
RNase 494
RNA polymerase 475

Saliva 539
Schiff stain 473
Sephadex gels 356, 369, 377, 404–406, 410, 424, 454, 464, 493, 519
sialic acid 538
sodium dodecyl sulphate (SDS) 493, 507, 540
soft-laser densitometer 471
sorbitol 374, 376, 380, 4440, 453, 454
staphylokinase 476
stokes radius 493, 539
succinic acid 376
sucrose density gradient 378
sulfurous acid 450
sulphobetaine DLH 507
sulphosalicylic acid 467
Sudan Black B 473, 539

Taurine 507
tetraethylene pentamine (TEPA) 358–365, 368
N,N,N′,N′-tetramethylethylene diamine (TEMED) 398, 439
tetramethyl urea 506, 539
thermal pH gradients 374
thin-layer IEF 450
thioglycolate 443, 506
thyrotropin (TSH) 530
thyroxine 473
thyroxine-binding-globulin 473
transferrin 472, 544
transient state IEF (TRANSIF) 392, 498, 499
trichloroacetic acid (TCA) 448, 465, 466
triethyl amine 445
triethylene tetramine (TETA) 358, 360, 362362
trinitrobenzensulphonate 542
Triton X-100 539, 543
tRNA 565
Tween 80 539
tyrosine 506, 520, 560

U-columns 397
urea 440, 494, 497, 506, 513, 522, 540, 542, 565, 570

Vinyl sulfonate 361
vitamin B_{12}-binding protein 376, 539

Zeins 522
zone convection IEF 367, 393, 397
zymograms 472–482